Structural Mechanics

Structural Mechanics provides an in-depth coverage of mechanics of structures with an approach that departs from the conventional strength of materials approach. It proceeds from the theory of elasticity to analyze the three-dimensional response to mechanical loads with a focus on slender structural elements: bars or beams, and assemblages thereof, known as trusses, frames, or ossatures.

Written with a rigorous perspective based on Saint-Venant's theory, this book develops practical solutions to structures problems. It enables readers to classify structural members and identify the conditions under which a structural component can be idealized for the analysis of load-bearing capacity. This book covers the three-dimensional theory of beams of any cross-sectional shape or form, including flexible rods, rings, tubes, multi-cell sections, and thin-walled structures.

This book is intended for upper-level undergraduate mechanical, aerospace, and civil engineering students taking courses in Structural Mechanics, Structural Analysis, and Mechanics of Aerospace Structures.

Instructors will be able to utilize a solutions manual and figure slides for their course.

A. Amine Benzerga is the General Dynamics Professor of Aerospace Engineering at Texas A&M University, USA, where he joined in 2004. He holds an engineering diploma from SUPAERO (Toulouse, 1995) and a Ph.D. in Materials Science and Engineering from Mines de Paris in 2000. Prior to joining Texas A&M, he was a Research Engineer at the National Research Laboratory of Gaz de France (now NaTran) on leave at Brown University (Providence, USA) where he later pursued his post-doctoral studies in Solid Mechanics. His research interests include the deformation and fracture of structural and functional materials. Dr. Benzerga was the recipient of a CAREER award from the U.S. National Science Foundation in 2008 and a Fulbright U.S. Scholar Award in 2022. He was the Edward Pete Aldridge Associate Professor of Aerospace Engineering at Texas A&M (2011–2014) and was elected as ASME Fellow in 2024. Until recently, he was the Director of the Center for intelligent Multifunctional Materials and Structures (CiMMS) and currently holds a joint appointment in the Department of Materials Science and Engineering at Texas A&M.

Structural Mechanics

A. Amine Benzerga

CRC Press
Taylor & Francis Group
Boca Raton London New York

CRC Press is an imprint of the
Taylor & Francis Group, an **informa** business

Designed cover image: A. Amine Benzerga

First edition published 2026
by CRC Press
2385 NW Executive Center Drive, Suite 320, Boca Raton FL 33431

and by CRC Press
4 Park Square, Milton Park, Abingdon, Oxon, OX14 4RN

CRC Press is an imprint of Taylor & Francis Group, LLC

Library of Congress Cataloging-in-Publication Data
Names: Benzerga, A. A. (A. Amine) author
Title: Structural mechanics / Amine A. Benzerga.
Description: First edition. | Boca Raton, FL : CRC Press, 2026. | Includes bibliographical references and index.
Identifiers: LCCN 2025046389 (print) | LCCN 2025046390 (ebook) | ISBN 9781032883274 hardback | ISBN 9781032883281 paperback | ISBN 9781003537229 ebook
Subjects: LCSH: Structural analysis (Engineering) | Structural engineering | LCGFT: Textbooks
Classification: LCC TA645 .B3964 2026 (print) | LCC TA645 (ebook)
LC record available at https://lccn.loc.gov/2025046389
LC ebook record available at https://lccn.loc.gov/2025046390

ISBN: 978-1-032-88327-4 (hbk)
ISBN: 978-1-032-88328-1 (pbk)
ISBN: 978-1-003-53722-9 (ebk)

DOI: 10.1201/9781003537229

Typeset in Times
by codeMantra

Access the Support Material: Routledge.com/9781032883274

Dedication

To my beloved mother and late father

Contents

Part I Elastic Solid

List of Figures

List of Tables

Preface

This book is concerned with the mechanics of structures, which specialize the general equations of solid mechanics by taking advantage of geometry-dictated simplifications. The aim of structural mechanics is to empower students and early-career professionals with a culture that allows them to interact with specialty (aerospace, mechanical, and civil) engineers and understand the complex approaches to dimensioning of load-bearing structures.

The text grew from the teaching experience of the author over the past 20 years at Texas A&M University. The basis are lecture notes in French by Serge Laroze at the Ecole Nationale Supérieure de l'Aéronautique et de l'Espace, better known as SUPAERO, which the author attended between 1992 and 1995. The original text was critically examined, augmented with proofs where omissions were noted, with examples added where those of the original were insufficient or too complex, with many more illustrations and figures to make difficult concepts or calculation results more accessible. In addition, a general introduction to the theory of elasticity and an original chapter on experimental measurements were, respectively, added at the beginning and at the end of this book. The result is a twofold increase in content volume without losing the intended focus and coherence.

Why a new book on structures? At the risk of being schematic, I would say that teaching structures has been approached in one of two ways: the "strength of materials" way or using the "theory of elasticity". The strength of materials approach has deep historical roots going, in Western literature, as far back as Galileo. The approach predates the introduction of such fundamental notions as those of stress and strain and eventually settled with Claude-Louis Navier's (1785–1836) *Résumé des Lecons*. A century later, this approach was advocated and popularized by Stephen Timoshenko (1878–1972) in the first edition of his *Strength of Materials* (1930). This approach focuses on approximate solutions and practical applications by simplifying all derivations in order to make the subject accessible to a student with the "usual preparation in mathematics". Arguments in favor of this approach are quite sensible until the shortcomings are critically examined. Further simplified in the second edition of 1940, Timoshenko's way of teaching the subject has become the de facto way, with many subsequent editions to this day under the title of Mechanics of Materials, including by many other authors. This is by far still the dominant way of teaching the subject.

In contrast to this elementary approach, the theory of elasticity approach was introduced by Barré de Saint-Venant in the mid-1850s, starting with his celebrated memoirs presented to the French Academy of Sciences in 1853 and 1855 and eventually settling in his commentary of Clebsch's book in 1883. In between, Saint-Venant published in 1864 a third edition of Navier's *Résumé des Lecons* augmenting it with a commentary that increased the volume from 120 to over 850 pages! The theory of elasticity approach to structures consists of solving boundary-value problems for all practical types of loading of prismatic beams, then generalizing the solutions to arbitrary beams. The solutions that are obtained are exact under well-specified conditions.

It would seem, therefore, that, at least from the instructor's point of view, the choice lies between a way that is approximate but simple and one that is exact but cumbersome. In scanning the textbook literature, one sees a continuum of choices between these two extremes. There have been quite a few attempts to inject more content from the theory of elasticity, but they remain rather timid on the whole. It would be fair to say that the subject has developed and matured, thanks to civil engineers. Historically speaking, many of the early developments were to accompany the design and safe operation of civil infrastructure. Interestingly, however, most attempts to break away from the traditional strength of materials approach have been made by, or destined for, professionals outside of civil engineering. In that vein, one notes the valuable attempts of Megson, Allen and Haisler, Donaldson, Bauchau, and Craig, among others.

Is the choice indeed between simple approximation and exact complication? In fact, there is more to the matter. Approximation is an art and, one may say, a science. The most disturbing issue with the elementary approach is that it employs gratuitous, and often wrong, assumptions to arrive at true results in some particular cases. These assumptions pertain to the manner in which things happen, for example, the conservation of the planar shape of beam cross-sections during deformation. Additional key assumptions are made, often unbeknownst to the readers, regarding the manner in which loads are applied. The student is then left to believe that such assumptions, at least those explicitly stated, are essential to the final result. This is damaging!

When the strength of materials was in its formative years, not much was known about stress and strain as we know them today. It made sense then to make assumptions about how things happen because we had no clue how they actually did happen. Consider the exemplary assumption of "plane sections remain plane", the bedrock of the so-called Euler–Bernoulli beam theory. We know today exactly and completely the circumstances under which sections remain plane and those under which they do not. When I say today, I mean we have known this for well over 170 years! Assuming that which we know is true is inelegant. Assuming that which we know is false is neglectful. The elementary approach is thus both inelegant and neglectful. On the other hand, the theory of elasticity approach is complete and insightful.

Fortunately for the utilitarian, how the subject is approached and taught does not affect how a wide range of practical problems are solved. Therefore, the basic training of students can be left intact. There are, however, other consequences of the choice made about how to teach the subject. In the elementary approach, the prevalence of often unrelated approximations across subtopics unavoidably leads to a dichotomized subject. The theory of elasticity, on the other hand, offers a unified way of covering the subject matter. This unification of subtopics helps develop patterns in the mind of the student; the regularity of patterns allows for compressibility of knowledge, and this, in turn, helps with retention.

Inelegance and neglect versus completeness and insight; juxtaposition of topics versus their unification are just some of the contrasts between the elementary approach and theory of elasticity approach and there is more. Strength of materials, as taught by the elementary approach, is essentially a terminal subject. What is there after the student has learned the 'flea' equation in extension, the moment–curvature relation in bending, or the torque versus angle of twist in torsion? How does that prepare him for the next course in a logical sequence culminating in, say a finite element course or the like? Not much. This is so because other than exercises in statics, which are taught in a course prior, nothing else is needed other than limited geometric considerations that live within this course and are absolutely useless elsewhere. By way of contrast, the theory of elasticity approach frames the mind of the student on how to formulate a well-posed boundary-value problem, illustrates the complexity of obtaining exact solutions to generic problems and therefore prepares him for the next course and the one following that, and so on. In passing, this approach definitely lowers the barrier to entry for graduate-level courses, which, in the US at least, are far too advanced in comparison to how mechanics is taught to undergraduates.

It would be pretentious to claim that the above arguments for or against either approach have escaped educators and other stakeholders for nearly a century. Then, why has the theory of elasticity approach not been fully developed and widely implemented? First, one cannot underestimate the role of inertia in any field, especially when the focus in academe is on new developments rather than refining old subjects. Second, initial conditions are important, and some pretty hard ones were set by Timoshenko's advocacy of the elementary approach. Two elements plead in favor: (i) the lack of student preparedness in mathematics, and (ii) feedback from prior attempts at introducing elasticity in structures, which were deemed as failures. These are two reasons that transpire in many of Timoshenko's writings, in particular in his History Granted that were good reasons then, are they still valid now?

It may be argued that the average level of student preparation in mathematics has declined. With the advent of powerful symbolic math software (Mathematica, Maple, and MATLAB), some laziness in mathematical reasoning, processing, and technique has slowly crept among instructors and students alike. To this latent overall weakening, add the might of artificial intelligence. The rise of AI is too important to be ignored, but the belief that it is sufficient is too naive to be left unchallenged. If anything, for engineers not to become mediocre consumers of AI, they need to strengthen their math, which, together with computer science, is the foundation par excellence of AI. Coming down to earth, so to speak, I believe that the average level of students in math in the US in 1940 (when Timoshenko published his second edition) was much lower than it is today. An analysis of this trend merits study in itself. My belief stems from the fact that the US of 1930–1940, between the two wars and Timoshenko's first two editions, was not yet the center of gravity of mechanics or of science in general. There is no need to elaborate further on the tremendous changes in the scientific landscape to this day.

There remains the need to analyze previous attempts at incorporating the theory of elasticity in structures education. These are of two kinds: those deemed as failures by Timoshenko prior to his time, and those that emerged since, which I have qualified above as timid. I must immediately point out the anachronistic character of Timoshenko's own assessment of the situation. He was fully aware that Saint-Venant had worked toward a theory of elasticity approach to the strength of materials in his 1864 edition of Navier's *Résumé des Lecons*. He noted and commended Saint-Venant's didactic efforts. However, he deemed other attempts, notably by Grashof and Winkler in Germany, as failures because of terse style and excessive abstraction. When he arrived at the conclusion that an elementary approach was the most suitable, he believed that Navier and Bresse, whom he held in high regard, would do the same. Clearly, Timoshenko did not seem to be aware of developments that were taking place in France around the very same time he published his second edition. The year 1934 appears as a critically important year after which the theory of elasticity approach was finally introduced at Ecole des Ponts et Chaussés, where Cauchy, Navier, and Bresse had taught. This was done by Gaston Pigeaud, then professor at the school, and seems to have prevailed, for the most part, to this day. The teaching of structures at SUPAERO essentially followed suit. The school was founded in 1909 in Paris and was moved to Toulouse in 1968. In some ways, education in structures was even more 'liberal' there, given the absence of the heavy legacy of civil engineering.

As for the attempts of the past few decades, the reader will understand better how they differ from the proposed approach when I outline various chapters of this book further below.

The aim of this book is thus to provide a way to teach structures entirely from the perspective of the theory of elasticity. It supports those instructors who have grown uneasy with the elementary approach by providing a coherent, economical, uncompromising, and naturally flowing description where various components are logically connected.

I must point out, at this juncture, that this book would be inappropriate in programs where the curriculum is limited to a single course on strength of materials. Using this text presupposes an ideal sequence of three courses, where this book may be used as an intermediate between a general introduction to the mechanics of solids and an introduction to the finite element method.

The reader is expected to be familiar with general concepts from statics, as well as notions of strain, stress, and stress–strain relations. On strain, the reader is expected to be familiar with the geometric meaning of extensional and shear strains as well as express mathematically the displacement–strain relationships describing the resulting motion. On stress, the student is expected to be familiar with the notion of traction and how it relates to stress as well as express stress equilibrium mathematically. On stress–strain relations, the student is expected to be familiar with the most basic of all, Hooke's law for linear elasticity, as well as fundamental implications, such as the superposition principle. Part I, with its four chapters, reviews the above fundamentals, which are typically covered in a first course on deformable solids. This part is written in a self-contained manner, generally without proofs. It contains a limited number of illustrations and examples aimed at refreshing previously seen material, perhaps in a new way. Likewise, a limited number of problems are offered at the end of each chapter. One important concept emphasized in this part is the Saint-Venant principle and a definition of a Saint-Venant problem, hinted at as a double infinity of boundary-value problems. The uniaxial loading of a bar and the axisymmetric torsion problem are used as examples to illustrate key issues. The entire part has a length comparable with a long chapter from the second part of this book.

Part II is the book. It is concerned with the three-dimensional theory of beams of any form or cross-sectional shape, including flexible rods, rings, tubes, multi-cell sections, and thin-walled structures.

Chapter 5 deals with all things geometric, once and for all. A formalism is developed for beams with double curvature. Two frames of reference bear importance: one associated with the centroidal axis, and the other with the principal second moments of area of the cross-section. In practice, the most advanced examples and problems are provided for beams admitting a median plane and loaded in that plane. This allows for a comprehensive coverage of rings and arches throughout the text. The second half of the chapter deals with beam statics. A daring, yet concise term of 'screwer' (from the French 'visseur') is introduced to mean the set of internal resultants (forces and moments) with respect to the centroid of the current cross-section. Differential equations of beam equilibrium are then derived in terms of the screwer, which has six components. In intrinsic (vector) form, the differential equations are as simple as they can be, regardless of double curvature. They only become complicated (to the naked eye) when projected onto one of the two frames of description.

There is no sign convention *per se*, only a convention about what is called action and what is called reaction in the sense of Newton's third law. Standard textbooks default on this issue in a way that confuses students.

The screwer has six components: three forces and three moments. Physically, however, it encompasses four types of solicitation: normal force, shear force, torque, and bending moment. Chapters 6 to 9 deal with these four types, not in this order, but rather in increasing order of difficulty. Essential results are obtained by formulating and solving a pedagogically designed sequence of Saint-Venant problems. It is the difficulty of the corresponding Saint-Venant problem that dictates the order in which various solicitations are covered. The difficulty goes crescendo from normal force, to bending, torsion then to shear force. The latter is de facto a general theory of bending, since among all screwer components, only the shear force cannot be conceived to act alone on all sections of a prismatic beam. Of particular interest is the analysis of thin-walled structures in Chapters 8 and 9. It is complete and, in the spirit of the book, presented in a unifying, naturally flowing manner. Two essential concepts and theorems enable a seamless transition from the coverage of solid sections to thin-walled tubes: the flux and circulation of shear tractions, having to do, respectively, with conditions of equilibrium and geometric fit (or strain compatibility). This enables original derivations of governing equations that are simpler and more elegant than what the literature provides in this regard. An example pertains to the Bredt formula in torsion. In this book, the Bredt equation refers to applying the flux theorem to an arbitrary wall domain under transverse shear, as a tribute to the German engineer and crane industry businessman, Rudolph Bredt (1842–1900).

Chapter 10 deals with combined loadings, as induced for example by mere curvature of the median line. Bresse's equations for ossatures are developed with some limited applications to frames and rings, showing how they can be used in hyperstatic problems. The second part of the chapter deals with stress concentrations if only to highlight one limitation of beam theory, yet demonstrate how the latter can still serve as a reference nominal solution. Some case studies are selected from the book of Rudolph Peterson, which is an important resource for the engineer.

Chapter 11 deals with the overall problem of structural failure. After an introductory section on discrete structures, focus is laid on the buckling of beams by compression-induced plane bending (the so-called column buckling). Then, the problem of bending-induced buckling by combined torsion and secondary bending is covered (so-called lateral buckling). Finally, the buckling of arches and rings is covered to some extent. Extensive tables are provided at the end of the chapter.

This book closes with Chapter 12 on experimental measurements. After recalling some basics about strain gauges and Wheatstone bridges, emphasis is laid on measuring the components of the screwer. When only one component is present, the methods are applicable to arbitrary cross-sections. Under combined loading, however, sufficient conditions must be sought to ensure the independence of various measurements. These conditions pertain to symmetries imposed on the cross-section. Independent measurements are presented as case studies in the final section of the chapter.

In each chapter, special emphasis is placed on synthesizing the results of a theory of elasticity analysis. Problems at the end of each chapter test the student's understanding of fundamental aspects, but most problems are geared toward applications and make use of the synthesis part of the chapter.

Part II of this book, Chapter 12 excluded, is intended to be used as the reference text for a semester-long course on the mechanics of structures. For this reason, many topics are left out, for example, thermal loading, dynamic loading, anisotropic elasticity, heterogeneous beams, plates and shells, energy methods, limit design, and limit load analysis.

A word of caution in the end. A book does not teach. If a book is not used appropriately, it will not be effective. My advocacy for a theory of elasticity perspective should not be mistaken for teaching continuum mechanics instead of structures. That would be detrimental. All content of this book that centers on the theory of elasticity constitutes lecture material at best. In particular, it should not be used as homework material and certainly not as exam material. Otherwise, this would take away valuable time from empowering students with skills for solving practical problems, which demand their own right of training.

Special care has been given to preparing the illustrations in this book. Part of the reason behind this effort is precisely to circumvent the inherent difficulty of the subject in general, and when it is approached with the mathematical theory of elasticity in particular. All figures were made from scratch using `PSTricks`, a LaTeX package.

As a visual aid, a color code is used throughout. Externally applied loads are shown in red and internal resultants in blue, while physical objects are shown using shades of maroon. Figures in Part I are made in black and white, with shades of gray, except for the red color for loads. This color code is commensurate with a flashback to a past course, as opposed to a colorful present! A solutions manual will be made available to instructors with answers provided to all problems. In addition, typical problems are worked out in detail in the manual with illustrations and figures using the same color code as in the main text. Google Gemini (version 2.5 Flash) was used to generate some of the alt text in this book.

A preliminary, much shorter edition of this text has been used as a course pack since the fall of 2019. It has enabled the author and two other colleagues to experiment with presenting the material. I thank members of the Texas A&M aerospace engineering classes since then who used the preliminary edition and provided valuable feedback. Special thanks to my colleague J.-B. le Graverend who has used the course pack multiple times and to G. Stevenson who provided many comments based on a student survey. I also wish to thank the many readers of various chapters of this book: M. Naraghi, A. Needleman, M. Vasoya, and N. Thomas on Part I; X. Poulain on Chapter 5; M. Torki on Chapter 7; A. Cruzado on Chapter 8; P.J. Guruprasad and R. Vignesh on Chapter 9; K. Chowdhury on Chapter 10; S. Keralavarma on Chapter 11; and K. Cundiff on Chapter 12. Their collective comments and criticisms have helped me make this book better than it otherwise would have been.

A. Amine Benzerga

Part I

Elastic Solid

1 Strain

1.1 DISPLACEMENT AND STRAIN

The concept of displacement, on the one hand, is central to the kinematics of solids, which is a branch of general mechanics. The concept of strain, on the other hand, is central to the kinematics of deformable solids, which is a branch of continuum solid mechanics. Displacement is more general than strain. It is necessary to strain but not sufficient for it. In other words, there can be no strain if there is no displacement, but mere rigid body displacement (translation or rotation) does not entail deformation of the body. Deformation implies some sort of distortion in which the positions of material particles within the body change relative to each other. All bodies are deformable. A rigid body is an abstract concept and an approximation of reality. Rigid body dynamics is an important branch of mechanics. Stability aside, this book only deals with deformable solids.

Physically, only two independent types of strain are conceivable: a relative change in length of a reference material line, Figure 1.1a, or a relative change in angle between two reference material lines, Figure 1.1b. The former is called normal strain, while the latter is called shear strain.

The normal strain in Figure 1.1a is *extensional* ($\varepsilon_x > 0$); it corresponds to an increase in the length of a line segment of length dx before deformation. Contraction would correspond to $\varepsilon_x < 0$. Figure 1.1a illustrates how the normal strain results from the variation of displacement along the direction of the displacement, i.e.,

$$\varepsilon_x = \frac{\partial u}{\partial x} \tag{1.1}$$

The sign of the shear strain bears no physical meaning. However, if the angle between

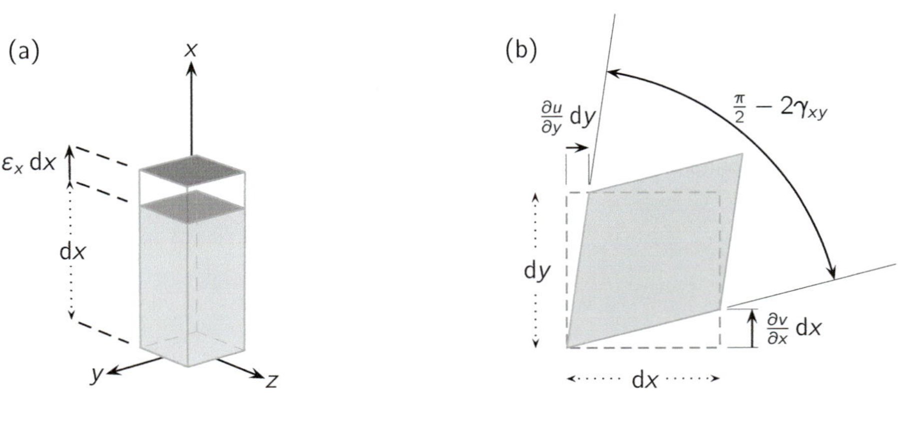

Figure 1.1 Concepts of (a) normal strain and (b) shear strain.

DOI: 10.1201/9781003537229-2

a pair of initially perpendicular line elements, say dx and dy, decreases, then $\gamma_{xy} > 0$, as in Figure 1.1b. If the angle increases, then $\gamma_{xy} < 0$. For small changes in angles, the angle is identified with its tangent, and the drawing in Figure 1.1b shows that:

$$2\gamma_{xy} = \frac{\partial u}{\partial y} + \frac{\partial v}{\partial x} \tag{1.2}$$

Because the body lies in a Euclidean space endowed with three dimensions, there are three independent normal strains and three independent shear strains. This presumes that the normal strain along an arbitrary direction can be related to the three independent normal strains and, as a matter of fact, to the three independent shear strains. Likewise, the shear strain in an arbitrary plane is related not only to all three independent shear strains but also to the three independent normal strains.

For our purposes, all information about the deformation of a given solid is contained in the concept of strain. Strain and displacement are related by[1]

$$\boxed{2\boldsymbol{\varepsilon} = \nabla\mathbf{u} + \nabla^T\mathbf{u}} \tag{1.3}$$

where $\boldsymbol{\varepsilon}$ denotes the strain tensor, $\mathbf{u}$ is the displacement vector, and the superscript T denotes the transpose. Thus, the strain tensor is the symmetric part of the displacement gradient $\nabla\mathbf{u}$.

In some cases, the skew-symmetric part of $\nabla\mathbf{u}$ contains additional information, e.g., rotation of cross-sections in beams. Rotation and displacement are related by

$$2\boldsymbol{\Omega} = \nabla\mathbf{u} - \nabla^T\mathbf{u} \tag{1.4}$$

where $\boldsymbol{\Omega}$ is the (infinitesimal) rotation tensor.

The above expressions are general, but the component-to-component relations depend on the chosen coordinate system. Let u, v, and w denote the displacement components in a Cartesian coordinate system with reference to some fixed basis, i.e.,

$$\{\mathbf{u}\} = \begin{Bmatrix} u(x,y,z) \\ v(x,y,z) \\ w(x,y,z) \end{Bmatrix} \tag{1.5}$$

Then, the components of the displacement gradient $\nabla\mathbf{u}$ are given by the matrix:

$$[\nabla\boldsymbol{u}] = \begin{bmatrix} \frac{\partial u}{\partial x} & \frac{\partial u}{\partial y} & \frac{\partial u}{\partial z} \\ \frac{\partial v}{\partial x} & \frac{\partial v}{\partial y} & \frac{\partial v}{\partial z} \\ \frac{\partial w}{\partial x} & \frac{\partial w}{\partial y} & \frac{\partial w}{\partial z} \end{bmatrix} \tag{1.6}$$

where the braces in $\{\mathbf{u}\}$ in Equation (1.5) and the brackets in $[\nabla\boldsymbol{u}]$ in Equation (1.6) are to emphasize that the given matrices are just representations of what the braces or brackets contain. There is an infinity of ways to represent, with components, the physically unique displacement and displacement gradient.

From their respective symmetry properties, the strain and rotation tensors take the general forms:

$$[\boldsymbol{\varepsilon}] = \begin{bmatrix} \varepsilon_x & \gamma_{xy} & \gamma_{xz} \\ \gamma_{xy} & \varepsilon_y & \gamma_{yz} \\ \gamma_{xz} & \gamma_{yz} & \varepsilon_z \end{bmatrix} \qquad [\boldsymbol{\Omega}] = \begin{bmatrix} 0 & \Omega_{xy} & \Omega_{xz} \\ -\Omega_{xy} & 0 & \Omega_{yz} \\ -\Omega_{xz} & -\Omega_{yz} & 0 \end{bmatrix} \tag{1.7}$$

Since it is symmetric, the strain tensor contains six independent components. Physically, there are two types of strain, as stated above. Normal strains describe the extension or contraction of material line elements. These are represented by the diagonal elements of $[\boldsymbol{\varepsilon}]$, namely ε_x, ε_y, and ε_z. Shear strains describe the relative change in angle between two reference material line elements. In Equation (1.7), these are represented by the off-diagonal elements γ_{xy}, γ_{xz}, and γ_{yz}. In advanced texts, the same Greek symbol ε is used for both extensional and shear strains with two indices for each. Here, a different choice is made if only to distinguish, throughout this chapter, the only two types of strain: normal (or extensional) with the symbol ε, and shear with the symbol γ.

On the other hand, the rotation tensor $\boldsymbol{\Omega}$ is skew-symmetric so that only three components are independent. Their physical meaning is given below.

In component form, the strain–displacement relations, Equation (1.3), become

$$\varepsilon_x = \frac{\partial u}{\partial x}; \qquad \varepsilon_y = \frac{\partial v}{\partial y}; \qquad \varepsilon_z = \frac{\partial w}{\partial z} \tag{1.8}$$

for the normal strains and

$$2\gamma_{xy} = \frac{\partial u}{\partial y} + \frac{\partial v}{\partial x}; \qquad 2\gamma_{xz} = \frac{\partial u}{\partial z} + \frac{\partial w}{\partial x}; \qquad 2\gamma_{yz} = \frac{\partial v}{\partial z} + \frac{\partial w}{\partial y} \tag{1.9}$$

for the shear strains, where use has been made of Equation (1.6) and the definition of a transpose. Also, the rotation–displacement relations, Equation (1.4), become

$$2\Omega_{xy} = \frac{\partial u}{\partial y} - \frac{\partial v}{\partial x}; \qquad 2\Omega_{xz} = \frac{\partial u}{\partial z} - \frac{\partial w}{\partial x}; \qquad 2\Omega_{yz} = \frac{\partial v}{\partial z} - \frac{\partial w}{\partial y} \tag{1.10}$$

Remark 1.1 *The strain–displacement relations and rotation–displacement relations are sometimes referred to as kinematic relations. When rotations are not relevant, kinematic relations simply refer to the first set of relations, i.e., Equation* (1.3) *in intrinsic form or Equations* (1.8) *and* (1.9) *in component form.* *

Remark 1.2 *The factor 2 appearing in the shear strain relations, Equation* (1.9), *is for mere convenience so that the tensorial form of Equation* (1.3) *holds for both shear and normal strains. Omitting factor 2 in Equation* (1.9) *would define the engineering (shear) strains. One cannot write a general tensorial relation for engineering strains as in Equation* (1.3). *

Remark 1.3 *The extensional strain along some arbitrary direction defined by unit vector* **n** *is given by*

$$\varepsilon_n = \mathbf{n} \cdot \boldsymbol{\varepsilon}\mathbf{n} \tag{1.11}$$

where the $\cdot$ *symbol refers to the usual vector scalar product and the term* $\boldsymbol{\varepsilon}\mathbf{n}$ *is evaluated in the usual way a matrix is multiplied by a vector. Also, the shear strain in some arbitrary plane defined by unit vectors* **n** *and* **m**, *taken to be orthogonal to each other, is*

$$\gamma_{nm} = \mathbf{n} \cdot \boldsymbol{\varepsilon}\mathbf{m} = \mathbf{m} \cdot \boldsymbol{\varepsilon}\mathbf{n} = \gamma_{mn} \tag{1.12}$$

since the strain tensor is symmetric. *

Remark 1.4 *The strain tensor has three invariants that enter its characteristic equation. One of these, the trace, has a physical meaning and corresponds to a relative change in volume due to deformation. Recall that the trace of a matrix is the sum of its diagonal components. If an initial chunk of material occupying volume V changes after deformation by an amount* δV, *then*

$$\frac{\delta V}{V} = \operatorname{tr} \boldsymbol{\varepsilon} = \varepsilon_x + \varepsilon_y + \varepsilon_z \tag{1.13}$$

The relative change in volume is also known as the dilatation. *

Finally, since only six strain components and three rotation components are independent, it is sometimes useful to store all strain information in a six-vector (i.e., a vector with six components), $\bar{\boldsymbol{\varepsilon}}$, and all rotations in a three-vector, $\boldsymbol{\omega}$, known as the pseudo-rotation vector. For strains, the order in which information is stored matters. For rotations, the signs of various terms also matter. Specifically,

$$\begin{aligned} [\bar{\boldsymbol{\varepsilon}}]^T &= \begin{bmatrix} \varepsilon_x & \varepsilon_y & \varepsilon_z & \sqrt{2}\gamma_{yz} & \sqrt{2}\gamma_{xz} & \sqrt{2}\gamma_{xy} \end{bmatrix} \\ &\equiv \begin{bmatrix} \varepsilon_1 & \varepsilon_2 & \varepsilon_3 & \varepsilon_4 & \varepsilon_5 & \varepsilon_6 \end{bmatrix} \end{aligned} \tag{1.14}$$

and

$$[\boldsymbol{\omega}]^T = \begin{bmatrix} -\Omega_{yz} & \Omega_{xz} & -\Omega_{xy} \end{bmatrix} \equiv \begin{bmatrix} \omega_x & \omega_y & \omega_z \end{bmatrix} \tag{1.15}$$

The above notation condensation is known as Voigt's notation. Equation (1.14) plays a role in writing constitutive relations in matrix form, which is more suitable for computer operations. The rationale for the $\sqrt{2}$ factor in the shear terms is proper accounting of energy and work. An illustration of that will be given in context in Section 3.5. Equation (1.15) is typically used in evaluating cross-section rotations in beam theory.

1.2 COMPATIBILITY EQUATIONS

When a deformable solid undergoes some strain, material points move relative to each other. In a deformable solid modeled as a continuum, two infinitely close material points must move "in concert" in such a way that no interpenetration of matter occurs and no spontaneous separation leading to void formation occurs either. This is the physical meaning of strain compatibility.

Mathematically, compatibility may be understood as follows. Consider a displacement field $\mathbf{u}$ where each component of the displacement, i.e., $u(x,y,z)$, $v(x,y,z)$, or $w(x,y,z)$, is a function of position $[\mathbf{x}]^T = [x,y,z]$. Starting from a continuous displacement field $\mathbf{u}(\mathbf{x})$, obtaining the corresponding strain field from the kinematic relations, Equation (1.3), is straightforward as this involves differentiation using the explicit relations in Equations (1.8) and (1.9). Conversely, imagine that it is the strain field that is known. Obtaining the corresponding displacement field would now require integrating kinematic relations (1.3). Without getting into much detail, there are some conditions that guarantee the integrability of Equation (1.3) to obtain a continuous displacement field. Such conditions are known as the compatibility equations.

There are several ways of expressing the compatibility equations. One essential aspect is that only six equations are independent. For example, in component form, they may be written as follows:

$$\frac{\partial^2 \varepsilon_x}{\partial y^2} + \frac{\partial^2 \varepsilon_y}{\partial x^2} = 2\frac{\partial^2 \gamma_{xy}}{\partial x \partial y} \tag{1.16a}$$

$$\frac{\partial^2 \varepsilon_y}{\partial z^2} + \frac{\partial^2 \varepsilon_z}{\partial y^2} = 2\frac{\partial^2 \gamma_{yz}}{\partial y \partial z} \tag{1.16b}$$

$$\frac{\partial^2 \varepsilon_z}{\partial x^2} + \frac{\partial^2 \varepsilon_x}{\partial z^2} = 2\frac{\partial^2 \gamma_{zx}}{\partial z \partial x} \tag{1.16c}$$

$$\frac{\partial^2 \varepsilon_x}{\partial y \partial z} = \frac{\partial}{\partial x}\left(-\frac{\partial \gamma_{yz}}{\partial x} + \frac{\partial \gamma_{zx}}{\partial y} + \frac{\partial \gamma_{xy}}{\partial z}\right) \tag{1.16d}$$

$$\frac{\partial^2 \varepsilon_y}{\partial z \partial x} = \frac{\partial}{\partial y}\left(-\frac{\partial \gamma_{zx}}{\partial y} + \frac{\partial \gamma_{xy}}{\partial z} + \frac{\partial \gamma_{yz}}{\partial x}\right) \tag{1.16e}$$

$$\frac{\partial^2 \varepsilon_z}{\partial x \partial y} = \frac{\partial}{\partial z}\left(-\frac{\partial \gamma_{xy}}{\partial z} + \frac{\partial \gamma_{yz}}{\partial x} + \frac{\partial \gamma_{zx}}{\partial y}\right) \tag{1.16f}$$

The above conditions are equivalent to the conditions of integrability of Equation (1.3), i.e., to the existence of a single-valued continuous displacement field, under some additional restrictions that go beyond the scope of this summary. For our purposes, therefore, the above six compatibility equations are equivalent to the six kinematic relations underlying Equations (1.8) and (1.9). It is easier to prove that the above conditions are necessary than to prove that they are sufficient. Both proofs are typically covered in an advanced course on solid mechanics.

The specific way in which the compatibility equations may be needed in this course will be clarified in Chapter 4 and illustrated in two important topics of structural mechanics in Part II. Specifically, the compatibility equations will be central in developing the general theory of torsion of solid sections in Chapter 8 and in developing the general theory of bending with transverse shear in Chapter 9.

Remark 1.5 *The above compatibility relations contain two types of relations. All others may be deduced by circular permutation of indices. Yet, they are rather complex, and typically one would need to look them up.* ∗

Remark 1.6 *One way to remember how to write the compatibility equations in a general form that encompasses the above equations as well as other forms is explained below.*[2]

In order to implement this mnemonic way, introduce the following surrogate notation for partial derivative:

$$D_i\varepsilon = \frac{\partial \varepsilon}{\partial i}$$

where i stands for one of the coordinates x, y, z. Also, denote normal strains as ε_{xx}*, etc. and shear strains as* ε_{xy}*, etc. Then, formally one may write successively*

$$DD\varepsilon + DD\varepsilon - DD\varepsilon - DD\varepsilon = 0 \tag{1.17a}$$

$$D_iD\varepsilon_j + DD\varepsilon - D_iD\varepsilon_j - DD\varepsilon = 0 \tag{1.17b}$$

$$D_iD\varepsilon_j + D_jD\varepsilon_i - D_iD\varepsilon_j - D_jD\varepsilon_i = 0 \tag{1.17c}$$

$$D_iD_k\varepsilon_{jl} + D_jD_l\varepsilon_{ik} - D_iD_k\varepsilon_{jl} - D_jD_l\varepsilon_{ik} = 0 \tag{1.17d}$$

to finally obtain

$$\boxed{D_iD_k\varepsilon_{jl} + D_jD_l\varepsilon_{ik} - D_iD_l\varepsilon_{jk} - D_jD_k\varepsilon_{il} = 0} \tag{1.18}$$

The first three statements are mathematically meaningless, yet formally valid. In the statement (1.17c), the indices from left to right read: i j ji i j ji. If the same pattern were kept when indices k and l are introduced, one would get Equation (1.17d)*, which is identically verified for any combination of indices. To break symmetry, one introduces the latter two indices following the sequence: kl lk lk kl, which then delivers the desired compatibility relations in Equation* (1.18)*. Equation (1.17d) is kept for didactic purposes and is grayed to indicate that it is to be dismissed.* ∗

1.3 PRINCIPAL STRAINS

Let λ be an eigenvalue of the strain matrix $[\boldsymbol{\varepsilon}]$. There is a direction, say along unit vector $\mathbf{n}$, such that

$$\boldsymbol{\varepsilon}\mathbf{n} = \lambda\mathbf{n} \tag{1.19}$$

The above relation is the very definition of an eigenvalue associated with eigenvector $\mathbf{n}$. Using Equation (1.12), the shear strain on a face of normal $\mathbf{n}$ in any direction of unit vector $\mathbf{m}$ is obtained as follows:

$$\mathbf{m}\cdot\boldsymbol{\varepsilon}\mathbf{n} = \mathbf{m}\cdot(\lambda\mathbf{n}) = \lambda\mathbf{m}\cdot\mathbf{n} = 0 \tag{1.20}$$

since $\mathbf{m}$ is perpendicular to $\mathbf{n}$. Hence, the face defined by an eigenvector of the strain matrix does not undergo any shear. This means that elementary material fibers that are oriented perpendicular to that face are only subjected to extensional strain: they can either dilate or contract. The amount of extensional strain is directly obtained from Equation (1.11) as follows:

$$\mathbf{n} \cdot \boldsymbol{\varepsilon}\mathbf{n} = \mathbf{n} \cdot (\lambda \mathbf{n}) = \lambda \mathbf{n} \cdot \mathbf{n} = \lambda \tag{1.21}$$

since $\mathbf{n}$ is a unit vector. The eigenvalue λ is said to be a principal strain.

The strain matrix $[\boldsymbol{\varepsilon}]$ in Equation $(1.7)_1$ is symmetric. Applying the spectral decomposition theorem of linear algebra, the strain matrix can be diagonalized. Specifically, the strain tensor $\boldsymbol{\varepsilon}$ has three real eigenvalues, $\varepsilon_1, \varepsilon_2, \varepsilon_3$, corresponding to three orthogonal eigenvectors, $\mathbf{e}_1, \mathbf{e}_2, \mathbf{e}_3$. The eigenvalues and eigenvectors of $\boldsymbol{\varepsilon}$ are called principal strains and principal strain directions, respectively. A plane containing two principal strain directions is called a principal plane for strain.

Remark 1.7 *There is no shear on a face parallel to a principal plane, since the normal to it is a principal direction; see Equation* (1.20). *Therefore, the angle between principal strain directions does not change upon deformation. The principal axes of strain remain perpendicular during deformation.* *

Remark 1.8 *The sum of the principal strains, which is the trace of the strain tensor, is the dilatation; see Remark 1.4. The value of the sum of the normal strains does not depend on the axes or basis used to express the components of the strain tensor:*

$$\operatorname{tr}\boldsymbol{\varepsilon} = \varepsilon_x + \varepsilon_y + \varepsilon_z = \varepsilon_1 + \varepsilon_2 + \varepsilon_3 \tag{1.22}$$

*

1.4 STRAIN TRANSFORMATIONS

1.4.1 PLANE TRANSFORMATION

We sometimes need to express the components of strain with respect to axes other than the default axes.

Assume, for the time being, plane deformations and that all straining occurs in the x–y plane. The only non-zero strain components are then ε_x, ε_y, and γ_{xy} with respect to an orthonormal basis $\{\mathbf{e}_x, \mathbf{e}_y, \mathbf{e}_z\}$. We wish to express the strain components in a new orthonormal basis $\{\mathbf{e}_x^*, \mathbf{e}_y^*, \mathbf{e}_z^*\}$, which is rotated by an angle θ about the z-axis (Figure 1.2a). The z-axis is common to both bases, i.e., $\mathbf{e}_z = \mathbf{e}_z^*$. The below expressions give the strain components in the starred basis in terms of those in the original basis and the angle θ:

$$\begin{cases} \varepsilon_x^* = \dfrac{\varepsilon_x - \varepsilon_y}{2}\cos 2\theta + \gamma_{xy}\sin 2\theta + \dfrac{\varepsilon_x + \varepsilon_y}{2} \\[2ex] \varepsilon_y^* = \dfrac{\varepsilon_y - \varepsilon_x}{2}\cos 2\theta - \gamma_{xy}\sin 2\theta + \dfrac{\varepsilon_x + \varepsilon_y}{2} \\[2ex] \gamma_{xy}^* = \dfrac{\varepsilon_y - \varepsilon_x}{2}\sin 2\theta + \gamma_{xy}\cos 2\theta \end{cases} \tag{1.23}$$

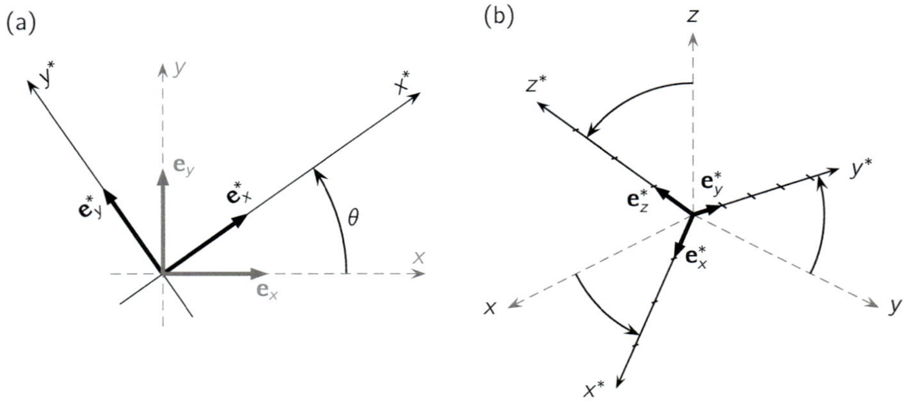

Figure 1.2 Original and rotated bases for (a) plane and (b) general transformations.

Adding term to term, the first two equations give $\varepsilon_x^* + \varepsilon_y^* = \varepsilon_x + \varepsilon_y$. Since the out-of-plane strain ε_z is zero, this sum is the dilatation, which must be conserved in any change of basis; see Remark 1.4. In turn, the invariance of the trace should serve as a check in keeping the signs correct when writing Equation (1.23).

The transformation relations in Equation (1.23) are often written for the state of plane strain defined above. They remain valid for a general deformation provided that the transformation is planar. To see this, it is necessary to introduce the most general transformation law for strain and then specialize the result to a plane transformation.

1.4.2 GENERAL TRANSFORMATION

For a general transformation, the new basis $\{\mathbf{e}_x^*, \mathbf{e}_y^*, \mathbf{e}_z^*\}$ is obtained from the original basis $\{\mathbf{e}_x, \mathbf{e}_y, \mathbf{e}_z\}$ by an arbitrary rotation (Figure 1.2b). Passing from the original basis to the rotated (starred) basis is accomplished using the matrix of direction cosines:

$$[\boldsymbol{R}] = \begin{bmatrix} \cos(\mathbf{e}_x, \mathbf{e}_x^*) & \cos(\mathbf{e}_x, \mathbf{e}_y^*) & \cos(\mathbf{e}_x, \mathbf{e}_z^*) \\ \cos(\mathbf{e}_y, \mathbf{e}_x^*) & \cos(\mathbf{e}_y, \mathbf{e}_y^*) & \cos(\mathbf{e}_y, \mathbf{e}_z^*) \\ \cos(\mathbf{e}_z, \mathbf{e}_x^*) & \cos(\mathbf{e}_z, \mathbf{e}_y^*) & \cos(\mathbf{e}_z, \mathbf{e}_z^*) \end{bmatrix} \tag{1.24}$$

This is the transformation matrix between the two bases. It is orthogonal, i.e., $\boldsymbol{R}^{-1} = \boldsymbol{R}^T$. In Figure 1.2b, only three of the nine angles are shown.

Then, any vector, for example, the displacement $\mathbf{u}$, transforms according to

$$\{\mathbf{u}\}^* = [\boldsymbol{R}]^T \{\mathbf{u}\} \tag{1.25}$$

Here, $\{\mathbf{u}\}^*$ denotes the (collection of) components of $\mathbf{u}$ in the starred basis and the superscript T is the transpose, as above.

Likewise, any tensor, for example the strain $\boldsymbol{\varepsilon}$, transforms according to:

$$[\boldsymbol{\varepsilon}]^* = [\boldsymbol{R}]^T [\boldsymbol{\varepsilon}][\boldsymbol{R}] \tag{1.26}$$

Example 1.1: Plane Transformation of an Arbitrary Deformation

For the plane transformation of Figure 1.2a, determine the strain components in the rotated basis, it being understood that no assumption is made about the state of strain itself being planar or not.

Solution:

The transformation matrix in Equation (1.24) specializes to

$$[\boldsymbol{R}] = \begin{bmatrix} \cos\theta & \cos(\theta+\frac{\pi}{2}) & 0 \\ \cos(\frac{\pi}{2}-\theta) & \cos\theta & 0 \\ 0 & 0 & 1 \end{bmatrix} = \begin{bmatrix} \cos\theta & -\sin\theta & 0 \\ \sin\theta & \cos\theta & 0 \\ 0 & 0 & 1 \end{bmatrix} \tag{1.27}$$

For a general state of strain, the strain matrix is given by Equation $(1.7)_1$. Then, a first calculation gives

$$[\boldsymbol{\varepsilon}][\boldsymbol{R}] = \begin{bmatrix} c\varepsilon_x + s\gamma_{xy} & -s\varepsilon_x + c\gamma_{xy} & \gamma_{xz} \\ c\gamma_{xy} + s\varepsilon_y & -s\gamma_{xy} + c\varepsilon_y & \gamma_{yz} \\ c\gamma_{xz} + s\gamma_{yz} & -s\gamma_{xz} + c\gamma_{yz} & \varepsilon_z \end{bmatrix} \tag{1.28}$$

using the shorthand notations: $c \equiv \cos\theta$ and $s \equiv \sin\theta$.

The strain components in the starred basis are then obtained using Equation (1.26) as follows:

$$\begin{cases} \varepsilon_x^* = \cos^2\theta\,\varepsilon_x + \sin^2\theta\,\varepsilon_y + 2\sin\theta\cos\theta\,\gamma_{xy} \\ \varepsilon_y^* = \sin^2\theta\,\varepsilon_x + \cos^2\theta\,\varepsilon_y - 2\sin\theta\cos\theta\,\gamma_{xy} \\ \gamma_{xy}^* = \sin\theta\cos\theta\,(\varepsilon_y - \varepsilon_x) + (\cos^2\theta - \sin^2\theta)\,\gamma_{xy} \end{cases} \tag{1.29}$$

for the in-plane components. Using standard trigonometric identities, these are exactly the same relations obtained under plane strain assumptions, Equation (1.23). The difference with plane strain is that there are three additional out-of-plane components, the expressions of which are eventually affected by the change of basis. These are given by the three remaining relations coming from Equation (1.26):

$$\begin{cases} \gamma_{xz}^* = \cos\theta\,\gamma_{xz} + \sin\theta\,\gamma_{yz} \\ \gamma_{yz}^* = -\sin\theta\,\gamma_{xz} + \cos\theta\,\gamma_{yz} \\ \varepsilon_z^* = \varepsilon_z \end{cases} \tag{1.30}$$

♡

1.5 GEOMETRIC REPRESENTATIONS OF STRAIN

Consider an elementary material fiber PQ along unit vector $\mathbf{n}$. Upon deformation, the fiber changes both its length and orientation, as depicted in Figure 1.3. The

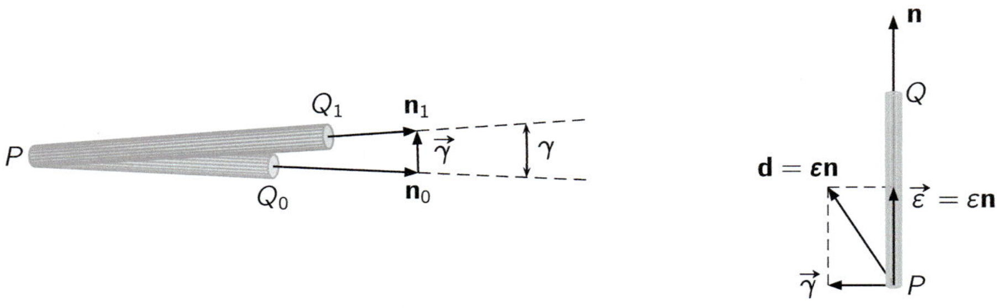

Figure 1.3 Decomposition of fiber deformation into stretch and glide.

strain tensor $\boldsymbol{\varepsilon}$, which is defined by Equation (1.3) in intrinsic form or by Equations (1.8) and (1.9) in component form, contains all information needed to describe these changes. This is accomplished through a "deformation vector" associated with the elementary fiber defined, relative to a frame tied to point P, by $\mathbf{d} = \boldsymbol{\varepsilon}\mathbf{n}$, and decomposed into "stretch" and "glide" terms as follows:

$$\mathbf{d} = \boldsymbol{\varepsilon}\mathbf{n} = \vec{\varepsilon} + \vec{\gamma} \tag{1.31}$$

Both terms are dimensionless and small in magnitude.

The stretch or relative elongation component contains information about the change in length:

$$\vec{\varepsilon} = \varepsilon\mathbf{n}; \qquad \varepsilon \equiv \frac{\delta PQ}{PQ} \tag{1.32}$$

where ε denotes the relative elongation along the fiber's direction $\mathbf{n}$ and is related to the notion of normal strain $\varepsilon_n = \mathbf{n} \cdot \boldsymbol{\varepsilon}\mathbf{n}$ introduced earlier; see Remark 1.3.

The glide component represents the fiber's reorientation:

$$\vec{\gamma} = \delta\mathbf{n} \tag{1.33}$$

where $\delta\mathbf{n} = \mathbf{n}_1 - \mathbf{n}_0$ is perpendicular to $\mathbf{n}$; see Figure 1.3. The length γ of $\vec{\gamma}$ is the measure in radian of the angle $(\mathbf{n}_0, \mathbf{n}_1)$.

The decomposition of $\mathbf{d}$ in Equation (1.31) into axial and transverse components is unique. The sign of ε is positive for extension and negative for contraction. On the other hand, the sign of γ has no meaning.

It is important to note that the decomposition in Equation (1.31) is unique for a given fiber orientation $\mathbf{n}$. However, from the same point P, one can draw arbitrarily oriented fibers and for each such orientation corresponds a different stretch-glide decomposition. This hints at the tensorial nature of strain. On the other hand, the strain tensor $\boldsymbol{\varepsilon}$ embodies all information about the state of strain at point P.

A graphical representation of the tensorial nature of strain is possible and results in what is known as a Mohr diagram. Let $\boldsymbol{\varepsilon}(P)$ be the strain tensor at a material point P. A fiber emanating from P and defined by unit vector $\mathbf{n}$ deforms in accordance with Equation (1.31). This deformation is uniquely decomposed into an axial (or normal) strain, $\vec{\varepsilon}$, and a transverse (or shear) strain, $\vec{\gamma}$, consistent with Equations (1.32) and (1.33). The fiber director and the magnitudes of the normal and shear strains of this fiber satisfy the following relationships:

$$\mathbf{n} \cdot \mathbf{n} = 1 \tag{1.34a}$$

$$\varepsilon \equiv \vec{\varepsilon} \cdot \mathbf{n} = \mathbf{d} \cdot \mathbf{n} = \mathbf{n} \cdot \boldsymbol{\varepsilon} \mathbf{n} \tag{1.34b}$$

$$\gamma^2 = \mathbf{d} \cdot \mathbf{d} - \varepsilon^2 \tag{1.34c}$$

An infinity of fibers may be made to emanate from point P, each with a different unit director $\mathbf{n}$. To each fiber corresponds a unique pair (ε, γ) of normal and shear strains, Figure 1.3. Thus, a fiber of director $\mathbf{n}$ may be represented by a point in a γ versus ε diagram, Figure 1.4a.

Three among all possible fiber directions that emanate from P are special in that they are only subjected to an axial strain and no shear. These are the fibers corresponding to the principal strain directions $\mathbf{e}_1, \mathbf{e}_2$, and $\mathbf{e}_3$ introduced in Section 1.3. The normal strains of these fibers are the principal strains $\varepsilon_1, \varepsilon_2, \varepsilon_3$. Since such fibers do not reorient upon deformation, their corresponding points in the γ–ε diagram lie on the horizontal axis, as depicted in Figure 1.4b where it is assumed that $\varepsilon_1 > \varepsilon_2 > \varepsilon_3$.

Let us find the domain swept by the representative (ε, γ) point when the director $\mathbf{n}$ takes all possible directions in space. With no loss of generality, we can work in the principal axes of strain as coordinate axes. The base vectors on these axes are $\mathbf{e}_1, \mathbf{e}_2$, and $\mathbf{e}_3$. The strain matrix $[\boldsymbol{\varepsilon}]$ is diagonal in its own principal axes. In these axes, the components of unit vector $\mathbf{n}$ and the deformation vector $\mathbf{d}$ are

$$[\mathbf{n}] = \begin{bmatrix} n_1 \\ n_2 \\ n_3 \end{bmatrix} \qquad [\mathbf{d}] = \begin{bmatrix} \varepsilon_1 n_1 \\ \varepsilon_2 n_2 \\ \varepsilon_3 n_3 \end{bmatrix} \tag{1.35}$$

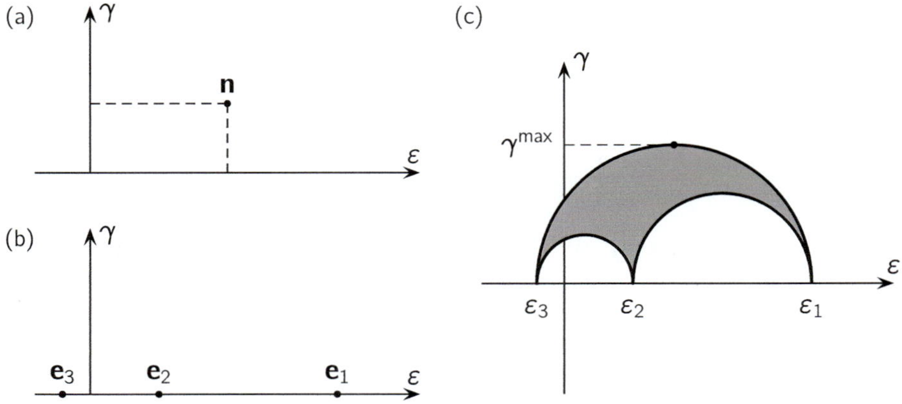

Figure 1.4 Mohr's diagram for strains. (a) Representative point for fiber along vector **n**. (b) Representative points for principal strain directions. (c) Full diagram.

using the definition of **d** in Equation (1.31). Then, the three relations in Equation (1.34c) become

$$\begin{cases} n_1^2+n_2^2+n_3^2=1 \\ \varepsilon_1 n_1^2+\varepsilon_2 n_2^2+\varepsilon_3 n_3^2=\varepsilon \\ \varepsilon_1^2 n_1^2+\varepsilon_2^2 n_2^2+\varepsilon_3^2 n_3^2=\varepsilon^2+\gamma^2 \end{cases} \tag{1.36}$$

This may be viewed as a linear system with respect to n_1^2, n_2^2, and n_3^2. The system in Equation (1.36) can be written in matrix form:

$$\begin{bmatrix} 1 & \varepsilon_1 & \varepsilon_1^2 \\ 1 & \varepsilon_2 & \varepsilon_2^2 \\ 1 & \varepsilon_3 & \varepsilon_3^2 \end{bmatrix}^T \begin{bmatrix} n_1^2 \\ n_2^2 \\ n_3^2 \end{bmatrix} = \begin{bmatrix} 1 \\ \varepsilon \\ \varepsilon^2+\gamma^2 \end{bmatrix} \tag{1.37}$$

The above 3×3 matrix is a Vandermonde matrix whose determinant is $(\varepsilon_2-\varepsilon_1)(\varepsilon_3-\varepsilon_1)(\varepsilon_3-\varepsilon_2)$. Therefore, provided that no two principal strains are equal, the above system is invertible and its solution is

$$\begin{cases} n_1^2=\dfrac{(\varepsilon-\varepsilon_2)(\varepsilon-\varepsilon_3)+\gamma^2}{(\varepsilon_1-\varepsilon_2)(\varepsilon_1-\varepsilon_3)} \\ n_2^2=\dfrac{(\varepsilon-\varepsilon_3)(\varepsilon-\varepsilon_1)+\gamma^2}{(\varepsilon_2-\varepsilon_3)(\varepsilon_2-\varepsilon_1)} \\ n_3^2=\dfrac{(\varepsilon-\varepsilon_1)(\varepsilon-\varepsilon_2)+\gamma^2}{(\varepsilon_3-\varepsilon_1)(\varepsilon_3-\varepsilon_2)} \end{cases} \tag{1.38}$$

Now, with no loss of generality, assume as in Figure 1.4 that the principal strains are ordered as $\varepsilon_1>\varepsilon_2>\varepsilon_3$. Then, Equation (1.38) implies that

$$\begin{cases} (\varepsilon-\varepsilon_2)(\varepsilon-\varepsilon_3)+\gamma^2\geq 0 \\ (\varepsilon-\varepsilon_3)(\varepsilon-\varepsilon_1)+\gamma^2\leq 0 \\ (\varepsilon-\varepsilon_1)(\varepsilon-\varepsilon_2)+\gamma^2\geq 0 \end{cases} \tag{1.39}$$

since $n_1^2\geq 0$, $n_2^2\geq 0$, and $n_3^2\geq 0$. The first inequality implies that the point (ε,γ) must lie outside the semicircle intersecting the horizontal axis at ε_2 and ε_3. The second inequality implies that the point (ε,γ) must lie inside the semicircle intersecting the horizontal axis at ε_1 and ε_3. Finally, the third inequality in Equation (1.39) implies that the point (ε,γ) must lie outside the semicircle passing through ε_1 and ε_2. In conclusion, the strain state $\boldsymbol{\varepsilon}(P)$ at material point P can be represented by the grayed region of Figure 1.4c.

Conversely, to each (ε,γ) point of Mohr's domain corresponds one value of the triplet (n_1^2,n_2^2,n_3^2), which means eight values of the triplet (n_1,n_2,n_3), hence eight vectors $\mathbf{n}$ in four pairs of opposites. At material point P, there are four distinct fibers that have the same state of strain (ε, γ).

The diagram in Figure 1.4c reveals that the maximum shear strain is given by

$$\gamma^{\max} = \frac{1}{2}(\varepsilon_1 - \varepsilon_3) \tag{1.40}$$

The normal strain, on the other hand, is bounded by the minor and major principal strains: $\varepsilon_3 \leq \varepsilon \leq \varepsilon_1$.

Some special cases are worth noting. If two principal strains are equal and different from the third, Mohr's diagram reduces to a single semicircle. If all three principal strains are equal, the diagram reduces to a point having Mohr coordinates $(e/3, 0)$ where $e \equiv \operatorname{tr}\boldsymbol{\varepsilon}$ is the dilatation; see Equation (1.13). The state of strain around P is then isotropic.

Three semicircles constitute the basis of Mohr's diagram, Figure 1.4c. The fact that these are semicircles and not full circles is because the sign of a normal strain has physical meaning (extension vs. contraction), whereas the sign of a shear strain has no physical meaning. An extension to full Mohr circles will be given in the context of Mohr's diagram for stresses in Chapter 2.

Example 1.2: Transformation from Diagonal Strain State

Consider a cylindrical bar deformed as shown in Figure 1.5a. Give the components of the strain tensor at a point P on the surface in a frame rotated by angle θ about the x-axis. Deformation is exaggerated for clarity.

Solution:

The bar deforms uniformly: all points have the same strain state. It expands axially and contracts radially. Therefore, the strain tensor takes everywhere the form:

$$[\boldsymbol{\varepsilon}] = \begin{bmatrix} \varepsilon_x & 0 & 0 \\ 0 & \varepsilon_y & 0 \\ 0 & 0 & \varepsilon_z \end{bmatrix} = \begin{bmatrix} \varepsilon_1 & 0 & 0 \\ 0 & \varepsilon_2 & 0 \\ 0 & 0 & \varepsilon_3 \end{bmatrix} \tag{1.41}$$

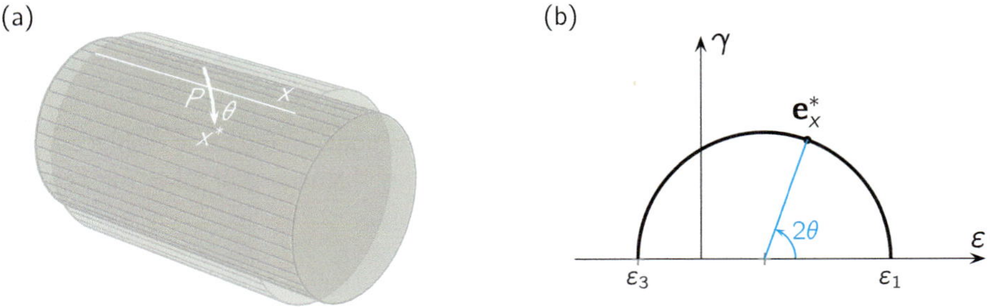

Figure 1.5 Principle of determining strains in any direction.

with respect to axes that contain the cylinder's x-axis and two arbitrary transverse and mutually perpendicular y- and z-axes. The strain matrix is diagonal. This means that the listed strains are the principal strains. Furthermore, axial symmetry imposes that $\varepsilon_2 = \varepsilon_3$.

To obtain the rotated strain matrix at surface point P, first consider the local frame tied to P. The tangent plane at P is defined by its normal, which is along the radial direction. The base vectors within the plane are given by $\mathbf{e}_x$ along the cylinder's axis and a second vector, say $\mathbf{e}_y$, along the hoop direction. These axes are not shown in Figure 1.5a to avoid clutter.

The rotation θ in the figure is about the radial axis. The new basis is given by $\{\mathbf{e}_x^*, \mathbf{e}_y^*, \mathbf{e}_z\}$. In this basis, the strain matrix is

$$[\boldsymbol{\varepsilon}]^* = \begin{bmatrix} \varepsilon_x^* & \gamma_{xy}^* & 0 \\ \gamma_{xy}^* & \varepsilon_y^* & 0 \\ 0 & 0 & \varepsilon_3 \end{bmatrix} \tag{1.42}$$

where

$$\begin{cases} \varepsilon_x^* = \dfrac{\varepsilon_1 - \varepsilon_2}{2}\cos 2\theta + \dfrac{\varepsilon_1 + \varepsilon_2}{2} \\[2ex] \varepsilon_y^* = \dfrac{\varepsilon_2 - \varepsilon_1}{2}\cos 2\theta + \dfrac{\varepsilon_1 + \varepsilon_2}{2} \\[2ex] \gamma_{xy}^* = \dfrac{\varepsilon_2 - \varepsilon_1}{2}\sin 2\theta \end{cases} \tag{1.43}$$

These relations specialize the general ones in Equation (1.23) when the reference frame is tied to the principal basis. Note that the strain state is not planar since $\varepsilon_z^* = \varepsilon_z = \varepsilon_3$; see Example 1.1.

Figure 1.5b shows the corresponding Mohr diagram. Because the eigenvalue ε_3 has multiplicity 2, the diagram reduces to a single semi-circle. The point in Mohr coordinates representative of the fiber at material point P oriented along $\mathbf{e}_x^*$ is such that the angle permitting its identification on the semicircle is twice the physical angle. Indeed, these Mohr coordinates are nothing but ε_x^* and γ_{xy}^*, given by the first and third relations in Equation (1.43). ♡

1.6 RIGID-BODY MOTIONS

When a body is in rigid motion, all material points undergo the same displacement. For infinitesimal strains and rotations, a rigid-body displacement takes the general form:

$$\mathbf{u} = \mathbf{u}_0 + \boldsymbol{\omega}_0 \times \mathbf{x} \tag{1.44}$$

where $\mathbf{x}$ is the position vector, as above, and $\mathbf{u}_0$, $\boldsymbol{\omega}_0$ are constants. Here, $\mathbf{u}_0$ represents a rigid-body translation with components a, b, c, and $\boldsymbol{\omega}_0$ is a rigid-body rotation about the origin having components α, β, γ.

In component form, Equation (1.44) reads

$$\begin{cases} u = \beta z - \gamma y + a \\ v = \gamma x - \alpha z + b \\ w = \alpha y - \beta x + c \end{cases} \tag{1.45}$$

It is straightforward to verify that the above displacement does not lead to any strain, by direct application of Equations (1.8) and (1.9). This is expected since deformation implies at least one of the following: (i) a relative change of length (between reference material points) and (ii) a relative change of angle (between reference material lines). On the other hand, rigid-body motion preserves both lengths and relative angles.

In structural mechanics, it is important to eliminate rigid-body motions as these do not lead to the kind of geometry changes that are of interest in the mechanics of deformable bodies. Of particular interest are situations where displacements are obtained by integration of the kinematic relations. Mathematically, integrating Equation (1.3), or equivalently their component forms in Equations (1.8) and (1.9), leads to integration constants. Physically, these constants represent rigid-body displacements.

In some structural mechanics problems, the rigid-body displacement may be used to infer salient features of the overall deformation of a structural member. For example, the overall deformation of a cross-section of a beam, viewed as a rigid entity, can be understood by carefully examining the rigid-body displacements that emerge from fully analytical solutions of beams treated as three-dimensional solids. Examples of this sort will be treated in Chapter 8 for establishing the notion of a center of twist and, most importantly, for analyzing the bending-induced rotation of cross-sections in Chapter 7.

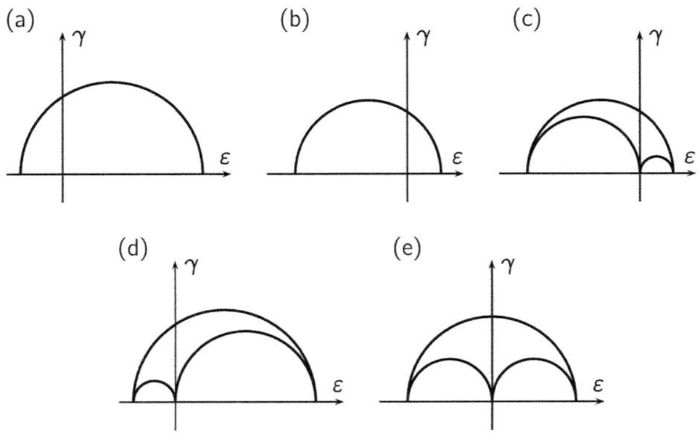

Figure 1.6 Mohr diagrams for various states of strain.

PROBLEMS

Problem 1.1 (On Shear Strain) *Strain gauge measurements lead to the following strain tensor:*

$$\boldsymbol{\varepsilon} = \begin{bmatrix} 0.85 & 0 & 0.65 \\ 0 & 0.2 & 3.9 \\ 0.65 & 3.9 & -0.45 \end{bmatrix} \textit{millistrain}; \quad \textit{1 millistrain} = 10^{-3}$$

By how much does the angle change between the lines that were along the y and z axes before deformation? ⋆

Problem 1.2 (On Strain Compatibility) *With reference to the compatibility equations, as stated in Equation* (1.18)*, answer the following: (i) How many equations are included in the single statement of Equation* (1.18)*? Hint: number of mappings from a set of cardinal 4 to a set of cardinal 3. (ii) How many equations among those written are identically verified? Hint: set $k = l$ or $i = j$. (iii) How many equations are duplicates on consideration of symmetry? Hint: the strain tensor is symmetric and the D operator commutes.* ⋆

Problem 1.3 (On Compatibility Equations) *What values should the indices i, j, k, l take in Equation* (1.18) *to obtain the six Equations (1.16a)–(1.16f)? Hint: try $i = x$, $j = y$, $k = y$, $l = z$ for Equation* (1.16e). ⋆

Problem 1.4 (On Principal Strains) *Consider the state of pure shear with $\gamma_{xy} = \gamma_{yx} = \gamma$, all other strain components being equal to zero. (i) Show that $\mathbf{e}_z$ is along a principal strain direction. (ii) Find by direct calculation the principal strains. (iii) Retrieve the results using a graphical method.* ⋆

Problem 1.5 (Mohr's Diagrams for Common Strain States) *Identify the strain state among the diagrams shown in Figure 1.6 out of: uniaxial compression, uniaxial tension, pure shear, plane strain compression, and plane strain tension. In each case, label the axes with the appropriate values (principal strains and maximum shear).* ⋆

Problem 1.6 (Integration of Kinematic Relations in Simple Case) *Determine the field of infinitesimal displacements corresponding to zero strain. To this end, follow these steps:*

1. *Show that all second derivatives of every displacement component are zero.*
2. *On that basis, justify using the displacement below as a trial field:*

   $$u(x,y,z) = u_1 yz + u_2 zx + u_3 xy + u_4 x + u_5 y + u_6 z + u_7$$

   with similar expressions for v and w in terms of seven constants each, $v_1 \cdots v_7$ and $w_1 \cdots w_7$, respectively.
3. *Using the "derivative" of Equation* $(1.9)_1$ *with respect to z, that of Equation* $(1.9)_2$ *with respect to y, and that of Equation* $(1.9)_3$ *with respect to x, show that $u_1 = v_2 = w_3 = 0$.*

4. *Show that the fact that the normal strains are zero implies that* $u_2 = u_3 = u_4 = 0$, *then that* $v_1 = v_3 = v_5 = 0$, *and that* $w_1 = w_2 = w_6 = 0$.
5. *Show that the fact that the shear strains are zero implies that* $u_5 + v_4 = 0$, *then that* $u_6 + w_4 = 0$, *and that* $v_6 + w_5 = 0$.
6. *Conclude that the displacement vector necessarily takes the form:*

$$\mathbf{u} = \boldsymbol{\omega} \times \mathbf{x} + \mathbf{u}_0$$

where $\boldsymbol{\omega}$ *and* $\mathbf{u}_0$ *are constant vectors to be determined in terms of* u_5, v_6 *and* w_4. ⋆

NOTES

1. The reader should be familiar with operators, such as the gradient, divergence, and Laplacian, which are also used in fluid dynamics. If not, jump directly to the component form in Equations (1.8) and (1.9).
2. This mnemonic way was “invented” by the author in 2004 when he first taught structural analysis.

2 Stress

The concept of traction (or stress vector) precedes that of stress (or stress tensor) and itself proceeds from the notion of (Newtonian) internal contact forces. A more fundamental theory is one that proceeds from (Newtonian) intermolecular (or interatomic) interactions, which are beyond the scope of a continuum theory.

2.1 TRACTION

Consider an elementary surface δS, not necessarily planar, centered about material point P, and locally separating the continuous body into two distinct parts. Each part exerts on the other via their contact surface δS a contact force $\pm\delta\mathbf{F}$. For definiteness, the contact surface δS is defined by its outer (unit) normal $\mathbf{n}$ so that the contact force is $+\delta\mathbf{F}$ when the action of the part outside on that inside the part is considered. Figure 2.1 depicts the elementary surface δS as a patch of a closed surface itself contained in the body, loaded as sketched. Depicting δS as a patch on a subpart allows to visualize an inside and outside for both $\mathbf{n}$ and $\delta\mathbf{F}$ to make sense.

As the surface area is made vanishingly smaller, physics demands that the magnitude of the internal contact force $\delta\mathbf{F}$ also becomes vanishingly small. As a consequence, the ratio of force per area is a priori undetermined in the limit. The Euler–Cauchy principle states that the limit exists does not depend on the way that limit is taken but only depends on the normal to the surface at P. That limit introduces the notion of traction, which is mathematically defined as follows:

$$\boxed{\mathbf{T} = \lim_{\delta S \to 0} \frac{\delta\mathbf{F}}{\delta S} = \frac{\mathrm{d}\mathbf{F}}{\mathrm{d}S}} \tag{2.1}$$

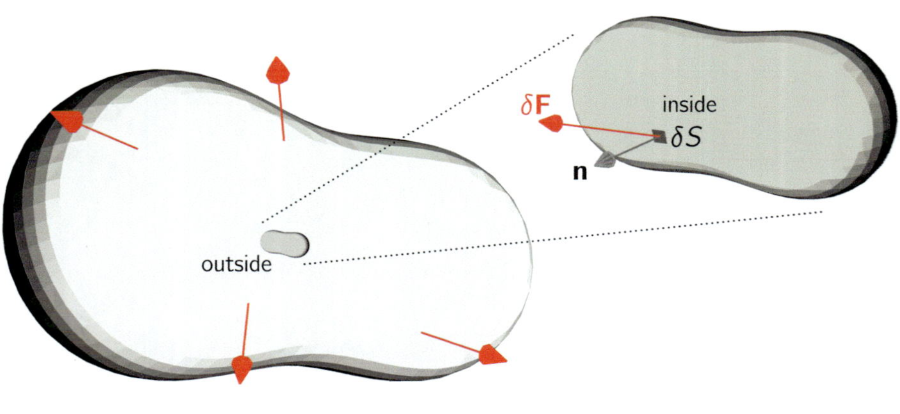

Figure 2.1 Concept of traction.

DOI: 10.1201/9781003537229-3

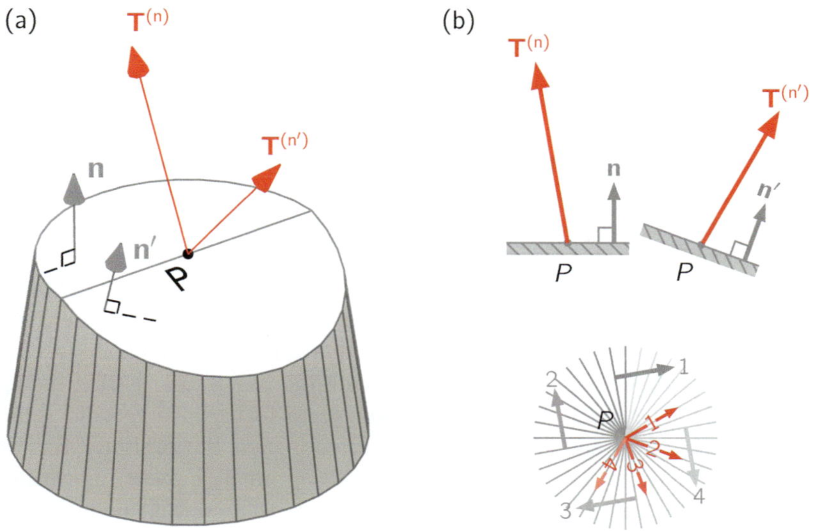

Figure 2.2 Limitation of concept of traction. (a) Two traction vectors at a point P lying on the common edge of two faces. (b) Schematics of the two faces (top) and of many more faces passing through P (bottom). Four faces numbered 1 to 4 are highlighted with the tractions acting on them shown in red.

where the surface area is identified with the surface δS itself, as usual. The traction is a vector, and its magnitude has units of pressure (Pa). The traction vector is also called the stress vector.

The notion of traction is valid at any material point. However, only at an exterior point or at some internal boundary (e.g., an interface) does it take a specific value since the normal is then fully specified. On the other hand, at an interior point, the value taken by the traction depends on the face it is resolved on. Figure 2.2a illustrates two traction vectors at the same interior point P. The notation for each traction vector includes the normal as a superscript, e.g., $\mathbf{T}^{(\mathbf{n})}$, to emphasize the dependence of the notion of traction on the face normal.

Since an infinite number of faces may be made to pass through the same point P, one can define an infinite number of tractions at P, Figure 2.2b. In the bottom sketch of Figure 2.2b, four faces out of many are indicated by their normals 1–4, with corresponding tractions shown in red. This warrants a better concept, the stress tensor.

2.2 STRESS

The traction is not a concept that one can work with to describe the state of internal contact forces at interior points of a given body. For this reason, a transformation $\boldsymbol{\sigma}$ is introduced which associates a given unit normal $\mathbf{n}$ with the traction $\mathbf{T}$ acting on the corresponding face. Hence, one may write

$$\boldsymbol{\sigma} : \mathbf{n} \mapsto \mathbf{T} \qquad \text{or} \quad \boldsymbol{\sigma}(\mathbf{n}) = \mathbf{T} \qquad \text{or simply} \quad \boxed{\boldsymbol{\sigma}\mathbf{n} = \mathbf{T}} \tag{2.2}$$

That a mapping of this sort can be stated is a matter of definition. What is of particular importance is whether the transformation is linear. Using his celebrated tetrahedron argument, Figure 2.3, Cauchy has shown that the transformation $\boldsymbol{\sigma}$ introduced above is indeed linear. This linear transformation is called the (Cauchy) stress tensor.

The argument begins with summing up the forces acting on the tetrahedron, including body and inertial forces. Just like the closed subpart in Figure 2.1, the tetrahedron is subjected to a traction distribution on each of its four faces. For simplicity, a right tetrahedron is considered with the base faces sharing the apex, taken as origin. The mean value theorem allows to focus on a specific value of traction on each face and relate the traction acting on the arbitrary face having normal $\mathbf{n}$, denoted $\mathbf{T}^{(\mathbf{n})}$, to the tractions acting on base faces, as illustrated in Figure 2.3. As the size of the tetrahedron is made to shrink, body and inertial forces go to zero faster than tractions, which are surface forces. All that remains, after taking the limit, is a relationship between $\mathbf{T}^{(\mathbf{n})}$ and the tractions $\mathbf{T}^{(-\mathbf{e}_x)}$, $\mathbf{T}^{(-\mathbf{e}_y)}$, $\mathbf{T}^{(-\mathbf{e}_z)}$ acting on the base faces. Figure 2.3a shows all tractions, but in Figure 2.3b, only base face tractions are shown one by one using a different perspective of the tetrahedron each time. The relationship between $\mathbf{T}^{(\mathbf{n})}$ and the other three tractions involves the ratios of surface area of normal $\mathbf{n}$ to the three areas of the base faces. These ratios are nothing but the components of the normal $\mathbf{n}$.

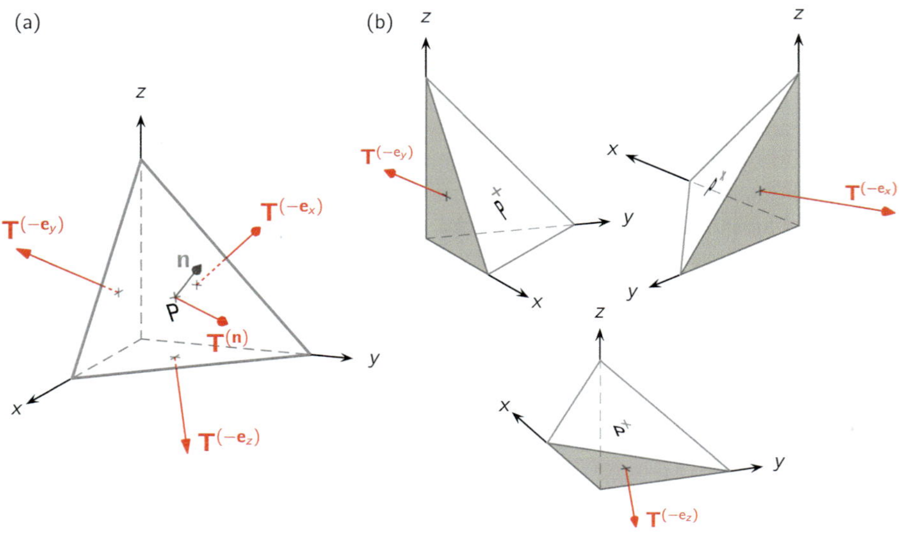

Figure 2.3 Cauchy's tetrahedron argument. (a) Right tetrahedron made up of one arbitrary face having normal $\mathbf{n}$ and three base faces. The traction acting at the center of each face is shown red. (b) Different views of the same tetrahedron each focusing on one base face (shown gray) with the traction acting on it.

That relationship constructs de facto the sought transformation and establishes its linearity.

In Cartesian coordinates, the component form of the stress versus traction relation, Equation (2.2), is

$$\begin{bmatrix} \sigma_x & \tau_{xy} & \tau_{xz} \\ \tau_{yx} & \sigma_y & \tau_{yz} \\ \tau_{zx} & \tau_{zy} & \sigma_z \end{bmatrix} \begin{bmatrix} n_x \\ n_y \\ n_z \end{bmatrix} = \begin{bmatrix} T_x \\ T_y \\ T_z \end{bmatrix} \tag{2.3}$$

In advanced texts, the same Greek symbol σ is used for both normal and shear stresses with two indices for each. Here, a different choice is made if only to distinguish throughout the text the only two types of stress: normal with the symbol σ and shear with the symbol τ.

For example, the traction acting on a face normal to the first base vector $\mathbf{e}_x$ has components: $T_x = \sigma_x$, $T_y = \tau_{yx}$, $T_z = \tau_{zx}$. The traction vector on that face may then be written as follows:

$$\mathbf{T} = \sigma_x \mathbf{e}_x + (\tau_{yx}\mathbf{e}_y + \tau_{zx}\mathbf{e}_z)$$

or, noting that the bracketed term is a vector contained in the y–z plane:

$$\mathbf{T} = \boldsymbol{\sigma}\mathbf{n} = \sigma\mathbf{n} + \tau\mathbf{m} \tag{2.4}$$

where $\mathbf{n} = \mathbf{e}_x$ is the unit normal and $\mathbf{m}$ is an in-plane unit vector in the direction of shear. Also, $\sigma = \sigma_x$ and τ are the magnitudes of the normal traction and shear traction, respectively.

More generally, the traction acting on an arbitrary face of normal $\mathbf{n}$ can uniquely be decomposed into a normal traction, $\vec{\sigma}$, and a shear traction, $\vec{\tau}$, Figure 2.4:

$$\mathbf{T} = \boldsymbol{\sigma}\mathbf{n} = \vec{\sigma} + \vec{\tau} = \sigma\mathbf{n} + \tau\mathbf{m} \tag{2.5}$$

The normal traction allows to define a state of tension versus compression following the sign of the scalar product $\vec{\sigma} \cdot \mathbf{n} = \sigma$. On the other hand, the sign of the shear

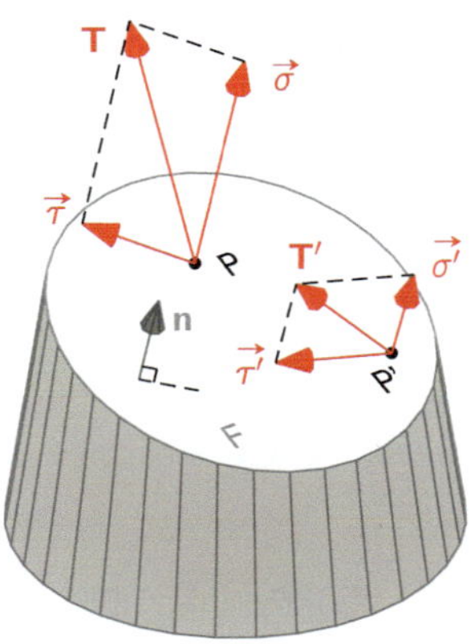

Figure 2.4 Decomposition of traction into normal and shear components.

traction τ has no physical meaning. In general, the traction varies pointwise and so does its decomposition into normal and shear tractions, as illustrated for the two points P and P' on face F in Figure 2.4.

2.3 EQUILIBRIUM

2.3.1 INSIDE THE BODY

The balance of angular momentum leads to symmetry of the stress tensor so that only six components are needed.

The statement of dynamic equilibrium follows from the balance of linear momentum and is then

$$\mathbf{div}\,\boldsymbol{\sigma} + \mathbf{b} = \rho \frac{\partial^2 \mathbf{u}}{\partial t^2} \tag{2.6}$$

where **div** is the divergence operator, **b** is the body force (per unit volume), ρ is the material density, **u** is the displacement, and t is the time. The full statement is needed when a transient response is sought, as in structural dynamics problems. Equation (2.6) is a compact statement of equilibrium. It provides an instantiation of Newton's second law ($\sum \mathbf{F} = m\mathbf{a}$) to a continuum, as opposed to a discrete particle or set of particles.

In the remainder of the course, focus is on structural statics with, in general, no body forces involved.[1] Under such circumstances, the equilibrium statement reduces to

$$\boxed{\mathbf{div}\,\boldsymbol{\sigma} = \mathbf{0}} \tag{2.7}$$

Equation (2.7) is a compact statement of equilibrium. It provides an instantiation of Newton's second law to continuum statics. It is so compact that it should never be forgotten.

Recall that the divergence of a tensor is a vector so that Equation (2.7) is equivalent to the three scalar equations:

$$\frac{\partial \sigma_x}{\partial x} + \frac{\partial \tau_{xy}}{\partial y} + \frac{\partial \tau_{xz}}{\partial z} = 0 \tag{2.8a}$$

$$\frac{\partial \tau_{xy}}{\partial x} + \frac{\partial \sigma_y}{\partial y} + \frac{\partial \tau_{yz}}{\partial z} = 0 \tag{2.8b}$$

$$\frac{\partial \tau_{xz}}{\partial x} + \frac{\partial \tau_{yz}}{\partial y} + \frac{\partial \sigma_z}{\partial z} = 0 \tag{2.8c}$$

Because the stress tensor is symmetric, Voigt's notation may be employed just like for the strain tensor, see Equation (1.14):

$$\begin{aligned}[\bar{\boldsymbol{\sigma}}]^T &= \begin{bmatrix}\sigma_x & \sigma_y & \sigma_z & \sqrt{2}\tau_{yz} & \sqrt{2}\tau_{xz} & \sqrt{2}\tau_{xy}\end{bmatrix} \\ &\equiv \begin{bmatrix}\sigma_1 & \sigma_2 & \sigma_3 & \sigma_4 & \sigma_5 & \sigma_6\end{bmatrix}\end{aligned} \tag{2.9}$$

Equation (2.9) plays a role in writing constitutive relations in matrix form, which is more suitable for computer operations. A rationale for the $\sqrt{2}$ factor in the shear terms will be given in Section 3.5.

2.3.2 EQUILIBRIUM AT A SURFACE

The three Equations (2.8a)–(2.8c) express the balance of forces along the x-, y-, and z-directions, respectively. The forces involved are the (local) internal contact forces that give rise to stress, and one is left wondering: where are the applied forces? What needs to be understood is that the statement of equilibrium, either in intrinsic form, Equation (2.7), or in component form, Equations (2.8a) to (2.8c), is valid locally, i.e., inside the body. Obviously, if the external surfaces of the body are not loaded in any way, then no stresses would arise inside the body, and the above equations would just be identically verified. This naturally leads us to consider equilibrium at a surface.

The general formulation of a boundary-value problem will be given in Chapter 4. Here, we are concerned with that part of the boundary condition that involves force-like quantities, i.e., tractions and stresses. Because the outer normal is fully specified on the external surface, only tractions can be imposed. Thus, if $\mathbf{n}$ denotes the surface outer normal, $\mathbf{T}_0$ is the specified traction on the surface, and $\boldsymbol{\sigma}$ is the stress in its immediate vicinity, one must have

$$\boldsymbol{\sigma}\mathbf{n} = \mathbf{T}_0 \tag{2.10}$$

If no tractions are imposed on a given portion of the boundary (i.e., if $\mathbf{T}_0 = \mathbf{0}$), then that portion is said to be traction-free. In solving problems, a traction-free boundary condition is just as important as any other boundary condition. For instance, in three-dimensional elasticity solutions for beams, only end sections are loaded and their lateral surfaces are typically traction-free.

2.4 PRINCIPAL STRESSES

Owing to the balance of angular momentum, the stress matrix $[\boldsymbol{\sigma}]$ in Equation (2.2) is symmetric. According to the spectral decomposition theorem, the stress matrix admits therefore three real eigenvalues, σ_1, σ_2, and σ_3, corresponding to three orthogonal eigenvectors, $\mathbf{E}_1, \mathbf{E}_2$, and $\mathbf{E}_3$. The eigenvalues and eigenvectors of $\boldsymbol{\sigma}$ are, respectively, called principal stresses and principal stress directions. The principal stress directions $\mathbf{E}_1, \mathbf{E}_2$, and $\mathbf{E}_3$ do not necessarily coincide with the principal strain directions $\mathbf{e}_1, \mathbf{e}_2$, and $\mathbf{e}_3$, introduced in Section 1.3.

2.5 MOHR'S DIAGRAM

Let $\boldsymbol{\sigma}(P)$ be the stress tensor at a material point P of the current configuration of a body. A face passing through P and defined by unit normal $\mathbf{n}$ is acted upon by the traction $\mathbf{T}$ in accordance with Cauchy's law, Equation (2.2). This traction can always be decomposed in a unique way into a normal traction, $\vec{\sigma}$, and a shear traction, $\vec{\tau}$, consistent with Equation (2.4) and Figure 2.4. The unit normal and the magnitudes of the normal and shear tractions acting on this face satisfy the following relationships:

$$\mathbf{n} \cdot \mathbf{n} = 1 \tag{2.11a}$$

$$\sigma \equiv \vec{\sigma} \cdot \mathbf{n} = \mathbf{T} \cdot \mathbf{n} = \mathbf{n} \cdot \boldsymbol{\sigma} \mathbf{n} \tag{2.11b}$$

$$\tau^2 = \mathbf{T} \cdot \mathbf{T} - \sigma^2 \tag{2.11c}$$

An infinity of faces may be made to pass through point P, each with a different normal $\mathbf{n}$, Figure 2.2. To each face corresponds a unique pair (σ, τ) of normal and shear stresses. Thus, a face of normal $\mathbf{n}$ may be represented by a point in a τ versus σ diagram, Figure 2.5a. Three among all possibles faces that pass through P are special in that they are only subjected to a normal stress and no shear stress. These are the faces corresponding to the principal stress directions $\mathbf{E}_1, \mathbf{E}_2, \mathbf{E}_3$ introduced in the previous section. The normal tractions acting on these faces are precisely the principal stresses $\sigma_1, \sigma_2, \sigma_3$. Since such faces have no shear, their corresponding points in the τ–σ diagram lie on the horizontal axis, as depicted in Figure 2.5b where it is assumed that $\sigma_1 > \sigma_2 > \sigma_3$.

A more general basic question is the following. $\mathcal{Q}_1$: given a stress state $\boldsymbol{\sigma}(P)$ at material point P, what domain is swept in the τ–σ diagram when the face (hence its normal $\mathbf{n}$) probes all possible orientations of the Euclidean space? To find that domain, choose the principal directions of $\boldsymbol{\sigma}$ as coordinate axes. In these axes, write the components of unit vector $\mathbf{n}$ and the traction $\mathbf{T}$:

$$[\mathbf{n}] = \begin{bmatrix} n_1 \\ n_2 \\ n_3 \end{bmatrix} \qquad [\mathbf{T}] = \begin{bmatrix} \sigma_1 n_1 \\ \sigma_2 n_2 \\ \sigma_3 n_3 \end{bmatrix} \tag{2.12}$$

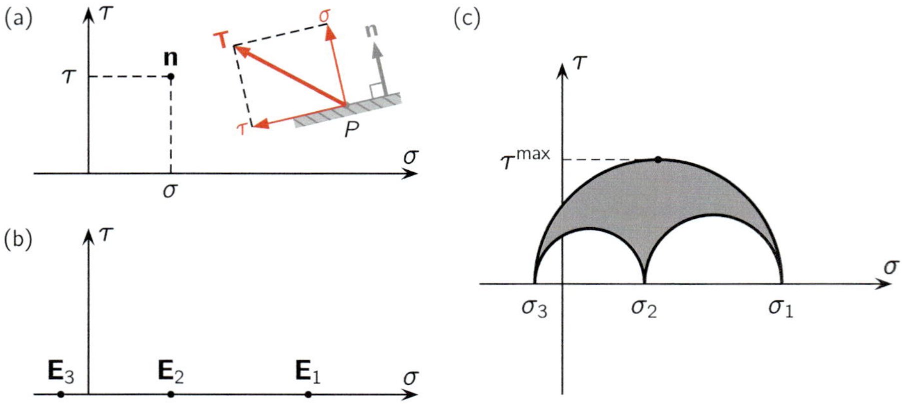

Figure 2.5 Mohr's diagram for stresses. (a) Representative point for a face having normal vector **n**. Inset shows decomposition of traction acting on the face. (b) Representative points for principal stress directions. (c) Full diagram.

The expression of $[\mathbf{T}]$ is simple because the matrix $[\boldsymbol{\sigma}]$ is diagonal in its own principal axes. Next, express Equation (2.11c) in component form:

$$\begin{cases} n_1^2 + n_2^2 + n_3^2 = 1 \\ \sigma_1 n_1^2 + \sigma_2 n_2^2 + \sigma_3 n_3^2 = \sigma \\ \sigma_1^2 n_1^2 + \sigma_2^2 n_2^2 + \sigma_3^2 n_3^2 = \sigma^2 + \tau^2 \end{cases} \tag{2.13}$$

Examining the above system of equations, a second question naturally arises; $\mathcal{Q}_2$: given a pair (σ, τ), what face passing through P, if any, is actually under the action of normal stress σ and shear stress τ? Finding that face amounts to determining its normal $\mathbf{n}$ through its components n_1, n_2, n_3. Thus, Equation (2.13) may be viewed as a system in the unknown components of $\mathbf{n}$ corresponding to fixed pair (σ, τ). To find that out, rewrite Equation (2.13) in matrix form:

$$\begin{bmatrix} 1 & \sigma_1 & \sigma_1^2 \\ 1 & \sigma_2 & \sigma_2^2 \\ 1 & \sigma_3 & \sigma_3^2 \end{bmatrix}^T \begin{bmatrix} n_1^2 \\ n_2^2 \\ n_3^2 \end{bmatrix} = \begin{bmatrix} 1 \\ \sigma \\ \sigma^2 + \tau^2 \end{bmatrix} \tag{2.14}$$

which is a linear system in n_1^2, n_2^2, n_3^2. The 3×3 matrix of the above system is a Vandermonde matrix whose determinant is $(\sigma_2 - \sigma_1)(\sigma_3 - \sigma_1)(\sigma_3 - \sigma_2)$. Therefore, provided that no two principal stresses are equal, the above system is invertible, and the solution may be written, after simplification, as follows:

$$\begin{cases} n_1^2 = \dfrac{(\sigma - \sigma_2)(\sigma - \sigma_3) + \tau^2}{(\sigma_1 - \sigma_2)(\sigma_1 - \sigma_3)} \\ n_2^2 = \dfrac{(\sigma - \sigma_3)(\sigma - \sigma_1) + \tau^2}{(\sigma_2 - \sigma_3)(\sigma_2 - \sigma_1)} \\ n_3^2 = \dfrac{(\sigma - \sigma_1)(\sigma - \sigma_2) + \tau^2}{(\sigma_3 - \sigma_1)(\sigma_3 - \sigma_2)} \end{cases} \tag{2.15}$$

The above result contains an answer to $\mathcal{Q}_2$ and implicitly shows that not all (σ, τ) pairs can be realized; see Problem 2.3. To see this more broadly, assume as above that the principal stresses are ordered as $\sigma_1 > \sigma_2 > \sigma_3$. Then, from $n_1^2 \geq 0$, etc., infer the following three inequalities from Equation (2.15):

$$\begin{cases} (\sigma - \sigma_2)(\sigma - \sigma_3) + \tau^2 \geq 0 \\ (\sigma - \sigma_3)(\sigma - \sigma_1) + \tau^2 \leq 0 \\ (\sigma - \sigma_1)(\sigma - \sigma_2) + \tau^2 \geq 0 \end{cases} \tag{2.16}$$

The first inequality implies that the point (σ, τ) representative of face $\mathbf{n}$ must lie outside the semicircle intersecting the horizontal axis at σ_2 and σ_3. The second inequality implies that the point (σ, τ) must lie inside the semicircle intersecting the horizontal axis at σ_1 and σ_3. Finally, the third inequality in Equation (2.16) implies that the point (σ, τ) must lie outside the semicircle passing through σ_1 and σ_2. It follows that the stress state $\boldsymbol{\sigma}(P)$ at material point P can be represented by the grayed region of Figure 2.5c. This completes the answer to question $\mathcal{Q}_1$.

This geometric representation of the stress state is known as Mohr's diagram and reflects the tensorial nature of the stress. There are other geometric representations of the stress tensor, such as Lamé's ellipsoid or Cauchy's quadrics, but these have fallen into disuse. Even Mohr's diagrams are not as frequently employed as they used to be. They are, however, very helpful in depicting stress (and strain) states and find use in strength theories, yield theories as well as strain gauge measurements, using strain transformations as shown in Chapter 1; also see Chapter 12.

Three semicircles constitute the basis of Mohr's diagram. The fact these are semicircles and not full circles is because the sign of a normal stress has physical meaning (tension versus compression) whereas the sign of a shear stress has no physical meaning. In all of the above equations, this fact manifests in that the shear stress always appears as a squared quantity. By way of extension, diagrams consisting of full Mohr circles can be constructed after adopting some sign convention for the shear stress, as explained further below.

It is important to keep in mind that the entire Mohr diagram in Figure 2.5 corresponds to a single material point P at a given instant of loading. The diagram reveals that the shear traction lies between 0 and a maximum value given by the relation

$$\tau^{\max} = \frac{1}{2}(\sigma_1 - \sigma_3) \tag{2.17}$$

On the contrary, the normal traction is bounded by the minor and major principal stresses: $\sigma_3 \leq \sigma \leq \sigma_1$. The diagram also reveals that not any value of the pair (σ, τ) in the above intervals can be realized. Only those within the grayed area are physically possible. One can determine the orientation of the faces where the maximum shear is realized; see Problem 2.4.

Some special cases are worth noting. A point on the semicircle passing through $(\sigma_1, 0)$ and $(\sigma_2, 0)$ corresponds to a face whose normal lies in the $\mathbf{E}_1 - \mathbf{E}_2$ principal plane. Likewise, a point on the semicircle passing through $(\sigma_2, 0)$ and $(\sigma_3, 0)$ corresponds to a face whose normal lies in the $\mathbf{E}_2 - \mathbf{E}_3$ principal plane, and so on.

It remains to determine graphically, if possible, the orientation of a given face when all one knows is the corresponding representative point in a Mohr diagram. First consider the case of a face in a principal plane, say $\mathbf{E}_1 - \mathbf{E}_3$ (i.e., a face whose normal $\mathbf{n}$ lies in that plane). The representative point necessarily lies on the semicircle intersecting the horizontal axis at σ_1 and σ_3, Figure 2.6a. As shown in the figure, let 2θ be the angle between the two radial lines joining the center of the semicircle and either the representative point of $\mathbf{n}$ or the representative point of $\mathbf{E}_1$, the latter being on the horizontal axis. Omitting the proof, it can be shown that this angle is exactly twice the angle θ between $\mathbf{n}$ and $\mathbf{E}_1$ in physical space, as illustrated in the inset of Figure 2.6a. For the sake of clarity, the face normal is recalled as a superscript, as in Figure 2.3. Note that the traction acting on the $\mathbf{E}_1$ face is normal to it, as expected for a principal face.

A special case is when the face normal $\mathbf{n}$ in Figure 2.6a is identified with the principal stress direction $\mathbf{E}_3$ corresponding to the minor principal stress. In that case the angle on the diagram would be π (i.e., 180°) since $\mathbf{E}_3$ is orthogonal to $\mathbf{E}_1$ in physical space. The same applies to any pair of principal directions, as illustrated in Figure 2.6b. The inset corresponds to principal directions $\mathbf{E}_1$ and $\mathbf{E}_2$ with the corresponding tractions having magnitudes commensurate with $\sigma_2 < \sigma_1$.

From the simple illustration in Figure 2.6b, it can be seen that two faces that are perpendicular to each other in physical space would be represented by diametrically opposite points on the diagram. This property can be fully taken advantage of provided that a sign is assigned to the shear stress. The process is illustrated in Figure 2.6c. A face of normal $\mathbf{m}$, perpendicular to $\mathbf{n}$, is now represented on a full Mohr circle as diametrically opposite to $\mathbf{n}$. The two insets serve to show that the tangential projection of the two tractions are equal in magnitude but the normal tractions are different. The sign convention consists of taking the shear traction to be positive if its action tends to rotate the face anti-clockwise (case of face $\mathbf{n}$) and negative otherwise (case of face $\mathbf{m}$).

This review chapter is closed with an example illustrating key stress-related concepts.

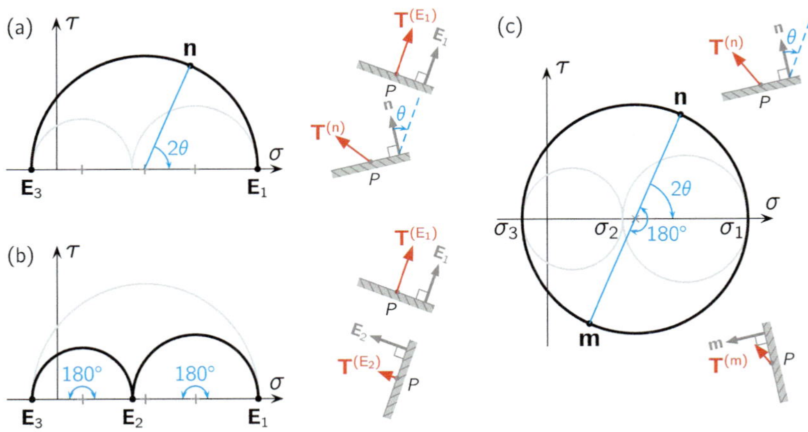

Figure 2.6 From Mohr's semicircles to full circles and the connection to face orientation in physical space. (a) The angle θ between face normal $\mathbf{n}$ and principal stress direction $\mathbf{E}_1$ (inset) becomes 2θ in Mohr's diagram. (b) Special cases of principal stress directions. (c) Construction of full circle knowing two points representative of two orthogonal faces $\mathbf{n}$ and $\mathbf{m}$.

Example 2.1: Mohr's Circles: Construction and Applications

A thin sheet is loaded in its plane. Measurements at some point P of the sheet lead to the following stress matrix:

$$[\boldsymbol{\sigma}] = \begin{bmatrix} 6 & 2 & 0 \\ 2 & 2 & 0 \\ 0 & 0 & 0 \end{bmatrix} \cdot 10^7 \text{Pa}$$

expressed in an orthonormal base $\{\mathbf{e}_x, \mathbf{e}_y, \mathbf{e}_z\}$, the first two base vectors being in the plane of the sheet and the last one through the thickness, Figure 2.7.

The following is required:

1. Determine mathematically the principal stresses acting at P and their associated principal directions.
2. Recover these results by constructing a Mohr diagram at P.
3. Find the orientation(s) of the face(s) passing through P on which the shear is maximum. What is the value of maximum shear?

Solution:

1. Principal stresses are the eigenvalues of $[\boldsymbol{\sigma}]$ and principal stress directions are their associated eigenvectors. λ is said to be an eigenvalue of $\boldsymbol{\sigma}$ if and only if

$$\exists \mathbf{n} \quad | \quad \boldsymbol{\sigma}\mathbf{n} = \lambda \mathbf{n} \tag{2.18}$$

or equivalently,

$$\exists \mathbf{n} \quad | \quad (\boldsymbol{\sigma} - \lambda \boldsymbol{I})\mathbf{n} = \mathbf{0} \tag{2.19}$$

where $\boldsymbol{I}$ denotes the identity. If the matrix $[\boldsymbol{\sigma} - \lambda \boldsymbol{I}]$ were invertible, the only solution would be $\mathbf{n} = \mathbf{0}$. For nontrivial solutions to exist, the matrix $[\boldsymbol{\sigma} - \lambda \boldsymbol{I}]$ must be singular, i.e., non-invertible. A necessary and sufficient condition for that is that its determinant be nil:

$$|\boldsymbol{\sigma} - \lambda \boldsymbol{I}| = 0 \tag{2.20}$$

It is common to start from Equation (2.20) to solve an eigenvalue problem, but it is important to keep in mind Equation (2.18), which constitutes the primitive definition of an eigenvalue λ and its associated eigenvector $\mathbf{n}$.

In the case of the thin sheet, the problem is further simplified by noting that 0 is an evident eigenvalue associated with the through-thickness direction, since

$$\boldsymbol{\sigma}\mathbf{e}_z = \mathbf{0} = 0\,\mathbf{e}_z \tag{2.21}$$

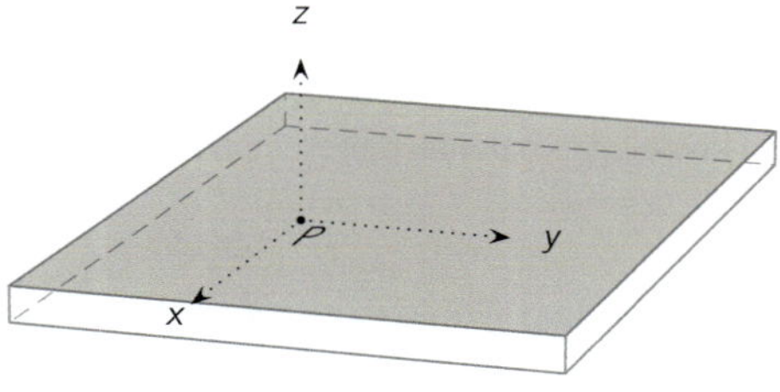

Figure 2.7 Thin sheet.

The problem, therefore, reduces to diagonalizing the upper left 2×2 matrix:

$$\begin{bmatrix} 6 & 2 \\ 2 & 2 \end{bmatrix} \quad \text{for which Equation (2.20) becomes} \quad \begin{vmatrix} 6-\lambda & 2 \\ 2 & 2-\lambda \end{vmatrix} = 0$$

The solution gives two positive roots. If principal stresses are ordered as $\sigma_1 > \sigma_2 > \sigma_3$, then

$$\sigma_1 = 2(2+\sqrt{2}) \cdot 10^7\,\text{Pa}, \quad \sigma_2 = 2(2-\sqrt{2}) \cdot 10^7\,\text{Pa}, \quad \sigma_3 = 0 \tag{2.22}$$

In order to find the principal stress directions, first note from Equation (2.21) that the eigenvector associated with σ_3 is $\mathbf{E}_3 = \mathbf{e}_z$. To find the other two, $\mathbf{E}_1$ and $\mathbf{E}_2$, specialize Equation (2.19) to the corresponding eigenvalues σ_1 and σ_2. For example, for σ_1 obtain

$$\begin{bmatrix} 6-\sigma_1 & 2 \\ 2 & 2-\sigma_1 \end{bmatrix} \begin{bmatrix} E_1^x \\ E_1^y \end{bmatrix} = \begin{bmatrix} 0 \\ 0 \end{bmatrix} \tag{2.23}$$

where E_1^x and E_1^y are the sought components of $\mathbf{E}_1$ in the $\{\mathbf{e}_x, \mathbf{e}_y\}$ base of Figure 2.7. Equation (2.23) is obviously not invertible. One can find a principal direction by choosing any value of a component and determining the other. For example, set $E_1^y = 1$; then, the first equation obtained by developing Equation (2.23) gives

$$(6-\sigma_1)E_1^x + 2 = 0 \quad \Rightarrow \quad E_1^x = -\frac{2}{6-2(2+\sqrt{2})} = -\frac{1}{1-\sqrt{2}} = 1+\sqrt{2} \tag{2.24}$$

Applying the same procedure to find $\mathbf{E}_2$, one finally gets

$$\{\mathbf{E}_1\} = \begin{Bmatrix} 1+\sqrt{2} \\ 1 \\ 0 \end{Bmatrix} \qquad \{\mathbf{E}_2\} = \begin{Bmatrix} 1-\sqrt{2} \\ 1 \\ 0 \end{Bmatrix} \qquad \{\mathbf{E}_3\} = \begin{Bmatrix} 0 \\ 0 \\ 1 \end{Bmatrix} \tag{2.25}$$

Noting that the stress matrix is symmetric, one knew beforehand that it admits three real eigenvalues, σ_1, σ_2, and σ_3, associated with three orthogonal eigenvectors, $\mathbf{E}_1$, $\mathbf{E}_2$, and $\mathbf{E}_3$. It can easily be verified that $\mathbf{E}_1 \cdot \mathbf{E}_2 = 0$.

2. The shear traction on a face of normal $\mathbf{e}_z$ is zero. Therefore, such a face is a principal one. What matters, therefore, is to determine Mohr's circle in the x–y plane. To this end, one takes advantage of the general property introduced in Figure 2.6c, namely that two physically perpendicular faces are represented by two diametrically opposite points in Mohr's diagram. The center of the circle is then the midpoint and its radius half the distance between them. This can be accomplished with any pair of perpendicular faces. In practice, the procedure is most conveniently carried out using base vectors $\mathbf{e}_x$ and $\mathbf{e}_y$.

Figure 2.8 illustrates the step-by-step process with brief explanations.

Point Associated with $\mathbf{e}_x$ Face

The traction acting on the $\mathbf{e}_x$ face is

$$\begin{bmatrix} 6 & 2 & 0 \\ 2 & 2 & 0 \\ 0 & 0 & 0 \end{bmatrix} \begin{Bmatrix} 1 \\ 0 \\ 0 \end{Bmatrix} \cdot 10^7\,\text{Pa} = \begin{Bmatrix} 6 \\ 2 \\ 0 \end{Bmatrix} \cdot 10^7\,\text{Pa} \tag{2.26}$$

It is represented in Figure 2.8a with its normal component $\sigma = 6$ and tangential (or shear) component $\tau = 2$, all given in units of 10^7 Pa. The corresponding point is reported on a Mohr diagram, as shown in Figure 2.8b. The shear stress is here counted positively since its action tends to rotate the face counterclockwise.

Point Associated with $\mathbf{e}_y$ Face

Likewise, the traction acting on the $\mathbf{e}_y$ face is

$$\begin{bmatrix} 6 & 2 & 0 \\ 2 & 2 & 0 \\ 0 & 0 & 0 \end{bmatrix} \begin{Bmatrix} 0 \\ 1 \\ 0 \end{Bmatrix} \cdot 10^7\text{Pa} = \begin{Bmatrix} 2 \\ 2 \\ 0 \end{Bmatrix} \cdot 10^7\text{Pa} \tag{2.27}$$

It is represented in Figure 2.8c with its normal component $\sigma = 2$ and tangential component $\tau = 2$, both given in units of 10^7 Pa. The corresponding point is reported on a Mohr diagram in Figure 2.8d. The shear stress is now counted negatively because its action tends to rotate the face clockwise.

Next, the two points corresponding to the $\mathbf{e}_x$ and $\mathbf{e}_y$ faces, which are denoted M_x and M_y in Figure 2.8e, allow to determine the center and diameter of Mohr's circle associated with the x–y principal plane. The symmetry of the stress tensor and the sign convention for the shear stress imply that the two shear tractions corresponding to M_x and M_y are necessarily equal in magnitude and opposite in sign. Hence, the midpoint of $[M_xM_y]$ must lie on the σ-axis in a Mohr diagram. The abscissa of the center is simply half the sum of the abscissas of M_x and M_y.

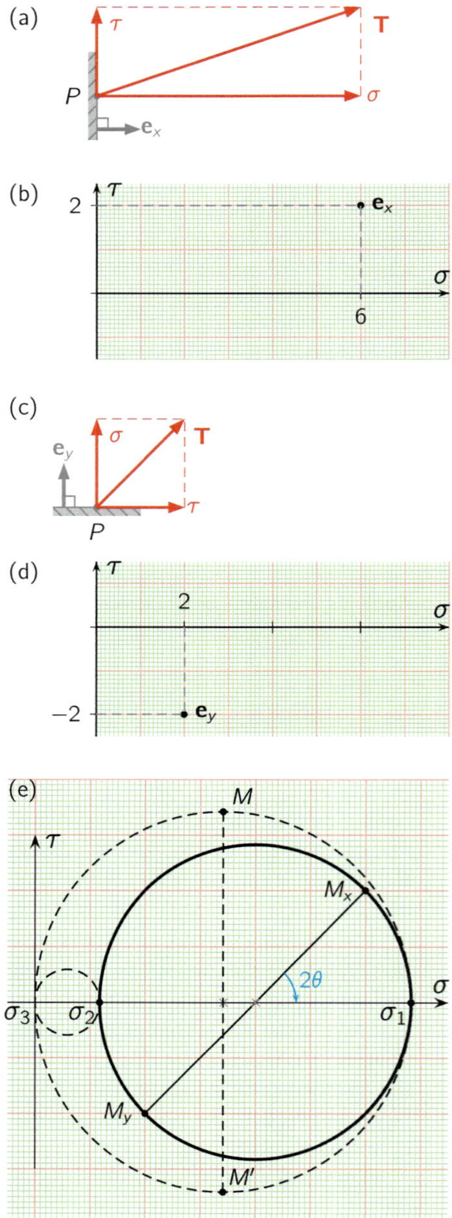

Figure 2.8 Step by step construction of Mohr's circles.

On the other hand, the diameter of the circle is simply obtained from the distance between them, i.e. the norm of $\overrightarrow{M_xM_y}$. As a result, one gets

$$\begin{cases} \text{Center: } \begin{pmatrix} 4 \\ 0 \end{pmatrix} \cdot 10^7\,\text{Pa} \\ \text{Radius: } 2\sqrt{2} \cdot 10^7\,\text{Pa} \end{cases} \tag{2.28}$$

The intersection of Mohr's circle with the horizontal axis gives the two principal stresses: σ_1 as the center's abscissa plus the radius and σ_2 as the center's abscissa minus the radius, hence the same result given in Equation (2.22).

The principal direction associated with the major principal stress σ_1 lies in the x–y plane and makes an angle θ with the x-axis. From Figure 2.8e, obtain

$$\tan 2\theta = 1 \quad \Rightarrow \quad \theta = \frac{\pi}{8} \tag{2.29}$$

Likewise, the principal direction associated with the intermediate principal stress σ_2 lies in the x–y plane and makes an angle $\theta = \pi/8$ with the y-axis. In other words, the principal directions $\mathbf{E}_1$ and $\mathbf{E}_2$ are, respectively, at 22.5° from the x-axis and y-axis in the x–y plane. It can be checked that these are the same directions as given by Equation (2.25) using the identity:

$$\frac{2}{\tan 2\theta} = \frac{1}{\tan\theta} - \tan\theta$$

and the fact that $\tan\theta = 1/(1+\sqrt{2})$ for the first principal stress for example.

3. To find the maximum shear, one needs to draw the entire Mohr diagram with its three circles. The other two circles are shown dashed in Figure 2.8e. It is evident from the largest circle that the maximum shear stress corresponds to the directions associated with points M and M'. By inspection, the maximum shear stress is half the major principal stress, consistent with Equation (2.17):

$$\tau^{\max} = (2+\sqrt{2}) \cdot 10^7\ \text{Pa} \tag{2.30}$$

The corresponding faces are at ±45° from the major principal stress direction $\mathbf{E}_1$ in the $\mathbf{E}_1 - \mathbf{E}_3$ plane, which is the $\mathbf{E}_1 - \mathbf{e}_z$ plane. ♡

Remark 2.1 *Clockwise and anti-clockwise directions are obviously arbitrary choices and must be interpreted for the observer of the page as is. An observer looking from into the page would have those two notions swapped. This serves to reinforce the idea that the sign of a shear stress has no physical meaning. See Problem 2.5 for more insight.* ∗

PROBLEMS

Problem 2.1 (Stress State at a Point) *At a point P of a solid, the principal stress directions are known and the principal stresses are* $\sigma_1 = 20$ *MPa,* $\sigma_2 = 10$ *MPa, and*

$\sigma_3 = -30$ *MPa. (i) Find the directions of maximum shear. (ii) Consider an elementary face centered about P with directors* $(-3, 12, 4)$. *Determine the components of the traction acting on that face, the normal stress, the shear stress vector* $\vec{\tau}$ *and the shear intensity* τ. ⋆

Problem 2.2 (Resultant of Hoop Tractions) *Let* $\vec{\tau} = \tau(r)\,\mathbf{e}_\theta$ *be a distribution of shear tractions acting in the hoop direction of a given face, with* $\tau(r)$ *an arbitrary function. Show that the resultant force on the face is zero.* ⋆

Problem 2.3 (On Possible Tractions) *(i) Starting from the first of Equation* (2.15) *show that the normal stress (or traction) acting on any face passing through a material point M cannot exceed the maximum principal stress. (ii) Likewise, starting from the third of Equation* (2.15) *show that the normal stress cannot fall below the minimum principal stress.* ⋆

Problem 2.4 (Faces of Maximum Shear) *Show that the faces of maximum shear are generally contained in the planes defined by : (i) the orientation of the intermediate principal stress and (ii) the bisectors of the plane defined by the minor and major principal stresses. Exhibit these faces by drawing them on a plate assuming that the intermediate stress direction is through the plate thickness.* ⋆

Problem 2.5 (Shear Sign Convention) *(i) In Example 2.1, draw the tractions acting on the faces of normals* $-\mathbf{e}_x$ *and* $-\mathbf{e}_y$, *and then determine the points on Mohr's diagram representing the two faces. Show that these points are* M_x *and* M_y, *unchanged from Figure 2.8, and explain why. (ii) In the same example, now change the shear stress convention by counting as positive the shear that tends to rotate the face clockwise. Determine the new locations of points* M_x *and* M_y *representative of faces* $\mathbf{e}_x$ *and* $\mathbf{e}_y$. *Do the new locations affect the corresponding Mohr circle? Conclude.* ⋆

Problem 2.6 (Determining Principal Stresses) *At a point P of a solid, the stress state is described by the following components:* $\sigma_x = 100$ *MPa,* $\sigma_y = 60$ *MPa,* $\tau_{xy} = -20$ *MPa in some axes Px, Py, Pz. Other stress components are zero. (i) Determine by direct calculation, then using Mohr's diagram, the principal stresses and principal stress directions at P. (ii) Find the maximum shear stress and the corresponding directions.* ⋆

Problem 2.7 (Analysis of Stress State) *At a point P of a solid, the stress state is described by the following matrix relative to an orthonormal frame* $\{O, xyz\}$:

$$[\boldsymbol{\sigma}] = \begin{bmatrix} 20 & 0 & 10 \\ 0 & 20 & -10 \\ 10 & -10 & 10 \end{bmatrix} \ MPa.$$

(i) Consider an elementary face centered about P with directors $(-1, 1, 2)$. *Determine the components of the traction acting on that face: the normal stress and the shear stress. (ii) Show that the state of stress at P is one of plane stress. (iii) Find the maximum shear stress and the corresponding directions.* ⋆

Problem 2.8 (Construction of a Stress Matrix) *A gravity dam, considered to be straight and indefinitely long, has the cross-section shown in Figure 2.9. Under the action of water pressure, the dam is under a state of plane stress in the plane of the cross-section (Oxy). At a point P on the upstream abutment side, one knows:*

- *the pressure p exerted by water*
- *the intensity q of the compressive stress prevailing along the vertical y direction.*

Determine in P the principal directions, the principal stresses, the shear stress acting on a face of normal y and the stress matrix in the x–y axes. ⋆

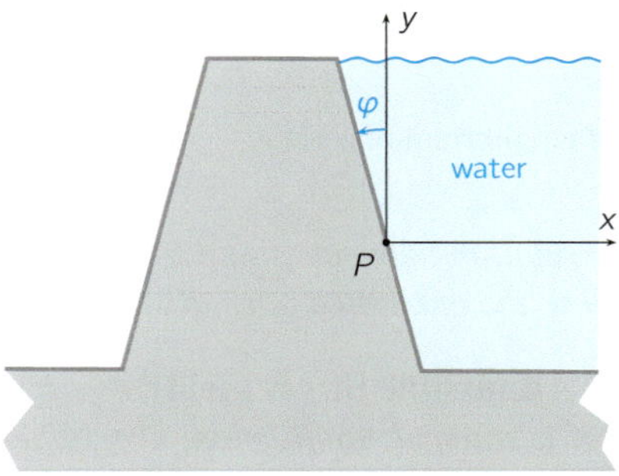

Figure 2.9 Dam under own weight and water pressure.

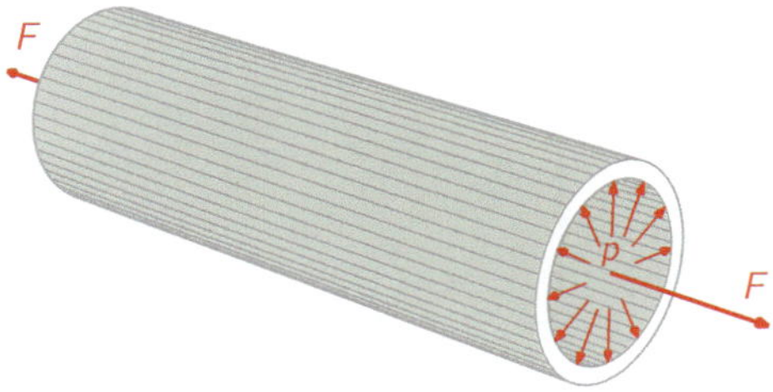

Figure 2.10 Tube under internal pressure and axial force.

Problem 2.9 (Pressurized Tube under Axial Force) *The thin-walled cylinder of Figure 2.10 has mean radius R and wall thickness $t \ll R$. It is subjected to an internal pressure p in addition to an axial force F, as shown.*

1. *Show that the radial (r), hoop (θ), and axial (z) directions are principal stress directions.*
2. *Show that the principal stresses are given by:*

$$\sigma_r = 0; \quad \sigma_\theta = \frac{pR}{t}; \quad \sigma_z = \frac{F}{2\pi Rt}$$

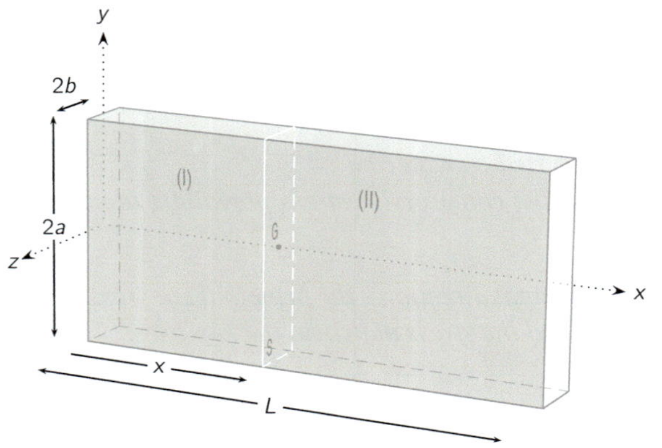

Figure 2.11 Conditions of equilibrium of a plate.

3. *What should be the relation between p and F for the maximum shear stress in the wall to be equal to the maximum normal stress?* ★

Problem 2.10 (Statically Admissible Stress Field) *A stress field $\boldsymbol{\sigma}(\mathbf{x})$ is said to be statically admissible if it satisfies equilibrium everywhere in the structure, i.e., $\mathbf{div}\,\boldsymbol{\sigma} + \mathbf{b} = 0$, which is Equation (2.6) without the inertial term, as well as any boundary conditions on the stresses of the form of Equation (2.10).*

Consider the rectangular plate shown in Figure 2.11.

1. *State the conditions under which the following stress tensor is statically admissible in the plate:*

$$[\boldsymbol{\sigma}] = \begin{bmatrix} \sigma_x & \tau_{xy} & 0 \\ \tau_{xy} & 0 & 0 \\ 0 & 0 & 0 \end{bmatrix} \text{ with } \begin{cases} \sigma_x = -\dfrac{M}{I}y \\ \tau_{xy} = \dfrac{V}{2I}(a^2 - y^2) \end{cases} \text{ where } \begin{cases} V = p(L-x) \\ M = \dfrac{1}{2}p(L-x)^2 \end{cases} \tag{2.31}$$

   Here, p is a constant having units of force per unit length and $I = \frac{4}{3}a^3 b$.
2. *Show that the body force acting on a slice of the plate between x and $x + \mathrm{d}x$ is $p\,\mathrm{d}x$ along the y-axis.*
3. *A cross-section S of the plate located at x separates part (I) upstream from part (II) downstream. Show that the surface forces applied by (II) on (I) following S admit as net resultant force $\mathbf{V} = V\mathbf{e}_y$ and net resultant moment about the center of S: $\mathbf{M} = M\mathbf{e}_z$, where $\mathbf{e}_y$ and $\mathbf{e}_z$ denote unit vectors along the y-axis and z-axis, respectively.* ★

Problem 2.11 (Mohr's Diagrams for Common Stress States) *Identify the stress state among the diagrams shown in Figure 2.12 out of: uniaxial tension, uniaxial compression, pure shear, and plane stress biaxial tension. In each case, label the axes with the appropriate values.* ★

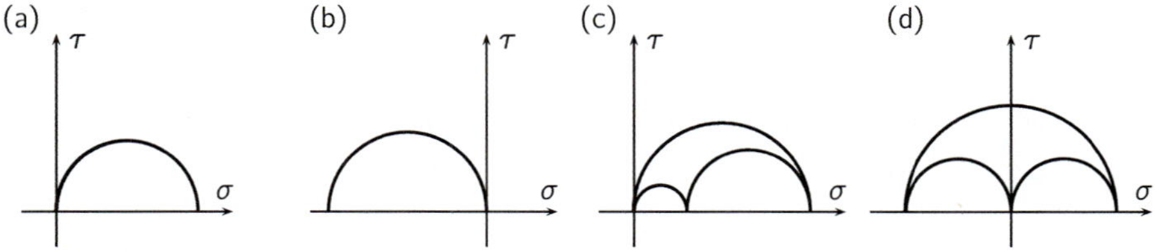

Figure 2.12 Mohr diagrams for various states of stress.

NOTE

1. When the weight of beams is considered, it is introduced as an applied force per unit length of the beam's span.

3 Stress–Strain Relations

Solid mechanics is a sub-field of continuum mechanics concerned with the stresses and deformation of solid bodies. Ignoring boundary conditions in a first step, the general field problem, in its simplest form, involves determining three fields: the displacement $\mathbf{u}$, the strain $\boldsymbol{\varepsilon}$, and the stress $\boldsymbol{\sigma}$. These concepts and their mathematical embodiment in tensor fields were summarized in Chapters 1 and 2. The number of unknowns in any such solid mechanics field problem is therefore fifteen, encompassing three displacement components, six strain components, and six stress components.

These fields are related by governing equations. So far, we have seen the kinematic relations, or equivalently the compatibility equations in Chapter 1, as well as the equilibrium equations in Chapter 2. There are six kinematic relations given by Equations (1.8) and (1.9) and three scalar equations of equilibrium, Equations (2.8a) to (2.8c). This makes a current total of nine equations. Closure of the system of equations in the unknown $\mathbf{u}$, $\boldsymbol{\varepsilon}$, and $\boldsymbol{\sigma}$ requires an additional set of six equations. Thus, a purely mathematical argument elucidates the need for six additional equations.

The missing six (scalar) equations are provided by a constitutive relation between stress and strain (involving more generally their rates for more complex behaviors.) The simplest of all constitutive relations is that for linear elasticity, known as Hooke's law. In this book, we will exclusively use this law to obtain solutions to structural mechanics problems.

The constitutive law rests on experimental facts. Historically, the law attributed to Hooke (17th century) has evolved to settle on two elastic constants: a tensile modulus named after Young (early 19th century) and a dimensionless coefficient after Poisson (early 19th century). The law can also be written in terms of two constants named after Lamé (mid-19th century). This serves as a reminder that, even for the relatively simple case of a homogeneous isotropic linear elastic material, much analytical thinking was required over a sustained period of time, well beyond experimental facts, to settle on the law of linear elasticity.

3.1 EXPERIMENTAL FACTS

A basic experiment to characterize material behavior is simple tension. The specimen has the shape shown in Figure 3.1 and loading ends for grips. Loading is along the x-axis. Under load, the stress state in the gauge section, i.e., far away from the grips, is uniform and uniaxial with σ_x being the only stress component:

$$\sigma_x = \frac{F}{S} \tag{3.1}$$

where F is the applied force and S is the cross-sectional area. The force is typically measured using a load cell or a dynamometer.

 DOI: 10.1201/9781003537229-4

A linear strain gauge, g_x, is typically glued to the front surface along the loading direction or on the side surface; see g'_x in Figure 3.1b. Longitudinal fibers elongate by the amount:

$$\varepsilon_x = \frac{\delta L}{L} \tag{3.2}$$

where L is the gauge length and δL is the elongation. This is typically measured using the strain gauge.

A plot of uniaxial stress versus longitudinal strain can then be produced, Figure 3.1. The stress versus strain response exhibits an initial part where the stress is proportional to the strain. The proportionality is expressed using

$$\sigma_x = E\varepsilon_x \tag{3.3}$$

where the slope of the curve (Figure 3.1a) is Young's modulus.

If the material is loaded to a point on the thick line and then unloaded, the reading from the strain gauge would show no strain after unloading. The deformation process is reversible, and there is no residual (or permanent) deformation. Such behavior is said to be linearly elastic.

Figure 3.2 illustrates the Poisson effect whereby cross-sections either shrink (in tension) or expand (in compression). In both cases, the loading is applied perpendicular to the plane of the figure, and the initial cross-section is shown dashed. For the

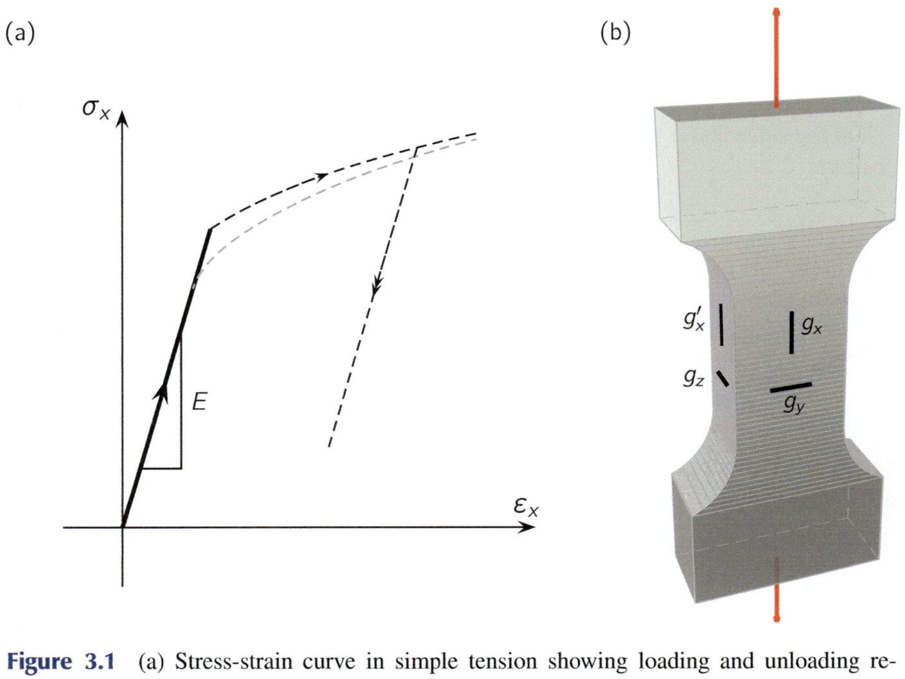

Figure 3.1 (a) Stress-strain curve in simple tension showing loading and unloading responses. (b) Tensile specimen with four strain gauges.

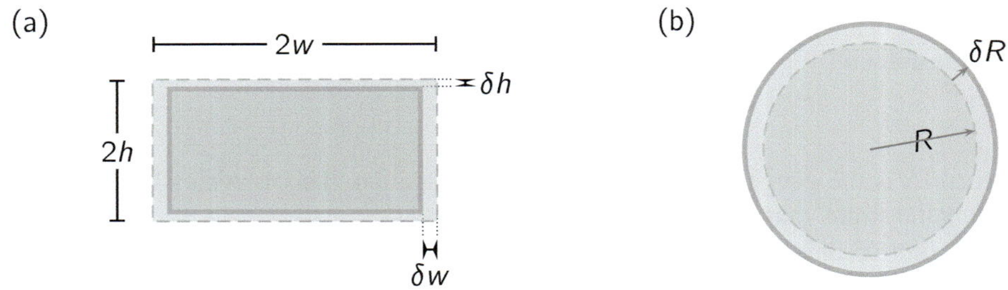

Figure 3.2 Illustration of the Poisson effect in (a) tension of a rectangular bar and (b) compression of a cylindrical pin.

rectangular section of Figure 3.2a, the transverse strains are:

$$\varepsilon_y = -\frac{\delta w}{w}, \quad \varepsilon_z = -\frac{\delta h}{h}$$

where the y- and z- axes are horizontal and vertical in the figure. Both strains are negative implying contraction. If the material is isotropic, the two transverse strains are equal: $\varepsilon_y = \varepsilon_z$. Poisson's ratio is then defined as follows:

$$\nu \equiv -\frac{\varepsilon_y}{\varepsilon_x} = -\frac{\varepsilon_z}{\varepsilon_x} \tag{3.4}$$

If two additional strain gauges are attached to the specimen, as shown in Figure 3.1b, then the transverse strains along the width and thickness of the specimen can be measured and an estimate of Poisson's ratio can be made using Equation (3.4).

On the other hand, for the cylindrical pin of Figure 3.2b, the transverse strain is positive:

$$\varepsilon_r = \varepsilon_\theta = \frac{\delta R}{R}$$

This implies expansion, consistent with the minus sign in Equation (3.4).

Another basic experiment is the torsion of a thin-walled tube, Figure 3.3. The tube has initial length L, mean radius R and wall thickness t. The shear stress in the hoop direction is given by

$$\tau_{x\theta} = \frac{M^t}{RS} \tag{3.5}$$

where M^t denotes the applied torque and $S = 2\pi Rt$ is the cross-sectional area. The shear strain is given by

$$\gamma_{x\theta} = \frac{D}{L}\omega \tag{3.6}$$

where $D = 2R$ is the diameter and ω is the angle of twist. The shear strain can be (indirectly) measured using a strain gauge at $\pm 45°$ from the tube's axis, as sketched in Figure 3.3b.

Figure 3.3a gives the outline of a shear stress versus shear strain curve from a torsion experiment. The overall response exhibits features similar to those of the

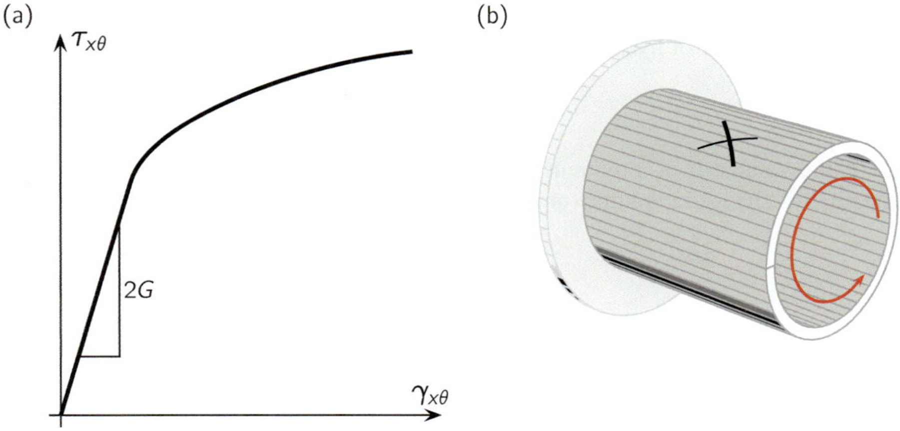

Figure 3.3 (a) Stress–strain curve in torsion. (b) Torsion tube and two strain gauges on its surface.

tensile test in Figure 3.1. However, the slope of the linear elastic part is now given by a different constant defined through

$$\tau_{x\theta} = 2G\gamma_{x\theta} \tag{3.7}$$

where G is called the shear modulus and the factor of 2 is for convenience and will be justified later. Unlike tension, torsion does not involve any changes in the radial or through-thickness dimensions of the thin-walled tube.

A less common, yet possible experiment is isostatic compression. Ideally, it consists of subjecting a material piece to hydrostatic pressure. A related thought experiment is to imagine a solid material immersed in a fluid pressurized to pressure p. The stress state applied to the material is then represented by a diagonal matrix where each diagonal term is equal to $-p$. This state is sketched in Figure 3.4.

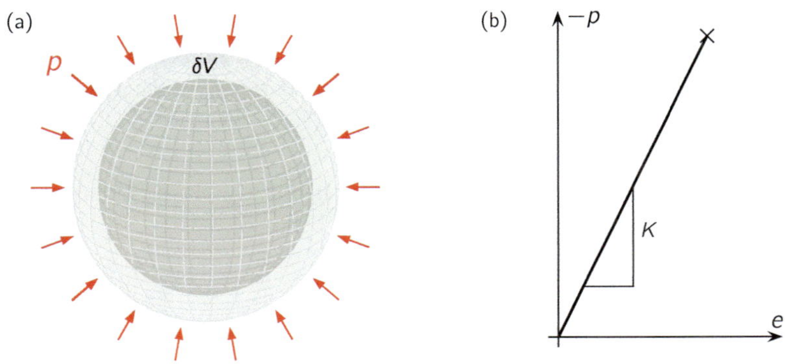

Figure 3.4 Notion of bulk modulus. (a) Hydrostatic loading. (b) Pressure versus dilatation response.

Under this state of hydrostatic pressure, we have

$$\operatorname{tr}\boldsymbol{\sigma} = -3p \tag{3.8}$$

Also, the volume V of a reference sphere decreases by an amount δV. As a result, the relative change in volume (or dilatation) is (see Remark 1.4)

$$e \equiv \operatorname{tr}\boldsymbol{\varepsilon} = \frac{\delta V}{V} \tag{3.9}$$

Figure 3.4b shows how the pressure versus dilatation response would look like should volumetric measurements be made. The relationship is linear and reversible; it may be expressed using

$$-p = Ke \tag{3.10}$$

where the slope of the linear elastic response is known as the bulk modulus.

If the material is made of a homogeneous isotropic material then a description based on the two elastic constants E and ν is sufficient. The portion shown dashed in the stress–strain curve of Figure 3.1 corresponds to the behavior beyond the elastic limit, also known as the proportional limit or yield stress of the material. That quantity is not as fundamental as the elastic constants but constitutes a de facto material property for the structural analyst. The more accurate the instrument of measurement, the earlier the detection of yielding, as illustrated in Figure 3.1a. In engineering practice, a yield stress defined at an offset strain (typically 0.2%) is used. Beyond this point, the stress–strain response becomes nonlinear. If the tensile specimen is deformed beyond the elastic limit, then unloaded, the unloading response is typically linear, as shown using the double arrow in Figure 3.1a. The unloading response has the same slope E as the initial linear elastic part. If unloading is complete, a permanent deformation is noted.

Remark 3.1 *The pressure versus volume change response in Figure 3.4b does not exhibit a nonlinear portion. For a standard material, yielding cannot occur under pure pressure loading, unless the material has internal porosity. In principle, the material would deform elastically until break, as marked by the symbol $\times$ in the figure.* $*$

The essence of structural analysis focuses on linearly elastic behavior and this book shall confine all attention to that regime of behavior.

3.2 HOOKE'S LAW

It is the constitutive law for linear isotropic elasticity. The strain is given as a (state) function of stress:

$$\boxed{\boldsymbol{\varepsilon} = \frac{1+\nu}{E}\boldsymbol{\sigma} - \frac{\nu}{E}(\operatorname{tr}\boldsymbol{\sigma})\boldsymbol{I}} \tag{3.11}$$

Here, E is Young's modulus, ν is Poisson's ratio, tr denotes the trace operator giving the sum of the diagonal components of a tensor, which is an invariant, and $\boldsymbol{I}$ is the identity tensor.

If Cartesian coordinates are used, the above equation may be written in matrix form as follows:

$$\begin{bmatrix} \varepsilon_x & \gamma_{xy} & \gamma_{xz} \\ \gamma_{xy} & \varepsilon_y & \gamma_{yz} \\ \gamma_{xz} & \gamma_{yz} & \varepsilon_z \end{bmatrix} = \frac{1+\nu}{E} \begin{bmatrix} \sigma_x & \tau_{xy} & \tau_{xz} \\ \tau_{xy} & \sigma_y & \tau_{yz} \\ \tau_{xz} & \tau_{yz} & \sigma_z \end{bmatrix} - \frac{\nu}{E}(\sigma_x + \sigma_y + \sigma_z) \begin{bmatrix} 1 & 0 & 0 \\ 0 & 1 & 0 \\ 0 & 0 & 1 \end{bmatrix} \tag{3.12}$$

from which the following six scalar equations are obtained after rearranging some terms:

$$\varepsilon_x = \frac{1}{E}[\sigma_x - \nu(\sigma_y + \sigma_z)] \tag{3.13a}$$

$$\varepsilon_y = \frac{1}{E}[\sigma_y - \nu(\sigma_z + \sigma_x)] \tag{3.13b}$$

$$\varepsilon_z = \frac{1}{E}[\sigma_z - \nu(\sigma_x + \sigma_y)] \tag{3.13c}$$

$$\gamma_{xy} = \frac{1+\nu}{E}\tau_{xy} \tag{3.13d}$$

$$\gamma_{xz} = \frac{1+\nu}{E}\tau_{xz} \tag{3.13e}$$

$$\gamma_{yz} = \frac{1+\nu}{E}\tau_{yz} \tag{3.13f}$$

Remark 3.2 *The intrinsic (or tensor) form of an equation is always more compact than its component counterpart. Consider, for instance, Equation* (1.3) *versus Equations* (1.8) *and* (1.9) *for kinematic relations, Equation* (2.7) *versus Equation (2.8) for equilibrium, and Equation* (3.11) *versus Equation (3.13) for Hooke's law. It may be argued that each of the scalar equations (3.13a)–(3.13f), taken individually, is simple enough to be remembered. However, there are other forms for simpler stress or strain states (namely, plane strain or plane stress) that resemble, but are yet different from Equations* (3.13a) *to* (3.13f)*. On the other hand, Equation* (3.11) *is a general form from which all other forms may be derived.* *

It is also common to write Hooke's law in matrix form, using Voigt's notations in Equations (1.14) and (2.9) for the strains and stresses, respectively, so that

$$\begin{bmatrix} \varepsilon_1 \\ \varepsilon_2 \\ \varepsilon_3 \\ \varepsilon_4 \\ \varepsilon_5 \\ \varepsilon_6 \end{bmatrix} = \frac{1}{E} \left[\begin{array}{ccc|ccc} 1 & -\nu & -\nu & & & \\ -\nu & 1 & -\nu & & 0 & \\ -\nu & -\nu & 1 & & & \\ \hline & & & 1+\nu & 0 & 0 \\ & 0 & & 0 & 1+\nu & 0 \\ & & & 0 & 0 & 1+\nu \end{array} \right] \begin{bmatrix} \sigma_1 \\ \sigma_2 \\ \sigma_3 \\ \sigma_4 \\ \sigma_5 \\ \sigma_6 \end{bmatrix} \tag{3.14}$$

For example, developing the first row gives: $\varepsilon_1 = [\sigma_1 - \nu(\sigma_2 + \sigma_3)]/E$, which is equivalent to Equation (3.13a) with the identifications $\varepsilon_1 = \varepsilon_x$ and $\sigma_1 = \sigma_x$, etc. from Equations (1.14) and (2.9). Likewise, developing the fourth row gives: $\varepsilon_4 = (1+\nu)\sigma_4/E$, which is the same as Equation (3.13f) with the identifications $\varepsilon_4 = \sqrt{2}\gamma_{yz}$ and $\sigma_4 = \sqrt{2}\tau_{yz}$.

Remark 3.3 *One advantage of the matrix format in Equation* (3.14) *is that it is amenable to generalization to the case of anisotropic materials, such as composites (where properties are orientation-dependent).* $*$

3.3 INVERTED FORM

When Hooke's law is written as in Equation (3.11), or any of its equivalent forms in the previous section, it is assumed that stresses are known so that the law provides the strains. In a class of problems, one actually proceeds from displacements, then the strains. In this case, the constitutive relation needs to be written so as to provide the stresses in terms of the strains. This involves inverting Equation (3.11). Details aside, the inverted form of Hooke's law reads

$$\boldsymbol{\sigma} = \lambda(\mathrm{tr}\,\boldsymbol{\varepsilon})\boldsymbol{I} + 2\mu\,\boldsymbol{\varepsilon} \tag{3.15}$$

Here, λ and μ are Lamé's constants, which are related to E and ν through:

$$\lambda = \frac{\nu E}{(1+\nu)(1-2\nu)} \tag{3.16}$$

$$\mu = \frac{E}{2(1+\nu)} \tag{3.17}$$

The second Lamé constant also has the meaning of a shear modulus and the notation $G \equiv \mu$ is often used, including in Part II.

Equation (3.15) may be written in various other forms, just like Equation (3.11), which was rewritten in component form, Equation (3.13), and in matrix form, Equation (3.14). The component and matrix forms of Equation (3.15) will not be used in the remainder of this text. For brevity, therefore, they are omitted here, but can be found in Section 4.A.

Remark 3.4 *In the literature, either form, Equation* (3.11) *or Equation* (3.15)*, may be referred to as Hooke's law. In other words, the terminology "inverted form" is just a convention. In the stress versus strain relation, Equation* (3.15)*, stiffness constants appear, whereas in the strain versus stress relation, Equation* (3.11)*, compliance constants appear. For example, when the matrix format is used, say in Equation* (3.14)*, the matrix multiplying the stress is the compliance matrix; its components have units of an inverse of pressure (recall that E is given in units of Pa, whereas ν is dimensionless). On the other hand, the Lamé constants used in Equation* (3.15) *have units of a stiffness, i.e., Pa.* ∗

Remark 3.5 *Another stiffness measure that is of general interest in elasticity is the bulk modulus. It is the stiffness of a solid when subjected to pure pressure (positive or negative) loading. Obviously, the bulk modulus must be related to previously introduced elastic constants, either E and ν in Equation* (3.11) *or λ and μ in Equation* (3.15)*. To see this, observe that under a pure pressure loading (of magnitude p) the stress tensor is diagonal ($-p\mathbf{I}$) and its trace is $-3p$. Then, from Equation* (3.11)*, the trace of the strain tensor (dilatation) is: $(\frac{1+\nu}{E} - 3\frac{\nu}{E})\mathrm{tr}\,\boldsymbol{\sigma}$. The bulk modulus K is nothing but the ratio of the pressure to the dilatation; hence,*

$$K = \frac{E}{3(1-2\nu)} \tag{3.18}$$

∗

3.4 SPECIAL CASES

3.4.1 PLANE STRAIN

In thick specimens, the change in thickness due to deformation is so small relative to the thickness that the overall through-thickness strain is negligible compared with other strain components. An example of such a specimen is shown in Figure 3.5a. This overall state of plane strain is encountered in each longitudinal section of the specimen sufficiently far from the front and back free surfaces.

Mathematically, a solid is said to be in a state of plane strain if there exists a frame where the displacement vector takes the following form:

$$\{\mathbf{u}\} = \left\{ \begin{array}{c} u(x,y) \\ v(x,y) \\ w(z) = Cz \end{array} \right\} \tag{3.19}$$

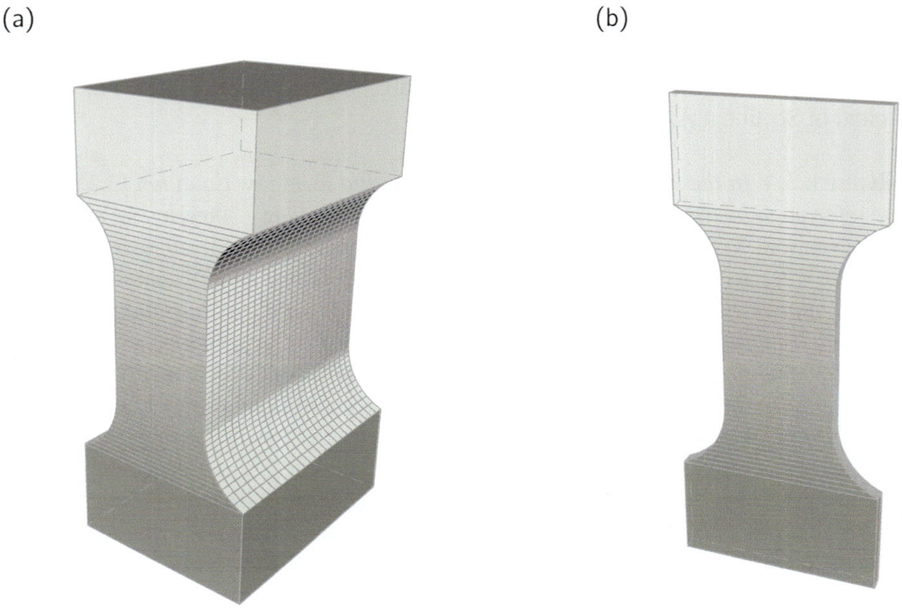

Figure 3.5 Example specimens that realize a state of (a) plane strain and (b) plane stress.

where C is a constant, most often equal to zero. The kinematic relations, Equations (1.8) and (1.9), then reduce to

$$\varepsilon_x = \frac{\partial u}{\partial x}; \qquad \varepsilon_y = \frac{\partial v}{\partial y}; \qquad \varepsilon_z = C \tag{3.20}$$

for the normal strains and

$$2\gamma_{xy} = \frac{\partial u}{\partial y} + \frac{\partial v}{\partial x}; \qquad \gamma_{xz} = 0; \qquad \gamma_{yz} = 0 \tag{3.21}$$

for the shear strains.

The strain and stress matrices then take the following form:

$$[\boldsymbol{\varepsilon}] = \begin{bmatrix} \varepsilon_x & \gamma_{xy} & 0 \\ \gamma_{xy} & \varepsilon_y & 0 \\ 0 & 0 & C \end{bmatrix}; \qquad [\boldsymbol{\sigma}] = \begin{bmatrix} \sigma_x & \tau_{xy} & 0 \\ \tau_{xy} & \sigma_y & 0 \\ 0 & 0 & \sigma_z \end{bmatrix} \tag{3.22}$$

Remark 3.6 *It is common to refer to the case $C = 0$ as plane strain and all other cases of $C \neq 0$ as generalized plane strain.* ∗

The intrinsic form of Hooke's law, Equation (3.11), is the same. The component form, Equation (3.13f), simplifies into:

$$\varepsilon_x = \frac{1+\nu}{E}[\sigma_x - \nu(\sigma_x + \sigma_y)] - \nu C \tag{3.23a}$$

$$\varepsilon_y = \frac{1+\nu}{E}[\sigma_y - \nu(\sigma_x + \sigma_y)] - \nu C \tag{3.23b}$$

$$\gamma_{xy} = \frac{1+\nu}{E}\tau_{xy} \tag{3.23c}$$

The inverted form is, in terms of E and ν (see Problem 3.3):

$$\sigma_x = \frac{E}{(1+\nu)(1-2\nu)}[(1-\nu)\varepsilon_x + \nu(\varepsilon_y + C)] \tag{3.24a}$$

$$\sigma_y = \frac{E}{(1+\nu)(1-2\nu)}[(1-\nu)\varepsilon_y + \nu(\varepsilon_x + C)] \tag{3.24b}$$

$$\sigma_z = \frac{E}{(1+\nu)(1-2\nu)}[\nu(\varepsilon_x + \varepsilon_y) + (1-\nu)C] \tag{3.24c}$$

$$\tau_{xy} = \frac{E}{1+\nu}\gamma_{xy} \tag{3.24d}$$

The following relation is to be noted (see Problem 3.4):

$$\sigma_z = \nu(\sigma_x + \sigma_y) + EC \tag{3.25}$$

3.4.2 PLANE STRESS

In thin specimens, Figure 3.5b, the relative change in thickness cannot be neglected. However, since the front and back surfaces are traction-free and the thickness is small, the variations of all tractions that are zero on those surfaces may be neglected.

Mathematically, a solid is said to be in a state of plane stress if there exists a frame where the stress and strain matrices take the following form:

$$[\boldsymbol{\sigma}] = \begin{bmatrix} \sigma_x & \tau_{xy} & 0 \\ \tau_{xy} & \sigma_y & 0 \\ 0 & 0 & 0 \end{bmatrix}; \qquad [\boldsymbol{\varepsilon}] = \begin{bmatrix} \varepsilon_x & \gamma_{xy} & 0 \\ \gamma_{xy} & \varepsilon_y & 0 \\ 0 & 0 & \varepsilon_z \end{bmatrix} \tag{3.26}$$

The component form of Hooke's law, Equation (3.13), simplifies into:

$$\varepsilon_x = \frac{1}{E}(\sigma_x - \nu\sigma_y) \tag{3.27a}$$

$$\varepsilon_y = \frac{1}{E}(\sigma_y - \nu\sigma_x) \tag{3.27b}$$

$$\varepsilon_z = -\frac{\nu}{E}(\sigma_x + \sigma_y) \tag{3.27c}$$

$$\gamma_{xy} = \frac{1+\nu}{E}\tau_{xy} \tag{3.27d}$$

The inverted form is, in terms of E and ν (see Problem 3.5):

$$\sigma_x = \frac{E}{1-\nu^2}(\varepsilon_x + \nu\varepsilon_y) \tag{3.28a}$$

$$\sigma_y = \frac{E}{1-\nu^2}(\varepsilon_y + \nu\varepsilon_x) \tag{3.28b}$$

$$\tau_{xy} = \frac{E}{1+\nu}\gamma_{xy} \tag{3.28c}$$

The following relation is to be noted (see Problem 3.6):

$$\varepsilon_z = -\frac{\nu}{1-\nu}(\varepsilon_x + \varepsilon_y) \tag{3.29}$$

3.5 ELASTIC POTENTIAL ENERGY

In a solid, stresses induce deformations, hence displacements. Therefore, they perform work between the initial and final equilibrium states. As a result, the solid stores this strain energy, which is called the elastic potential energy.

A classical calculation in a principal frame leads to the general expression for the elastic potential energy W per unit volume:

$$\frac{dW}{dV} = \frac{1}{2}\boldsymbol{\sigma} : \boldsymbol{\varepsilon} \tag{3.30}$$

where the symbol ":" indicates a double contraction (a sort of scalar product between tensors), just like the scalar product between vectors denoted by "·". In component form, we have

$$\frac{dW}{dV} = \frac{1}{2}[\sigma_x\varepsilon_x + \sigma_y\varepsilon_y + \sigma_z\varepsilon_z + 2(\tau_{xy}\gamma_{xy} + \tau_{yz}\gamma_{yz} + \tau_{zx}\gamma_{zx})] \tag{3.31}$$

Remark 3.7 *One may also write:* $\boldsymbol{\sigma} : \boldsymbol{\varepsilon} = \bar{\boldsymbol{\sigma}} \cdot \bar{\boldsymbol{\varepsilon}}$ *using Voigt's condensation, Equation* (1.14), *where the standard scalar product is used between the 6-vectors* $\bar{\boldsymbol{\sigma}}$ *and* $\bar{\boldsymbol{\varepsilon}}$. *This explains the* $\sqrt{2}$ *factor appearing in Equations* (1.14) *and* (2.9) *so that the shear terms, which appear twice in the double contraction, are properly accounted for.* *

Remark 3.8 *There is a variant of Voigt's notation whereby the shear strains* ε_4, *etc. in Equation* (3.14) *are defined using a factor 2 instead of* $\sqrt{2}$. *When this is the case, the* $\sqrt{2}$ *is dropped from the shear stresses* σ_4, *etc. This amounts to properly accounting for work-like or energy terms, as in Equation* (3.31). *If such a convention were to be used, then the lower-right block matrix in Equation* (3.14) *would have to be multiplied by a factor of 2.* *

PROBLEMS

Problem 3.1 (How to Invert Hooke's Law by Hand) *Starting from the strain versus stress form of Hooke's law, Equation* (3.11), *obtain by direct inversion Equation* (3.15). *Likewise, starting from Equation* (3.15), *obtain Equation* (3.11). ⋆

Problem 3.2 (Relation between Shear Modulus and Young's Modulus) *Show that the shear modulus G is not independent of the other two elastic constants E and ν. To this end, consider a state of pure shear and establish Equation* (3.17) *using either strain/stress transformations or appropriate Mohr diagrams.* ⋆

Problem 3.3 (Hooke's Law under Plane Strain) *Establish Equation* (3.23c) *and Equation* (3.24d) *by inverting the matrix form of Hooke's law, Equation* (3.14). *We will take $C = 0$.* ⋆

Problem 3.4 (Out-of-Plane Stress in Plane Strain) *Derive Equation* (3.25). ⋆

Problem 3.5 (Hooke's Law under Plane Stress) *Establish Equation* (3.27d) *and Equation* (3.28c) *by inverting the matrix form of Hooke's law, Equation* (3.14). ⋆

Problem 3.6 (Out-of-Plane Strain in Plane Stress) *Derive Equation* (3.29). ⋆

Problem 3.7 (On Elastic Constant Calculations) *Measurements lead to the following stress and strain tensors:*

$$\boldsymbol{\sigma} = \begin{bmatrix} 2 & 0 & 1 \\ 0 & 1 & 6 \\ 1 & 6 & 0 \end{bmatrix} 10^2\, MPa; \quad \boldsymbol{\varepsilon} = \begin{bmatrix} 0.85 & 0 & 0.65 \\ 0 & 0.2 & 3.9 \\ 0.65 & 3.9 & -0.45 \end{bmatrix} \textit{millistrain}$$

(i) By how much does the angle change between the lines that were along the y and z axes before deformation? (ii) What is Young's modulus E of the material? ⋆ ⋆

Problem 3.8 (Specific Energy Calculations) *Calculate the elastic potential energy for the three physical experiments of Figures 3.1, 3.3 and 3.4.* ⋆

4 The General Problem of Elastostatics

4.1 FIELD EQUATIONS

For ease of reference, the governing equations of a field elasticity problem are gathered in this section. With inertial forces disregarded, one deals with a problem of elastostatics.

Kinematics: The kinematic relations comprise the strain–displacement relations given by Equation (1.3), which are equivalent (under some conditions) to the compatibility equations given by (1.16a)–(1.16f):

$$2\boldsymbol{\varepsilon} = \nabla \mathbf{u} + \nabla^T \mathbf{u}, \tag{K}$$

as well as the rotation–displacement relations given by Equation (1.4), rewritten here:

$$2\boldsymbol{\Omega} = \nabla \mathbf{u} - \nabla^T \mathbf{u} \tag{K'}$$

Their component forms are given by Equations (1.8) to (1.10).

Equilibrium: The statement of local equilibrium is given by Equation (2.7) (no body force):

$$\mathbf{div}\,\boldsymbol{\sigma} = \mathbf{0} \tag{E}$$

The component form is given by Equations (2.8a) to (2.8c).

Constitution: For isotropic linear elasticity, Hooke's law reads:

$$\boldsymbol{\varepsilon} = \frac{1+\nu}{E}\boldsymbol{\sigma} - \frac{\nu}{E}(\mathrm{tr}\,\boldsymbol{\sigma})\boldsymbol{I} \quad \Longleftrightarrow \quad \boldsymbol{\sigma} = \lambda(\mathrm{tr}\,\boldsymbol{\varepsilon})\boldsymbol{I} + 2\mu\,\boldsymbol{\varepsilon} \tag{H}$$

with λ and μ given by Equations (3.16) and (3.17) in terms of E and ν.

The general problem defined by field equations {(K), (E) and (H)} involves 15 scalar equations, six in Equation (K), three in Equation (E) and six in Equation (H). It also involves 15 unknowns, which comprise the three components of the displacement $\mathbf{u}$, the six components of the strain $\boldsymbol{\varepsilon}$, and the six components of the stress $\boldsymbol{\sigma}$. Thus, there are as many scalar unknowns as there are equations. Once the displacements are determined, the rotations may be obtained directly from Equation (K').

The component form of the governing system {(K), (E), (H)} is given in Section 4.A using both Cartesian and cylindrical coordinates.

DOI: 10.1201/9781003537229-5

4.2 SOLUTION METHODS

4.2.1 DISPLACEMENT METHOD

One way to approach the general solution to the system of equations $\{(K), (E), (H)\}$ is to treat the displacement **u** as the main unknown and seek a single (vectorial) governing equation in **u**. This is possible and may be seen as follows. Substitute for the strains, as given by Equation (K), into the inverted Hooke's law, Equation $(H)_2$, then inject the resulting expression of the stress into the equilibrium statement, Equation (E). The result will be a second-order partial differential equation in the sole unknown **u**. That equation is known as Navier's equation, which is the counterpart of the Navier–Stokes equations (for an incompressible linearly viscous fluid).

Navier's equation has many fundamental implications in solid mechanics, but, in this course, we will make no use of it. This does not mean that we will not employ the displacement method as a conceptual approach to solving some fundamental problems. For example, the general solution to the pure bending of a prismatic beam is most easily approached using a displacement method.

4.2.2 FORCE METHOD

An alternative approach to solving system $\{(K), (E), (H)\}$ consists of taking the stress $\boldsymbol{\sigma}$ as the main unknown and seeking a single (tensorial) governing equation in $\boldsymbol{\sigma}$. To obtain the latter, substitute for the strain as given by the compliance form of Hooke's law, Equation $(H)_1$, into the compatibility equations (formally, this would best be done using the "mnemonic" format in Equation (1.18)), then use equilibrium, Equation (E), to simplify the final equation. The result will be a second-order partial differential equation in the sole unknown $\boldsymbol{\sigma}$. That equation is known as the Beltrami equation.

Just like the Navier equation, the Beltrami equation has far-reaching implications for linear elasticity. While it will not be used as such in the remainder of the text, the underlying stress formulation (or force method) will be extensively used in deriving general solutions to the torsion of prisms in Chapter 8 and their bending with shear in Chapter 9. For this reason, the compatibility relations play a central role in developing an exact theory of beams.

4.3 BOUNDARY-VALUE PROBLEM

The constitutive relation, Equation (H), is a state equation. However, one must be cognizant of the fact that Equations (K) and (E) are differential equations, and as such, they require boundary conditions. Even in an unbounded medium field problem, some constraints on the unknowns may be needed at infinity to obtain physically reasonable solutions. Specifically, Equations (K) and (E) are partial differential equations where each unknown is a field, i.e., a quantity that depends on x, y, z. Therefore, solution to the system of equations $\{(K), (E), (H)\}$ requires spatial boundary conditions.

Note that in elastodynamics (i.e., replacing Equation (E) with Equation (2.6) for fully dynamic equilibrium in the system $\{(K), (E), (H)\}$) both initial and boundary conditions would be required (or mathematically speaking, boundary conditions in the spatio-temporal domain).

All elastostatic problems obviously share the same governing field equations. It is through the boundary conditions that two problems are distinguished from each other. Defining boundary conditions requires defining two things: (i) the boundary of the body itself and (ii) how the boundary is loaded.

On a smooth surface boundary, the normal is well defined so that imposing a tensorial quantity on such a boundary is meaningless. Thus, boundary conditions are either specified in terms of displacements or tractions, as illustrated in Figure 4.1. In a general formulation of a boundary-value problem of elastostatics, a portion of the boundary may be subjected to imposed displacements, while another portion would be subjected to imposed tractions. In Figure 4.1, these portions are denoted S_u and S_t, respectively.

In sum, the governing equations of a boundary-value problem in elastostatics comprise the system (K), (E), (H) for any point inside the body along with two types of boundary conditions on the surface S:

- Displacement boundary conditions: For displacements $\mathbf{u}_0$ imposed on a portion S_u of the surface S,

$$\mathbf{u} = \mathbf{u}_0 \tag{4.1}$$

  This includes fixed portions for which $\mathbf{u} = \mathbf{0}$ (supports, fixtures).
- Traction boundary conditions: For tractions $\mathbf{T}_0$ imposed on a portion S_t of the surface,

$$\boldsymbol{\sigma}\mathbf{n} = \mathbf{T}_0 \tag{4.2}$$

  This includes traction-free portions for which $\boldsymbol{\sigma}\mathbf{n} = \mathbf{0}$ (free surfaces).

Of particular interest are problems with all-around imposed displacements or all-around imposed tractions. In the former case, when analytical solutions are sought or possible, the Navier equation (see Section 4.2.1) would be a natural starting point. In the latter, the Beltrami equation (see Section 4.2.2) would be the natural starting point.

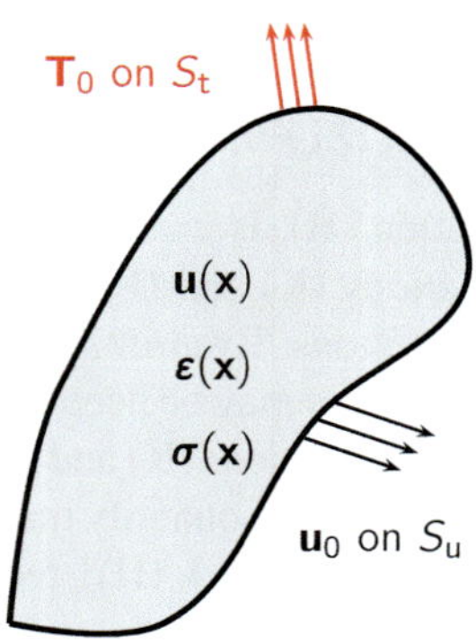

Figure 4.1 Boundary-value problem in solid mechanics.

4.4 BASIC PRINCIPLES

4.4.1 EXISTENCE AND UNIQUENESS OF SOLUTIONS

The boundary-value problem of elastostatics defined by system (K), (E), (H) completed with boundary conditions specified by Equations (4.1) and (4.2) admits a unique solution provided that the stiffness matrix is positive definite. This is guaranteed if both the shear modulus $G \equiv \mu$ given by Equation (3.17) and the bulk modulus given by Equation (3.18) are positive. This leads to the following restriction on Poisson's ratio:

$$-1 < \nu < \frac{1}{2} \tag{4.3}$$

These bounds are typically verified by all materials. Furthermore, the vast majority of materials of interest to the aerospace industry, among others, have $\nu > 0$. Materials with a negative Poisson ratio are known as auxetics. They typically have an internal structure that enables lateral expansion in simple tension.

Remark 4.1 *One should be appreciative of the availability of such a profound result in elastostatics (and more generally linear elasticity). Issues of existence and uniqueness are among the toughest in mathematics. For example, in fluid mechanics, there are no formal proofs to date of the existence of solutions, or their uniqueness when they exist, to the Navier–Stokes equations. Both existence and uniqueness are important concepts, but proofs of existence are context-dependent and typically much harder to establish.* $\ast$

Remark 4.2 *The existence and uniqueness of solutions in linear elasticity, under the not-so-restrictive condition of Equation* (4.3)*, has far-reaching implications. Indeed, if a trial solution satisfies all governing equations, then it is the solution. In particular, this applies when implementing some of the solution methods expounded in Section 4.2. For example, if a trial stress field is shown to satisfy the Beltrami equation and all boundary conditions are satisfied, then the solution is complete.* $\ast$

4.4.2 SAINT-VENANT'S PRINCIPLE

In some traction-formulated problems, the tractions are not specified pointwise on parts of the boundary. Instead, they are imposed on average so that a resultant force or a resultant moment is imposed. Such situations are far more practical than initially thought. For instance, the specific distribution of the imposed tractions is not exactly known in some basic mechanical tests, such as in simple tension. Yet, one can be assured that the net imposed load consists exclusively of, say a tensile force.

Figure 4.2 illustrates the situation for a bar under three loading scenarios. Lateral surfaces are free of any tractions while end sections are loaded such that global equilibrium (balance of linear and angular momenta) is satisfied. The three loading scenarios have in common the same net force, say F_0, on the top boundary. In loading scenario 1, the traction distribution is uniform and equal to $T_0 = F_0/S$ with S the cross-section area. In loading scenario 2, the traction distribution is nonuniform but

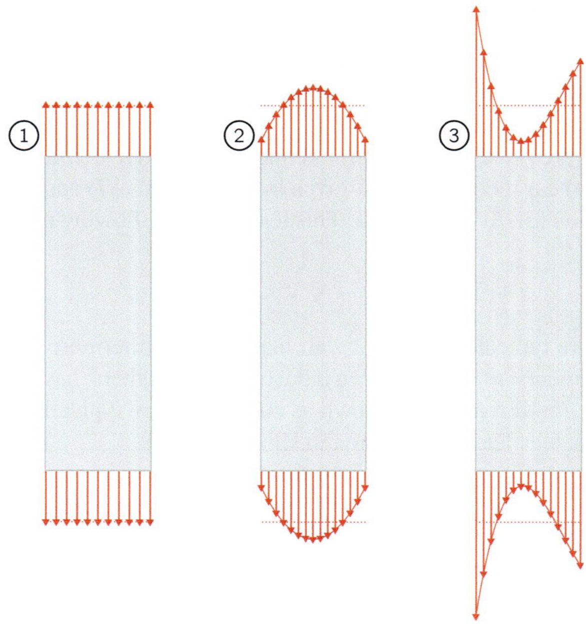

Figure 4.2 Bar under three loading scenarios sharing the same net force.

symmetric about the bar's vertical axis of symmetry. As a result, there is no net moment associated with this traction distribution. The average traction, shown dotted, is also equal to T_0. Thus, loading scenarios 1 and 2 not only share the same net force F_0 but also the same moment (zero in both cases). Loading systems 1 and 2 are said to be statically equivalent. On the other hand, loading scenario 3 has a nonuniform traction distribution that generates the same net force F_0 but also a non-zero net moment because the distribution is not symmetric. The resultant moment at the top boundary balances that on the bottom boundary. Loading system 3 is not statically equivalent to systems 1 and 2.

For the more general body of Figure 4.1, traction boundary conditions may be specified on average through

$$\int_{\bar{S}_t} \mathbf{T}\,\mathrm{d}S = \mathbf{F}_0 \tag{4.4}$$

and

$$\int_{\bar{S}_t} \overrightarrow{OP} \times \mathbf{T}\,\mathrm{d}S = \mathbf{M}_0 \tag{4.5}$$

Here, $\bar{S}_t$ denotes the surface of the body over which the exact distribution of tractions is not known or specified. Also, $\mathbf{F}_0$ is an imposed net force, $\mathbf{M}_0$ an imposed net moment about the origin O and P is the current material point within $\bar{S}_t$.

It must be realized that a boundary-value problem of elastostatics where boundary conditions such as Equations (4.4) and (4.5) hold on a portion of the surface of the body cannot admit a unique solution, since the tractions are not fully specified on the surface. However, an immensely important result known as the Saint-Venant principle enables the following to be stated.

Theorem 4.1: Saint-Venant's Principle

The boundary-value problem defined by the fundamental system $\{(K), (E), (H)\}$ inside the body, traction boundary conditions of the type of Equation (4.2) on a portion, S_t, of the boundary and *average* traction boundary conditions of the type in Equations (4.4) and (4.5) on a *complementary* portion, $\bar{S}_t$, of the boundary (i.e., such that $S_t \cup \bar{S}_t = S$) admits a two-term solution:

1. a near-field term that depends on the exact distribution of tractions on $\bar{S}_t$.
2. a far-field term that only depends on $\mathbf{F}_0$ and $\mathbf{M}_0$.

The near-field solution, which is valid in the vicinity of $\bar{S}_t$ is undetermined since the exact distribution of tractions on $\bar{S}_t$ is not known. The far-field solution is unique. ◊

Applying the above result to the bar example of Figure 4.2, it follows that, sufficiently far from the top and bottom boundaries, Problems 4.1 and 4.2 admit exactly the same solution, in that the displacement, strain, and stress fields are indistinguishable for the two loading scenarios. On the other hand, Problem 4.3 has a different solution away from the boundary because the traction distribution has an additional resultant moment, which the first two loading scenarios do not have.

A variant of Theorem 4.1 was first stated as a principle by the French elastician Barré de Saint-Venant in the mid-1800s and was proved over a century later. So it has the status of a theorem.

The theorem is all the more significant that the decay of the near-field solution is shown to be exponential. In other words, the problem-specific solution is valid only locally, close to where the boundary conditions are applied. Sufficiently far away (which is actually quite close), the far-field solution remains the same regardless of details about how the average force $\mathbf{F}_0$ and/or moment $\mathbf{M}_0$ are applied.

The family of boundary-value problems defined in the statement of Theorem 4.1 is known as a Saint-Venant problem. It is of general interest in mechanics but has particular grounding in structural mechanics as we will see in Part II.

Remark 4.3 *While the focus of an introductory structural mechanics course is on determining the far-field solution, analysis of the near-field solution is of much practical importance. It may be viewed as a stress concentration problem. A basic introduction to this topic will be found in Chapter 10.* $*$

4.4.3 SUPERPOSITION PRINCIPLE

In elastostatics, and more generally linear elasticity as well as other branches of physics, solutions can be superposed because the governing equations, including boundary conditions, are linear. This may be formally stated as follows.

Theorem 4.2: Superposition Principle

Let $\{\mathbf{u}^{(1)}, \boldsymbol{\varepsilon}^{(1)}, \boldsymbol{\sigma}^{(1)}\}$ be the solution to the boundary-value problem defined by (K), (E), (H) inside the body and boundary conditions Equations (4.1) and (4.2) on the surface, specialized for applied displacements $\mathbf{u}_0^{(1)}$ and applied tractions $\mathbf{T}_0^{(1)}$. Likewise, let $\{\mathbf{u}^{(2)}, \boldsymbol{\varepsilon}^{(2)}, \boldsymbol{\sigma}^{(2)}\}$ be the solution to a similar boundary-value problem but now with applied displacements $\mathbf{u}_0^{(2)}$ and applied tractions $\mathbf{T}_0^{(2)}$. Let α, β be two arbitrary scalars and define $\mathbf{u} = \alpha\mathbf{u}^{(1)} + \beta\mathbf{u}^{(2)}$, etc. Then, the set $\{\mathbf{u}, \boldsymbol{\varepsilon}, \boldsymbol{\sigma}\}$ is the solution to the boundary-value problem corresponding to applied displacements $\alpha\mathbf{u}_0^{(1)} + \beta\mathbf{u}_0^{(2)}$ and applied tractions $\alpha\mathbf{T}_0^{(1)} + \beta\mathbf{T}_0^{(2)}$. ◇

The superposition principle finds many applications in solid and structural mechanics. Since beam theory may be viewed as a reduced order theory of linear elasticity for slender members, the principle is naturally applied to the elements of reduction of the theory, such as beam deflections and internal resultants, as we will see in Part II. For example, the deflections of a cantilever beam are well known. If such a beam is subjected to more complex loading, involving a distributed load, say due to pressure distributions from aerodynamic loads, then the complex problem may be decomposed into two or more simpler problems. Furthermore, the original problem may be hyperstatic (i.e., statically indeterminate) , but its decomposition into smaller problems may result into isostatic (i.e., statically determinate) problems, which are easier to solve analytically. An example of this sort is given in Chapter 7; see Example 7.3.

4.5 EXAMPLES

The governing equations of linear elasticity given in Section 4.1 are solved for two basic problems: uniaxial loading of a bar and axisymmetric torsion. Both are formulated in terms of traction boundary conditions, as explained in Section 4.3. Therefore, the force method sketched in Section 4.2 is used. For definiteness, specific traction distributions are considered from the outset, it being understood that the obtained solutions remain valid for other statically equivalent traction distributions, as per the Saint-Venant principle of Section 4.4.2.

Example 4.1: Uniaxial Loading of a Bar

A rectangular bar of length L and cross-section $S = wh$, made up of a material with elastic constants E and ν, is loaded by uniform tractions on end sections, as shown in Figure 4.3. The bar is reported to a Cartesian coordinate system with $\mathbf{e}_x, \mathbf{e}_y, \mathbf{e}_z$ the base vectors. Interpret this loading by computing traction resultants, and then determine the stresses, strains, displacements, and rotations everywhere in the bar.

Solution:

Let T_0 denote the magnitude of the applied uniform tractions. Begin with mathematically expressing the traction boundary conditions as follows:

- On end face S_0, where the outer unit normal $\mathbf{n} = -\mathbf{e}_x$:

$$\boldsymbol{\sigma}\mathbf{n} = -T_0\mathbf{e}_x \quad \text{with} \quad T_0 = \frac{F}{S} \tag{4.6}$$

- On end face S_1, where $\mathbf{n} = \mathbf{e}_x$:

$$\boldsymbol{\sigma}\mathbf{n} = T_0\mathbf{e}_x = \frac{F}{S}\mathbf{e}_x \tag{4.7}$$

- On lateral surface S_2, where $\mathbf{n} = \pm\mathbf{e}_y$ or $\mathbf{n} = \pm\mathbf{e}_z$ (see inset in Figure 4.3):

$$\boldsymbol{\sigma}\mathbf{n} = 0 \tag{4.8}$$

That $T_0 = F/S$ is valid in Equations (4.6) and (4.7) follows from the definition of tractions, Equation (2.1), which gives, for example on S_1:

$$\mathbf{F} = \int_{S_1} \mathbf{T}\,\mathrm{d}S = \int_{S_1} T_0\mathbf{e}_x\,\mathrm{d}S = T_0\mathbf{e}_x \int_{S_1} \mathrm{d}S = T_0 S\mathbf{e}_x \tag{4.9}$$

This shows that the net force is oriented along the bar's axis and has magnitude $F = T_0 S$. Furthermore, the assumed uniform traction distribution has zero moment. The net moment with respect to, say the centroid G of the section is

$$\mathbf{M} = \int_{S_1} \overrightarrow{GP} \times \mathbf{T}\,\mathrm{d}S = T_0 \int_{S_1} \overrightarrow{GP} \times \mathbf{e}_x\,\mathrm{d}S \tag{4.10}$$

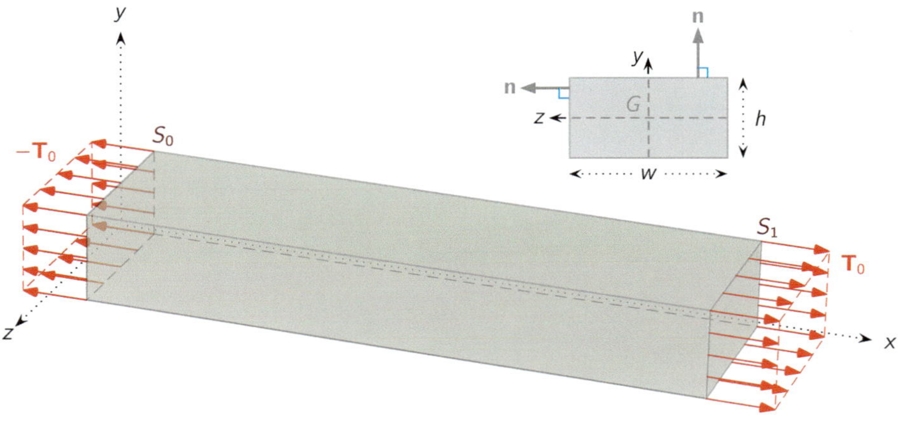

Figure 4.3 Bar under uniformly distributed tractions on end sections.

where P denotes some arbitrary point within S_1 such that $\overrightarrow{GP} = y\mathbf{e}_y + z\mathbf{e}_z$. Thus,

$$\mathbf{M} = T_0 \int_{S_1} (z\mathbf{e}_y - y\mathbf{e}_z)\,\mathrm{d}S = T_0 \left(\int_{S_1} z\,\mathrm{d}S \right) \mathbf{e}_y - T_0 \left(\int_{S_1} y\,\mathrm{d}S \right) \mathbf{e}_z \tag{4.11}$$

where the bracketed integrals are nil by definition of the centroid. Hence, $\mathbf{M} = \mathbf{0}$. The prescribed loading amounts, therefore, to a single axial net force $\mathbf{F} = F\mathbf{e}_x$ on face S_1 and an opposite force $-\mathbf{F}$ on face S_0, Figure 4.4. The bar is either under tension ($F > 0$) or compression ($F < 0$).

The so defined problem admits a unique solution; see Section 4.4.1. Thus, if a trial solution satisfies all governing equations, then it is *the* solution. For the uniaxial bar, it makes sense to attempt for the stress tensor the following matrix:

$$[\boldsymbol{\sigma}] = \begin{bmatrix} \frac{F}{S} & 0 & 0 \\ 0 & 0 & 0 \\ 0 & 0 & 0 \end{bmatrix} \tag{4.12}$$

such that the boundary conditions, Equations (4.6) to (4.8), are satisfied.

Anywhere in the bar, the governing equations are the system $\{$(K), (E), (H)$\}$ introduced in Section 4.1. These equations are expressions of kinematics, equilibrium and constitution, respectively. The tensor $\boldsymbol{\sigma}$ in Equation (4.12) being constant, it trivially satisfies equilibrium, i.e., Equation (E). From Hooke's law, Equation (H), the strain matrix necessarily takes the following form:

$$[\boldsymbol{\varepsilon}] = \frac{F}{ES} \begin{bmatrix} 1 & 0 & 0 \\ 0 & -\nu & 0 \\ 0 & 0 & -\nu \end{bmatrix} \tag{4.13}$$

Also, refer to the component form of Hooke's law, Equation (3.13f).

The technically tedious part consists of integrating the kinematic relations, Equation (K), to obtain the displacements. The derivation is given here for completeness, and may be skipped on first reading.

The displacements are related to the strains by Equation (K), which on account

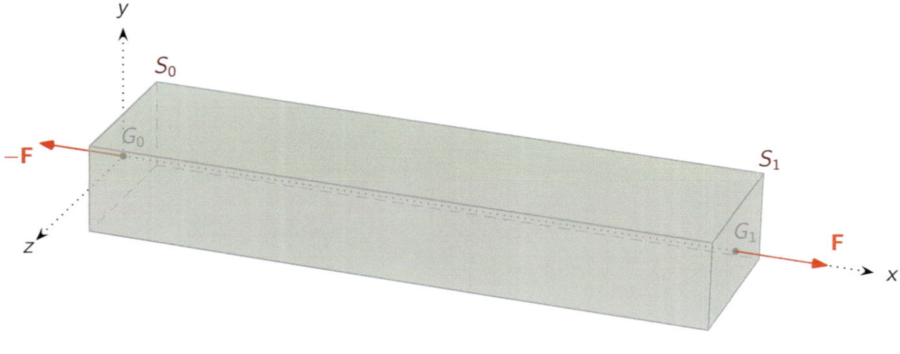

Figure 4.4 Bar under uniaxial loading.

of Equation (4.13) gives

$$\begin{cases} \dfrac{\partial u}{\partial x} = \dfrac{F}{ES} \\ \dfrac{\partial v}{\partial y} = -\nu \dfrac{F}{ES} \\ \dfrac{\partial w}{\partial z} = -\nu \dfrac{F}{ES} \end{cases} \tag{4.14}$$

for the normal strains using the component form of Equation (K), i.e., Equation (1.8) and

$$\begin{cases} \dfrac{\partial u}{\partial y} + \dfrac{\partial v}{\partial x} = 0 \\ \dfrac{\partial u}{\partial z} + \dfrac{\partial w}{\partial x} = 0 \\ \dfrac{\partial v}{\partial z} + \dfrac{\partial w}{\partial y} = 0 \end{cases} \tag{4.15}$$

for the shear strains using the component form in Equation (1.9). Equations (4.14) and (4.15) are integrable because compatibility in this problem is trivially satisfied. Integration of Equation (4.14) yields

$$\begin{cases} u = \dfrac{F}{ES} x + U(y,z) \\ v = -\nu \dfrac{F}{ES} y + V(x,z) \\ w = -\nu \dfrac{F}{ES} z + W(x,y) \end{cases} \tag{4.16}$$

where U is a 'constant' of partial integration with respect to x, and is hence possibly a function of y and z; likewise for integration functions V and W. The absence of any shears, Equation (4.15), implies constraints on these integration functions:

$$\begin{cases} \dfrac{\partial U}{\partial y} = -\dfrac{\partial V}{\partial x} = f_3(z) \\ \dfrac{\partial U}{\partial z} = -\dfrac{\partial W}{\partial x} = f_2(y) \\ \dfrac{\partial V}{\partial z} = -\dfrac{\partial W}{\partial y} = f_1(x) \end{cases} \tag{4.17}$$

Indeed, given that U in Equation $(4.16)_1$ is a function of y and z, its derivative with respect to y in Equation $(4.17)_1$ is also function of y and z. But the constraint implied by the vanishing of γ_{xy} makes it equal to a function of x and z, which is the derivative of $-V(x,z)$. This implies that both derivatives in Equation $(4.17)_1$ are at best functions of z alone (i.e., functions of the common variable). Apply the same reasoning to complete Equation $(4.17)_2$ and Equation $(4.17)_3$.

Each of the three unknown functions U, V, W appears twice in Equation (4.17). Integration of these now gives

$$\begin{cases} U(y,z) = f_3(z)y + g_3(z) = f_2(y)z + h_2(y) \\ V(x,z) = f_1(x)z + g_1(x) = -f_3(z)x + h_3(z) \\ W(x,y) = -f_2(y)x + g_2(y) = -f_1(x)y + h_1(x) \end{cases} \tag{4.18}$$

where the three g_i's and the three h_i's are integration functions of their respective variables.

It is easily shown that each of the f_i, g_i and h_i functions is linear. Indeed, assume that $f_3(z)$, for example, contains powers higher than 1. Since the right-hand side of Equation $(4.18)_1$ is linear in z, the only possibility is that such powers would cancel out with those of $g_3(z)$. But the latter are not multiplied by y, unlike $f_3(z)$. The reasoning holds for any nonlinear terms in $f_3(z)$. Hence, $f_3(z)$ must be a linear function. A similar reasoning applies to all functions in all three statements of Equation (4.18).

It follows that Equation (4.18) may be written as follows:

$$\begin{cases} U(y,z) = u_1 zy + u_2 y + u_3 z + u_4 \\ V(x,z) = v_1 xz + v_2 z + v_3 x + v_4 \\ W(x,y) = w_1 yx + w_2 x + w_3 y + w_4 \end{cases} \tag{4.19}$$

where u_i, v_i, w_i are constants. These 12 constants are not independent. Indeed, the constraints by Equation (4.17), not yet fully utilized, imply

$$\begin{cases} u_1 z + u_2 = -v_1 x - v_3,\ \forall x,z \quad \Rightarrow \quad u_1 = 0, v_1 = 0, u_2 = -v_3 \\ u_1 y + u_3 = -w_1 y - w_2,\ \forall y \quad \Rightarrow \quad w_1 = -u_1 (=0), u_3 = -w_2 \\ v_1 x + v_2 = -w_1 x - w_3,\ \forall x \quad \Rightarrow \quad v_2 = -w_3 \end{cases} \tag{4.20}$$

Equation (4.19) may then be simplified as follows:

$$\begin{cases} U(y,z) = u_2 y - w_2 z + u_4 \\ V(x,z) = v_2 z - u_2 x + v_4 \\ W(x,y) = w_2 x - v_2 y + w_4 \end{cases} \tag{4.21}$$

then recast in the following form:

$$\begin{Bmatrix} U \\ V \\ W \end{Bmatrix} = \begin{bmatrix} 0 & u_2 & -w_2 \\ -u_2 & 0 & v_2 \\ w_2 & -v_2 & 0 \end{bmatrix} \begin{Bmatrix} x \\ y \\ z \end{Bmatrix} + \begin{Bmatrix} u_4 \\ v_4 \\ w_4 \end{Bmatrix} = \begin{Bmatrix} u_2 \\ v_2 \\ w_2 \end{Bmatrix} \times \begin{Bmatrix} x \\ y \\ z \end{Bmatrix} + \begin{Bmatrix} u_4 \\ v_4 \\ w_4 \end{Bmatrix} \tag{4.22}$$

Equation (4.22) clearly shows that the functions U, V, W that emerged upon first integration in Equation (4.16) are nothing but rigid-body displacements; see Section 1.6. On the connection between the antisymmetric rotation matrix and the pseudo-rotation vector; see Equation (1.15).

As indicated in Section 1.6, it is only natural for rigid-body terms to emerge from integrating the kinematic relations. If a point, say the origin O, is fixed then $u_4 = v_4 = w_4 = 0$. If, in addition, the rotation of a neighborhood of O is zero then $u_2 = v_2 = w_2 = 0$. The only deformation induced displacements are therefore the first terms in Equation (4.16). In particular, the deformations imply no rotations and $\boldsymbol{\omega} = \mathbf{0}$.

Note that much of the technical work above pertains to a physical interpretation of rigid-body displacements U, V, W in Equation (4.16). Another classical integration method to this effect is studied under Problem 1.6. ♡

Example 4.2: Axisymmetric Torsion of a Bar

A cylindrical bar of length L and radius R, made up of a material with elastic constants E and ν, is loaded exclusively on end sections by hoop tractions that linearly increase from the axis of the bar, as shown in Figure 4.5. The bar is reported to a local cylindrical coordinate system with $\mathbf{e}_r, \mathbf{e}_\theta, \mathbf{e}_z$ the base vectors. Interpret this loading by computing traction resultants, then determine the stresses, strains, displacements and rotations everywhere in the bar.

Solution:

The traction boundary conditions are as follows:

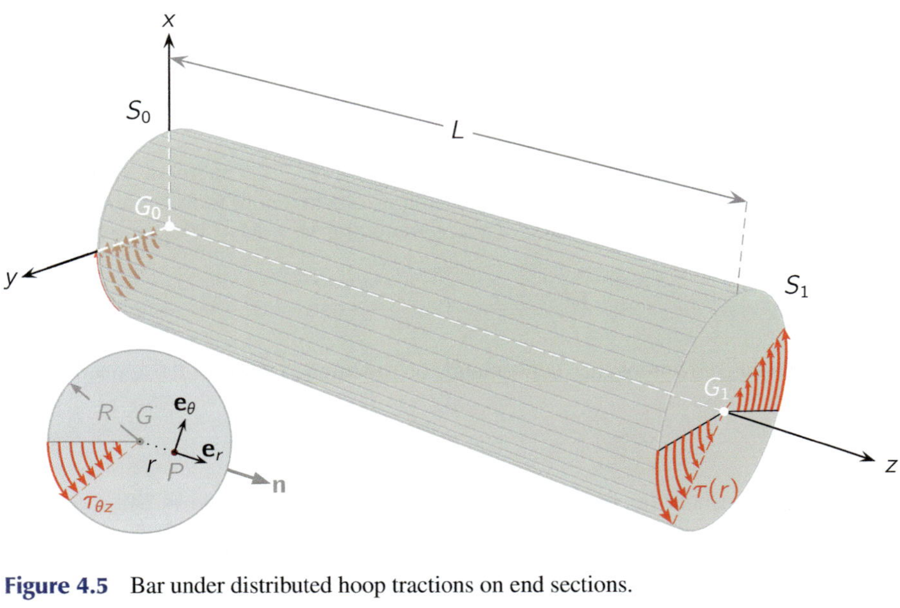

Figure 4.5 Bar under distributed hoop tractions on end sections.

- On end face S_1, where $\mathbf{n} = \mathbf{e}_z$:

$$\boldsymbol{\sigma}\mathbf{n} = \tau(r)\,\mathbf{e}_\theta = \tau_0 \frac{r}{R}\mathbf{e}_\theta \tag{4.23}$$

where τ_0 denotes the value of the maximum applied traction at the outer surface, i.e., $\tau_0 = \tau(R)$.

- On end face S_0, where the outer unit normal $\mathbf{n} = -\mathbf{e}_z$:

$$\boldsymbol{\sigma}\mathbf{n} = -\tau(r)\,\mathbf{e}_\theta \tag{4.24}$$

- On lateral surface S_2, where $\mathbf{n} = \mathbf{e}_r$ (see inset in Figure 4.5):

$$\boldsymbol{\sigma}\mathbf{n} = 0 \tag{4.25}$$

Consider the tractions applied to end section S_1. In terms of the normal–tangential traction decomposition in Equation (2.5), here one has: $\sigma = 0$ (no normal tractions) and $\mathbf{m} = \mathbf{e}_\theta$ (shear tractions are in the hoop direction) or equivalently $\vec{\sigma} = \vec{0}$ and $\vec{\tau} = \tau\mathbf{e}_\theta$. It can be shown that this traction distribution has no resultant force, i.e.,

$$\mathbf{F} = \int_{S_1} \mathbf{T}\,\mathrm{d}S = \int_{S_1} \vec{\tau}\,\mathrm{d}S = \mathbf{0}; \tag{4.26}$$

see Problem 2.2. The distribution of tractions on face S_1 only results in a net moment given by

$$\mathbf{M} = \int_{S_1} \overrightarrow{GP} \times \mathbf{T}\,\mathrm{d}S = \int_{S_1} \overrightarrow{GP} \times \vec{\tau}\,\mathrm{d}S = \int_{S_1} r\mathbf{e}_r \times \tau\mathbf{e}_\theta\,\mathrm{d}S = \left(\int_{S_1} r\tau(r)\,\mathrm{d}S\right)\mathbf{e}_z \tag{4.27}$$

Using the expression of $\tau(r)$ in Equation (4.23), one gets

$$\mathbf{M} = \frac{\tau_0}{R}\left(\int_{S_1} r^2\,\mathrm{d}S\right)\mathbf{e}_z \tag{4.28}$$

The integral in the above equation is a constant $J = \pi R^4/2$, property of the cross-section. It appears, therefore, that the loading results in a pure moment about the bar's axis, which has the meaning of a torque. Letting $\mathbf{M}^t = \mathbf{M} = M^t\mathbf{e}_z$, one has

$$M^t = \frac{\tau_0}{R}J \quad \Rightarrow \quad \tau_0 = \frac{M^t R}{J} \tag{4.29}$$

and the applied tractions may conveniently be rewritten in terms of the applied net torque as follows:

$$\tau(r) = \frac{M^t}{J}r \tag{4.30}$$

Likewise, the shear tractions applied to end section S_0, Equation (4.24), have a zero resultant force and a resultant moment equal to $-\mathbf{M}^t$. Thus, the boundary-value problem defined by the traction distributions shown in Figure 4.5 is statically equivalent to two opposite torques, equal in magnitude, applied to the end sections, as depicted in Figure 4.6.

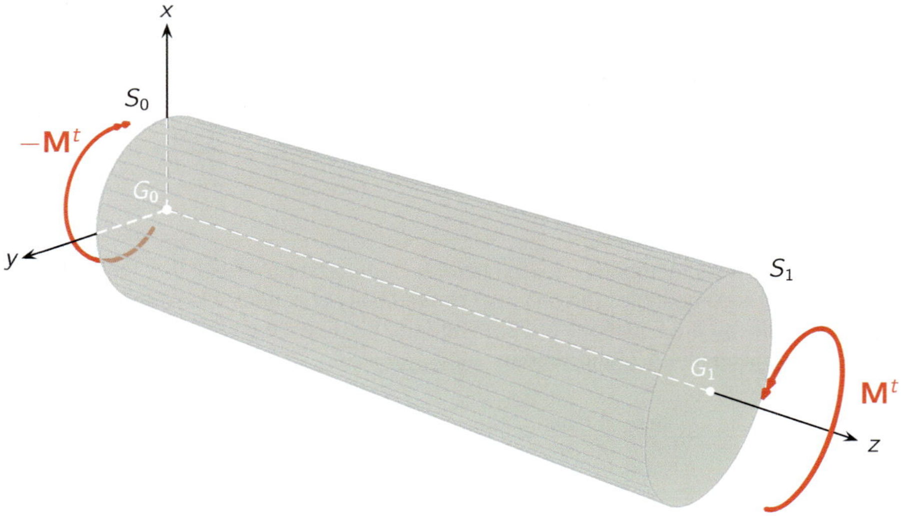

Figure 4.6 Bar under torsion.

The problem is axially symmetric: the cylindrical geometry admits the z axis as an axis of symmetry and the applied loading respects that symmetry. By way of consequence, the solution is independent of angle θ.

Invoking the uniqueness of solutions in elastostatics, Section 4.4.1, a trial solution approach is used, as in Example 4.1. One may attempt the following stress matrix:

$$[\boldsymbol{\sigma}] = \begin{bmatrix} 0 & 0 & 0 \\ 0 & 0 & \tau_{\theta z} \\ 0 & \tau_{z\theta} & 0 \end{bmatrix} \tag{4.31}$$

which mimics the traction distribution applied to end faces. Indeed, the applied tractions act on a face of normal $\mathbf{e}_z$ in the hoop direction $\mathbf{e}_\theta$. Furthermore, one may attempt exactly the same linear variation in r for the solution, i.e.,

$$\tau_{z\theta}(r) = \tau_0 \frac{r}{R} = \frac{M^t}{J} r \tag{4.32}$$

such that the boundary conditions, Equations (4.23) to (4.25), are satisfied. This trial stress field is sketched in the inset of Figure 4.5 on the current cross-section of the bar centered about point G. The stress field given by Equations (4.31) and (4.32) satisfies local equilibrium. Indeed, in cylindrical coordinates, Equation (E) reduces to the two equations (see Section 4.A.2):

$$\begin{cases} \dfrac{\partial \tau_{\theta z}}{\partial z} = 0 \\ \dfrac{\partial \tau_{z\theta}}{\partial \theta} = 0 \end{cases} \tag{4.33}$$

The first of these is valid because the shear stress in Equation (4.32) is taken to be independent of z. The second equation is valid because of axial symmetry.

Next, using Hooke's law, Equation (H), the strain matrix takes the following form:

$$[\boldsymbol{\varepsilon}] = \gamma_{\theta z} \begin{bmatrix} 0 & 0 & 0 \\ 0 & 0 & 1 \\ 0 & 1 & 0 \end{bmatrix} \quad \text{where} \quad \gamma_{\theta z} = \frac{1+\nu}{E}\tau_{\theta z} = \frac{1}{2G}\frac{M^t}{J}r \tag{4.34}$$

and G is the shear modulus.

Since a stress method is adopted here, it is important to check that the above strain field is compatible. The only two compatibility equations where appears the shear strain $\gamma_{\theta z}$ are Equations (4.52d) and (4.52f). They are given in the chapter's appendix, Section 4.A.2, and simplify into:

$$\frac{\partial^2 \gamma_{\theta z}}{\partial r^2} + \frac{1}{r}\frac{\partial \gamma_{\theta z}}{\partial r} - \frac{1}{r^2}\gamma_{\theta z} = \frac{\partial^2 \gamma_{\theta z}}{\partial r^2} + \frac{\partial}{\partial r}\left(\frac{\gamma_{\theta z}}{r}\right) = 0 \tag{4.35}$$

$$\frac{\partial}{\partial z}\left(\frac{\partial \gamma_{\theta z}}{\partial r}\right) - \frac{1}{r}\frac{\partial \gamma_{\theta z}}{\partial z} = 0 \tag{4.36}$$

They are both satisfied on account of the fact that $\gamma_{\theta z}$ is linear in r and does not depend on z.

In order to obtain the displacements, the component form of the kinematic relations, Equation (K), in cylindrical coordinates must be integrated. These are given by Equation (4.50f) in the Appendix (Section 4.A.2). Begin with the normal strains, which are all zero here, Equation (4.34). After setting to zero derivatives with respect to θ, the first three relations in Equation (4.50f) give

$$\begin{cases} \varepsilon_r = \dfrac{\partial u}{\partial r} = 0 & \Rightarrow \quad u = u(z) \\[2ex] \varepsilon_\theta = \dfrac{u}{r} + \dfrac{1}{r}\dfrac{\partial v}{\partial \theta} = \dfrac{u}{r} = 0 & \Rightarrow \quad u(z) = 0 \\[2ex] \varepsilon_z = \dfrac{\partial w}{\partial z} = 0 & \Rightarrow \quad w = w(r) \end{cases} \tag{4.37}$$

Setting now to zero all derivatives with respect to θ in the last three relations of Equation (4.50f) yields, on account of Equation (4.34):

$$\begin{cases} \gamma_{r\theta} = \dfrac{1}{2}\left(\dfrac{1}{r}\dfrac{\partial u}{\partial \theta} - \dfrac{v}{r} + \dfrac{\partial v}{\partial r}\right) = 0 \;\Rightarrow\; \dfrac{\partial v}{\partial r} = \dfrac{v}{r} \;\Rightarrow\; v(r,z) = f(z)r \\[2ex] \gamma_{rz} = \dfrac{1}{2}\left(\dfrac{\partial u}{\partial z} + \dfrac{\partial w}{\partial r}\right) = 0 \;\Rightarrow\; w'(r) = 0 \;\Rightarrow\; w(r) = w(0) = 0 \\[2ex] \gamma_{\theta z} = \dfrac{1}{2}\left(\dfrac{\partial v}{\partial z} + \dfrac{1}{r}\dfrac{\partial w}{\partial \theta}\right) = \dfrac{1}{2G}\dfrac{M^t}{J}r \;\Rightarrow\; \dfrac{\partial v}{\partial z} = \dfrac{M^t}{GJ}r \end{cases} \tag{4.38}$$

where $f(z)$ in the first equation is a function to be determined. Also, in the second equation rigid body displacement is eliminated by fixing the centroid of the initial section S_0. It appears, therefore, that both the radial displacement u and axial displacement w are zero everywhere in the bar. Combining the first and third relations in Equation (4.38) gives

$$f'(z)r = \frac{M^t}{GJ}r \qquad \Rightarrow \qquad f(z) = \frac{M^t}{GJ}z \tag{4.39}$$

where rigid body translation is again ignored. Thus, the displacement in the hoop direction is determined as follows:

$$v(r,z) = \frac{M^t}{GJ}rz \tag{4.40}$$

Having so determined the displacements, the rotations are inferred from Equation (K′) in cylindrical coordinates, given in Equation (4.51c) of the Appendix.

$$\begin{cases} \omega_r = -\Omega_{\theta z} = -\dfrac{1}{2}\left(\dfrac{\partial v}{\partial z} - \dfrac{1}{r}\dfrac{\partial w}{\partial \theta}\right) = -\dfrac{M^t}{2GJ}r \\ \omega_\theta = \Omega_{rz} = \dfrac{1}{2}\left(\dfrac{\partial u}{\partial z} - \dfrac{\partial w}{\partial r}\right) = 0 \\ \omega_z = -\Omega_{r\theta} = -\dfrac{1}{2}\left(\dfrac{1}{r}\dfrac{\partial u}{\partial \theta} - \dfrac{v}{r} - \dfrac{\partial v}{\partial r}\right) = \dfrac{v}{r} = \dfrac{M^t}{GJ}z \end{cases} \tag{4.41}$$

Of particular importance is the rotation ω_z, which is found to be independent of r. All points of a given cross-section, i.e., at some fixed value of z, have the same rotation. Thus, ω_z represents the rotation of the cross-section as a whole and represents the twist due to the applied torque M^t. The rotation ω_r is worth noting but has secondary importance in visualizing the overall deformation. ♡

PROBLEMS

Problem 4.1 (Saint-Venant Problem in Tension) *A bar of width $2a$, height $2b$, and unit thickness is loaded as shown in Figure 4.7. Each loading scenario consists of a different traction distribution on the top and bottom boundaries such that global equilibrium is guaranteed. Let T_0 denote the uniform traction in load system 1. (i) Determine the net force per unit thickness F_0 in 1. (ii) Find the traction distributions in 2 and 3 that would have the same resultant F_0. The distribution in 3 is quadratic. (iii) Which load systems are statically equivalent? (iv) Apply the Saint-Venant principle to the appropriate loading scenarios and conclude.* ⋆

Problem 4.2 (Saint-Venant Problem in Bending) *Three identical beams are subjected to pure bending using three different traction distributions on the end sections, Figure 4.8. All other surfaces are traction-free. Load system 1 consists of a piecewise constant traction, load system 2 of a linearly varying traction, and load system 3 of a quadratic variation of the traction, as depicted. If $\pm T_0$ is the piecewise uniform traction in 1, determine the traction profiles such that all beams are subjected to the same bending moment M_0.* ⋆

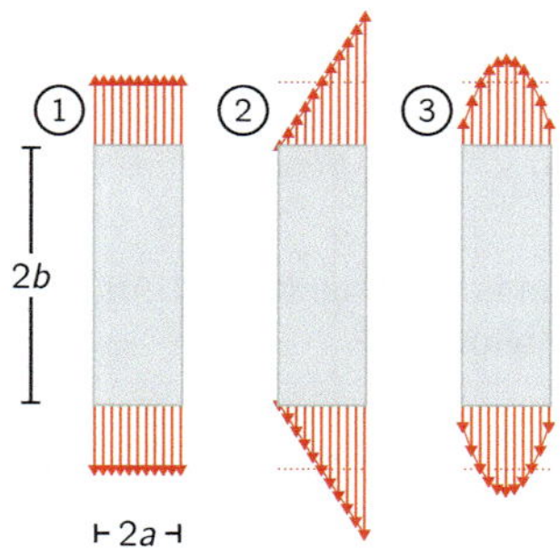

Figure 4.7 Boundary traction distributions.

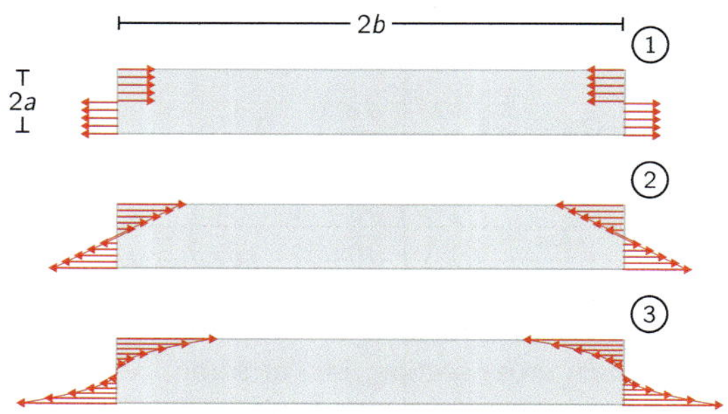

Figure 4.8 Beams loaded in pure bending.

Problem 4.3 (Torsion of a Cylindrical Bar) *A cylindrical bar of radius R is subjected at one end to the shear traction distribution:*

$$\vec{\tau} = \tau_0 \sin\left(\frac{\pi}{2R} r\right) \mathbf{e}_\theta$$

where $\mathbf{e}_\theta$ *denotes the base vector in the hoop direction, r the distance to the axis, and* τ_0 *a constant. The exact opposite tractions are applied to the other end. Lateral surfaces are left free of any tractions. At a distance to be characterized from end sections, determine the stresses, strains, displacements and rotations in the bar.* ⋆

APPENDIX

4.A SUMMARY OF GOVERNING EQUATIONS

4.A.1 CARTESIAN COORDINATES

4.A.1.1 Kinematic Relations

Equation (K) is equivalent to the six scalar equations:

$$\varepsilon_x = \frac{\partial u}{\partial x} \tag{4.42a}$$

$$\varepsilon_y = \frac{\partial v}{\partial y} \tag{4.42b}$$

$$\varepsilon_z = \frac{\partial w}{\partial z} \tag{4.42c}$$

$$\gamma_{xy} = \frac{1}{2}\left(\frac{\partial u}{\partial y} + \frac{\partial v}{\partial x}\right) \tag{4.42d}$$

$$\gamma_{xz} = \frac{1}{2}\left(\frac{\partial u}{\partial z} + \frac{\partial w}{\partial x}\right) \tag{4.42e}$$

$$\gamma_{yz} = \frac{1}{2}\left(\frac{\partial v}{\partial z} + \frac{\partial w}{\partial y}\right) \tag{4.42f}$$

Equation (K') is equivalent to the three scalar equations:

$$\Omega_{xy} = \frac{1}{2}\left(\frac{\partial u}{\partial y} - \frac{\partial v}{\partial x}\right) = -\omega_z \tag{4.43a}$$

$$\Omega_{xz} = \frac{1}{2}\left(\frac{\partial u}{\partial z} - \frac{\partial w}{\partial x}\right) = \omega_y \tag{4.43b}$$

$$\Omega_{yz} = \frac{1}{2}\left(\frac{\partial v}{\partial z} - \frac{\partial w}{\partial y}\right) = -\omega_x \tag{4.43c}$$

4.A.1.2 Compatibility Equations

Equation (1.16f) is reproduced here for ease of reference:

$$\frac{\partial^2 \varepsilon_x}{\partial y^2} + \frac{\partial^2 \varepsilon_y}{\partial x^2} = 2\frac{\partial^2 \gamma_{xy}}{\partial x \partial y} \tag{4.44a}$$

$$\frac{\partial^2 \varepsilon_y}{\partial z^2} + \frac{\partial^2 \varepsilon_z}{\partial y^2} = 2\frac{\partial^2 \gamma_{yz}}{\partial y \partial z} \tag{4.44b}$$

$$\frac{\partial^2 \varepsilon_z}{\partial x^2} + \frac{\partial^2 \varepsilon_x}{\partial z^2} = 2\frac{\partial^2 \gamma_{zx}}{\partial z \partial x} \tag{4.44c}$$

$$\frac{\partial^2 \varepsilon_x}{\partial y \partial z} = \frac{\partial}{\partial x}\left(-\frac{\partial \gamma_{yz}}{\partial x} + \frac{\partial \gamma_{zx}}{\partial y} + \frac{\partial \gamma_{xy}}{\partial z}\right) \tag{4.44d}$$

$$\frac{\partial^2 \varepsilon_y}{\partial z \partial x} = \frac{\partial}{\partial y}\left(-\frac{\partial \gamma_{zx}}{\partial y} + \frac{\partial \gamma_{xy}}{\partial z} + \frac{\partial \gamma_{yz}}{\partial x}\right) \tag{4.44e}$$

$$\frac{\partial^2 \varepsilon_z}{\partial x \partial y} = \frac{\partial}{\partial z}\left(-\frac{\partial \gamma_{xy}}{\partial z} + \frac{\partial \gamma_{yz}}{\partial x} + \frac{\partial \gamma_{zx}}{\partial y}\right) \tag{4.44f}$$

4.A.1.3 Equilibrium Equations

Equation (E) is equivalent to the three scalar equations:

$$\frac{\partial \sigma_x}{\partial x} + \frac{\partial \tau_{xy}}{\partial y} + \frac{\partial \tau_{xz}}{\partial z} = 0 \tag{4.45a}$$

$$\frac{\partial \tau_{xy}}{\partial x} + \frac{\partial \sigma_y}{\partial y} + \frac{\partial \tau_{yz}}{\partial z} = 0 \tag{4.45b}$$

$$\frac{\partial \tau_{xz}}{\partial x} + \frac{\partial \tau_{yz}}{\partial y} + \frac{\partial \sigma_z}{\partial z} = 0 \tag{4.45c}$$

4.A.1.4 Constitutive Relation

Equation (H) is equivalent to the six scalar equations:

$$\varepsilon_x = \frac{1}{E}\left[\sigma_x - \nu(\sigma_y + \sigma_z)\right] \tag{4.46a}$$

$$\varepsilon_y = \frac{1}{E}\left[\sigma_y - \nu(\sigma_z + \sigma_x)\right] \tag{4.46b}$$

$$\varepsilon_z = \frac{1}{E}\left[\sigma_z - \nu(\sigma_x + \sigma_y)\right] \tag{4.46c}$$

$$\gamma_{xy} = \frac{1+\nu}{E}\tau_{xy} \tag{4.46d}$$

$$\gamma_{xz} = \frac{1+\nu}{E}\tau_{xz} \tag{4.46e}$$

$$\gamma_{yz} = \frac{1+\nu}{E}\tau_{yz} \tag{4.46f}$$

The inverted form of Hooke's law, Equation (H), is equivalent to the six scalar equations:

$$\sigma_x = (\lambda + 2\mu)\varepsilon_x + \lambda\left[\varepsilon_y + \varepsilon_z\right] \tag{4.47a}$$

$$\sigma_y = (\lambda + 2\mu)\varepsilon_y + \lambda\left[\varepsilon_z + \varepsilon_x\right] \tag{4.47b}$$

$$\sigma_z = (\lambda + 2\mu)\varepsilon_z + \lambda\left[\varepsilon_x + \varepsilon_y\right] \tag{4.47c}$$

$$\tau_{xy} = 2\mu\gamma_{xy} \tag{4.47d}$$

$$\tau_{xz} = 2\mu\gamma_{xz} \tag{4.47e}$$

$$\tau_{yz} = 2\mu\gamma_{yz} \tag{4.47f}$$

The Lamé constants are related to Young's modulus E and Poisson's ratio ν by:

$$\lambda = \frac{\nu E}{(1+\nu)(1-2\nu)} \tag{4.48a}$$

$$\mu = \frac{E}{2(1+\nu)} \equiv G \tag{4.48b}$$

where the second Lamé constant μ is also the shear modulus G. Conversely, E and ν are given in terms of λ and μ through:

$$E = \mu \frac{3\lambda + 2\mu}{\lambda + \mu} \tag{4.49a}$$

$$\nu = \frac{\lambda}{2(\lambda + \mu)} \tag{4.49b}$$

4.A.2 CYLINDRICAL COORDINATES

4.A.2.1 Kinematic Relations

Equation (K) is equivalent to the six scalar equations:

$$\varepsilon_r = \frac{\partial u}{\partial r} \tag{4.50a}$$

$$\varepsilon_\theta = \frac{1}{r}\left(u + \frac{\partial v}{\partial \theta}\right) \tag{4.50b}$$

$$\varepsilon_z = \frac{\partial w}{\partial z} \tag{4.50c}$$

$$\gamma_{r\theta} = \frac{1}{2}\left(\frac{1}{r}\frac{\partial u}{\partial \theta} - \frac{v}{r} + \frac{\partial v}{\partial r}\right) \tag{4.50d}$$

$$\gamma_{rz} = \frac{1}{2}\left(\frac{\partial u}{\partial z} + \frac{\partial w}{\partial r}\right) \tag{4.50e}$$

$$\gamma_{\theta z} = \frac{1}{2}\left(\frac{\partial v}{\partial z} + \frac{1}{r}\frac{\partial w}{\partial \theta}\right) \tag{4.50f}$$

where $u(r,\theta,z)$, $v(r,\theta,z)$, and $w(r,\theta,z)$ denote the components of the displacement vector $\mathbf{u}$ on the basis $\{\mathbf{e}_r, \mathbf{e}_\theta, \mathbf{e}_z\}$.

Equation (K') is equivalent to the three scalar equations:

$$\Omega_{r\theta} = \frac{1}{2}\left(\frac{1}{r}\frac{\partial u}{\partial \theta} - \frac{v}{r} - \frac{\partial v}{\partial r}\right) = -\omega_z \tag{4.51a}$$

$$\Omega_{rz} = \frac{1}{2}\left(\frac{\partial u}{\partial z} - \frac{\partial w}{\partial r}\right) = \omega_\theta \tag{4.51b}$$

$$\Omega_{\theta z} = \frac{1}{2}\left(\frac{\partial v}{\partial z} - \frac{1}{r}\frac{\partial w}{\partial \theta}\right) = -\omega_r \tag{4.51c}$$

4.A.2.2 Compatibility Equations

For axisymmetric problems where $\partial(\cdot)/\partial\theta = 0$,

$$\frac{\partial^2 \varepsilon_\theta}{\partial r^2} - \frac{1}{r}\frac{\partial \varepsilon_r}{\partial r} + \frac{2}{r}\frac{\partial \varepsilon_\theta}{\partial r} = 0 \tag{4.52a}$$

$$\frac{\partial^2 \varepsilon_\theta}{\partial z^2} + \frac{1}{r}\frac{\partial \varepsilon_z}{\partial r} = \frac{2}{r}\frac{\partial \gamma_{rz}}{\partial z} \tag{4.52b}$$

$$\frac{\partial^2 \varepsilon_z}{\partial r^2} + \frac{\partial^2 \varepsilon_r}{\partial z^2} = 2\frac{\partial^2 \gamma_{zr}}{\partial z \partial r} \tag{4.52c}$$

$$\frac{\partial}{\partial r}\left(-\frac{\partial \gamma_{\theta z}}{\partial r} + \frac{\partial \gamma_{r\theta}}{\partial z}\right) + \frac{2}{r}\frac{\partial \gamma_{r\theta}}{\partial z} - \frac{1}{r}\frac{\partial \gamma_{\theta z}}{\partial r} + \frac{1}{r^2}\gamma_{\theta z} = 0 \tag{4.52d}$$

$$\frac{\partial^2 \varepsilon_\theta}{\partial z \partial r} + \frac{1}{r}\frac{\partial \varepsilon_\theta}{\partial z} - \frac{1}{r}\frac{\partial \varepsilon_r}{\partial z} = 0 \tag{4.52e}$$

$$\frac{\partial}{\partial z}\left(-\frac{\partial \gamma_{r\theta}}{\partial z} + \frac{\partial \gamma_{\theta z}}{\partial r}\right) - \frac{1}{r}\frac{\partial \gamma_{\theta z}}{\partial z} = 0 \tag{4.52f}$$

4.A.2.3 Equilibrium Equations

Equation (E) is equivalent to the three scalar equations:

$$\frac{\partial \sigma_r}{\partial r} + \frac{1}{r}\frac{\partial \tau_{r\theta}}{\partial \theta} + \frac{\sigma_r - \sigma_\theta}{r} + \frac{\partial \tau_{rz}}{\partial z} = 0 \tag{4.53a}$$

$$\frac{\partial \tau_{r\theta}}{\partial r} + \frac{1}{r}\frac{\partial \sigma_\theta}{\partial \theta} + 2\frac{\tau_{r\theta}}{r} + \frac{\partial \tau_{\theta z}}{\partial z} = 0 \tag{4.53b}$$

$$\frac{\partial \tau_{rz}}{\partial r} + \frac{1}{r}\frac{\partial \tau_{\theta z}}{\partial \theta} + \frac{\tau_{rz}}{r} + \frac{\partial \sigma_z}{\partial z} = 0 \tag{4.53c}$$

4.A.2.4 Constitutive Relation

Equation (H) is equivalent to the six scalar equations:

$$\varepsilon_r = \frac{1}{E}\left[\sigma_r - \nu(\sigma_\theta + \sigma_z)\right] \tag{4.54a}$$

$$\varepsilon_\theta = \frac{1}{E}\left[\sigma_\theta - \nu(\sigma_z + \sigma_r)\right] \tag{4.54b}$$

$$\varepsilon_z = \frac{1}{E}\left[\sigma_z - \nu(\sigma_r + \sigma_\theta)\right] \tag{4.54c}$$

$$\gamma_{r\theta} = \frac{1+\nu}{E}\tau_{r\theta} \tag{4.54d}$$

$$\gamma_{rz} = \frac{1+\nu}{E}\tau_{rz} \tag{4.54e}$$

$$\gamma_{\theta z} = \frac{1+\nu}{E}\tau_{\theta z} \tag{4.54f}$$

The inverted form of Hooke's law, Equation (H), is equivalent to the six scalar equations:

$$\sigma_r = (\lambda + 2\mu)\varepsilon_r + \lambda\,[\varepsilon_\theta + \varepsilon_z] \tag{4.55a}$$

$$\sigma_\theta = (\lambda + 2\mu)\varepsilon_\theta + \lambda\,[\varepsilon_z + \varepsilon_r] \tag{4.55b}$$

$$\sigma_z = (\lambda + 2\mu)\varepsilon_z + \lambda\,[\varepsilon_r + \varepsilon_\theta] \tag{4.55c}$$

$$\tau_{r\theta} = 2\mu\gamma_{r\theta} \tag{4.55d}$$

$$\tau_{rz} = 2\mu\gamma_{rz} \tag{4.55e}$$

$$\tau_{\theta z} = 2\mu\gamma_{\theta z} \tag{4.55f}$$

4.A.2.5 Compatibility Equations

More generally than implied by Equation (4.52f):

$$\frac{1}{r^2}\frac{\partial^2 \varepsilon_r}{\partial \theta^2} + \frac{\partial^2 \varepsilon_\theta}{\partial r^2} - \frac{1}{r}\frac{\partial \varepsilon_r}{\partial r} + \frac{2}{r}\frac{\partial \varepsilon_\theta}{\partial r} = \frac{2}{r}\frac{\partial^2 \gamma_{r\theta}}{\partial r \partial \theta} + \frac{2}{r^2}\frac{\partial \gamma_{r\theta}}{\partial \theta} \tag{4.56a}$$

$$\frac{\partial^2 \varepsilon_\theta}{\partial z^2} + \frac{1}{r^2}\frac{\partial^2 \varepsilon_z}{\partial \theta^2} + \frac{1}{r}\frac{\partial \varepsilon_z}{\partial r} = \frac{2}{r}\frac{\partial^2 \gamma_{\theta z}}{\partial \theta\, \partial z} + \frac{2}{r}\frac{\partial \gamma_{rz}}{\partial z} \tag{4.56b}$$

$$\frac{\partial^2 \varepsilon_z}{\partial r^2} + \frac{\partial^2 \varepsilon_r}{\partial z^2} = 2\frac{\partial^2 \gamma_{zr}}{\partial z \partial r} \tag{4.56c}$$

$$\frac{1}{r}\frac{\partial^2 \varepsilon_r}{\partial \theta\, \partial z} = \frac{\partial}{\partial r}\left(-\frac{\partial \gamma_{\theta z}}{\partial r} + \frac{1}{r}\frac{\partial \gamma_{zr}}{\partial \theta} + \frac{\partial \gamma_{r\theta}}{\partial z}\right) + \frac{2}{r}\frac{\partial \gamma_{r\theta}}{\partial z} - \frac{\partial}{\partial r}\left(\frac{\gamma_{\theta z}}{r}\right) \tag{4.56d}$$

$$\frac{\partial^2 \varepsilon_\theta}{\partial z \partial r} + \frac{1}{r}\frac{\partial \varepsilon_\theta}{\partial z} - \frac{1}{r}\frac{\partial \varepsilon_r}{\partial z} = \frac{1}{r}\frac{\partial}{\partial \theta}\left(-\frac{1}{r}\frac{\partial \gamma_{zr}}{\partial \theta} + \frac{\partial \gamma_{r\theta}}{\partial z} + \frac{\partial \gamma_{\theta z}}{\partial r}\right) + \frac{1}{r^2}\frac{\partial \gamma_{\theta z}}{\partial \theta} \tag{4.56e}$$

$$\frac{1}{r}\frac{\partial^2 \varepsilon_z}{\partial r \partial \theta} - \frac{1}{r^2}\frac{\partial \varepsilon_z}{\partial \theta} = \frac{\partial}{\partial z}\left(-\frac{\partial \gamma_{r\theta}}{\partial z} + \frac{\partial \gamma_{\theta z}}{\partial r} + \frac{1}{r}\frac{\partial \gamma_{zr}}{\partial \theta}\right) - \frac{1}{r}\frac{\partial \gamma_{\theta z}}{\partial z} \tag{4.56f}$$

Part II

Beams

5 Geometry and Statics of Beams

5.1 DEFINITIONS

DEFINITION 5.1 A beam is any solid generated by a planar surface S when its centroid G traces a curve $\widehat{G_0G_1}$, the plane of S being normal to that curve at G (Figure 5.1). The curve $\widehat{G_0G_1}$ is the beam's *median line* and S its *cross-section*. Furthermore, the following two conditions must be met:

1. Let the diameter D of S be the distance between its farthest two points; then the diameter of each cross-section must be much smaller than the length and the radii of curvature and torsion of median line $\widehat{G_0G_1}$.
2. Any variations in cross-section S (size, shape, built-in twist) with arc length s of G on $\widehat{G_0G_1}$ must be very slow. ♣

The median line is also called *centroidal axis* (Figure 5.1), but the former term is preferred when dealing with arbitrary beams.

A beam is said to be *straight* if its centroidal axis is a line segment. If, in addition, its cross-section S is constant (dimensions, shape, and built-in twist), then the beam is said to be prismatic (or cylindrical). A non-straight beam is a curved beam that may be planar or skewed.

A ring is a beam whose median line is a closed curve. In this case, the initial and terminal cross-sections are the same and may be arbitrarily chosen.

A tube is any hollow beam; it may contain one or more cavities.

A thin-walled beam has a cross-section formed by one or more thin band, that is a band whose thickness t is much smaller than the diameter D of S.

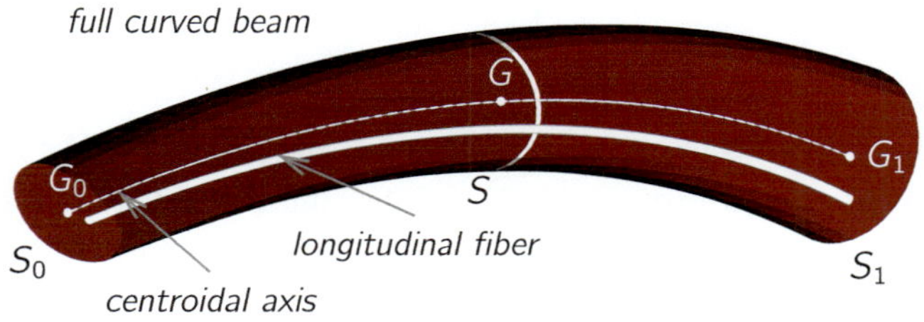

Figure 5.1 An arbitrary beam.

DOI: 10.1201/9781003537229-7

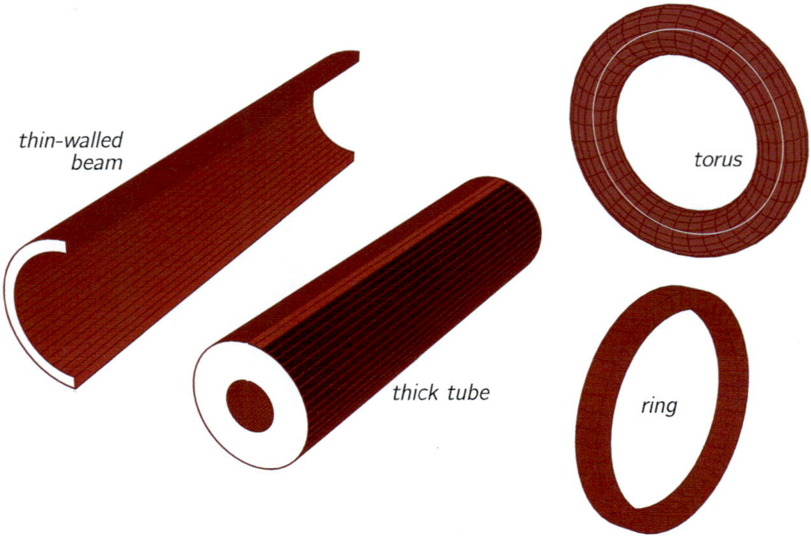

Figure 5.2 Examples of beams.

In an arbitrary beam, a longitudinal fiber is an infinitely released tube generated by an element dS of S when G traces $\widehat{G_0G_1}$ (Figure 5.1).

Figure 5.2 shows examples of beams meeting Definition 5.1. Unless otherwise noted, the material making up the beam is assumed to be homogeneous, elastic, and isotropic.

5.2 MEDIAN LINE

Let $\{O,xyz\}$ be a Cartesian frame with orthonormal basis $\{\mathbf{e}_1,\mathbf{e}_2,\mathbf{e}_3\}$. The median line $\widehat{G_0G_1}$ of the beam generally admits parametric representation:

$$G\begin{cases} x = x(u) \\ y = y(u) \\ z = z(u) \end{cases} \tag{5.1}$$

where G is the current centroid and $u \in [u_0,u_1]$ is the parameter. The mapping between the positions of G over arc $\widehat{G_0G_1}$ and the values of u between u_0 and u_1 is assumed to be bijective, i.e., one to one. Also, functions $x(u)$, $y(u)$ and $z(u)$ are taken to be sufficiently regular on $[u_0,u_1]$.

Taking the origin of arc lengths at some point of the median line (generally G_0), a positive orientation is introduced from G_0 to G_1, Figure 5.3. The mapping between current point G and arc length s is also bijective.

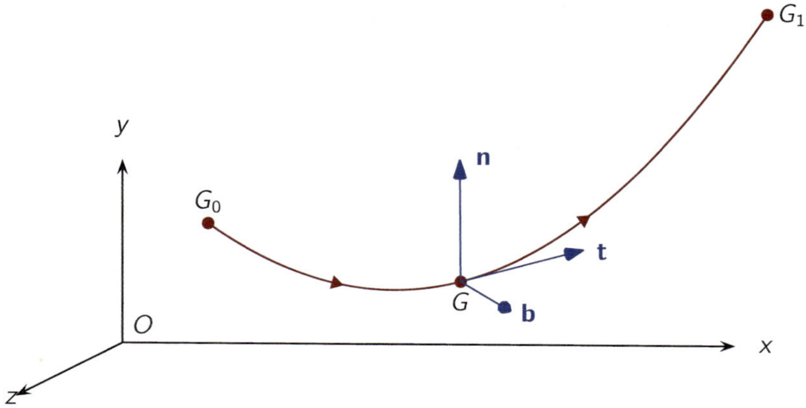

Figure 5.3 Median line and associated Frenet triad.

The derivative of position-vector $\overrightarrow{OG}$ with respect to s is the unit tangent in G to $\widehat{G_0G_1}$. It is denoted by $\mathbf{t}$ and is oriented in the direction of increasing arc lengths (Problem 5.2):

$$\mathbf{t} = \frac{\mathrm{d}\overrightarrow{OG}}{\mathrm{d}s} \begin{cases} \dfrac{\mathrm{d}x}{\mathrm{d}s} \\[2ex] \dfrac{\mathrm{d}y}{\mathrm{d}s} \\[2ex] \dfrac{\mathrm{d}z}{\mathrm{d}s} \end{cases} \tag{5.2}$$

It follows from Equation (5.2) that

$$s = \widehat{G_0G} = \pm \int_{u_0}^{u_1} \sqrt{x'^2 + y'^2 + z'^2}\,\mathrm{d}u \tag{5.3}$$

which is the practical formula for arc length s, Problem 5.3. In Equation (5.3), the + sign applies if parameter u increases when G traces the median line in the positive orientation chosen on $\widehat{G_0G_1}$ and *vice versa*.

5.2.1 THEOREM OF THE MOBILE FRAME

Let $\{\mathbf{e}_1, \mathbf{e}_2, \mathbf{e}_3\}$ be an orthonormal basis whose vectors are differentiable functions of some parameter u. Define A as the matrix that associates the three $\mathbf{e}_i$ vectors with their derivatives so that

$$\frac{\mathrm{d}}{\mathrm{d}u}\begin{Bmatrix} \mathbf{e}_1 \\ \mathbf{e}_2 \\ \mathbf{e}_3 \end{Bmatrix} = \begin{bmatrix} A_{11} & A_{12} & A_{13} \\ A_{21} & A_{22} & A_{23} \\ A_{31} & A_{32} & A_{33} \end{bmatrix} \begin{Bmatrix} \mathbf{e}_1 \\ \mathbf{e}_2 \\ \mathbf{e}_3 \end{Bmatrix} \tag{5.4}$$

Theorem 5.1

If basis $\{\mathbf{e}_i\}$ remains orthonormal when u varies, then matrix A is antisymmetric. ◇

Proof *Since $\{\mathbf{e}_i\}$ is orthonormal, $\mathbf{e}_i \cdot \mathbf{e}_j = \delta_{ij}$ with δ_{ij} the Kronecker symbol (=1 if $i = j$, 0 otherwise). Differentiating that equation gives:*

$$\frac{\mathrm{d}\mathbf{e}_i}{\mathrm{d}u} \cdot \mathbf{e}_j + \mathbf{e}_i \cdot \frac{\mathrm{d}\mathbf{e}_j}{\mathrm{d}u} = 0$$

By definition, $A_{ij} = \frac{\mathrm{d}\mathbf{e}_i}{\mathrm{d}u} \cdot \mathbf{e}_j$. Thus, the above equation simply states that $A_{ij} + A_{ji} = 0$. ■

5.2.2 FRENET'S FRAME

Define an orthonormal mobile frame associated with arc $\widehat{G_0G_1}$, the origin being at G. The first base vector is the unit tangent $\mathbf{t}$, Equation (5.2). This vector has constant magnitude, and its derivative (say with respect to the arc length) is normal to it. One may therefore write

$$\frac{\mathrm{d}\mathbf{t}}{\mathrm{d}s} = \frac{1}{R}\mathbf{n} \tag{5.5}$$

with

- $\mathbf{n}$ a unit vector normal to $\mathbf{t}$, called the *principal normal vector*, Figure 5.3. It is taken as second base vector; and
- $1/R$ a scalar with dimension L^{-1}. It defines the curvature of arc $\widehat{G_0G_1}$ at G.

The third base vector, defined by $\mathbf{b} = \mathbf{t} \times \mathbf{n}$, is the binormal. The right-handed orthonormal frame $\{G, \mathbf{t}, \mathbf{n}, \mathbf{b}\}$ is Frenet's frame associated with curve $\widehat{G_0G_1}$ in G, Figure 5.3.

5.2.3 FRENET'S FORMULAE

Apply Equation (5.4) to Frenet's mobile frame taken as function of the arc length s. With due account of the antisymmetry of the corresponding A matrix, one obtains

$$\frac{\mathrm{d}}{\mathrm{d}s}\begin{Bmatrix}\mathbf{t}\\ \mathbf{n}\\ \mathbf{b}\end{Bmatrix} = \begin{bmatrix}0 & \frac{1}{R} & 0\\ -\frac{1}{R} & 0 & \frac{1}{T}\\ 0 & -\frac{1}{T} & 0\end{bmatrix}\begin{Bmatrix}\mathbf{t}\\ \mathbf{n}\\ \mathbf{b}\end{Bmatrix} \tag{5.6}$$

Proof *Matrix A being antisymmetric, it is completely determined by its components A_{12}, A_{13} and A_{23}. Having posed $\frac{\mathrm{d}\mathbf{t}}{\mathrm{d}s} = \frac{1}{R}\mathbf{n}$, see Equation (5.5), one has determined $A_{12} = 1/R$ and $A_{13} = 0$, the latter following by necessity from the definition of* **n**. *One poses then $A_{23} = 1/T$, a scalar called* torsion *of the curve $\widehat{G_0G_1}$ in G.* ■

Remark 5.1 *The orientation of vector* **n** *is arbitrary; replacing* **n** *with* $-\mathbf{n}$ *changes* R *to* $-R$. *Thus, the sign of curvature has, in general, no geometric meaning.*

For plane curves, it is customary to choose **n** *directly perpendicular to* **t**, *that is* $\langle \mathbf{t}, \mathbf{n} \rangle = +\frac{\pi}{2}$. *With this choice,* R *is then a positive scalar if* **n** *is oriented toward the concavity of curve* $\widehat{G_0G_1}$, *negative otherwise. For skew curves, one generally orients* **n** *toward the concavity of* $\widehat{G_0G_1}$, *which amounts to taking* $R > 0$.

The center of curvature C^R *of curve* $\widehat{G_0G_1}$ *in* G *is defined by* $\overrightarrow{GC^R} = R\mathbf{n}$; R *is called the* radius of curvature *of arc* $\widehat{G_0G_1}$ *in* G. *

Remark 5.2 *The point* C^T *defined by* $\overrightarrow{GC^T} = T\mathbf{b}$ *is called the* center of torsion *of* $\widehat{G_0G_1}$ *in* G *and* T *the* radius of torsion. *The sign of* T *has the following geometric meaning: as* G *moves along the positive direction on* $\widehat{G_0G_1}$, *Frenet's frame turns around* **t** *in the positive direction if* T *is positive, and in the negative direction if* T *is negative.* *

Remark 5.3 *The three planes that contain* G *and are normal to* **t**, **n**, **b** *are called the* normal plane, rectifying plane, *and* osculating plane, *respectively.* *

Remark 5.4 *A curve has no torsion if and only if it is a plane curve. A curve has no curvature if and only if it is straight.* *

Remark 5.5 *At an inflexion point, one has* $\frac{d\mathbf{t}}{ds} = 0$ *so that the local curvature is nil and vector* **n** *is undetermined. However, it can be defined by continuity.*

On the other hand, at each point of a straight line, one has $\frac{1}{R} = 0$ *and* **n** *is undetermined. One then poses an osculating plane that contains the line and principal normals.* *

Example 5.1: Geometry of Helical Spring

The median line is a circular helix with parametric representation:

$$G \begin{cases} x = a\cos\theta \\ y = a\sin\theta \\ z = \lambda a\theta \end{cases} \tag{5.7}$$

in the axes of Figure 5.4. Here, a and λ denote the so-called radius and slope of the helix, respectively. Parameter θ is the polar angle of vector $\overrightarrow{OG'}$ with respect to the x-axis, G' being the projection of G onto the x–y plane, Figure 5.4. Helix $\widehat{G_0G_1}$ is oriented as shown, in the direction of increasing θ.

Using Equations (5.2) and (5.5) and the definition of the binormal, one obtains

$$\sqrt{1+\lambda^2}\,\mathbf{t} = \begin{cases} -\sin\theta \\ \cos\theta \\ \lambda \end{cases} \qquad \mathbf{n} = \begin{cases} -\cos\theta \\ -\sin\theta \\ 0 \end{cases} \qquad \sqrt{1+\lambda^2}\,\mathbf{b} = \begin{cases} \lambda\sin\theta \\ -\lambda\cos\theta \\ 1 \end{cases} \tag{5.8}$$

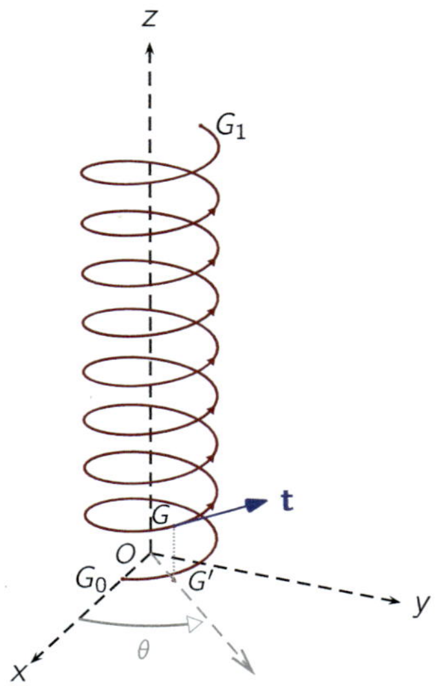

Figure 5.4 A helical spring.

In addition, the radii of curvature and torsion are found as follows:

$$R = a(1+\lambda^2) \qquad T = a\frac{1+\lambda^2}{\lambda} \tag{5.9}$$

Note that the (constant) radius of curvature of the helix is not equal to a which is abusively called radius. In fact, a is the radius of the cylinder that contains the helix. ♡

5.3 CROSS-SECTION

5.3.1 CENTROID

It is the center of mass G of cross-section S. Let Oy and Oz be two perpendicular axes in the plane of the section; then the coordinates of G are determined by

$$\boxed{y^G = \frac{1}{S}\iint_S y\,\mathrm{d}S \qquad z^G = \frac{1}{S}\iint_S z\,\mathrm{d}S} \tag{5.10}$$

y and z being the coordinates of current point P within S, center of elementary area dS (Figure 5.5). In vector form,

$$\overrightarrow{OG} = \frac{1}{S}\iint_S \overrightarrow{OP}\,\mathrm{d}S \tag{5.11}$$

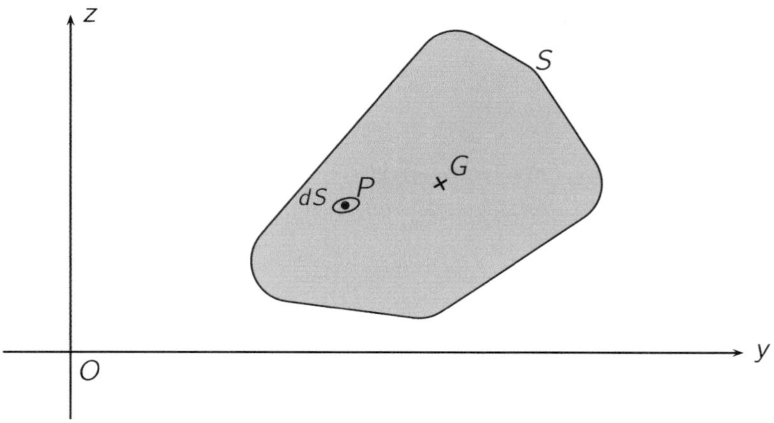

Figure 5.5 An arbitrary cross-section.

If section S admits an axis of symmetry, then G lies on that axis; if S admits two axes of symmetry then G is their intersection (center of symmetry); see Examples 5.4 and 5.6 further below.

In some cases, useful theorems permit straightforward calculations without evaluating the integrals in Equation (5.10).

Theorem 5.2: Guldin's First Theorem

If some arc having centroid G and length $\mathcal{L}$ is used as a meridian to generate a surface of revolution, then the surface area is given by

$$\mathcal{A} = 2\pi z^G \mathcal{L} \tag{5.12}$$

when the arc is drawn in the half-plane $z \geq 0$. ♢

Example 5.2: Centroid of Semi-Circular Thin Section

The thin-walled section has a semi-circular shape of thickness t and mean radius $R \gg t$ (Figure 5.6a). The cross-section is identified with a line with perimeter $\mathcal{L} = \pi R$. Revolving that line about the y-axis generates a sphere whose surface area is $\mathcal{A} = 4\pi R^2$. Guldin's first formula, Equation (5.12), applies with

$$4\pi R^2 = 2\pi z^G \cdot \pi R$$

from which one gets

$$z^G = \frac{2}{\pi} R \tag{5.13}$$

♡

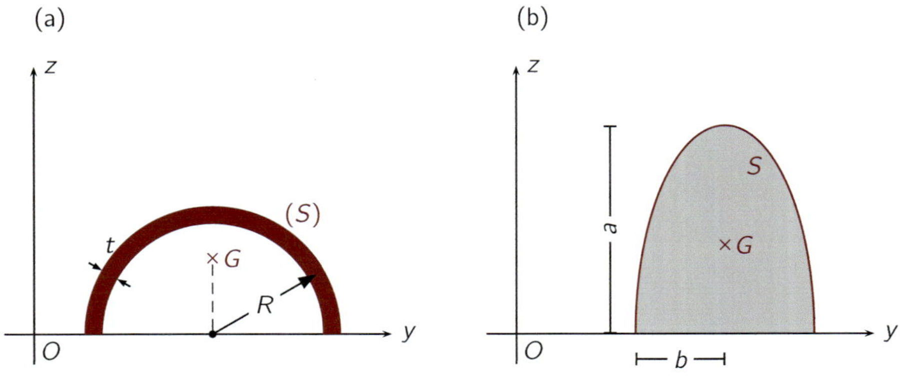

Figure 5.6 Application of Guldin's theorems. (a) Thin semi-circular profile. (b) Solid semi-elliptical section.

Theorem 5.3: Guldin's Second Theorem

If some planar domain having centroid G and area $\mathcal{S}$ is used as a meridian to generate a solid of revolution, then the volume of the solid is given by

$$\mathcal{V} = 2\pi z^G \mathcal{S} \tag{5.14}$$

when the domain is drawn in the half-plane $z \geq 0$. ◇

Example 5.3: Centroid of Semi-Elliptic Section

The section has a semi-elliptic shape with major and minor axes a and b along the z and y axes, respectively (Figure 5.6b). Guldin's second formula, Equation (5.14), applies with

$$\frac{4}{3}\pi a^2 b = 2\pi z^G \cdot \pi \frac{ab}{2}$$

from which one gets

$$z^G = \frac{4}{3\pi} a \tag{5.15}$$

The result is remarkably independent of the minor axis b. ♡

In some cases, the cross-section consists of several parts, and it is easier to determine the centroid of each part separately before applying a "composite" rule. Let us illustrate that for two such parts of the cross-section, for example a semi-circular part and a straight part for a D section (see Problem 5.15). Letting O denote the origin of coordinates, Equation (5.11) is algebraically written as follows:

$$S\overline{OG} = S_1\overline{OG_1} + S_2\overline{OG_2} \tag{5.16}$$

where S, S_1, and S_2 denote the cross-sectional areas of the whole section, part 1, and part 2, respectively. Likewise, G, G_1, and G_2 denote their respective centroids. The overbar refers to algebraic (signed) measure, which is important to keep track of the positioning of points relative to the origin.

5.3.2 SECOND MOMENTS OF AREA

DEFINITION 5.2 For a cross-section S lying in the Oyz plane, Figure 5.5, define:

- the second moment of area of S with respect to the Oy axis as the positive scalar:

$$\boxed{I_y = \iint_S z^2 \, \mathrm{d}S} \tag{5.17}$$

- the second moment of area of S with respect to the Oz axis as the positive scalar:

$$\boxed{I_z = \iint_S y^2 \, \mathrm{d}S} \tag{5.18}$$

- the second moment of area of S with respect to the Ox axis as the positive scalar:

$$I_x = \iint_S \overline{OP}^2 \, \mathrm{d}S = I_y + I_z \tag{5.19}$$

- the product moment of S with respect to the Oy and Oz axes as the scalar:

$$I_{yz} = \iint_S yz \, \mathrm{d}S \tag{5.20}$$

which can be positive, negative, or nil. ♣

Second moments of area are quadratic moments and have dimension L^4. They are equal to the moments of inertia defined in general mechanics when taking S as a thin planar plate of unit surface mass. I_x is also known as the polar moment of area.

Remark 5.6 *Let Gx, Gy, Gz denote the axes passing through the centroid and parallel to Ox, Oy, Oz. Upon a change of origin, Huygens' formulae apply:*

$$\begin{cases} I_{Oy} = I_{Gy} + S\, z^{G2} \\ I_{Oz} = I_{Gz} + S\, y^{G2} \\ I_{Ox} = I_{Gx} + S\,(y^{G2} + z^{G2}) \\ I_{Oyz} = I_{Gyz} + S\, y^G z^G \end{cases} \tag{5.21}$$

where $I_{Oy} \equiv I_y$ given by Equation (5.17), $I_{Oz} \equiv I_z$ *given by Equation* (5.18), *etc. A variant of these equations is commonly known as the* parallel axis theorem. *From the definition of G and the second moments of area, Equations* (5.17) *to* (5.20), *the proof is straightforward.* *

Example 5.4: (Second Moments of Area of Basic Sections)

Calculate the second moments of area I_y, I_z, I_x, I_{yz} for (a) a rectangle of width w and height h; (b) a disk of radius R; (c) a tube of inner and outer radii R_1 and R_2, respectively; and (d) an elliptical section of semi-axes a, b with $a > b$, Figure 5.7.

For ease of reference, the results are summarized in Table 5.1. All expressions must scale as a length to the fourth power. Key steps involved in the calculations are provided below. Details may be worked out under Problem 5.13.

All sections have two axes of symmetry so that G is identified with O, the origin of coordinates and $I_{yz} = 0$. Also, since $I_x = I_y + I_z$, Equation (5.19), it suffices to determine I_y and I_z. Furthermore, for the disk and the tube, the results are invariant upon changing the axes so that $I_y = I_z$. On the other hand, for the rectangle and ellipse, I_y is deducible from I_z and vice versa. In the end, only three direct calculations are needed. Results for the tube, and thin tube can be obtained without integration.

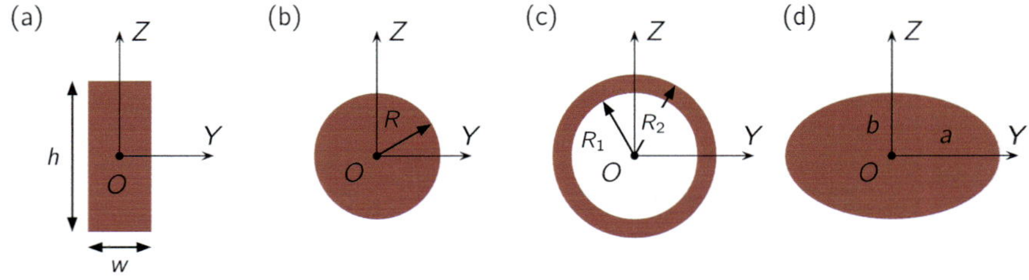

Figure 5.7 Symmetric cross-sections.

Table 5.1
Second Moments of Area of Sections in Figure 5.7

Case	I_y	I_z	I_x
(a) Rectangle	$\frac{wh^3}{12}$	$\frac{w^3h}{12}$	$\frac{wh}{12}(w^2+h^2)$
(b) Disk	$\frac{\pi R^4}{4}$	$\frac{\pi R^4}{4}$	$\frac{\pi R^4}{2}$
(c) Tube	$\frac{\pi}{4}(R_2^4 - R_1^4)$	$\frac{\pi}{4}(R_2^4 - R_1^4)$	$\frac{\pi}{2}(R_2^4 - R_1^4)$
thin tube	$\pi R^3 t$	$\pi R^3 t$	$2\pi R^3 t$
(d) Ellipse	$\frac{\pi ab^3}{4}$	$\frac{\pi a^3 b}{4}$	$\frac{\pi ab}{4}(a^2+b^2)$

(a) For the rectangular section, Figure 5.7a, from Equation (5.18)

$$I_z = \iint\limits_S y^2\,\mathrm{d}S = \int_{y=-\frac{w}{2}}^{\frac{w}{2}} \int_{z=-\frac{h}{2}}^{\frac{h}{2}} y^2\,\mathrm{d}y\,\mathrm{d}z = \int_{-\frac{w}{2}}^{\frac{w}{2}} y^2\,\mathrm{d}y \int_{-\frac{h}{2}}^{\frac{h}{2}} \mathrm{d}z = \left[\frac{y^3}{3}\right]_{-\frac{w}{2}}^{\frac{w}{2}} h$$
$$= \frac{1}{12} w^3 h$$

Deduce that $I_y = \frac{1}{12}h^3 w$ by interchanging w and h. The highest power appears for the dimension that corresponds to the distance to the axis about which the moment is calculated. More on that will follow in Section 5.3.3 further below.

(b) For the disk, Figure 5.7b, using polar coordinates is appropriate with, for example, $y = r\cos\theta$

$$I_z = \int_{\theta=0}^{2\pi} \int_{r=0}^{R} (r\cos\theta)^2 r\,\mathrm{d}r\,\mathrm{d}\theta = \left(\int_{\theta=0}^{2\pi} \cos^2\theta\,\mathrm{d}\theta\right)\left(\int_{r=0}^{R} r^3\,\mathrm{d}r\right) = \pi\frac{R^4}{4}$$

where use was made of the identity $\cos^2\theta = \frac{1}{2} + \frac{1}{2}\cos 2\theta$, the integral of $\cos 2\theta$ over a period being zero.

(c) For the tube, Figure 5.7c, it suffices to note that for any integrable function $f(r)$

$$\int_{R_1}^{R_2} f(r)\,\mathrm{d}r = \int_0^{R_2} f(r)\,\mathrm{d}r - \int_0^{R_1} f(r)\,\mathrm{d}r \tag{5.22}$$

thereby allowing to obtain the I_z for the tube from that of the disk.

(d) For the elliptical section, Figure 5.7d, one may use horizontal or vertical strips of infinitesimal thickness while noting that the bounds of one integral depend on the other variable (see solution to Problem 5.13). Alternatively, the change of variable

$$\begin{cases} y = ar\cos\theta \\ z = br\sin\theta \end{cases} \tag{5.23}$$

is appropriate with $0 \leq \theta \leq 2\pi$ and $0 \leq r \leq 1$ so that, for example,

$$I_z = \int_{\theta=0}^{2\pi} \int_{r=0}^{1} (ar\cos\theta)^2 abr\,\mathrm{d}r\,\mathrm{d}\theta = a^3 b\left(\int_{\theta=0}^{2\pi} \cos^2\theta\,\mathrm{d}\theta\right)\left(\int_{r=0}^{1} r^3\,\mathrm{d}r\right) = \pi\frac{a^3 b}{4}$$

♡

Remark 5.7 *One may write*

$$I_y = S\rho_y^2, \qquad I_z = S\rho_z^2, \qquad I_x = S\rho_x^2$$

thus defining the radii of gyration ρ_y, ρ_z, ρ_x. *

5.3.3 TENSOR OF SECOND MOMENTS OF AREA

The arbitrary cross-section of Figure 5.5 is replotted in Figure 5.8. Consider a line $\mathcal{L}$ within the plane of S, passing through the origin O. Let $\mathbf{u}$ be a unit vector on $\mathcal{L}$ with polar angle α from the Oy axis.

DEFINITION 5.3 The second moment of area of a plane section S with respect to a line $\mathcal{L}$ lying in its plane is

$$I_{\mathcal{L}} = \iint_S d^2(P,\mathcal{L})\,\mathrm{d}S \tag{5.24}$$

where $d(P,\mathcal{L})$ stands for the distance between current point $P \in S$ and line $\mathcal{L}$. ♣

The above definition generalizes those given by Equations (5.17) to (5.19). Using the projection H of P onto the line, Fig. 5.8, Equation (5.24) is expressed as follows:

$$I_{\mathcal{L}} = \iint_S \overline{PH}^2\,\mathrm{d}S = \iint_S (\mathbf{u}\times\overrightarrow{OP})^2\,\mathrm{d}S = \iint_S (z\cos\alpha - y\sin\alpha)^2\,\mathrm{d}S$$

Developing the latter equation leads to:

$$I_{\mathcal{L}} = I_y\cos^2\alpha - 2I_{yz}\sin\alpha\cos\alpha + I_z\sin^2\alpha$$

The above equation introduces a positive definite quadratic form, which may be written in matrix form:

$$I_{\mathcal{L}} = \begin{Bmatrix}\cos\alpha & \sin\alpha\end{Bmatrix}\begin{bmatrix} I_y & -I_{yz} \\ -I_{yz} & I_z\end{bmatrix}\begin{Bmatrix}\cos\alpha \\ \sin\alpha\end{Bmatrix} \tag{5.25}$$

Equation (5.25) thus introduces the symmetric square matrix of the tensor of quadratic moments of S in O. This matrix can be diagonalized by rotating the axes by some angle θ given by (up to a $\pi/2$ period)

$$\tan 2\theta = \frac{-2I_{yz}}{I_y - I_z} \tag{5.26}$$

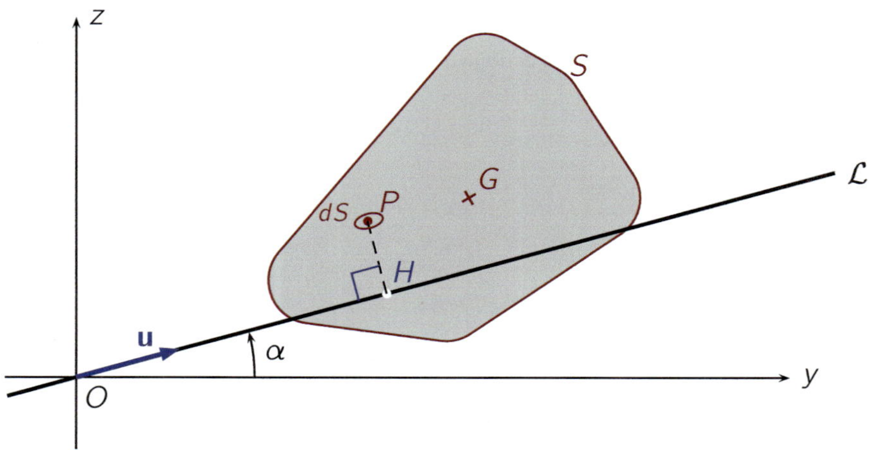

Figure 5.8 Second moment of area with respect to arbitrary axis $\mathcal{L}$.

The so-rotated axes are the *principal axes of inertia* of cross-section S. The eigenvalues of the above matrix are the *principal second moments of area*, i.e., the second moments of area with respect to the principal axes of inertia, and are given by

$$\frac{1}{2}(I_y + I_z) \pm \sqrt{(I_y - I_z)^2 + 4I_{yz}^2} \tag{5.27}$$

In this book, we shall exclusively work within the *central principal axes of inertia*, that is using the centroid as the origin. These axes are denoted GX, GY, GZ. The corresponding *central and principal second moments of area* are denoted I_X, I_Y, I_Z.

Remark 5.8 *To avoid notation clutter, the principal second moments of area about O, Equation* (5.27), *are not given a notation. One obtains the* central *principal second moments of area by applying the appropriate Huygens' formulae, namely the first three relations in Equation* (5.21) *and principal axes.* $*$

Remark 5.9 *When cross-section S admits an axis of symmetry, the latter is a central principal axis of inertia.* $*$

Remark 5.10 *Examination of Equation* (5.25) *reveals that the moments of inertia do not have the same transformation law as stress or strain components (see Part I). This is because the tensor of second moments of area and the stress or strain tensors have different definitions.* $*$

Example 5.5: Second Moments of Area for Common Sections

The central principal second moments of area of frequently encountered cross-sections are given below for quick reference (Figure 5.9).

For the thin tube, one may start from the thick tube result in Table 5.1 and carry out a Taylor expansion in the small parameter t/R. Alternatively, write, for example,

$$I_y = \int_{\theta=0}^{2\pi} (R\sin\theta)^2 tR\,\mathrm{d}\theta \tag{5.28}$$

where integration through the thickness is already accounted for by considering $r = R$ to the first order. ♡

Example 5.6: Geometric Properties of Common Cross-Sections

Calculate the position of the centroid, and then the second moments of area for the sections shown in Figure 5.28 of Problem 5.14. For the I-section, consider the thin wall limit when $t \ll h$ and $t' \ll h'$.

All sections have their centroid on OZ, axis of symmetry, so that only z^G is needed ($y^G = 0$). The parallel axis (Huygens) theorem, Equation (5.21), must be used where appropriate, including for the I-section when dealing with the top and bottom flanges. The results are reported in Table 5.2.

Note that $I_X = I_Y + I_Z$ is not provided explicitly in the table. ♡

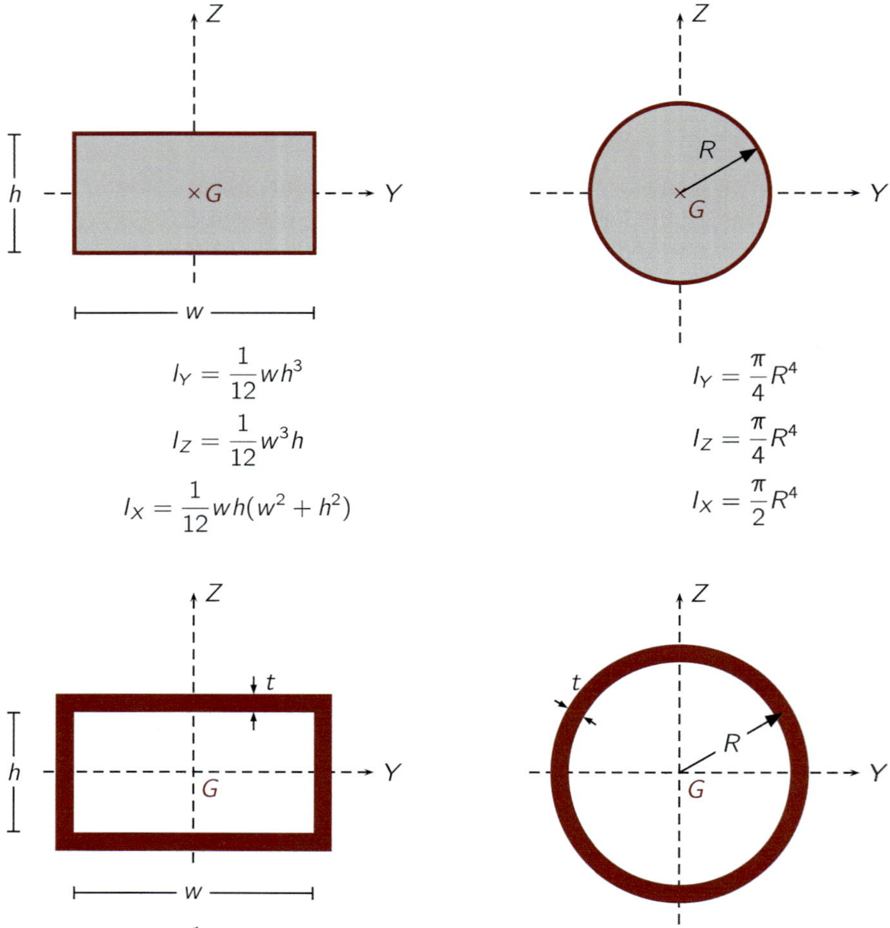

$$I_Y = \frac{1}{12}wh^3 \qquad I_Z = \frac{1}{12}w^3h \qquad I_X = \frac{1}{12}wh(w^2+h^2)$$

$$I_Y = \frac{\pi}{4}R^4 \qquad I_Z = \frac{\pi}{4}R^4 \qquad I_X = \frac{\pi}{2}R^4$$

$$I_Y = \frac{1}{6}h^2(3w+h)t \qquad I_Z = \frac{1}{6}w^2(3h+w)t \qquad I_X = \frac{1}{6}(w+h)^3t$$

$$I_Y = \pi R^3 t \qquad I_Z = \pi R^3 t \qquad I_X = 2\pi R^3 t$$

Figure 5.9 Some central principal second moments of area.

5.3.4 FRENET'S FRAME AND THE PRINCIPAL FRAME OF INERTIA

At a point G of the median line, centroid of cross-section S, two orthonormal frames have been defined in Sections 5.2.2 and 5.3.3. They share the same origin, G, and the same first axis GX of unit vector $\mathbf{t}$.

The principal frame of inertia $\{G, XYZ\}$ of Section 5.3.3 is thus obtained from

Table 5.2
Second Moments of Area and Centroids of Common Sections

Case	z^G	I_Y	I_Z
(a) I section	0	$\frac{th^3}{12}+\frac{2t'^3h'}{3}+\frac{t'h'h^2}{2}+t'^2h'h$	$\frac{t^3h}{12}+\frac{t'h'^3}{6}$
thin section	0	$\frac{th^3}{12}+\frac{t'h'h^2}{2}$	$\frac{t'h'^3}{6}$
(b) Semi-circle	$\frac{2}{\pi}R$	$\left(\frac{\pi}{2}-\frac{4}{\pi}\right)R^3t$	$\frac{\pi R^3t}{2}$
(c) Triangle	$\frac{h}{3}$	$\frac{Bh^3}{36}$	$\frac{B^3h}{48}$
(d) Half-disk	$\frac{4}{3\pi}R$	$\left(\frac{\pi}{8}-\frac{8}{9\pi}\right)R^4$	$\frac{\pi R^4}{8}$

Frenet's frame $\{G, \mathbf{t}, \mathbf{n}, \mathbf{b}\}$ of Section 5.2.2 by a rotation of angle φ about the longitudinal axis GX, Figure 5.10. With reference to aircraft wing or compressor blade design terminology, the built-in twist angle φ may be termed the "washout" angle.

A change of frame can be carried out using the formulae in Table 5.3. Taking into account Frenet's formulae, Equation (5.6), the transformation matrix of Table 5.3 and the mobile frame theorem, Section 5.2.1, the derivatives of the principal base vectors with respect to the arc length are given by (see Problem 5.9)

$$\frac{\mathrm{d}}{\mathrm{d}s}\begin{Bmatrix}\mathbf{I}\\\mathbf{J}\\\mathbf{K}\end{Bmatrix}=\begin{bmatrix}0 & \dfrac{\cos\varphi}{R} & -\dfrac{\sin\varphi}{R}\\ -\dfrac{\cos\varphi}{R} & 0 & \dfrac{1}{T}+\dfrac{\mathrm{d}\varphi}{\mathrm{d}s}\\ \dfrac{\sin\varphi}{R} & -\dfrac{1}{T}-\dfrac{\mathrm{d}\varphi}{\mathrm{d}s} & 0\end{bmatrix}\begin{Bmatrix}\mathbf{I}\\\mathbf{J}\\\mathbf{K}\end{Bmatrix} \tag{5.29}$$

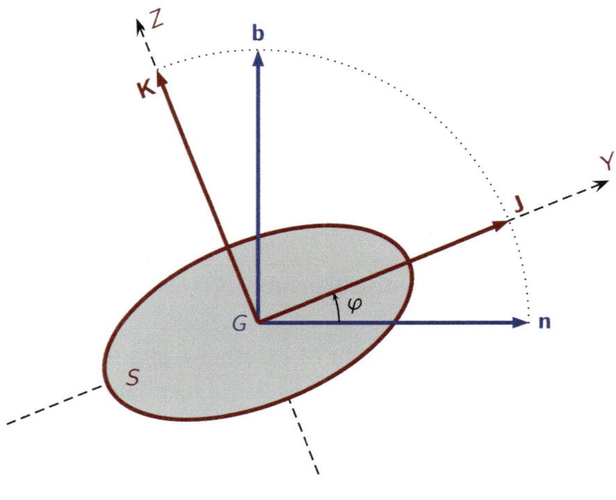

Figure 5.10 The washout angle.

Table 5.3
Transformation Matrix between the Principal and Frenet Frames

	I	**J**	**K**
t	1	0	0
n	0	$\cos\varphi$	$-\sin\varphi$
b	0	$\sin\varphi$	$\cos\varphi$

Remark 5.11 *In most cases, the two frames coincide for all cross-sections and the washout angle φ is identically nil.* ∗

Remark 5.12 *When second moments of area I_Y and I_Z are equal, say in circular or square sections, any straight line passing through G is a principal axis of inertia. In such cases, the axes GY and GZ are arbitrary and one can take $\varphi = 0$.* ∗

DEFINITION 5.4 A beam is said to admit a median plane (Π) if

- its median line $\widehat{G_0G_1}$ is contained within plane (Π); and
- one of the principal axes, say GY, remains in (Π) as G traces the arc $\widehat{G_0G_1}$. ♣

Remark 5.13 *A beam that admits a median plane has no washout and $\varphi = 0$ for all sections.* ∗

5.4 CROSS-SECTIONAL LOADS

5.4.1 INTERNAL RESULTANTS (SCREWER)

Any cross-section S divides the beam in two parts:

- an upstream part (I) between the initial section S_0 and current section S;
- a downstream part (II) between current section S and the final section S_1.

The notions of upstream and downstream are with respect to the median line being oriented from G_0 to G_1, Figure 5.11.

The beam being subjected to a system of external – applied and reaction – loads, in equilibrium, part (II) exerts on part (I) through S a resultant force $\mathbf{R}$ and a resultant moment $\mathbf{M}$ in G. To visualize this, consider – as a thought experiment – cutting the beam at section S, thus revealing the two lips (S^-) and (S^+) of the cut, each limiting the upstream and downstream parts (I) and (II), respectively. For each part to be in equilibrium, it is therefore necessary to apply on (S^-) a system of loads equivalent

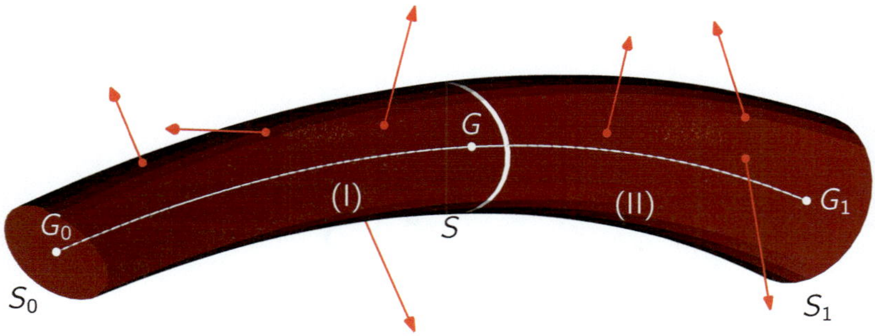

Figure 5.11 Current cross-section in a loaded beam.

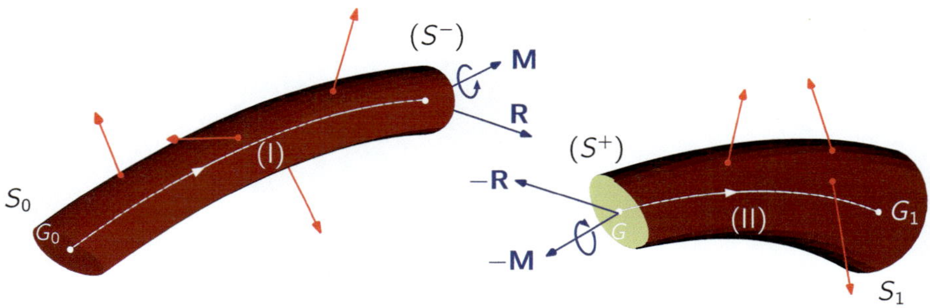

Figure 5.12 Definition of internal resultants in a beam, the "screwer".

to **R** and **M** and on (S^+) the opposite load system. The process is illustrated in Figure 5.12.

DEFINITION 5.5 The resultant force, **R**, and resultant moment in G, **M**, are called the internal resultants, or simply the "screwer" associated with current cross-section S ♣

The *screwer* has six components, three for **R** and three for **M**, Figure 5.13.

1. The projection of **R** on GX is the *normal* or *axial force* **N**, of algebraic measure N, which is positive in tension and negative in compression.
2. The projection of **R** onto the plane of cross-section S is the *shear force* **V**, with components V_Y and V_Z on the principal axes.
3. The projection of **M** on GX is the *longitudinal moment* $\mathbf{M}^X$, of algebraic measure M_X.
4. The projection of **M** onto the plane of cross-section S is the *bending moment* $\mathbf{M}^b$, with components M_Y and M_Z on the principal axes.

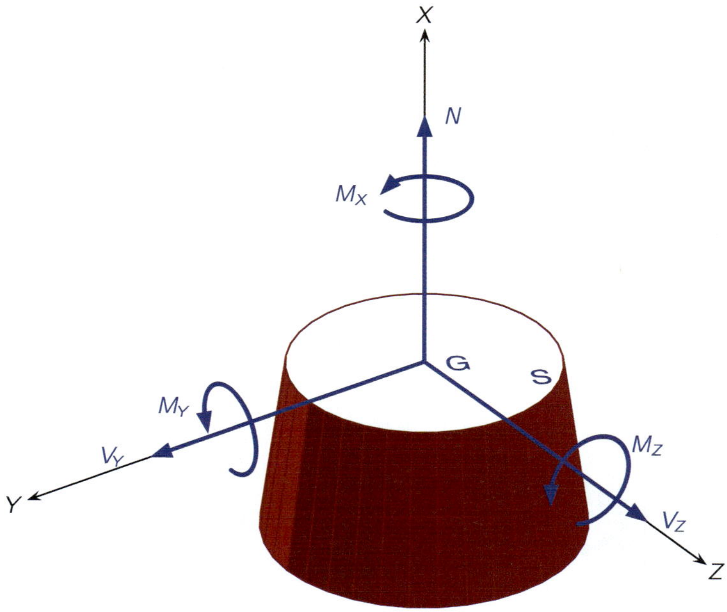

Figure 5.13 Screwer components, *alias* the internal resultants.

Remark 5.14 *The term "screwer" is not standard. It is an attempted translation of the French term "visseur". A more common term is "internal resultants" or "stress resultants". The new term is used sporadically in this text for its brevity and the historical precedence of the concept it contains over any notion of traction or stress, of which it is the resultant.* $*$

Remark 5.15 *In the language of torsors, it is both appropriate and convenient to think of the screwer as a force torsor with reduction elements* **R** *and* **M** *in G. It is denoted* $\{\mathcal{S}\}$ *and may be written as follows:*

$$\{\mathcal{S}\} = \begin{Bmatrix} \mathbf{R} \\ \mathbf{M} \end{Bmatrix}_G = \begin{Bmatrix} N & M_X \\ V_Y & M_Y \\ V_Z & M_Z \end{Bmatrix}_{G,XYZ} \tag{5.30}$$

$*$

Remark 5.16 *In some texts, it is the action of the upstream part (I) on the downstream part (II) that defines the internal resultants (or screwer in the new terminology), i.e., the action of* (S^-) *on* (S^+)*. One goes from one convention to the other by a simple change of sign. Note that this "sign convention" has fundamentally nothing to do with coordinates or components.* $*$

Remark 5.17 *The internal resultants are functions of arclength s span-wise along the median line. Considering a cross-section S other than that shown in Figure 5.11* *would change some, if not all, of the screwer components. Thus, what is to be determined in general are the functions* $\mathbf{R}(s)$ *and* $\mathbf{M}(s)$ *and the corresponding screwer-component diagrams* $s \mapsto \mathbf{R}(s)$ *and* $s \mapsto \mathbf{M}(s)$. $*$

5.4.2 DETERMINING THE SCREWER COMPONENTS

Since the screwer is fundamentally a force torsor, Equation (5.30), it is formally convenient to calculate it using torsor formalism. Let $\{\mathcal{F}\}$ denote the torsor of external forces, inclusive of reactions. For a given cross-section *S*, $\{\mathcal{F}\}$ is decomposed into a torsor $\{\mathcal{F}^{\mathrm{I}}\}$ applied upstream from *S* and a torsor $\{\mathcal{F}^{\mathrm{II}}\}$ applied downstream from it.

Consider the equilibrium of portion (II), Figure 5.12; the external forces acting on it are:

- on the one hand, those forming torsor $\{\mathcal{F}^{\mathrm{II}}\}$;
- on the other hand, those forming the screwer $\{-\mathcal{S}\}$.

It then follows that: $\{\mathcal{F}^{\mathrm{II}}\} + \{-\mathcal{S}\} = \{0\}$, i.e.,

$$\{\mathcal{S}\} = \{\mathcal{F}^{\mathrm{II}}\}. \tag{5.31}$$

One then enunciates the following:

Theorem 5.4: Fundamental Theorem of Beam Statics

The screwer acting on a beam cross-section equals the torsor of all external forces, inclusive of reactions, applied downstream from that section. $\diamondsuit$

The practical rule for calculating the six components of $\{\mathcal{S}\}$, which are the internal resultants depicted in Figure 5.13, is the following:

1. The resultant **R** equals the sum of all external forces, inclusive of reaction forces, applied to the beam downstream from *S*;
2. The resultant moment **M** equals the moment about *G*, centroid of *S*, of all external forces, inclusive of reaction forces, augmented with all external couples, inclusive of reaction moments, applied to the beam downstream from *S*.

Remark 5.18 *In some cases, it is more convenient to express the equilibrium of part (I) upstream from S. The resultants* **R** *and* **M** *are then, respectively, the opposite of*

1. *the resultant of external forces applied to part (I) of the beam;*
2. *the resultant moment in G of the same forces.*

In torsor terms, $\{\mathcal{S}\} = -\{\mathcal{F}^{\mathrm{I}}\}$. *This result and that of Equation* (5.31) *are, of course, equivalent. Indeed, by considering the equilibrium of the entire beam, as shown in Figure 5.11, one has* $\{\mathcal{F}^{\mathrm{II}}\} = -\{\mathcal{F}^{\mathrm{I}}\}$. $*$

Remark 5.19 *For the reader not familiar with torsors, Theorem 5.4 may be established by simply considering the free-body diagram of portion (II) in Figure 5.12. The free-body diagram of portion (I) would lead to relating the internal resultants with the forces and moments applied to part (I). Again, the two results are equivalent by considering the free-body diagram of the whole beam, Figure 5.11.* ∗

Remark 5.20 *Among the external forces included in torsors* $\{\mathcal{F}^{\mathrm{I}}\}$ *and* $\{\mathcal{F}^{\mathrm{II}}\}$ *there can be some hyperstatic unknowns, be it reaction forces – if the structure is externally hyperstatic – or internal forces and moments, as could arise in trusses, frames, or rings.* ∗

Example 5.7: Prismatic Beam under Its Own Weight

A prismatic beam of length $4L$, admitting XOY as a vertical plane of symmetry, is subjected to its own weight $P = 4pL$ and supported as shown in Figure 5.14 where p has the meaning of a weight per unit length. The reactions at the supports are equal: $R^A = R^B = P/2 = 2pL$.

Applying the fundamental theorem (Theorem 5.4) first shows that the only non-zero screwer components are the shear V_Y and the bending moment M_Z. This is because all external forces, whether applied (P) or reactions (R^A and R^B), are oriented along the Y principal axis. In addition, one finds

- For $L < X \leq 2L$:

$$V_Y = -p(2L - X) \qquad \text{and} \qquad M_Z = -\frac{1}{2}p(2L - X)^2 \tag{5.32}$$

- For $-L < X < L$,

$$V_Y = pX \qquad \text{and} \qquad M_Z = -\frac{1}{2}pX^2 \tag{5.33}$$

- For $-2L \leq X < -L$,

$$V_Y = p(2L + X) \qquad \text{and} \qquad M_Z = -\frac{1}{2}p(2L + X)^2 \tag{5.34}$$

The above equations enable the shear and bending moment diagrams, $X \mapsto \mathbf{V}_Y(X)$ and $X \mapsto \mathbf{M}_Z(X)$, to be drawn (see Figure 5.14). Note that:

- The reactions at the supports introduce discontinuities of the first kind in the shear diagram.
- The shear force is the derivative, with an opposite sign, of the bending moment with respect to X. These issues will be examined further later.

It is possible, and even recommended for the beginner, to retrieve the above results by first drawing an appropriately chosen part of the beam comprised between one end of the beam and a "cut", i.e., the current cross-section centered about G. Then, the free body diagram of that portion allows to obtain the internal resultants. The procedure is illustrated in Figure 5.15 where $V = V_Y$ and $M = M_Z$.

If the position x of the current cross-section S is such that $L < x < 2L$, then it is more efficient to consider the equilibrium of the part downstream from S. The orientation of the shear force V and bending moment M is consistent with the general convention in Figure 5.12. The length of the part shown being $2L - x$, force

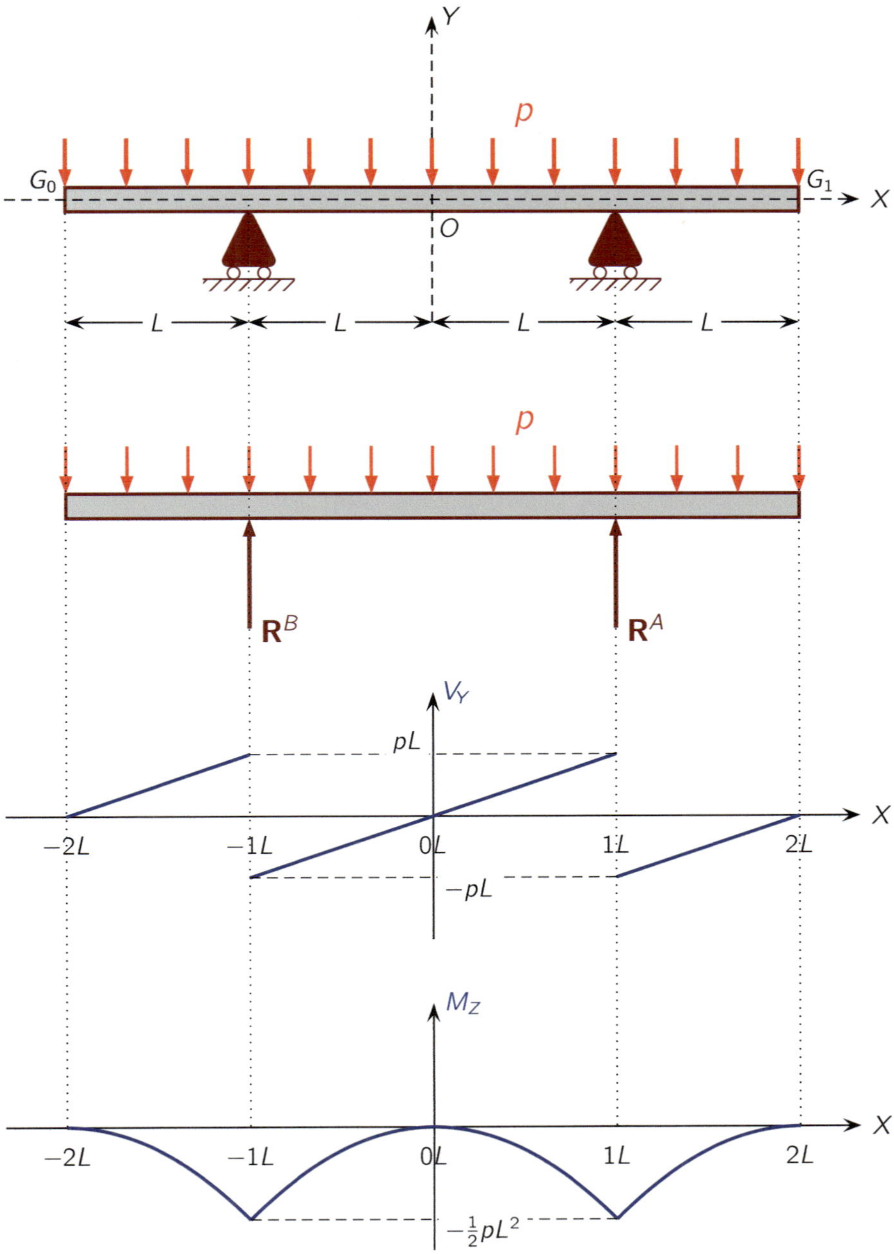

Figure 5.14 Simply supported beam under its own weight.

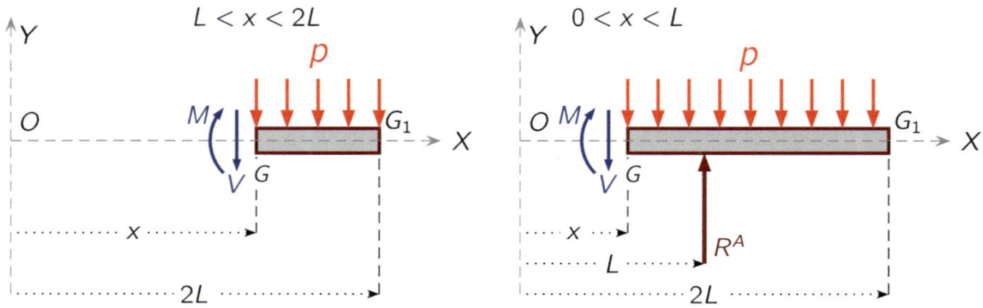

Figure 5.15 Method of sections to determine internal resultants.

equilibrium along the vertical gives

$$-V_Y - p(2L - X) = 0 \tag{5.35}$$

which is the same result as in Equation $(5.32)_1$. Moment equilibrium about G, on the other hand, gives

$$-M_Z - p(2L - X)\left[\frac{1}{2}(2L - X)\right] = 0 \tag{5.36}$$

where the net downward force $p(2L - X)$ has an armlength of $\frac{1}{2}(2L - X)$. This recovers Equation $(5.32)_2$.

If now the cut is located between the two supports, for example, if $0 < x < L$ as in the right panel of Figure 5.15, then the free-body diagram now includes the reaction force R^A in addition to the weight. Force equilibrium now gives

$$-V_Y - p(2L - X) + R^A = 0, \tag{5.37}$$

which implies Equation $(5.33)_1$ since $R^A = 2pL$ from global equilibrium. Likewise, moment equilibrium about G gives

$$-M_Z - \frac{1}{2}p(2L - X)^2 + R^A(L - x) = 0 \tag{5.38}$$

which gives Equation $(5.33)_2$ after simplification.

The solution is completed for $-2L \leq X < -L$ by employing the symmetry relations:

$$V(-X) = -V(X) \qquad \text{and} \qquad M(-X) = M(X) \tag{5.39}$$

which provide the same result as in Equation (5.34). Alternatively, a part may be chosen between the left end at G_0 and the current cross-section (the cut), that is upstream from the cut. The shear force and bending moment would have to be appropriately oriented consistent with Figure 5.12. In doing so, the result of Equation (5.34) follows in a straightforward manner. ♡

Example 5.8: Helical Spring in Tension

The geometry of the spring was treated in Example 5.1. The origin G_0 is fixed to the x axis. The terminal section at G_1 is welded to a rigid bar G_1C of length a with

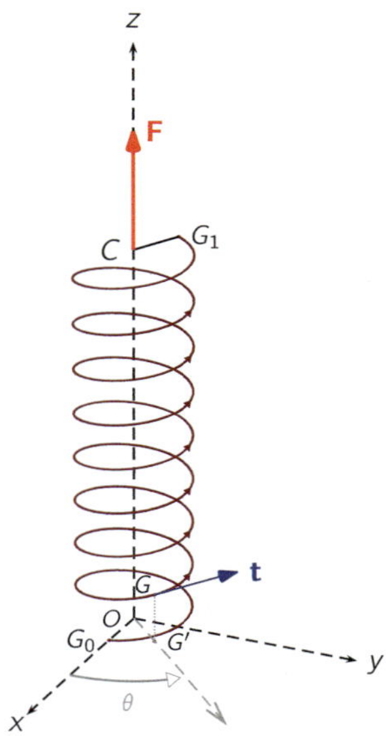

Figure 5.16 A helical spring in tension.

point C lying on the z axis, Figure 5.16. A tensile load $\mathbf{F}$ is applied at C along the z direction.

Applying the fundamental theorem (Theorem 5.4) allows the calculation of the screwer components on the Frenet axes. On the current cross-section parameterized by θ, one finds

$$\mathbf{R} = \begin{cases} N = \mathbf{F}\cdot\mathbf{t} = \dfrac{\lambda}{\sqrt{1+\lambda^2}}F \\ V_n = \mathbf{F}\cdot\mathbf{n} = 0 \\ V_b = \mathbf{F}\cdot\mathbf{b} = \dfrac{1}{\sqrt{1+\lambda^2}}F \end{cases} \tag{5.40}$$

$$\text{and}\quad \mathbf{M} = \begin{cases} M_X = (\overrightarrow{GC}, \mathbf{F}, \mathbf{t}) = \dfrac{a}{\sqrt{1+\lambda^2}}F \\ M_n = (\overrightarrow{GC}, \mathbf{F}, \mathbf{n}) = 0 \\ M_b = (\overrightarrow{GC}, \mathbf{F}, \mathbf{b}) = -\dfrac{a\lambda}{\sqrt{1+\lambda^2}}F \end{cases} \tag{5.41}$$

where, for example, the notation $(\overrightarrow{GC}, \mathbf{F}, \mathbf{t})$ stands for $(\overrightarrow{GC}\times\mathbf{F})\cdot\mathbf{t}$.

Hence, each internal resultant is the same in all cross-sections. Their expressions in the global frame, $Oxyz$, would not only be cumbersome in terms of θ but also would hide this essential result. Furthermore, note that a spring in tension is actually under combined loading that includes axial and shearing forces, as well as bending and torsion moments. ♡

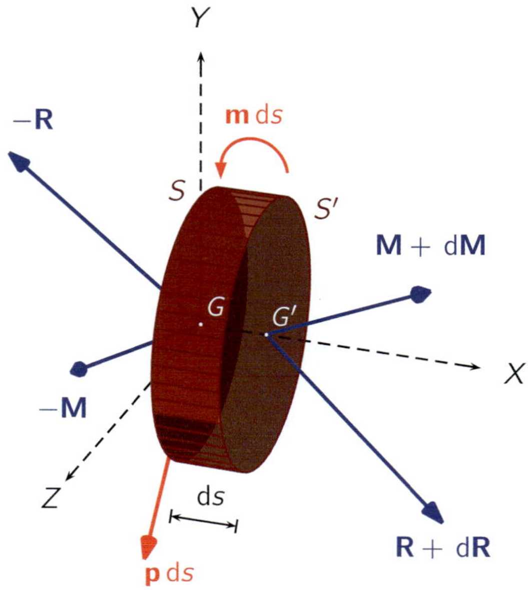

Figure 5.17 Slice of arbitrary beam under load.

5.5 EQUILIBRIUM EQUATIONS FOR BEAMS

5.5.1 GENERAL CASE

Consider an elementary slice of a beam, comprised between two neighboring cross-sections S and S', with arclengths s and $s+\mathrm{d}s$ and centered about G and G', respectively, as shown in Figure 5.17.

Suppose that all external forces and couples applied to the beam at the level of this slice are distributed. Let $\mathbf{p}\,\mathrm{d}s$ and $\mathbf{m}\,\mathrm{d}s$ be, respectively, their general resultant and resultant moment in G.

The internal resultants on (S^+) are $-\mathbf{R}(s)$ and $-\mathbf{M}(s)$, while those acting on (S'^-) are $\mathbf{R}(s)+\mathrm{d}\mathbf{R}$ and $\mathbf{M}(s)+\mathrm{d}\mathbf{M}$. The notation is consistent with the definitions in Section 5.4.1 and Figure 5.12.

Equilibrium of the slice is then written as follows:

$$\left\{\begin{aligned} &-\mathbf{R}+\left(\mathbf{R}+\frac{\mathrm{d}\mathbf{R}}{\mathrm{d}s}\mathrm{d}s\right)+\mathbf{p}\,\mathrm{d}s=\mathbf{0}\\ &-\mathbf{M}+\left(\mathbf{M}+\frac{\mathrm{d}\mathbf{M}}{\mathrm{d}s}\mathrm{d}s\right)+\overrightarrow{GG'}\times\mathbf{R}+\mathbf{m}\,\mathrm{d}s=\mathbf{0}\end{aligned}\right. \tag{5.42}$$

where $\overrightarrow{GG'}=\mathbf{t}\,\mathrm{d}s$ by definition of the unit tangent. In Equation $(5.42)_2$, moments are taken with respect to the center G of section S. Thus, the third term corresponds to the moment of $\mathbf{R}+\frac{\mathrm{d}\mathbf{R}}{\mathrm{d}s}\,\mathrm{d}s$, which acts at G'. The equation is obtained by neglecting the second-order term (in $\mathrm{d}s^2$) before first-order ones (in $\mathrm{d}s$). Simplifying,

$$\boxed{\mathbf{p}+\frac{\mathrm{d}\mathbf{R}}{\mathrm{d}s}=\mathbf{0} \qquad \text{and} \qquad \mathbf{m}+\frac{\mathrm{d}\mathbf{M}}{\mathrm{d}s}+\mathbf{t}\times\mathbf{R}=\mathbf{0}} \tag{5.43}$$

In vector form, the differential equations of beam equilibrium are thus quite simple. These two vectorial equilibrium equations for the elementary slice are equivalent to six scalar equations that will now be written in the central principal frame $\{G,XYZ\}$. On that basis, let:

$$\mathbf{p}=\begin{Bmatrix} p_X \\ p_Y \\ p_Z \end{Bmatrix} \qquad \mathbf{m}=\begin{Bmatrix} m_X \\ m_Y \\ m_Z \end{Bmatrix} \qquad \mathbf{R}=\begin{Bmatrix} N \\ V_Y \\ V_Z \end{Bmatrix} \qquad \mathbf{M}=\begin{Bmatrix} M_X \\ M_Y \\ M_Z \end{Bmatrix}$$

Projecting force and moment equilibrium, Equation (5.43), with due account of derivation formulae for the $\{G,XYZ\}$ frame, Equation (5.29), gives

$$\begin{cases} p_X+\dfrac{\mathrm{d}N}{\mathrm{d}s}-V_Y\dfrac{\cos\varphi}{R}+V_Z\dfrac{\sin\varphi}{R}=0 \\[2ex] p_Y+N\dfrac{\cos\varphi}{R}+\dfrac{\mathrm{d}V_Y}{\mathrm{d}s}-V_Z\left(\dfrac{1}{T}+\dfrac{\mathrm{d}\varphi}{\mathrm{d}s}\right)=0 \\[2ex] p_Z-N\dfrac{\sin\varphi}{R}+\dfrac{\mathrm{d}V_Z}{\mathrm{d}s}+V_Y\left(\dfrac{1}{T}+\dfrac{\mathrm{d}\varphi}{\mathrm{d}s}\right)=0 \end{cases} \tag{5.44}$$

$$\begin{cases} m_X+\dfrac{\mathrm{d}M_X}{\mathrm{d}s}-M_Y\dfrac{\cos\varphi}{R}+M_Z\dfrac{\sin\varphi}{R}=0 \\[2ex] m_Y+M_X\dfrac{\cos\varphi}{R}+\dfrac{\mathrm{d}M_Y}{\mathrm{d}s}-M_Z\left(\dfrac{1}{T}+\dfrac{\mathrm{d}\varphi}{\mathrm{d}s}\right)-V_Z=0 \\[2ex] m_Z-M_X\dfrac{\sin\varphi}{R}+\dfrac{\mathrm{d}M_Z}{\mathrm{d}s}+M_Y\left(\dfrac{1}{T}+\dfrac{\mathrm{d}\varphi}{\mathrm{d}s}\right)+V_Y=0 \end{cases} \tag{5.45}$$

In general, therefore, mere (double) curvature of the median line of a beam leads to coupling among all screwer components: bending (M_Y, M_Z), torsion (M_X), extension (N) and shear (V_Y, V_Z) are all intertwined.

5.5.2 BEAM WITH MEDIAN PLANE LOADED IN THAT PLANE

Let xOy be the median plane of a curvilinear member that meets Definition 5.4. In this case, the cross-section has no washout ($\varphi = 0$) and the median line may have curvature but no torsion ($1/T = 0$), as sketched in Figure 5.18. If such a beam is loaded in its median plane, then

$$p_Z=0, \quad m_X=0, \quad m_Y=0, \quad V_Z=0, \quad M_X=0, \quad M_Y=0$$

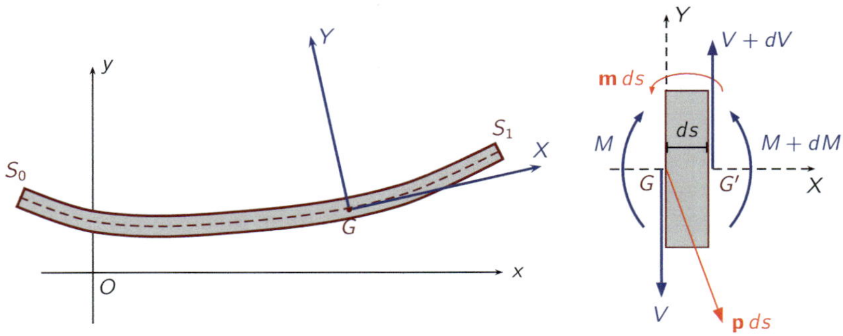

Figure 5.18 Beam with a median plane loaded in that plane.

Three equations of the set (5.44) and (5.45) are then identically verified. The remaining set consists of force equilibrium along *GX* and *GY* and moment equilibrium about *GZ*:

$$\begin{cases} p_X + \dfrac{\mathrm{d}N}{\mathrm{d}s} - \dfrac{V}{R} = 0 \\ p_Y + \dfrac{\mathrm{d}V}{\mathrm{d}s} + \dfrac{N}{R} = 0 \\ m_Z + \dfrac{\mathrm{d}M}{\mathrm{d}s} + V = 0 \end{cases} \tag{5.46}$$

with $V = V_Y$ and $M = M_Z$.

Remark 5.21 *In the vast majority of cases, beams having a median plane admit that plane as a plane of symmetry.* *

5.5.3 CIRCULAR ARCHES AND RINGS

Arches are structures made of curvilinear members. When the arch admits a median plane and is loaded in that plane, its equilibrium can be analyzed using Equation (5.46). If, in addition, the arch has constant curvature, it is circular, and these equations can be further simplified and partially solved. Rings are another practical

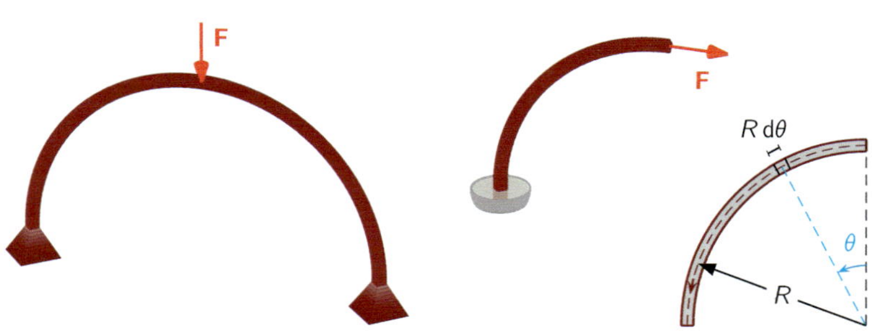

Figure 5.19 Examples of circular arches.

example of circular beams. Let R denote the constant radius of curvature. The element of arclength is then $\mathrm{d}s = R\,\mathrm{d}\theta$ where θ is the angle identifying the current cross-section from some reference section, Figure 5.19.

Using Equation (5.46), the equilibrium of an arch is governed by

$$\begin{cases} Rp_X + \dfrac{\mathrm{d}N}{\mathrm{d}\theta} - V = 0 \\ Rp_Y + \dfrac{\mathrm{d}V}{\mathrm{d}\theta} + N = 0 \\ \dfrac{\mathrm{d}M}{\mathrm{d}\theta} + RV = 0 \end{cases} \iff \begin{cases} V = \dfrac{\mathrm{d}N}{\mathrm{d}\theta} + Rp_X \\ \dfrac{\mathrm{d}^2N}{\mathrm{d}\theta^2} + N = -R\left(\dfrac{\mathrm{d}p_X}{\mathrm{d}\theta} + p_Y\right) \\ \dfrac{\mathrm{d}M}{\mathrm{d}\theta} = -RV \end{cases} \tag{5.47}$$

in the absence of distributed couples ($m_Z = 0$). The second statement in Equation $(5.47)_2$ is obtained by eliminating the shear V from the first two equations to the left of the equivalence.

In any arch equilibrium problem, the distributed loads p_X and p_Y are given. Equation $(5.47)_2$ then constitutes a second-order ordinary differential equation (linear with constant coefficients) in the sole unknown $N(\theta)$. The general solution is then given as follows:

$$N(\theta) = A\cos\theta + B\sin\theta + N_{\mathrm{p}}(\theta) \tag{5.48}$$

where A and B are integration constants to be determined from boundary conditions, and $N_{\mathrm{p}}(\theta)$ is a particular solution to the differential equation, and depends on the loading via p_X and p_Y.

The shear force is then obtained from Equation $(5.47)_1$ by differentiation:

$$V(\theta) = -A\sin\theta + B\cos\theta + \frac{\mathrm{d}N_{\mathrm{p}}}{\mathrm{d}\theta} + Rp_X \tag{5.49}$$

and the bending moment from Equation $(5.47)_3$ by integration:

$$M(\theta) = -RA\cos\theta - RB\sin\theta - RN_{\mathrm{p}}(\theta) - R^2\int_0^\theta p_X\,\mathrm{d}\phi \tag{5.50}$$

Note that the distributed load p_Y does not enter explicitly the above expressions but affects all results through the particular solution term $N_{\mathrm{p}}(\theta)$.

Remark 5.22 *Arches are used in large-span structures, such as airplane hangars, long-span bridges, and architectural members (arched domes, vaults, etc.) Unlike straight beams, the design of which is governed by bending moments and shearing forces (see Example 5.7), a distinguishing feature of an arch is the development of horizontal thrusts at the supports, in addition to the vertical reactions, even in the absence of a horizontal load. As seen in Equation* (5.47)*, the screwer in any section of an arch includes axial compression, shear force, and bending moment. The bending moment and shear force are comparatively smaller than those of a straight beam of the same span due to the presence of the horizontal thrusts. The latter significantly reduce the moments and shear forces at any section of the arch, which results in reduced member size and a more economical design compared to other structures. In addition, arches are aesthetically pleasant.* *

Remark 5.23 *Rings are common structural components in aerospace structures, such as fuselage frames, landing gears, and jet engines. Webs and skins of semi-monocoque construction are thin and cannot withstand large concentrated loads. At the locations of concentrated loads, stiffeners are introduced in the design of structures, which serve to "diffuse" the load to the attached skins or webs. For example, fuselage frames typically comprise slender-curved members and thin rings. A fuselage structure has closely spaced bulkheads or frames that resist transverse loads, while the fuselage shell resist longitudinal loads, i.e., those acting backward and forward. Typically, fuselage frames are modeled as rings subjected to in-plane loading.* *

Example 5.9: Semi-Circular Beam under a Mid-Span Point Load

Consider a simply supported semi-circular beam admitting xOy as a median plane and loaded in compression, as shown in Figure 5.20. The figure also depicts the principal frame GXY and the angle θ, which parameterizes the location of the current cross-section. The shape of the cross-section need not be specified; it can be anything that admits a median plane as per Definition 5.4.

The portions corresponding to $\theta > 0$ and $\theta < 0$ must be treated separately because of the point force. On each interval, Equations 5.46 are integrable and, using $p_X = p_Y = 0 = m_Z$ and $\mathrm{d}s = R\,\mathrm{d}\theta$, apply as follows:

$$\frac{\mathrm{d}N}{\mathrm{d}\theta} - V = 0, \qquad N + \frac{\mathrm{d}V}{\mathrm{d}\theta} = 0, \qquad \frac{\mathrm{d}M}{\mathrm{d}\theta} + RV = 0 \tag{5.51}$$

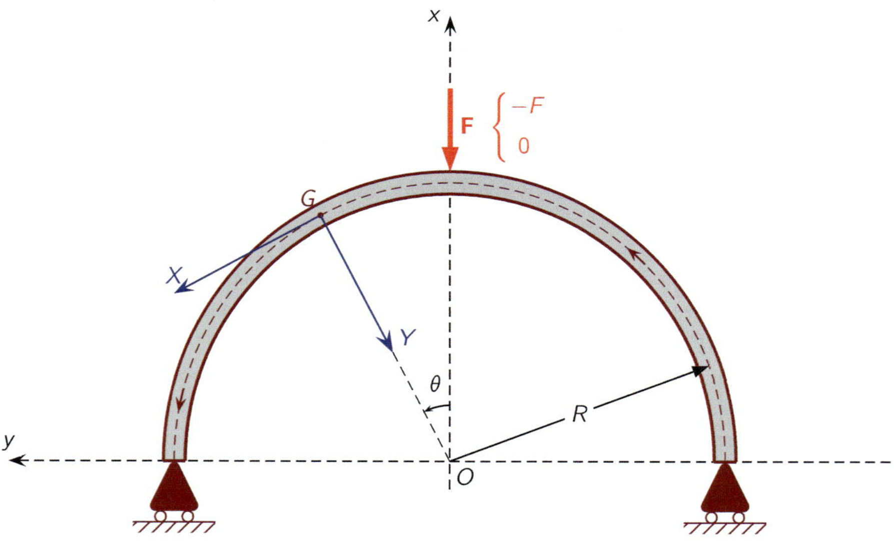

Figure 5.20 Semi-circular beam with a median plane.

Eliminating V between the first two leads to the equivalent system:

$$\frac{d^2N}{d\theta^2} + N = 0, \qquad V = \frac{dN}{d\theta}, \qquad \frac{dM}{d\theta} = -RV \tag{5.52}$$

which is the practical system of equations to solve for any circular beam, in the absence of a distributed load. Equation (5.52) may also be obtained directly from Equation (5.47).

The reactions at the support are equal to $F/2$. Here, the solution of Equation (5.52) is to be determined subject to boundary conditions:

$$\text{For } \theta = \frac{\pi}{2}: \quad N = -\frac{F}{2}, \quad V = 0, \quad M = 0 \tag{5.53}$$

$$\text{For } \theta = -\frac{\pi}{2}: \quad N = -\frac{F}{2}, \quad V = 0, \quad M = 0 \tag{5.54}$$

and to the following patching:

$$\text{For } \theta = 0: \quad [\![V]\!] = -F, \quad N \text{ and } M \text{ continuous}$$

where the notation $[\![.]\!]$ stands for a jump discontinuity. To arrive at Equation (5.53), it suffices to consider the free-body diagram of a slice comprised between the section at the left support (where $\theta = \pi/2$) and an infinitely close cut, Figure 5.21. Likewise, to establish Equation (5.54), consider the free-body diagram of a slice comprised between the section at the right support ($\theta = -\pi/2$) and an infinitely close cut, Figure 5.21.

Solving Equation (5.52) gives:

$$N(\theta) = \begin{cases} a\cos\theta + b\sin\theta & \text{for } \theta > 0 \\ a'\cos\theta + b'\sin\theta & \text{for } \theta < 0 \end{cases} \qquad V(\theta) = \begin{cases} -a\sin\theta + b\cos\theta & \text{for } \theta > 0 \\ -a'\sin\theta + b'\cos\theta & \text{for } \theta < 0 \end{cases}$$

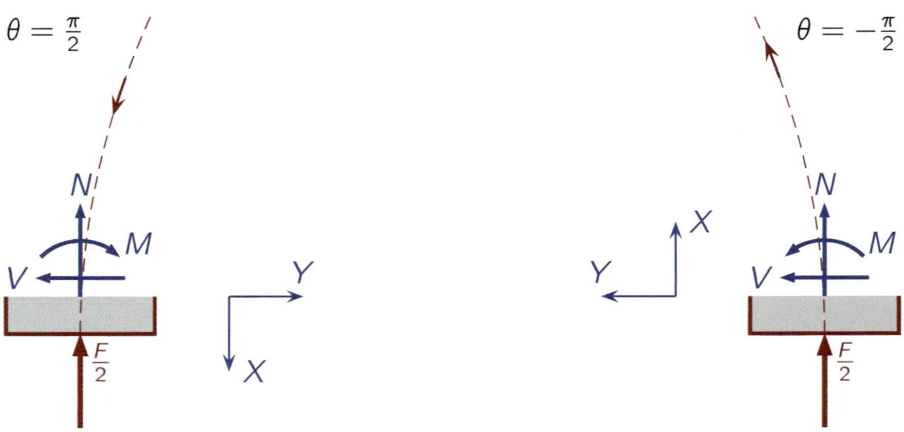

Figure 5.21 Boundary conditions on internal resultants at the two supports.

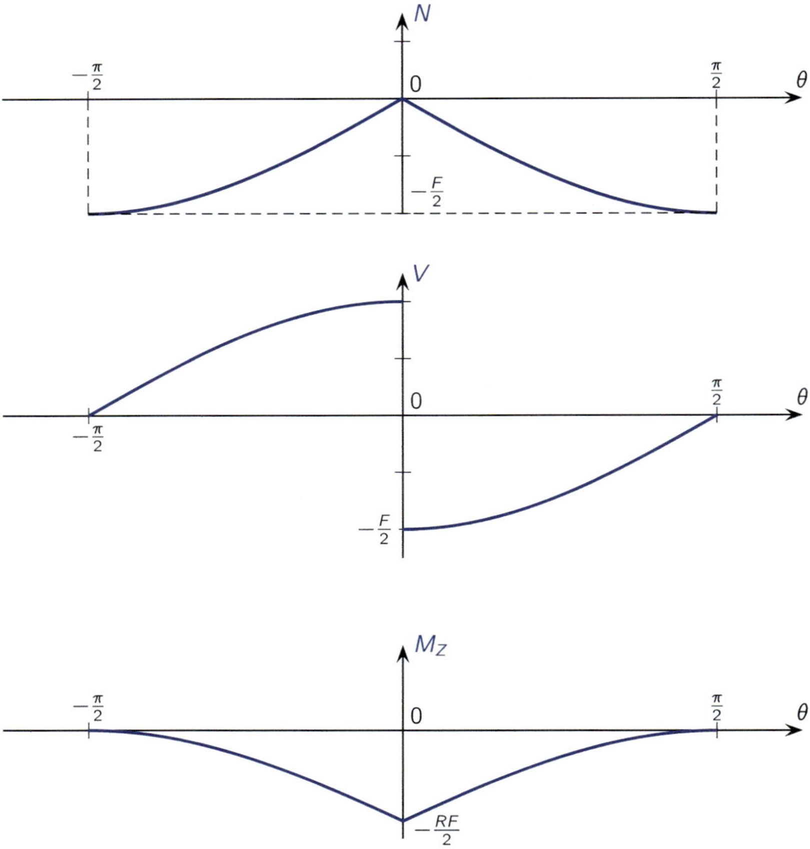

Figure 5.22 Internal resultants in the circular beam of Figure 5.20.

and

$$M(\theta) = \begin{cases} -R(a\cos\theta + b\sin\theta + c) & \text{for } \theta > 0 \\ -R(a'\cos\theta + b'\sin\theta + c') & \text{for } \theta < 0 \end{cases}$$

where integration constants a, a', b, b', c, c' are determined from boundary conditions, Equations (5.53) and (5.54):

$$a = 0, \quad b = -\frac{F}{2}, \quad c = \frac{F}{2} \quad \text{and} \quad a' = 0, \quad b' = \frac{F}{2}, \quad c' = \frac{F}{2}$$

The patching at $\theta = 0$ is here identically verified. The sought functions are then

$$\begin{cases} \theta < 0: \quad N = \frac{F}{2}\sin\theta; \qquad V = \frac{F}{2}\cos\theta; \qquad M = -\frac{RF}{2}(1 + \sin\theta) \\ \\ \theta > 0: \quad N = -\frac{F}{2}\sin\theta; \qquad V = -\frac{F}{2}\cos\theta; \qquad M = -\frac{RF}{2}(1 - \sin\theta) \end{cases} \tag{5.55}$$

and the corresponding diagrams are shown in Figure 5.22. ♡

5.5.4 STRAIGHT BEAM WITH MEDIAN PLANE

In this case, the arclength is identified with the X coordinate along the beam's axis, Figure 5.23, and the product $\mathbf{t} \times \mathbf{R}$ in the general statement of equilibrium, Equation (5.43), is identified with $\mathbf{I} \times \mathbf{R}$ where $\mathbf{I}$ is the unit vector along the beam's axis.

Thus, one obtains the much simpler equations:

$$\left\{\begin{array}{l} p_X + \dfrac{\mathrm{d}N}{\mathrm{d}x} = 0 \\[2ex] p_Y + \dfrac{\mathrm{d}V_Y}{\mathrm{d}x} = 0 \\[2ex] p_Z + \dfrac{\mathrm{d}V_Z}{\mathrm{d}x} = 0 \end{array}\right. \qquad \left\{\begin{array}{l} m_X + \dfrac{\mathrm{d}M_X}{\mathrm{d}x} = 0 \\[2ex] m_Y + \dfrac{\mathrm{d}M_Y}{\mathrm{d}x} - V_Z = 0 \\[2ex] m_Z + \dfrac{\mathrm{d}M_Z}{\mathrm{d}x} + V_Y = 0 \end{array}\right. \tag{5.56}$$

The above equations could also be obtained from Equations (5.44) and (5.45) by setting $\varphi = 0$, $1/R = 0$ and $1/T = 0$ (no washout, no curvature, no torsion).

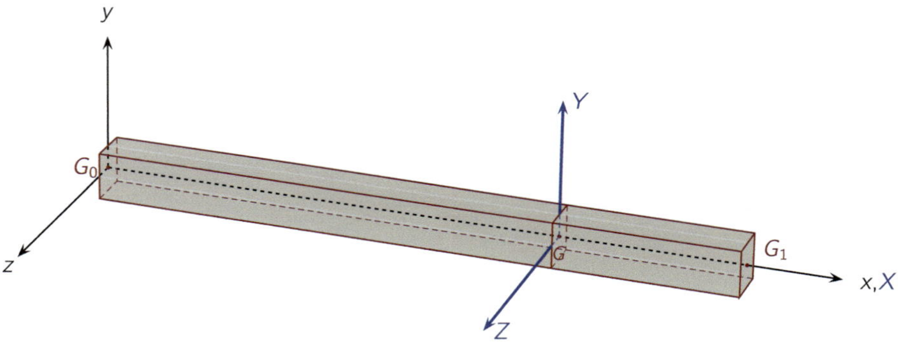

Figure 5.23 Straight beam with a median plane.

Example 5.10: Screwer in a Straight Beam by Integration

A simply supported prismatic beam admitting xOy as a median plane, is subjected to a uniform load over half of its span as shown in Figure 5.24 where q is a force per unit length.

The following is required:

1. Determine the non-zero components of the screwer using the 'section-cut method', as in Example 5.7.
2. Recover these results by integrating the different equations of equilibrium.
3. Plot the screwer diagrams $x \mapsto \mathbf{R}(x)$ and $x \mapsto \mathbf{M}(x)$.

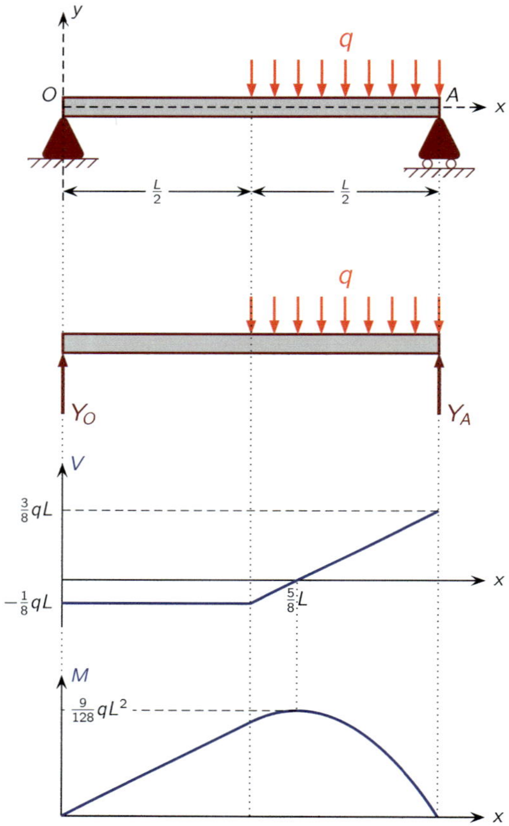

Figure 5.24 Uniformly distributed load over half the span.

Solution:

1. Section-cut Method:
   a. Global equilibrium: Referring to the free-body diagram in Figure 5.24, the balance of forces and balance of moments about the origin O read:

$$\left\{\begin{array}{l} Y_O+Y_A+\left(-q\dfrac{L}{2}\right)=0 \\ Y_AL-q\dfrac{L}{2}\left(\dfrac{3L}{4}\right)=0 \end{array}\right. \quad\Longrightarrow\quad \left\{\begin{array}{l} Y_O=\dfrac{1}{8}qL \\ Y_A=\dfrac{3}{8}qL \end{array}\right. \tag{5.57}$$

   b. Section located at x with $0<x<L/2$: The shear force V and bending moment M are the only non-zero components of the screwer. One can obtain V and M in two ways:

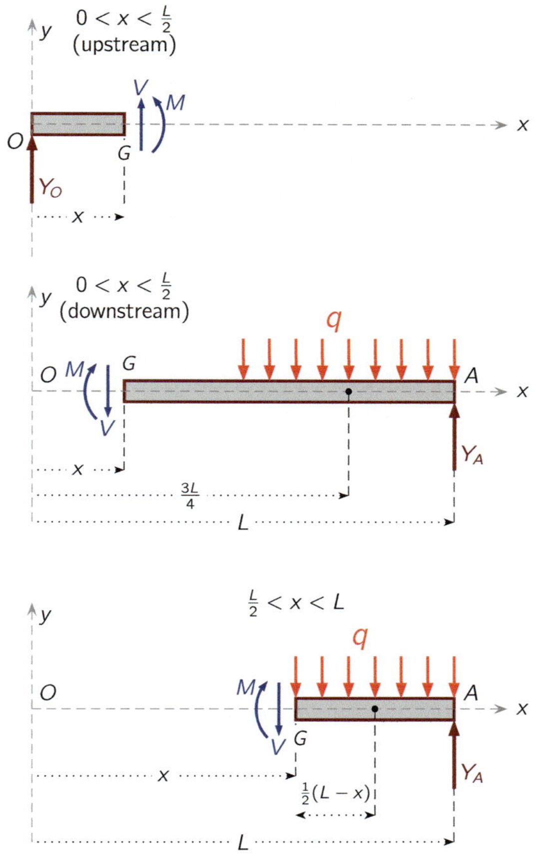

Figure 5.25 Free-body diagrams for determining V and M.

- Looking upstream from the section: Referring to the corresponding free-body diagram in Figure 5.25,

$$\begin{cases} V + Y_O = 0 \\ M - Y_O x = 0 \end{cases} \implies \begin{cases} V = -\frac{1}{8}qL \\ M = \frac{1}{8}qLx \end{cases} \tag{5.58}$$

where the moments are taken with respect to the center G of the current cross-section (to eliminate V).

- Looking downstream from the section: Referring to the second free-body diagram in Figure 5.25,

$$\begin{cases} -V - q\dfrac{L}{2} + \dfrac{3}{8}qL = 0 \\ -M - q\dfrac{L}{2}\left(\dfrac{3L}{4} - x\right) + \dfrac{3}{8}qL(L-x) = 0 \end{cases} \implies \tag{5.59}$$

$$\begin{cases} V = \dfrac{3}{8}qL - q\dfrac{L}{2} = -\dfrac{1}{8}qL \\ M = -\dfrac{3}{8}qL^2 + \dfrac{1}{2}qLx + \dfrac{3}{8}qL^2 - \dfrac{3}{8}qLx = \dfrac{1}{8}qLx \end{cases} \tag{5.60}$$

which is the same result as Equation (5.58), but with more work.

c. Section located at x with $L/2 < x < L$: Now, consider only the part downstream from the section (see Figure 5.25):

$$\begin{cases} -V - q(L-x) + \dfrac{3}{8}qL = 0 \\ -M - q(L-x)\dfrac{L-x}{2} + \dfrac{3}{8}qL(L-x) = 0 \end{cases} \implies \tag{5.61}$$

$$\begin{cases} V = qx - \dfrac{5}{8}qL \\ M = -\dfrac{1}{2}qx^2 + \dfrac{5}{8}qLx - \dfrac{1}{8}qL^2 \end{cases} \tag{5.62}$$

2. Method of Integration: The starting point is now a set of differential equations. Given the loading, the only non-zero component of the general resultant $\mathbf{R}$ is V_Y and the only non-zero component of the resultant moment $\mathbf{M}$ is M_Z. To lighten the notation, these will be denoted V and M, respectively. Then, Equation (5.56) reduces to two equations:

$$\begin{cases} \dfrac{dV}{dx} + p_Y = 0 \\ \dfrac{dM}{dx} + V = 0 \end{cases} \quad \text{with} \quad p_Y = \begin{cases} 0 & \text{for} \quad 0 < x < \dfrac{L}{2} \\ -q & \text{for} \quad \dfrac{L}{2} < x < L \end{cases} \tag{5.63}$$

a. From global equilibrium, the reactions Y_O and Y_A at the supports are as given by Equation (5.57).

b. Case with $0 < x < L/2$: Integrating the first equation in (5.63) gives

$$\frac{dV}{dx} = 0 \quad \implies \quad V = V(0) \tag{5.64}$$

and integrating the second gives

$$\frac{dM}{dx} = -V(0) \quad \implies \quad M = -V(0)x + M(0) \tag{5.65}$$

where $V(0)$ and $M(0)$ are constants of integration, to be determined from boundary conditions as follows. Consider the free-body diagram of a slice comprised between the section at the left support (where $x = 0$) and an infinitely close cut, Figure 5.26. From this, one gets $V(0) = -Y_O = -\frac{1}{8}qL$ and $M(0) = 0$. Inserting these values in the integrated equations above, one recovers the expressions for V and M in Equation (5.58).

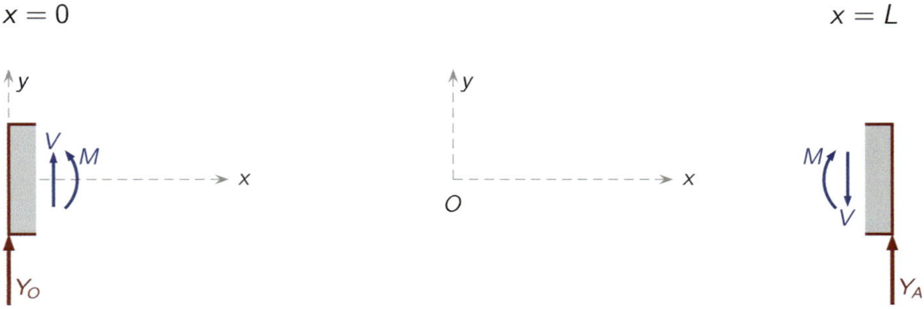

Figure 5.26 Boundary conditions on V and M at the two supports in Example 5.10.

c. Case with $L/2 < x < L$: Integrating the first equation in (5.63) now gives

$$\frac{\mathrm{d}V}{\mathrm{d}x} = q \qquad \Longrightarrow \qquad V = qx + A \tag{5.66}$$

and integrating the second gives

$$\frac{\mathrm{d}M}{\mathrm{d}x} = -qx - A \qquad \Longrightarrow \qquad M = -q\frac{x^2}{2} - Ax + B \tag{5.67}$$

where A and B are constants of integration, to be determined from boundary conditions at $x = L$. Consider the free-body diagram of a slice comprised between the section at the right support (where $x = L$) and an infinitely close cut, Figure 5.26. Because the slice is infinitely thin, the net distributed force is vanishingly small and does not contribute. From this, one gets $V(L) = Y_A = \frac{3}{8}qL$ and $M(L) = 0$, which can be further expanded, using Equations (5.66) and (5.67), as follows:

$$\left\{ \begin{array}{l} qL + A = \frac{3}{8}qL \quad \Longrightarrow \quad A = -\frac{5}{8}qL \\ -\frac{1}{2}qL^2 - AL + B = 0 \end{array} \right. \quad \Longrightarrow \quad B = -\frac{1}{8}qL^2 \tag{5.68}$$

By inserting these values in Equations (5.66) and (5.67), the expressions for V and M in Equation (5.62) are recovered.

3. The shear and bending moment diagrams are plotted in Figure 5.24. ♡

Remark 5.24 *In most structural mechanics problems, the distributed couple is nil* ($\mathbf{m} = \mathbf{0}$)*. Under such circumstances, the last two moment equilibrium equations in (5.56) yield*

$$V_Z = \frac{\mathrm{d}M_Y}{\mathrm{d}x} \qquad \textit{and} \qquad V_Y = -\frac{\mathrm{d}M_Z}{\mathrm{d}x}$$

and the last of Equation (5.46) *becomes:*

$$V_Y = -\frac{\mathrm{d}M_Z}{\mathrm{d}s}.$$

As first noted in Example 5.7, this property is, in fact, general for straight beams. ∗

Remark 5.25 *On a section where is applied a point load (force or couple), certain screwer components undergo a jump (a discontinuity of the first kind) and are not differentiable. This was the case for the shear force on supported sections in Figure 5.14 of Example 5.7.* ∗

Remark 5.26 *A direct method of evaluation of the screwer was given in Section 5.4.2 and illustrated in Examples 5.7 and 5.8. Integrating the differential equations of equilibriumconstitutes an alternative method.*

These equations are integrable on any portion of the beam where no point loads (forces or couples) are applied. If such point loads are applied, then one integrates over the various intervals of integrability and "patches" the solutions together while taking discontinuities into account. ∗

5.6 INTERNAL RESULTANTS AND STRESS

The screwer constitutes a system of surface forces acting on internal section S; see Figure 5.12. Its definition is reminiscent of that of surface tractions (see Section 2.1 in Part I), the difference being that the screwer is a net torsor that acts on the entire cross-section, whereas tractions act pointwise in that section or any other surface. Therefore, there must be a direct relationship between internal resultants and cross-sectional tractions, and that relationship is one of integration.

Let $\mathrm{d}S$ be an area element within S centered about current point P, Figure 5.27. The elementary surface (or contact) force acting on it is precisely $\mathbf{T}\,\mathrm{d}S$ where $\mathbf{T}$ is the traction vector (see Equation (2.1) in Part I).

Defining $\vec{\sigma}$ and $\vec{\tau}$ as the normal and tangential components of $\mathbf{T}$ with respect to cross-section S, the following relationships between internal resultants and stresses (or tractions) hold invariably:

$$\begin{aligned} \mathbf{N} &= \iint_S \vec{\sigma}\,\mathrm{d}S; \quad \mathbf{V} = \iint_S \vec{\tau}\,\mathrm{d}S \\ \mathbf{M}^{\times} &= \iint_S \overrightarrow{GP} \times \vec{\tau}\mathrm{d}S; \quad \mathbf{M}^{b} = \iint_S \overrightarrow{GP} \times \vec{\sigma}\mathrm{d}S \end{aligned} \tag{5.69}$$

where the color code matches that of Figure 5.27. Here, $\mathbf{V}$ has components V_Y, V_Z in the plane of the cross-section. Likewise, $\mathbf{M}^b$ has components M_Y, M_Z in the same plane.

Traction components $\vec{\sigma}$ and $\vec{\tau}$, respectively, represent the normal stress (or traction) and shear stress (or traction) acting on the cross-section. The notation $\vec{\sigma}$ is used instead of $\boldsymbol{\sigma}$ to avoid confusion between the normal stress vector and the stress tensor. In fact, the beam being the peculiar solid described in Definition 5.1 the normal on any cross-section is identified with $\mathbf{t}$, the unit tangent to its median line and parallel to GX. Thus, one may write: $\mathbf{T} = \boldsymbol{\sigma}\mathbf{t}$ so that the above equations can be explicitly written in terms of stress components in the $\{G, XYZ\}$ frame. For example,

$$N = \iint\limits_S \sigma_{XX}\,\mathrm{d}S; \qquad V_Y = \iint\limits_S \sigma_{XY}\,\mathrm{d}S; \qquad V_Z = \iint\limits_S \sigma_{XZ}\,\mathrm{d}S \tag{5.70}$$

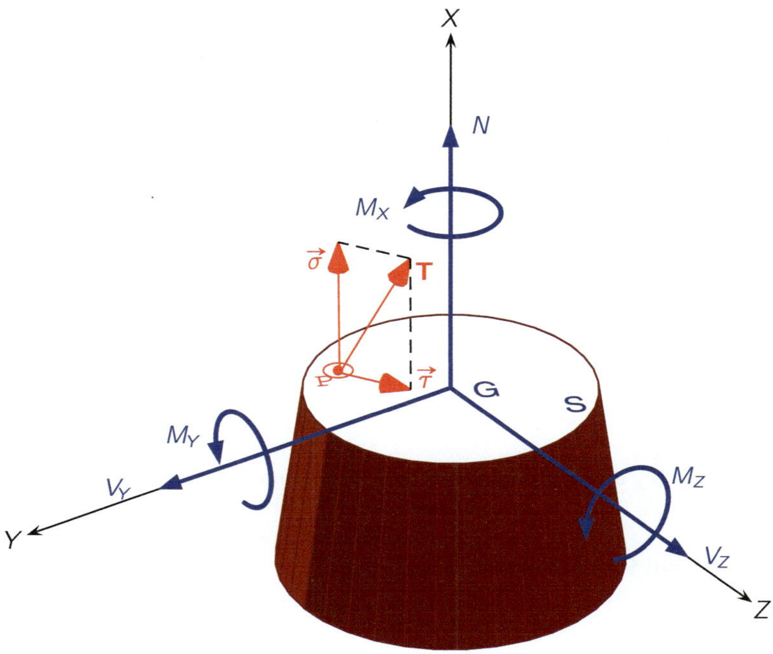

Figure 5.27 Connection between tractions and the screwer.

Similar relations may be obtained for M_X, M_Y and M_Z; see Problem 5.12.

The internal resultants acting on a given cross-section S can be viewed as the superposition of four simple types of loading: the normal (or axial) force **N**, the shear force **V**, the longitudinal moment $\mathbf{M}^X$, and the bending moment $\mathbf{M}^b$. The analysis of stresses, strains and displacements due to these loadings is the subject of the next four chapters. Taken by increasing order of complexity, the normal force **N** is treated in Chapter 6, bending $\mathbf{M}^b$ in Chapter 7, torsion $\mathbf{M}^X$ in Chapter 8 and transverse shear **V** in Chapter 9. The latter constitutes the general theory of bending.

PROBLEMS

Problem 5.1 (Beam or Not Beam) *For each of the solid objects shown in Figure 5.2, state if it meets Definition 5.1 of a beam. Justify your answer.* ⋆

Problem 5.2 (Unit Tangent to a Space Curve) *Show that the vector* **t** *defined by Equation* (5.2) *has indeed the properties enunciated in that proposition, namely (i) it is tangent to the median line in G, (ii) it is a unit vector, and (iii) it is oriented in the direction of increasing arc lengths.* ⋆

Problem 5.3 (Arc Length Equation) *Starting from Equation* (5.2) *show the arc length formula, Equation* (5.3). ⋆

Problem 5.4 (On Frenet's Mobile Frame)

1. *With point G lying on a space curve, the first derivative of the position vector* $\overrightarrow{OG}$ *with respect to the arc length is a unit vector, but its second derivative is not; why?*
2. *Show that* $\dfrac{d\mathbf{t}}{ds}$ *is normal to* **t**.
3. *Show that the curvature and torsion can be expressed using:*

$$\frac{1}{R} = \left\| \frac{d^2\mathbf{r}}{ds^2} \right\|$$

$$\frac{1}{T} = R^2 \left(\frac{d\mathbf{r}}{ds} \times \frac{d^2\mathbf{r}}{ds^2} \right) \cdot \frac{d^3\mathbf{r}}{ds^3}$$

where $\mathbf{r} \equiv \overrightarrow{OG}$ *is the position vector.* ⋆

Problem 5.5 (Geometry of a Coil Spring) *Prove Equations* (5.8) *and* (5.9) *of Example 5.1.* ⋆

Problem 5.6 (Centroid Calculations)

1. *Without any calculation and using Example 5.3 find the centroid of a semi-circular disk.*
2. *Find the centroid of a semi-elliptic thin section.* ⋆

Problem 5.7 (Parallel Axis Theorem) *Prove Huygens' formulae, Equation (5.21).* ⋆

Problem 5.8 (Principal Second Moments of Area) *Derive the relations giving the central principal second moments of area shown in Figure 5.9.* ⋆

Problem 5.9 (Principal Axes of Inertia) *Work out derivation details to establish Equation* (5.29). ⋆

Problem 5.10 (On the Concept of Screwer) *In Figure 5.12, change the orientation of the median line to be from* G_1 *to* G_0. *Would that affect the definition of the screwer (or internal resultants)?* ⋆

Problem 5.11 (Beam Equilibrium Equations) *Work out derivation details to establish the six equations of equilibrium (5.44)–(5.45).* ⋆

Problem 5.12 (Relation between Screwer and Stress Components) *The general relationships established in Section 5.6, Equation (5.35), can be specialized to express the internal resultants in terms of specific stress components. An example was given for N,* V_Y, *and* V_Z. *Obtain similar relations for moments* M_X, M_Y, *and* M_Z. ⋆

Problem 5.13 (Second Moments of Area of Basic Sections) *(i) Calculate the second moments of area* I_Y, I_Z, I_X *for the sections studied under Example 5.4 in terms of the dimensions shown in Figure 5.7. (ii) For case (c), consider the limit of a thin-walled tube with* $t \ll R$ *where* $t = R_2 - R_1$ *and* $R = (R_1 + R_2)/2$. ⋆

Problem 5.14 (Geometric Properties of Common Cross-Sections) *(i) Determine the position of the centroid and the second moments of area for the sections shown in Figure 5.28. (ii) For the I-section in (a), consider the thin wall limit when $t \ll h$ and $t' \ll h'$.* ⋆

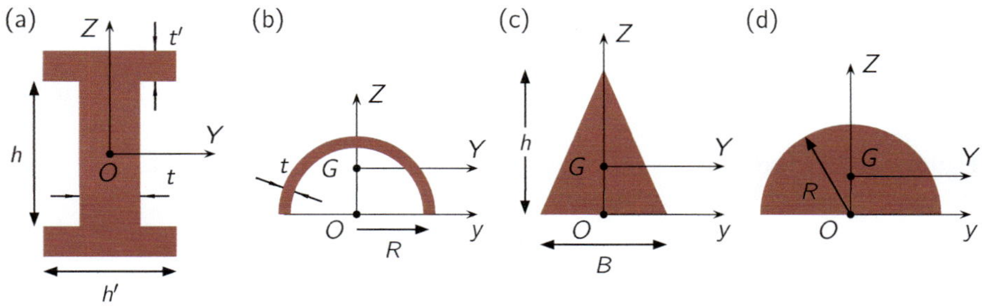

Figure 5.28 Common cross-sections of beams.

Problem 5.15 (Geometric Properties of D Section) *Calculate the cross-sectional area, the position of the centroid, then the second moments of area for a thin-walled D section having a semi-circular part of radius R and thickness t and a straight part of length $2R$ and thickness t', Figure 5.29.* ⋆

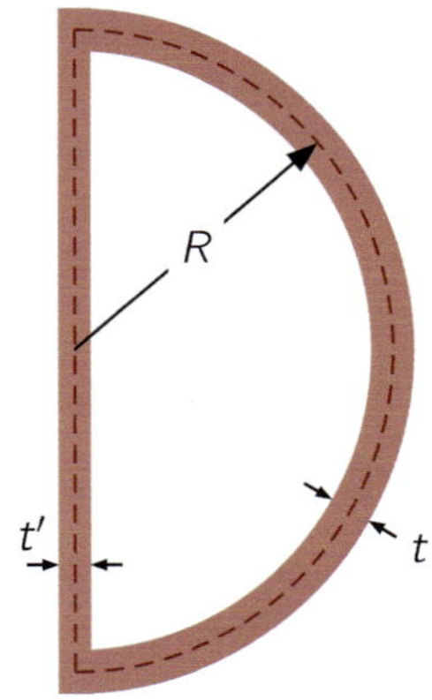

Figure 5.29 Thin-walled D section.

Problem 5.16 (Arch with Constant Reduced Inertia) *A parabolic arch having span $2a$ and rise r is shown in Figure 5.30. The reduced second moment of area is defined as $I^* = I \cos\theta$ where θ is the angle between the tangent to the arch at the current point and the horizontal. The arch is designed for stability. To this end, it is*

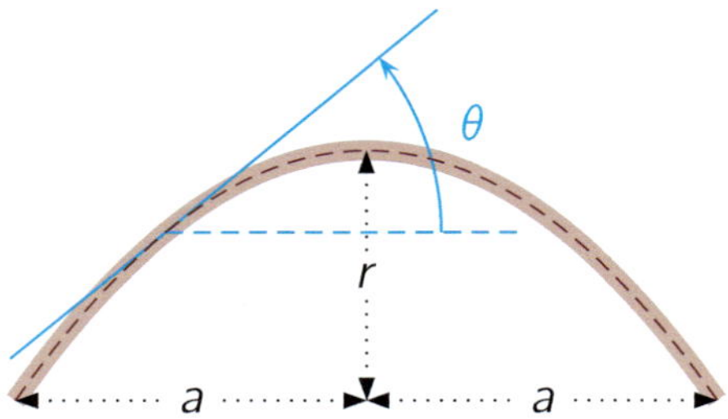

Figure 5.30 Parabolic arch.

desired to keep I^* constant along the arch's span. (i) For a rectangular cross-section of fixed width w and varying height h, find the span-wise law of evolution of the height h so that the reduced inertia does remain constant. (ii) Repeat the question for a thin-walled rectangular section of uniform thickness t. (iii) Draw the three-dimensional shape of the parabolic arch with the variable section in each case. ⋆

Problem 5.17 (Isostatic Straight Beam) *A straight beam pinned at O, simply supported at A is loaded as shown in Figure 5.31 with q a given force per unit length. Draw the shear and bending moment diagrams.* ⋆

Problem 5.18 (Shear and Bending Moment Diagrams) *A beam OAB is subjected to a distributed force of intensity q over the left half of its span and a point couple* **C** *having magnitude C along the z-axis at its right end, Figure 5.32. (i) Determine the reactions at the supports. (ii) Give in each bay the shear force $V = V_Y$ and the bending moment $M = M_Z$ as a function of the spanwise position x. (iii) Using $\ell = 1$ m, $q = 600$ N/m and $C = 400$ N m, draw the shear and bending moment diagrams.* ⋆

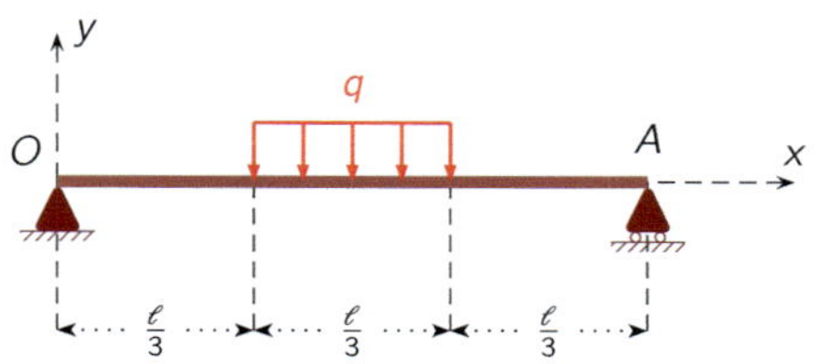

Figure 5.31 Simply supported beam under distributed load over part of the span.

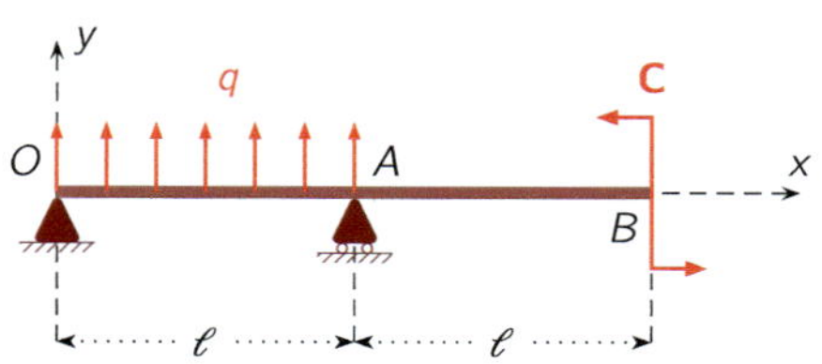

Figure 5.32 Beam under distributed load and a point couple at one end.

Problem 5.19 (Shear and Bending Moment Diagrams) *Consider a straight beam of length 3ℓ sitting on two supports at both ends, and loaded as shown in Figure 5.33; the beam's weight is neglected.*

(i) Calculate the transverse shear $V_Y = V(x)$ and the bending moment $M_Z = M(x)$ on the current cross-section at abscissa x. (ii) Draw the corresponding diagrams. Take $\Gamma = F\ell > 0$ where Γ is the algebraic (signed) length of vector $\mathbf{\Gamma}$ on Oz. ⋆

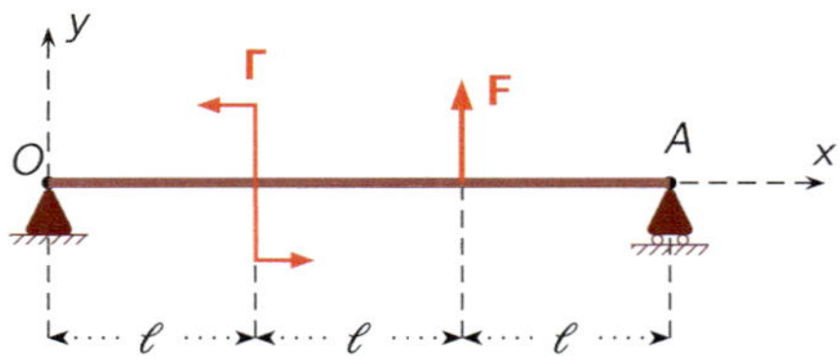

Figure 5.33 Doubly supported beam subjected to a point force and a point couple.

Problem 5.20 (Shear and Bending Moment Diagrams) *A structure is made up of two beams OA and AB articulated together by a pin-joint at A. The structure (OAB) is straight and admits xOy as median plane, Figure 5.34. The structure is clamped at O and simply supported at B (horizontal support plane). A point load* **F** *with components X and Y is applied at section C located at a distance c from the origin. In addition, a uniform load q pet unit length of the beam is applied to part AB, as depicted. Calculate the components of the screwer on all cross-sections of the structure and draw the corresponding diagrams.* ⋆

Problem 5.21 (Curved Beam: Screwer, Diagrams) *A beam with median plane xOy, of constant cross-section S, has a semi-circular median line of radius R, Figure 5.35. The beam lies in a vertical plane (Oy being the ascending vertical) and is simply supported, as shown. It is subjected to its own weight with p denoting the weight per unit length (p is depicted on the current section only). (i) Calculate on the current cross-section, located at angle θ, the normal force N, the transverse shear $V = V_Y$ and the bending moment $M = M_Z$. (ii) Draw the corresponding diagrams.* ⋆

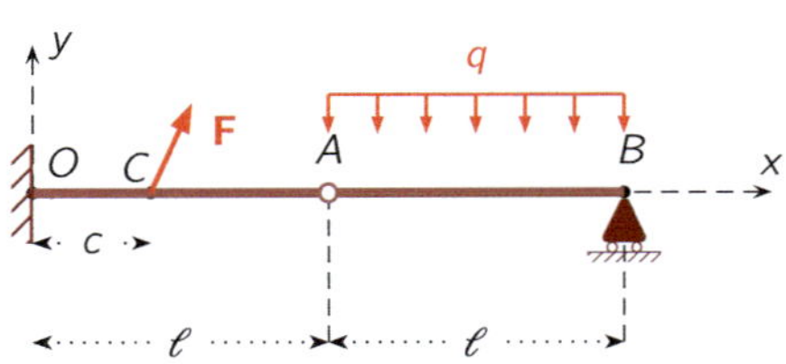

Figure 5.34 Structure subjected to a point load and a distributed load.

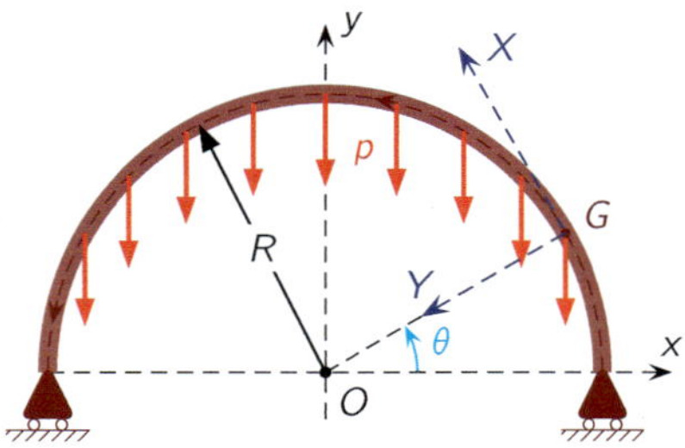

Figure 5.35 Doubly supported semi-circular beam under its own weight.

Problem 5.22 (Curved Beam: Screwer, Diagrams) *A simply supported semi-circular beam with median plane xOy, of constant cross-section S, is under the action of a couple Γ at mid-span, as depicted in Figure 5.36. (i) Determine the reactions at the supports. (ii) Calculate the non-zero components of the screwer acting on the current cross-section. (iii) Draw the diagrams of the normal force, shear force, and bending moment.* ⋆

Problem 5.23 (Semi-Circular Beam: Screwer, Diagrams) *A semi-circular beam of constant cross-section is pinned at one end and simply supported at the other, as shown in Figure 5.37. The beam is under the action of a horizontal point force F at mid-span. (i) Determine the reactions at the supports. (ii) Calculate the non-zero components of the screwer acting on the current cross-section. (iii) Draw the corresponding diagrams.* ⋆

Problem 5.24 (Planar Gantry: Screwer and Diagrams) *Consider the gantry in Figure 5.38 consisting of three beams AB, BC, and CD, welded together. End section A is a pivot, while the other end D is simply supported on a horizontal plane.*

This structure is under the action of a uniformly distributed load of intensity q on the ascending part AB. To describe the structure, point A is taken as the origin of arclengths.

Focus is placed on the plane problem. (i) Determine the reactions. (ii) Determine

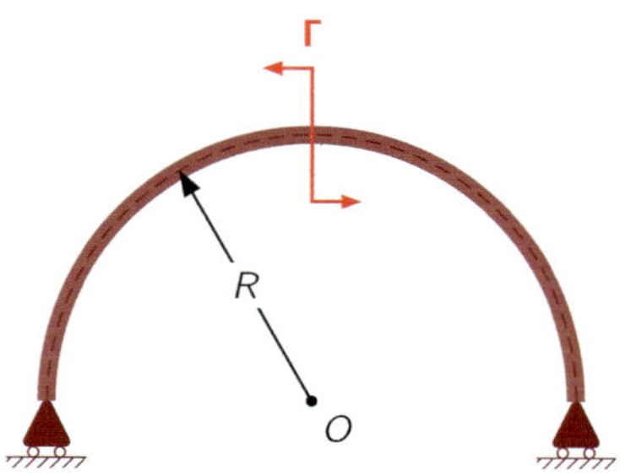

Figure 5.36 Doubly supported semi-circular beam under a point couple.

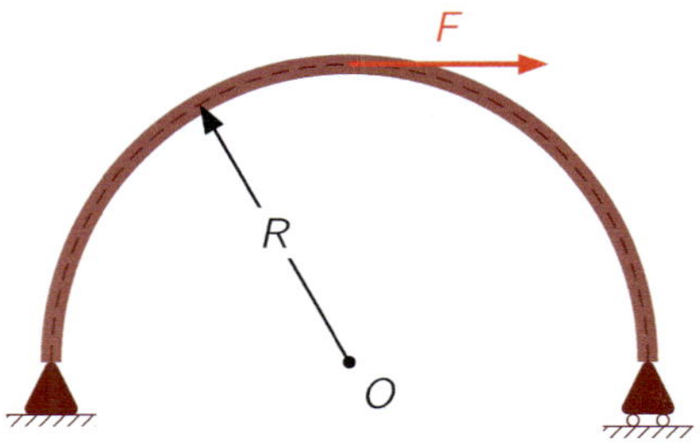

Figure 5.37 Curved beam under a horizontal force at mid-span.

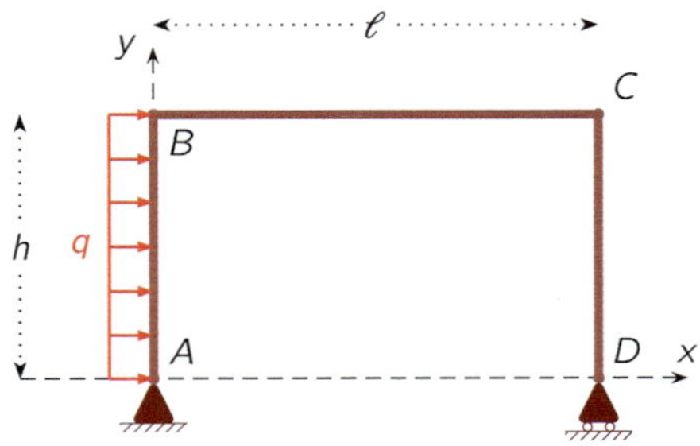

Figure 5.38 Gantry subjected to a uniform distributed load.

on each cross-section the components N, V_Y, and M_Z of the screwer. (iii) Draw the corresponding diagrams. ⋆

Problem 5.25 (Screwer in a Helical Spring) *Consider a helical spring with radius*

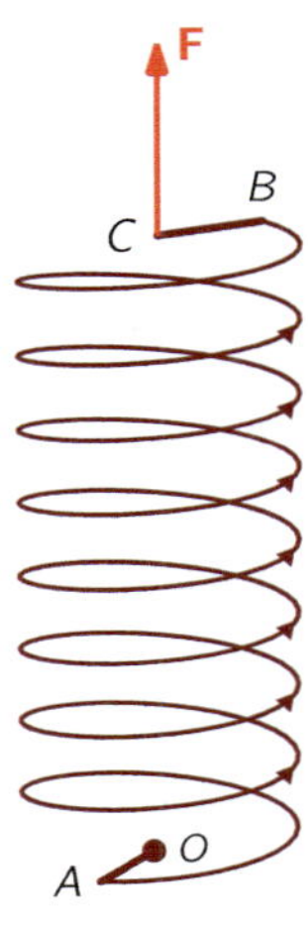

Figure 5.39 Helical spring under tension.

R and slope α connected through a ball joint at O and subjected to a force F at C, Figure 5.39. (i) Calculate on each cross-section the components of the screwer. (ii) Verify that they satisfy the differential equations of equilibrium. ⋆

Problem 5.26 (Curved versus Straight Beam) *Consider a semicircular beam under a single downward force at mid-span, Figure 5.40. Compare the bending moments and shear forces in the curved beam to those acting under three-point bending of a straight beam of the same span, as shown in the figure.* ⋆

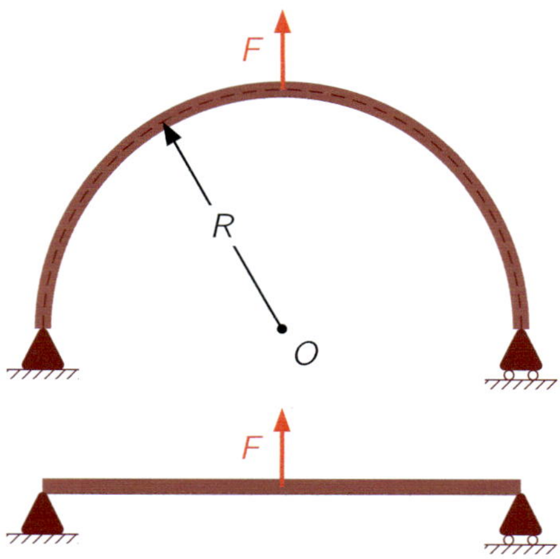

Figure 5.40 Three-point bending in curved versus straight beams.

6 Normal Force

6.1 BARS

Consider a prismatic beam of length L and limited by cross-sections S_0 and S_1, Figure 6.1.

Assume that uniformly distributed loads are applied on S_0 and S_1 with net axial forces $-\mathbf{F}$ and $\mathbf{F}$, respectively. Let the x-axis be along the centroidal axis and the y- and z-axes associated with S_0 and passing through its centroid G_0. Such a beam, so loaded, is called a *bar*.

DEFINITION 6.1 A bar is a prismatic beam loaded by net axial forces at either end. ♣

On current cross-section S, the screwer reduces to the normal force $N = F$ (positive in tension, negative in compression). According to the simple tension experiment and the corresponding Saint-Venant problem described in Example 4.1, the following results hold for an elastic material having Young's modulus E and Poisson's ratio ν:

- In terms of stress:

$$[\boldsymbol{\sigma}] = \begin{bmatrix} \sigma_x & 0 & 0 \\ 0 & 0 & 0 \\ 0 & 0 & 0 \end{bmatrix} \quad \text{with} \quad \sigma_x = \frac{F}{S} \tag{6.1}$$

- In terms of strains: using Hooke's law in component form, Equation (3.13f),

$$[\boldsymbol{\varepsilon}] = \begin{bmatrix} \varepsilon_x & 0 & 0 \\ 0 & \varepsilon_y & 0 \\ 0 & 0 & \varepsilon_z \end{bmatrix} \quad \text{with} \quad \begin{cases} \varepsilon_x = \dfrac{\sigma_x}{S} = \dfrac{F}{ES} \\ \varepsilon_y = \varepsilon_z = -\nu\varepsilon_x \end{cases} \tag{6.2}$$

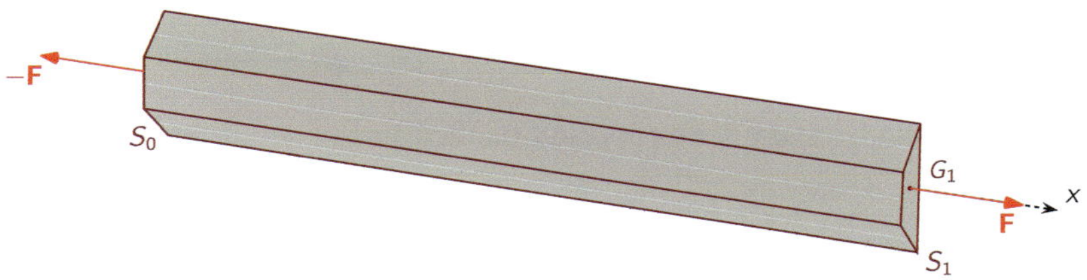

Figure 6.1 A prismatic beam.

DOI: 10.1201/9781003537229-8

- In terms of displacements and rotations, with respect to a frame tied to S_0:

$$\{\mathbf{u}\} = \begin{cases} u = \dfrac{F}{ES}x \\ v = -\dfrac{\nu F}{ES}y \\ w = -\dfrac{\nu F}{ES}z \end{cases} \quad \text{and} \quad \boldsymbol{\omega} = \mathbf{0} \tag{6.3}$$

The potential elastic energy stored in the slice comprised between sections S and S', located at x and $x + \mathrm{d}x$, is $\mathrm{d}W = \frac{1}{2}N\,\mathrm{d}u = \frac{1}{2}\frac{N^2}{ES}\,\mathrm{d}x$ so that

$$\frac{\mathrm{d}W}{\mathrm{d}x} = \frac{1}{2}\frac{N^2}{ES} \quad \text{and} \quad W = \frac{1}{2}\frac{N^2L}{ES} \tag{6.4}$$

The ratio between the force F and the elongation $\varepsilon_x L$ is the overall stiffness, $\frac{ES}{L}$, of the bar; its inverse $\frac{L}{ES}$ is the flexibility. Also, ES is the *lineal stiffness*, that is the stiffness per unit length of the bar.

Remark 6.1 *In general, forces* **F** *and* −**F** *are applied to the bar by means of grip fixtures, as illustrated in Figure 6.2. According to the Saint-Venant principle (see Theorem 4.1 in Section 4.1), the previous results are valid, except in the neighborhood of the grip attachments where the stress field is perturbed.* ∗

6.2 TRUSSES

DEFINITION 6.2 A truss is any assemblage of bars connected at their ends by ball joints that form the nodes. In addition, it is assumed that:

1. all connections to surroundings are also made of ball joints;
2. external forces are exclusively applied to nodes;
3. the weight of each bar is negligible in comparison with external forces. ♣

Under such circumstances, each bar undergoes either pure tension or pure compression. Indeed, consider the equilibrium of bar AB. Let $\mathbf{F}^A$ and $\mathbf{F}^B$ be the net external forces applied at A and B, then:

$$\mathbf{F}^A + \mathbf{F}^B = \mathbf{0}, \qquad \overrightarrow{AB} \times \mathbf{F}^B = \mathbf{0}$$

which implies that $\mathbf{F}^A$ and $\mathbf{F}^B$ are opposites on their line of action AB.

Figure 6.2 A tie-rod.

Remark 6.2 *Numerous structures are made by assembling prismatic beams, the connections being realized by welding, bolting, or riveting. This is the case, for instance, of cranes, aircraft engine mounts, metallic bridges and viaducts, powerline pylons, antenna masts, etc. Such structures do not strictly meet the above criteria for a truss since the node connections are not ball joints and can thus generate couples. However, calculations show that the effects of the normal force (strains, stresses, and displacements) prevail over those of other loads. Such structures can therefore be analyzed as trusses to a very good degree of approximation.* $*$

Remark 6.3 *A tensegrity structure is a type of truss with some members in compression (bars) and others prestressed in tension (cables).* $*$

Remark 6.4 *At the design stage of a structure, one should attempt the truss idealization as a first option. This is so because it is light, simple, and economical.* $*$

6.3 PLANAR TRUSSES

A truss is said to be planar if the median lines of all constituent bars and the applied forces lie within the same plane, say xOy.

6.3.1 ISOSTATIC AND HYPERSTATIC CHARACTER

Consider a planar truss, freed from all connections with its surroundings. Let n be the number of nodes and b the number of bars.

If the truss is internally statically determinate, or *isostatic*, it constitutes a solid. Immobilization of the bars relative to each other is assured and removal of any bar would make the truss *hypostatic*. The positioning of the truss in its plane, defined by three independent parameters, may also be described by positioning its n nodes, that is by $2n$ parameters related by b independent relations. Thus, one may write:

$$2n - b = 3 \quad \text{or} \quad b = 2n - 3$$

If now the truss is internally statically indeterminate, or *hyperstatic* of degree p, it contains p redundant bars, and one has

$$b = 2n - 3 + p \quad \Longrightarrow \quad p = b - 2n + 3 \tag{6.5}$$

Regarding the system of connections with the exterior, it is isostatic if it constrains three nodal degrees of freedom. Such constraints would introduce three scalar reactions for three equilibrium equations. The truss is externally hyperstatic of degree q if the external connections constrain $(3+q)$ nodal degrees of freedom, causing $(3+q)$ unknown reactions for still only three equilibrium equations.

6.3.2 NODAL EQUILIBRIUM

Let A be a node of the truss related to other nodes A^i by bars AA^i, Figure 6.3. Let $\mathbf{F}$ be the resultant of external forces acting on node A with components F_x and F_y. This force includes applied forces as well as any possible reactions. Also, let φ^i be the polar angle of $\overrightarrow{AA^i}$ with respect to the x axis, and N^i the (algebraic) normal force in bar AA^i. If $\mathbf{N}^i$ is the force applied by bar AA^i on node A, one verifies that N^i is its algebraic measure on a unit vector along $\overrightarrow{AA^i}$.

Force equilibrium at node A writes: $\mathbf{F}+\sum \mathbf{N}^i = \mathbf{0}$, the summation being extended over all bars sharing node A. Projecting onto the x and y axes yields

$$\begin{cases} F_x+\sum N^i \cos\varphi^i = 0 & \text{along } Ox \\ F_y+\sum N^i \sin\varphi^i = 0 & \text{along } Oy \end{cases} \tag{6.6}$$

Thus, by expressing the equilibrium of all n nodes, one obtains a linear system of $2n$ equations.

6.3.3 NODAL DISPLACEMENTS

Consider now bar A^iA^j, characterized by length L^{ij}, cross-sectional area S^{ij}, Young's modulus E^{ij}, and let φ^{ij} be the polar angle of $\overrightarrow{A^iA^j}$ with respect to Ox.

In the initial state, the Pythagorean theorem gives:

$$(x^j-x^i)^2+(y^j-y^i)^2 = L^{ij2}$$

where x^i, y^i and x^j, y^j denote the coordinates of nodes A^i and A^j, respectively. Under the action of loading, these coordinates undergo variations u^i, v^i and u^j, v^j, small in

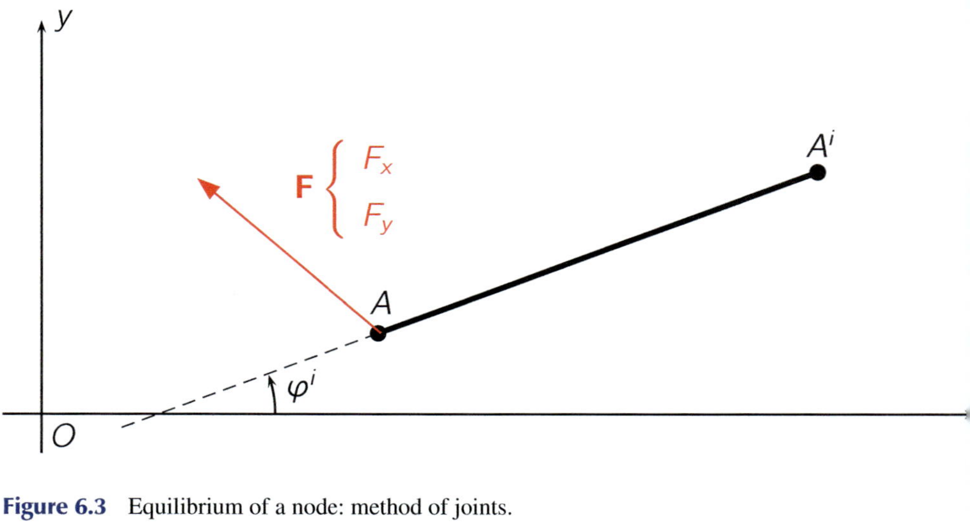

Figure 6.3 Equilibrium of a node: method of joints.

comparison with L^{ij}, such that by differentiation of the above relation one gets

$$(x^j - x^i)(u^j - u^i) + (y^j - y^i)(v^j - v^i) = L^{ij}\delta L^{ij} \tag{6.7}$$

From the component form of Hooke's law, Equation (3.13a), or directly from Equation $(6.2)_2$, given that ε_x is nothing but the relative elongation:

$$\frac{\delta L^{ij}}{L^{ij}} = \left(\frac{N}{ES}\right)^{ij} \quad \Rightarrow \quad \delta L^{ij} = \left(\frac{NL}{ES}\right)^{ij}$$

Thus, dividing both sides of Equation (6.7) by L^{ij}, one obtains, account being taken of Hooke's law:

$$\boxed{\cos\varphi^{ij}\,(u^j - u^i) + \sin\varphi^{ij}\,(v^j - v^i) = \left(\frac{NL}{ES}\right)^{ij}} \tag{6.8}$$

One such equation may be written for each bar of the truss joining nodes A^i and A^j.

6.3.4 METHOD OF RESOLUTION

Assume the following is given:

- the truss geometry, in particular node coordinates, bar lengths and cross-sectional areas;
- Young's moduli of the constituent materials; and
- the links to the exterior.

Denoting by p and q the degrees of internal and external hyperstatic character, respectively, the governing equations consist of:

- the $2n$ nodal equilibrium equations;
- the b nodal displacement equations; and
- the $(3+q)$ conditions imposed on the nodal displacements by the external connections/links.

Resolution of the linear system of $(2n+b+3+q)$ equations permits to calculate, in all cases, the $(2n+b+3+q)$ unknowns of the problem, namely:

- the $2n$ nodal displacements;
- the b normal forces in the bars; and
- the $(3+q)$ scalar reactions due to external links.

Once these unknowns are determined, one infers the stresses, strains, and displacements at any point of the truss.

Example 6.1: Hyperstatic Planar Truss

Consider the planar truss of Figure 6.4. The bars have length L or $L\sqrt{2}$, cross-sectional area S and Young's modulus E. The global axes, external links, and loading are as specified in Figure 6.4. Also, nodes and bars are numbered as shown.

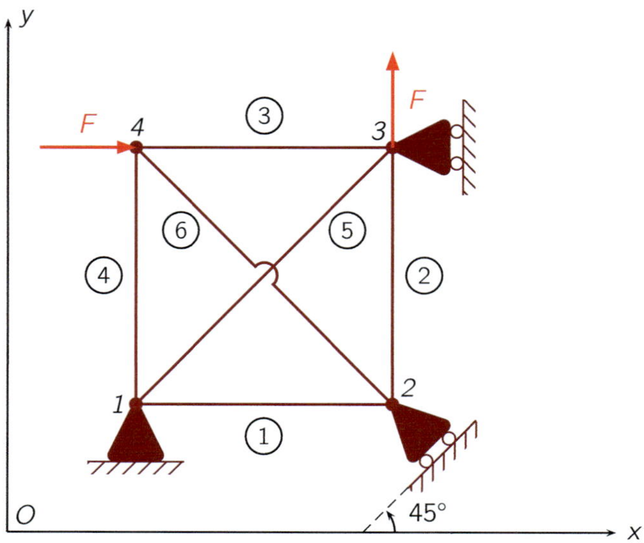

Figure 6.4 A hyperstatic planar truss ($p = 1$, $q = 1$).

The truss is externally hyperstatic of degree 1 ($q = 1$). One degree of freedom is blocked in excess of what is needed for a planar structure. Said otherwise, there is one redundant reaction resulting from blocking that additional degree of freedom.

Likewise, the truss is internally hyperstatic of degree 1 ($p = 1$). To see this, it suffices to apply Equation (6.5) using $b = 6$ and $n = 4$.

Where applicable, let X_i and Y_i denote the reactions at node i along the x and y directions, respectively. The nodal equilibrium equations read:

$$\text{Node 1}\left\{\begin{array}{l} X_1 + N_1 + N_5\dfrac{\sqrt{2}}{2} = 0 \\ Y_1 + N_4 + N_5\dfrac{\sqrt{2}}{2} = 0 \end{array}\right. \tag{6.9}$$

$$\text{Node 2}\left\{\begin{array}{l} X_2 - N_1 - N_6\dfrac{\sqrt{2}}{2} = 0 \\ Y_2 + N_2 + N_6\dfrac{\sqrt{2}}{2} = 0 \end{array}\right. \tag{6.10}$$

$$\text{Node 3}\left\{\begin{array}{l} X_3 - N_3 - N_5\dfrac{\sqrt{2}}{2} = 0 \\ F - N_2 - N_5\dfrac{\sqrt{2}}{2} = 0 \end{array}\right. \tag{6.11}$$

$$\text{Node 4}\left\{\begin{array}{l} F + N_3 + N_6\dfrac{\sqrt{2}}{2} = 0 \\ -N_4 - N_6\dfrac{\sqrt{2}}{2} = 0 \end{array}\right. \tag{6.12}$$

These equations directly follow from applying Equation (6.6) or using the method of joints by drawing separate free-body diagrams for all nodes. The latter is better for the beginner.

The nodal displacement equations follow from applying Equation (6.8) and write:

$$\text{Bar 1:} \quad u_2 - u_1 = \frac{N_1 L}{ES} \tag{6.13}$$

$$\text{Bar 2:} \quad v_3 - v_2 = \frac{N_2 L}{ES} \tag{6.14}$$

$$\text{Bar 3:} \quad u_3 - u_4 = \frac{N_3 L}{ES} \tag{6.15}$$

$$\text{Bar 4:} \quad v_4 - v_1 = \frac{N_4 L}{ES} \tag{6.16}$$

$$\text{Bar 5:} \quad \frac{\sqrt{2}}{2}(u_3 - u_1) + \frac{\sqrt{2}}{2}(v_3 - v_1) = \frac{N_5 L\sqrt{2}}{ES} \tag{6.17}$$

$$\text{Bar 6:} \quad -\frac{\sqrt{2}}{2}(u_4 - u_2) + \frac{\sqrt{2}}{2}(v_4 - v_2) = \frac{N_6 L\sqrt{2}}{ES} \tag{6.18}$$

The boundary conditions constitute the equations imposed by external links:

$$\text{Node 1:} \quad u_1 = 0 \qquad v_1 = 0 \tag{6.19}$$

$$\text{Node 2:} \quad u_2 = v_2 \tag{6.20}$$

$$\text{Node 3:} \quad u_3 = 0 \tag{6.21}$$

Note that Equation (6.20) follows from writing that the displacement perpendicular to the support is nil.

Solving the above system involves three generic steps. First, use nodal equilibrium to express all internal forces in terms of two of them, say N_5 and N_6. The problem being hyperstatic of overall degree $p+q=2$, one knows beforehand that this is possible. Thus, the equations that do not involve external reactions give

$$\left\{\begin{array}{lll} \text{Equation } (6.11)_2 & \Longrightarrow & N_2 = F - N_5\dfrac{\sqrt{2}}{2} \\ \text{Equation } (6.12)_1 & \Longrightarrow & N_3 = -F - N_6\dfrac{\sqrt{2}}{2} \\ \text{Equation } (6.12)_2 & \Longrightarrow & N_4 = -N_6\dfrac{\sqrt{2}}{2} \end{array}\right. \tag{6.22}$$

The equations that involve reactions Y_1, Y_2 and X_3 give

$$\left\{\begin{array}{lll} \text{Equation } (6.9)_2 & \Longrightarrow & Y_1 = -N_4 - N_5\dfrac{\sqrt{2}}{2} = \dfrac{\sqrt{2}}{2}(N_6 - N_5) \\ \text{Equation } (6.10)_2 & \Longrightarrow & Y_2 = -N_2 - N_6\dfrac{\sqrt{2}}{2} = -F + \dfrac{\sqrt{2}}{2}(N_5 - N_6) \\ \text{Equation } (6.11)_1 & \Longrightarrow & X_3 = N_3 + N_5\dfrac{\sqrt{2}}{2} = -F + \dfrac{\sqrt{2}}{2}(N_5 - N_6) \end{array}\right. \tag{6.23}$$

where the results in Equation (6.22) have been accounted for. Note that $X_3 = Y_2$ by inspection. On the other hand, the reaction components on the inclined support are necessarily related by $X_2 = -Y_2$. The remaining reaction X_1 may be obtained by eliminating N_1 between Equation $(6.9)_1$ and Equation $(6.10)_1$:

$$\text{Equation } (6.9)_1 + \text{Equation } (6.10)_1 \quad \Longrightarrow \quad X_1 + X_2 = \frac{\sqrt{2}}{2}(N_6 - N_5) \tag{6.24}$$

On account of $X_2 = -Y_2$ and Equation $(6.23)_2$, obtain $X_1 = -F$, which is a remarkably simple result. Inserting this result in the equilibrium of node 1 delivers the last internal force N_1 in terms of the hyperstatic unknowns:

$$\text{Equation } (6.9)_1 \quad \Longrightarrow \quad N_1 = -X_1 - \frac{\sqrt{2}}{2}N_5 = F - \frac{\sqrt{2}}{2}N_5 \tag{6.25}$$

This completes step 1.

The second step consists of simplifying the nodal displacement equations by (i) incorporating boundary conditions, and (ii) substituting for $N_1 \cdots N_4$ their expressions in terms of N_5, N_6 from step 1. This is putative; actual substitution may be delayed until the final step to avoid cluttered expressions. Thus, per bar:

$$\begin{aligned}
&\text{Equation } (6.13) \quad \Longrightarrow \quad u_2 = \frac{N_1 L}{ES} \\
&\text{Equation } (6.14) \quad \Longrightarrow \quad v_3 = u_2 + \frac{N_2 L}{ES} = (N_1 + N_2)\frac{L}{ES} \qquad (6.26) \\
&\text{Equation } (6.15) \quad \Longrightarrow \quad u_4 = -\frac{N_3 L}{ES} \\
&\text{Equation } (6.16) \quad \Longrightarrow \quad v_4 = \frac{N_4 L}{ES}
\end{aligned}$$

$$\text{Equation } (6.17) \quad \Longrightarrow \quad v_3 = 2\frac{N_5 L}{ES} \tag{6.27}$$

$$\text{Equation } (6.18) \quad \Longrightarrow \quad \frac{\sqrt{2}}{2}(u_2 - u_4) + \frac{\sqrt{2}}{2}(v_4 - u_2) = \frac{N_6 L\sqrt{2}}{ES} \tag{6.28}$$

The final step consists of solving. It becomes immediately clear that v_3 is given by two separate routes, Equations (6.26) and (6.27). Equating these two gives

$$N_1 + N_2 = 2N_5 \tag{6.29}$$

Also, Equation (6.28) involves displacements that are already available using the unnumbered expressions implied by Equations (6.13), (6.15) and (6.16) such that

$$\frac{\sqrt{2}}{2}(N_1 + N_3) + \frac{\sqrt{2}}{2}(N_4 - N_1) = \frac{N_6 L\sqrt{2}}{ES}$$

which readily simplifies into

$$N_3 + N_4 = 2N_6 \tag{6.30}$$

After inserting their expressions for $N_1 \cdots N_4$ from Equations (6.22) and (6.25) into the above two equations, obtain

$$\begin{cases}
F - N_5\dfrac{\sqrt{2}}{2} + F - \dfrac{\sqrt{2}}{2}N_5 = 2N_5 \quad \Longrightarrow \quad N_5 = \dfrac{2F}{2+\sqrt{2}} \\[2ex]
-F - N_6\dfrac{\sqrt{2}}{2} - \dfrac{\sqrt{2}}{2}N_6 = 2N_6 \quad \Longrightarrow \quad N_6 = -\dfrac{F}{2+\sqrt{2}}
\end{cases} \tag{6.31}$$

The first gives N_5, and the second N_6 gives in terms of the applied force F.

Having so determined the two hyperstatic unknowns, N_5 and N_6, the remaining internal forces are obtained using Equations (6.22) and (6.25), the (yet to be determined) reactions are obtained using Equation (6.23), and the nodal displacements using the results from step 2. Rearranging finally gives:

$$\text{Reactions:} \begin{cases} X_1 = -F \qquad Y_1 = -\dfrac{3}{2}(\sqrt{2}-1)F \\ X_2 = \dfrac{5-3\sqrt{2}}{2}F \qquad Y_2 = -X_2 \\ X_3 = -\dfrac{5-3\sqrt{2}}{2}F \end{cases}$$

$$\text{Normal forces:} \begin{cases} N_1 = (2-\sqrt{2})F \qquad N_2 = (2-\sqrt{2})F \\ N_3 = -\dfrac{3-\sqrt{2}}{2}F \qquad N_4 = \dfrac{\sqrt{2}-1}{2}F \\ N_5 = (2-\sqrt{2})F \qquad N_6 = -\dfrac{2-\sqrt{2}}{2}F \end{cases}$$

$$\text{Nodal displacements:} \begin{cases} u_1 = 0, \quad v_1 = 0 \\ u_2 = (2-\sqrt{2})\dfrac{FL}{ES}, \quad v_2 = (2-\sqrt{2})\dfrac{FL}{ES} \\ u_3 = 0, \quad v_3 = 2(2-\sqrt{2})\dfrac{FL}{ES} \\ u_4 = \dfrac{3-\sqrt{2}}{2}\dfrac{FL}{ES}, \quad v_4 = \dfrac{\sqrt{2}-1}{2}\dfrac{FL}{ES} \end{cases}$$

The deformed configuration is depicted, exaggerated, in Figure 6.5.

It is important to identify bars under compression for these are prone to buckling, as will be analyzed in Chapter 11. Here, bars 3 and 6 are under compression. Interestingly, bars 1, 2, and 5 undergo the same level of tension. ♡

Remark 6.5 *In the above example, one could have expressed global equilibrium where internal forces do not appear. There are three equilibrium equations for four unknown reactions, consistent with* $q = 1$. *

Remark 6.6 *If the truss is hyperstatic, there is no hope in getting all forces (internal and reactions) from equilibrium alone. In the above example, assume one had used global equilibrium to express three of the reactions in terms of a hyperstatic reaction. Then nodal equilibrium, Equations* (6.9) *to* (6.12), *would have involved the internal forces* $N_1 \cdots N_6$ *in addition to one unknown reaction. This means eight equations for seven unknowns. One might naively think that the eight equilibrium equations would suffice to determine the seven unknowns. However, the problem being hyperstatic, the system of equations is degenerate. If one were to form its matrix, the latter would be singular, hence not invertible.* *

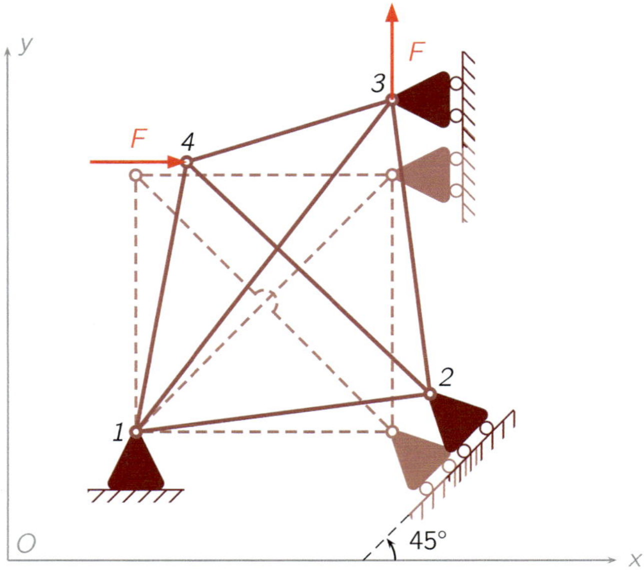

Figure 6.5 Truss of Figure 6.4 after deformation.

6.4 THREE-DIMENSIONAL TRUSSES

This section extends to three dimensions the results obtained above for planar trusses. Three-dimensional trusses are ubiquitous in space structures and satellite support systems, among others.

6.4.1 ISOSTATIC AND HYPERSTATIC CHARACTER

If the truss is internally hyperstatic of degree p ($p = 0$ for an isostatic truss), then the relationship $b = 3n - 6 + p$ holds between the number of bars b and the number of nodes n. Indeed, the positioning of any three-dimensional structure is defined by six independent parameters (the coordinates of any point, say the center of mass, and three angles). It may also be described by positioning its n nodes, that is by $3n$ parameters related by b independent relations.

If the truss is externally hyperstatic of degree q, then there are $(6+q)$ unknown scalar reactions for six global equilibrium equations. The external connections constrain $(6+q)$ nodal degrees of freedom.

6.4.2 NODAL EQUILIBRIUM

With $\{O, xyz\}$ an orthonormal frame associated with the truss, let:

- A be a node related to other nodes A^i by bars AA^i;

- **F** the resultant of external forces acting on node A with components F_x, F_y, and F_z, including applied forces as well as any possible reactions;
- α^i, β^i, and γ^i the direction cosines of $\overrightarrow{AA^i}$;
- N^i the (algebraic) normal force in bar AA^i.

Force equilibrium at node A writes

$$\begin{cases} F_x + \sum N^i \alpha^i = 0 & \text{along } Ox \\ F_y + \sum N^i \beta^i = 0 & \text{along } Oy \\ F_z + \sum N^i \gamma^i = 0 & \text{along } Oz, \end{cases} \tag{6.32}$$

the summation being extended over all bars sharing node A. Since there are n such nodes, one obtains a linear system of $3n$ equations.

6.4.3 NODAL DISPLACEMENTS

For a bar A^iA^j having length L^{ij}, cross-sectional area S^{ij}, and Young's modulus E^{ij}, let α^{ij}, β^{ij}, γ^{ij} be the direction cosines of vector $\overrightarrow{A^iA^j}$ and u^i, v^i, w^i the displacements of node A^i along Ox, Oy, Oz, respectively. The nodal displacement equations then read

$$\alpha^{ij}\left(u^j - u^i\right) + \beta^{ij}\left(v^j - v^i\right) + \gamma^{ij}\left(w^j - w^i\right) = \left(\frac{NL}{ES}\right)^{ij} \tag{6.33}$$

Proof *In the initial state, one has:*

$$(x^j - x^i)^2 + (y^j - y^i)^2 + (z^j - z^i)^2 = L^{ij2}$$

where x^i, y^i, z^i denote the coordinates of node A^i. Under the action of loading, these coordinates undergo variations u^i, v^i, w^i, small in comparison with L^{ij}. Differentiating the above relation gives:

$$(x^j - x^i)(u^j - u^i) + (y^j - y^i)(v^j - v^i) + (z^j - z^i)(w^j - w^i) = L^{ij}\delta L^{ij}$$

Dividing both sides by L^{ij}, one obtains Equation (6.33) *on account of Hooke's law: $\delta L^{ij} = \left(\frac{NL}{ES}\right)^{ij}$ and the definition of direction cosines.* ■

6.4.4 METHOD OF RESOLUTION

Given the truss geometry, Young's moduli of the constituent materials and the links to the exterior, the governing equations are:

- the $3n$ nodal equilibrium equations;
- the b nodal displacement equations; and
- the $(6+q)$ conditions on the nodal displacements due to external links.

Resolution of the linear system of $(3n+b+6+q)$ equations permits to calculate, in all cases, the $(3n+b+6+q)$ unknowns of the problem, namely:

- the $3n$ nodal displacements u^i, v^i, w^i;
- the b normal forces in the bars, N^i; and
- the $(6+q)$ scalar reactions due to external links.

Details of the resolution method depend on p and q, the degrees of internal and external hyperstatic character, respectively,

6.5 NORMAL FORCE IN ARBITRARY BEAMS

6.5.1 BASIC EQUATIONS

In the previous sections, bars and assemblages thereof were analyzed. Here, the objective is to extend Equations (6.1) to (6.4) to an arbitrary beam that is subjected to an arbitrary loading giving rise on current cross-section S, identified by arclength s, to a screwer $\{\mathcal{S}\}$, in particular to a normal force $N(s)$, Figure 6.6.

To do so, consider a thin slice of the beam, comprised between cross-sections S and S' located at arclengths s and $s + \mathrm{d}s$. This slice can be considered prismatic, given the assumptions underlying Definition 5.1 of a beam, namely that the diameter of each section S is small compared with the length of the median line and with the radii of curvature and torsion of that line in G. It follows that the results obtained for a bar can be extended to the slice. In the principal frame $\{G,XYZ\}$ associated with S, under the sole action of N, one thus has:

- The stress matrix:

$$[\boldsymbol{\sigma}] = \begin{bmatrix} \sigma_X & 0 & 0 \\ 0 & 0 & 0 \\ 0 & 0 & 0 \end{bmatrix} \qquad \text{with} \quad \sigma_X = \frac{N}{S} \tag{6.34}$$

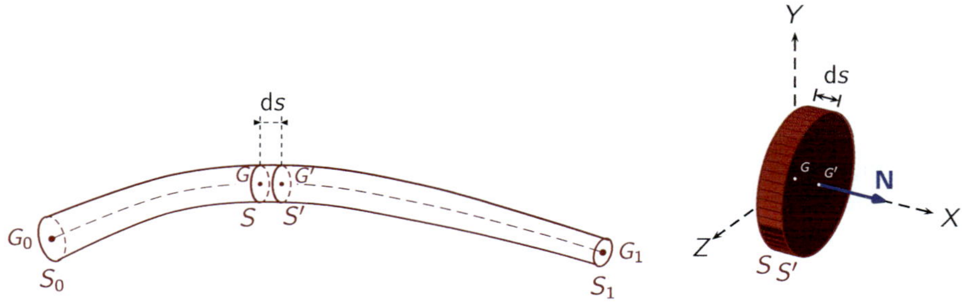

Figure 6.6 Axial load in an arbitrary beam.

- The strain matrix:

$$[\boldsymbol{\varepsilon}] = \begin{bmatrix} \varepsilon_X & 0 & 0 \\ 0 & \varepsilon_Y & 0 \\ 0 & 0 & \varepsilon_Z \end{bmatrix} \quad \text{with} \quad \begin{cases} \varepsilon_X = \dfrac{\sigma_X}{E} = \dfrac{N}{ES} \\ \varepsilon_Y = \varepsilon_Z = -\nu \dfrac{N}{ES} \end{cases} \tag{6.35}$$

- The displacement of S' with respect to S has a translation along director:

$$\mathrm{d}\boldsymbol{\Lambda} = \frac{N}{ES}\,\mathrm{d}s\,\mathbf{t} = \frac{\mathbf{N}}{ES}\,\mathrm{d}s \tag{6.36}$$

The potential elastic energy stored in that slice:

$$\frac{\mathrm{d}W}{\mathrm{d}s} = \frac{1}{2}\frac{N^2}{ES} \tag{6.37}$$

6.5.2 EXAMPLES

Example 6.2: Aircraft Propeller

Consider a blade modeled as a straight beam having variable cross-section $S(x)$ at a distance x from the rotor's axis (Figure 6.7) and made up of a material with mass density ρ. At sufficiently large rotation, with angular velocity ω, a centrifugal force arises in each blade. To obtain its expression, recall that the centrifugal force acting on a particle of mass m_p rotating about an axis is given by: $\mathbf{f} = -m_\mathrm{p}\boldsymbol{\omega} \times (\boldsymbol{\omega} \times \mathbf{r})$, where $\boldsymbol{\omega}$ denotes the angular velocity vector and $\mathbf{r}$ denotes the position vector of the particle, the origin being on the axis. The concept of centrifugal force is particularly useful in the stress analysis of rotating bodies that can be approximated as beams because internal resultants are defined with respect to a frame associated with the beam, which is a rotating frame, not an inertial frame.

Evaluating the above expression in the present case, the centrifugal force field in each blade, given as a body force per unit volume, is $f(x) = \rho\omega^2 x$. By way of consequence, a tensile normal force prevails in the blade, which can be evaluated using Theorem 5.4 in Section 5.4.2 as follows:

$$N(x) = \int_V f(\xi)\,\mathrm{d}V = \rho\omega^2 \int_{\xi=x}^{L} \xi\, S(\xi)\,\mathrm{d}\xi \tag{6.38}$$

where L denotes the distance between the rotor's axis and the tip of the blade and V the volume of the blade occupied by the portion between the sections at x and L.

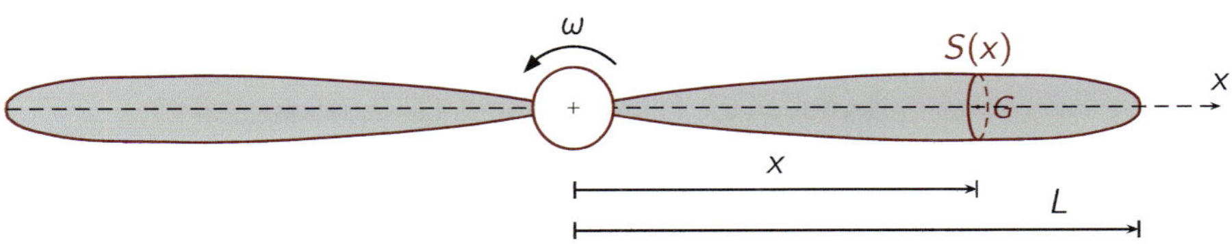

Figure 6.7 Normal force in a propeller.

Alternatively, one may obtain the same result by simply drawing the free-body diagram of the portion of the blade comprised between $S(x)$ and the tip. A slice (not shown) of thickness $d\xi$, located at abscissa ξ where $x < \xi < L$, contributes to the total internal force by $f(\xi)S d\xi$. Summing up all contributions, which are nonuniformly distributed between section S and the tip, leads to Equation (6.38).

Helicopter rotor blades typically have a constant cross-section. Thus,

$$N(x) = \frac{1}{2} m\omega^2 (L^2 - x^2) \tag{6.39}$$

where m denotes the blade's mass per unit length. ♡

Remark 6.7 *As noted under Remark 5.18, one can also evaluate the normal force N by analyzing the part upstream from the current cross-section S in Figure 6.7. In doing so, one must not forget the reaction force R at the origin. Indeed, global equilibrium of the blade, expressed in the rotating frame, yields: $R = -M\omega^2 L/2$ (case of constant cross-section) where $M = mL$ is the total mass of the blade. Applying the counterpart of Theorem 5.4 according to Remark 5.18 then gives*

$$N(x) = -\rho\omega^2 \int_0^x \xi\, S(\xi)\, d\xi - R$$

which delivers upon integration the same result as above. ∗

Remark 6.8 *In blades and wings, aerodynamic forces induce a bending moment, a torque and a shearing force.* ∗

Example 6.3: Water Tower

A weight W is placed atop a column of height h, axisymmetric (of axis Ox, vertical) and made up of a material having weight density ρg. To avoid overdesigning the column it is desirable that the normal stress be the same at all heights. How should the cross-section $S(x)$ evolve with x such that the compressive stress σ_x is uniform in the column, say equal to some (negative) value σ?

The normal force in the column at height x is $N(x) = S(x)\sigma$ and satisfies (refer to the free body diagram in Figure 6.8):

$$N(x) = -W - \rho g \int_x^h S(\xi)\, d\xi = S(x)\,\sigma \tag{6.40}$$

This relation gives by derivation the following differential equation:

$$\frac{dS}{S} = \frac{\rho g}{\sigma} dx \tag{6.41}$$

Hence,

$$S = S_0 \exp \frac{\rho g}{\sigma} x \tag{6.42}$$

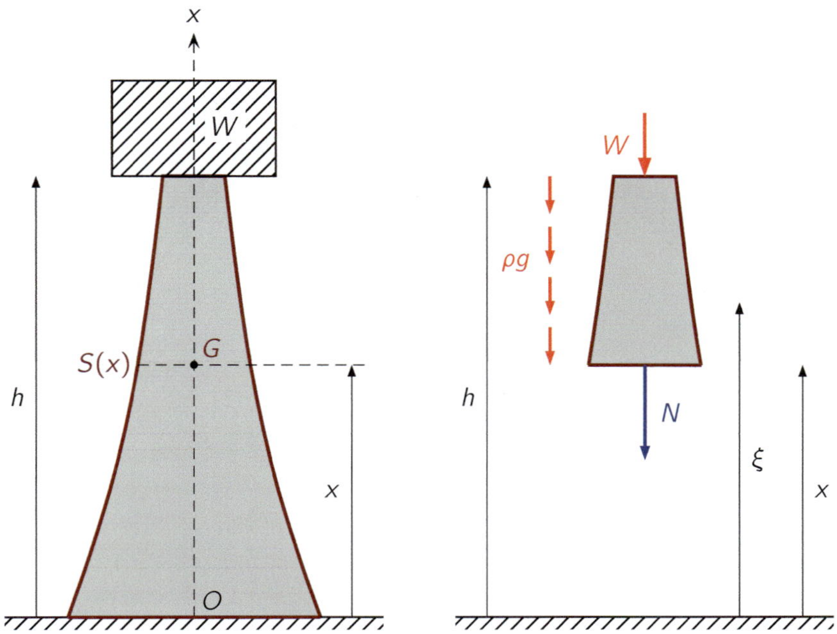

Figure 6.8 Water tower under weight.

where the base cross-section S_0 is determined from the boundary condition

$$N(h) = -W = S(h)\sigma \qquad \text{at } x = h \tag{6.43}$$

and is thus given by

$$S_0 = -\frac{W}{\sigma}\exp-\frac{\rho g h}{\sigma} \tag{6.44}$$

With a cross-section so designed, the net compressive stress is the same at any height. ♡

Example 6.4: Vault under Uniform Loading

Consider a beam with median plane (xOy), of constant cross-section S, of median line $\widehat{G_0G_1}$ (arc of radius R, centered at O, having length $L = 2R\alpha$, Figure 6.9). External links at both ends are simple hinges that allow for possible rotations of the cross-sections and for radial translations. The loading consists of a uniformly distributed centripetal radial force having intensity p per unit length of the median line.

Let us determine the reactions at the supports as well as the internal resultants acting on current cross-section S, located using polar angle θ, Figure 6.9.

Equilibrium equations (5.46) are written as follows, considering the constant

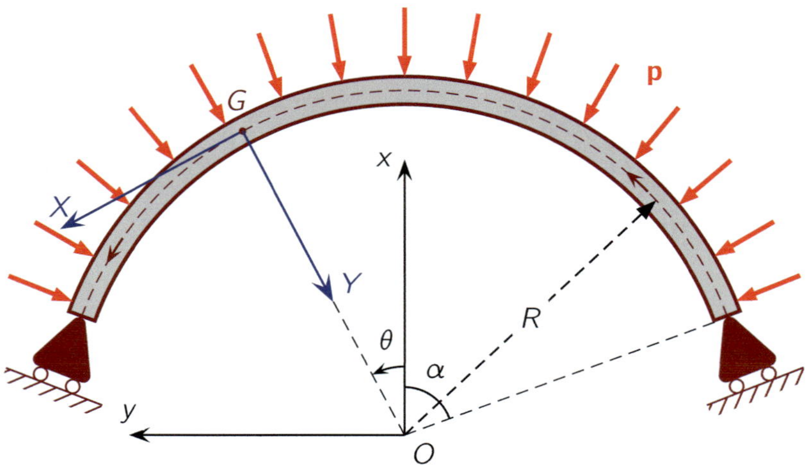

Figure 6.9 Circular arch under uniform pressure.

curvature:

$$\begin{cases} \dfrac{dN}{d\theta} - V = 0 \\ pR + N + \dfrac{dV}{d\theta} = 0 \\ \dfrac{dM}{d\theta} + RV = 0 \end{cases} \tag{6.45}$$

where N is the normal force, V is the shear, and M is the bending moment about G.

Integration of the above equations, as in Example 5.9, gives

$$\begin{cases} N = pR(A\cos\theta + B\sin\theta - 1) \\ V = pR(-A\sin\theta + B\cos\theta) \\ M = -pR^2(A\cos\theta + B\sin\theta + C) \end{cases} \tag{6.46}$$

with A, B, C constants to be determined; also see Section 5.5.3. The boundary conditions on the shear force and bending moment are (see Figure 6.10):

$$V = 0, \quad M = 0 \qquad \text{for}\ \theta = \pm\alpha$$

They can only be satisfied if $A = B = C = 0$ so that on each cross-section:

$$N = -pR, \quad V = 0, \quad M = 0$$

Thus, the vault works in pure compression, of uniform intensity pR. The reactions at the supports are also equal to pR. Indeed, using the last set of boundary conditions, now on N, one has $Q = -N$ for $\theta = \pm\alpha$; see Figure 6.10.

The length L and mean radius R of the beam undergo relative variations equal to $-\frac{pR}{ES}$. ♡

Figure 6.10 Boundary conditions on N, V, M at each support.

Remark 6.9 *Curvilinear members that work under pure tension or compression have special importance in structural analysis. The median curve of an arch working under tension or compression when subjected to a given pressure loading is said to be the funicular curve for that loading. The previous example shows that a circular arch is funicular for a uniform radial pressure. Likewise, a parabolic arch is funicular for uniform vertical pressure. This may be inferred from the equilibrium shape of a cable under uniformly distributed vertical load, as will be treated in the following section; see Example 6.7.* ∗

6.6 STATICS OF CABLES

DEFINITION 6.3 A cable is an infinitely flexible and twistable (i.e., torsionally flexible) beam satisfying the two conditions:

1. Of small extensibility (relative elongations under load very small before unity $\delta L/L \ll 1$); and
2. Of being loaded by forces (concentrated and distributed), exclusive of any couple. ♣

The above definition applies to what are commonly referred to as cables, ropes or strings. Under such conditions, at equilibrium, the net moment **M** of the screwer is identically nil so that the general equations of equilibrium (5.43) may be written:

$$\mathbf{p} + \frac{\mathrm{d}\mathbf{R}}{\mathrm{d}s} = \mathbf{0} \qquad \text{and} \qquad \mathbf{t} \times \mathbf{R} = \mathbf{0} \tag{6.47}$$

Equation $(6.47)_2$ implies that the resultant force is tangential to the cable. In other words, the screwer is reduced to the sole normal force **N**, which is determined from

$$\mathrm{d}\mathbf{N} + \mathbf{p}\,\mathrm{d}s = \mathbf{0} \tag{6.48}$$

In fact, the normal force N is necessarily positive and is called *tension*. Indeed, a cable does not support any compressive load.

The general problem of cable statics consists of determining, for given loading and a system of links:

- the curve that defines the median line;
- the tension N in each cross-section of the cable; and
- the reaction forces.

Remark 6.10 *A cable may be thought of as a beam that does not have an equilibrium shape at rest, and which takes, under the action of loads, an equilibrium shape such that the internal resultants are reduced, on each cross-section, to a pure tension.* $*$

The next three sections deal with three cases based on the type of loading (concentrated and/or distributed forces).

6.6.1 LOADING BY CONCENTRATED FORCES

In this case, $\mathbf{p} = \mathbf{0}$ everywhere. In particular, the weight of the cable is neglected. Along the cable, the points of application of the forces (external and reactions) are called nodes and denoted $A_0, A_1 \dots A_{n-1}, A_n$ with A_0 and A_n denoting the cable ends.

Between two consecutive nodes A_i and A_{i+1}, cable equilibrium, that is Equation (6.48), writes $\mathrm{d}\mathbf{N} = \mathbf{0}$, which implies that $\mathbf{N}$ is constant in direction and magnitude. Each portion of the cable A_iA_{i+1} is therefore straight, and the tension in it is constant.

Therefore, the generic problem of cables under point forces is posed so as to determine:

- the polygonal line $A_0A_1 \dots A_{n-1}A_n$;
- the tension in each portion comprised between two nodes; and
- the reaction forces.

To achieve this, one solves a system of equations obtained by expressing at each node:

1. force equilibrium; and
2. the imposed geometric constraints.

Example 6.5: Long Cable under Concentrated Forces

In vertical plane (xOy), where Oy denotes the ascending vertical, two frictionless pulleys are fixed with their axes parallel to Oz. Their diameters are taken to be sufficiently small, in comparison with the length of the cable, to be considered as points. The pulleys thus constitute the nodes A_1 and A_3 having specified coordinates (x_1, y_1) and (x_3, y_3), respectively; see Figure 6.11.

On the cable of total length L, three nodes are considered, in addition to nodes A_1 and A_3 above, which are constrained to pass through pulleys:

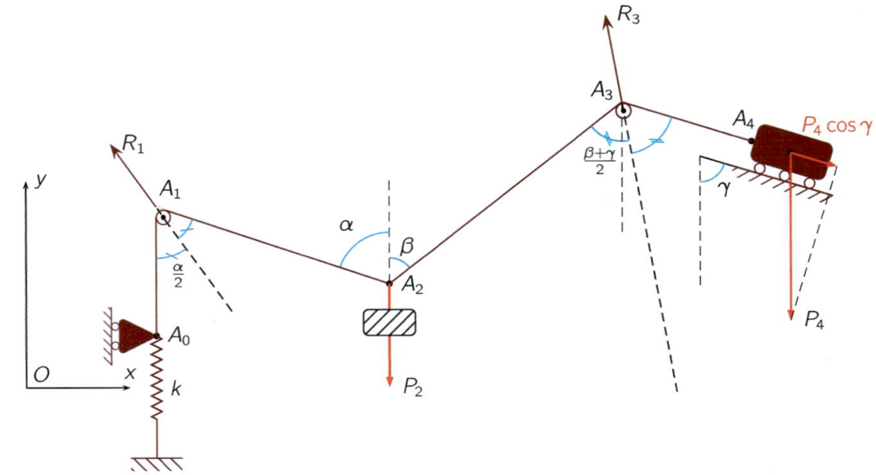

Figure 6.11 Long cable under concentrated forces.

- Node A_0, the origin of the cable, is fixed to the ground by means of a spring of stiffness k. When the spring is at rest, its tip at A_0 is taken to lie on the x-axis.
- Node A_2, located at a distance L_{02} from A_0 along the cable. A weight P_2 is suspended at A_2.
- Node A_4, the end of the cable located at a distance L_{24} from A_2 along the cable. This node is fixed to a cart having weight P_4, which is free to move on a rail contained within the (xOy) plane and inclined at an angle γ from the vertical. Segment A_3A_4 remains parallel to the rail.

Figure 6.11 shows the cable, as loaded. One seeks to determine, at equilibrium, the following:

- the polygonal line $A_0A_1A_2A_3A_4$, in particular the coordinates of A_0, A_2, and A_4 as well as angles α and β;
- the cable tensions N_{ii+1} in the various portions A_iA_{i+1} (four in total); and
- the reactions at the pulleys, R_1 and R_3.

The equations of nodal equilibrium write:

$$\text{For } A_0\text{:} \quad N_{01} = ky_0 \tag{6.49}$$

$$\text{For } A_1\text{:} \quad N_{01} = N_{12} \quad \text{and} \quad R_1 = 2N_{01}\cos\frac{\alpha}{2} \tag{6.50}$$

$$\text{For } A_2\text{:} \quad P_2 = N_{12}\cos\alpha + N_{23}\cos\beta \quad \text{and} \quad N_{12}\sin\alpha = N_{23}\sin\beta \tag{6.51}$$

$$\text{For } A_3\text{:} \quad N_{23} = N_{34} \quad \text{and} \quad R_3 = 2N_{23}\cos\frac{\beta+\gamma}{2} \tag{6.52}$$

$$\text{For } A_4\text{:} \quad N_{34} = P_4\cos\gamma \tag{6.53}$$

The geometric constraints on nodes are:

For A_0: set value for x_0 with $x_0 = x_1$; recall that x_1 is given as input.

For A_1: given values of x_1 and y_1.
For A_2: $A_0A_1 + A_1A_2 = L_{02}$ (given) and $A_2A_3 + A_3A_4 = L_{24}$ (given).
For A_3: given values of x_3 and y_3.
For A_4: set slope of segment A_3A_4.

Mathematically, the constraints on A_2 write:

$$y_1 - y_0 + \frac{y_1 - y_2}{\cos\alpha} = L_{02} \quad \text{with} \quad x_2 - x_1 = (y_1 - y_2)\tan\alpha \tag{6.54}$$

$$\frac{y_3 - y_2}{\cos\beta} + \frac{y_3 - y_4}{\cos\gamma} = L_{24} \quad \text{with} \quad x_3 - x_2 = (y_3 - y_2)\tan\beta \tag{6.55}$$

Also, the constraint on A_4 writes

$$x_4 - x_3 = (y_3 - y_4)\tan\gamma \tag{6.56}$$

The above conditions, static and kinematic, constitute a *non-linear* system that permits the calculation of all unknowns: the four tensions $N_{i\,i+1}$, the coordinates of nodes A_0, A_2, A_4, angles α and β, and reactions R_1, R_3 of the pulleys. In the process, one can also find the minimum value of weight P_4 below which there is no equilibrium configuration. ♡

6.6.2 LOADING BY DISTRIBUTED FORCES

Let $\mathbf{p}(s)$ be the distributed force per unit length, assumed to be continuous along the median line $\widehat{A_0A_1}$. The only point forces are the reactions at the two ends.

The only equilibrium condition, Equation (6.48), is therefore integrable over the entire arc $\widehat{A_0A_1}$. Projecting on the global axes x, y, z, one gets

$$\begin{cases} \mathrm{d}\left(N\dfrac{\mathrm{d}x}{\mathrm{d}s}\right) + p_x\,\mathrm{d}s = 0 \\ \mathrm{d}\left(N\dfrac{\mathrm{d}y}{\mathrm{d}s}\right) + p_y\,\mathrm{d}s = 0 \\ \mathrm{d}\left(N\dfrac{\mathrm{d}z}{\mathrm{d}s}\right) + p_z\,\mathrm{d}s = 0 \end{cases} \quad \text{with} \quad \mathbf{p}\begin{cases} p_x \\ p_y \\ p_z \end{cases} \quad \text{and} \quad \mathbf{t}\begin{cases} \dfrac{\mathrm{d}x}{\mathrm{d}s} \\ \dfrac{\mathrm{d}y}{\mathrm{d}s} \\ \dfrac{\mathrm{d}z}{\mathrm{d}s} \end{cases} \tag{6.57}$$

Supplementing the three scalar equations (6.57) with the following two equations:

$$\mathrm{d}s = \sqrt{\mathrm{d}x^2 + \mathrm{d}y^2 + \mathrm{d}z^2} \quad \text{and} \quad L = \int_{A_0}^{A_1} \mathrm{d}s \tag{6.58}$$

L being the cable's length, defines a system of equations, the integration of which deliversthe equilibrium shape and the tension, account being taken of boundary conditions at A_0 and A_1.

Remark 6.11 *In the Frenet frame* $\{G, \mathbf{t}, \mathbf{n}, \mathbf{b}\}$ *associated with centroid G, the vectorial equation of equilibrium, Equation* (6.48)*, admits the three projections:*

$$\begin{cases} dN + p_t\, ds = 0 \\ N + Rp_n = 0 \\ p_b = 0 \end{cases} \qquad \textit{with} \quad \mathbf{p} \begin{cases} p_t \\ p_n \\ p_b \end{cases} \tag{6.59}$$

and R being the radius of curvature in G. *

Example 6.6: The Catenary Problem

The two ends of a cable are anchored at two points A_0 and A_1 of the vertical (xOy) plane. The cable is subjected to its own weight. Let p be its weight per unit length and L its length, which is larger than the distance A_0A_1 with $x_0 < x_1$. The median line is oriented from A_0 to A_1, Figure 6.12.

In order to determine the shape of the cable in equilibrium and the tension N in it, start from the relevant equilibrium conditions, Equation (6.57):

$$d\left(N\frac{dx}{ds}\right) = 0 \qquad \text{and} \qquad d\left(N\frac{dy}{ds}\right) - p\,ds = 0$$

A first integration gives:

$$N\frac{dx}{ds} = N_x \qquad \text{and} \qquad N\frac{dy}{ds} = ps + C$$

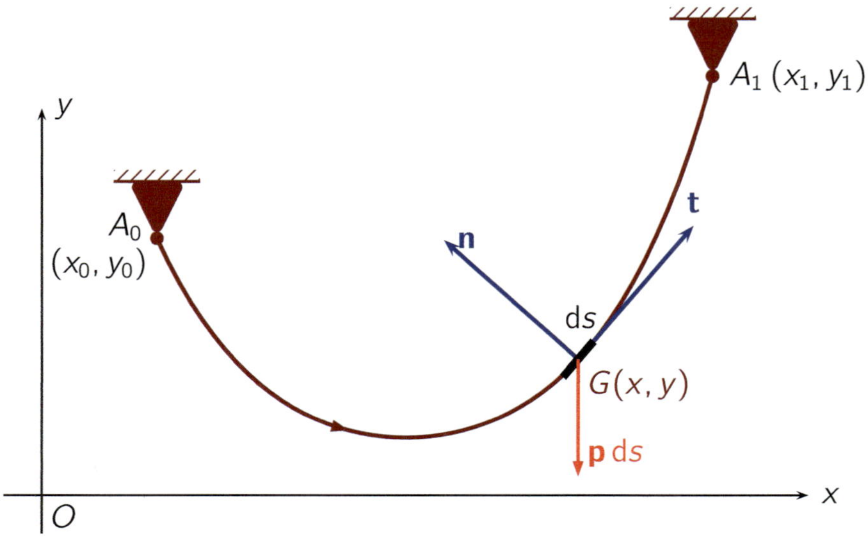

Figure 6.12 A cable under its own weight.

where N_x and C are constants, N_x being the (constant) projection of N on Ox.

Eliminating N between the above two equations gives next:

$$\frac{\mathrm{d}y}{\mathrm{d}x} = \frac{p}{N_x}s + \frac{C}{N_x}$$

then by derivation to discard the auxiliary constant:

$$\frac{\mathrm{d}^2y}{\mathrm{d}x^2} = \frac{p}{N_x}\frac{\mathrm{d}s}{\mathrm{d}x},$$

which on account of Equation (6.58) gives the governing differential equation for the equilibrium shape:

$$\frac{\mathrm{d}^2y}{\mathrm{d}x^2} = \frac{p}{N_x}\sqrt{1+\left(\frac{\mathrm{d}y}{\mathrm{d}x}\right)^2}$$

Finally, integrating twice the above equation yields

$$\frac{\mathrm{d}y}{\mathrm{d}x} = \sinh\left(\frac{p}{N_x}(x-a)\right) \quad \text{and} \quad y-b = \frac{N_x}{p}\cosh\left(\frac{p}{N_x}(x-a)\right) \tag{6.60}$$

where a and b are additional constants of integration. It follows that $\widehat{A_0A_1}$ is an arc of a catenary. Immediately, one has therefore

$$\frac{\mathrm{d}s}{\mathrm{d}x} = \cosh\left(\frac{p}{N_x}(x-a)\right) \quad \text{and} \quad N = N_x\cosh\left(\frac{p}{N_x}(x-a)\right) \tag{6.61}$$

Integration constants N_x, a, and b are determined using the relations:

$$\begin{cases} L = \displaystyle\int_{x_0}^{x_1} \cosh\left(\frac{p}{N_x}(x-a)\right)\mathrm{d}x = \frac{N_x}{p}\left[\sinh\left(\frac{p}{N_x}(x_1-a)\right) - \sinh\left(\frac{p}{N_x}(x_0-a)\right)\right] \\ y_0 - b = \dfrac{N_x}{p}\cosh\left(\dfrac{p}{N_x}(x_0-a)\right) \\ y_1 - b = \dfrac{N_x}{p}\cosh\left(\dfrac{p}{N_x}(x_1-a)\right) \end{cases}$$

which express that the catenary has length L and passes through points A_0 and A_1. Note that a and $b+N_x/p$ are the coordinates of the apex of the catenary (which could eventually fall outside of arc $\widehat{A_0A_1}$) where the radius of curvature takes the value N_x/p.

In summary, the shape of the catenary is that of hyperbolic function given by Equation $(6.60)_2$ and the tension in the cable varies with position and is given by Equation $(6.61)_2$. The specifics (minimum point on the catenary, maximum tension, etc.) depend on constants N_x, a and b, which are determined by the above relations and themselves depend on the relative positions of anchoring points A_0 and A_1. ♡

Example 6.7: Suspended Bridge

A cable is anchored at both ends and subjected to a downward distributed force with per unit length magnitude $p = f\frac{\mathrm{d}x}{\mathrm{d}s}$, f being constant. This is the case, for instance, of the holding cable of a suspended bridge, if the apron has weight density f and the suspensions, assumed to be infinitely close together, are all equally stretched. In addition, it is assumed that holding cable and suspensions have weights that are negligible in comparison with that of the apron, and $x_0 < x_1$.

The equilibrium of element ds (Figure 6.13) is given by Equation (6.57), specialized to:

$$\mathrm{d}\left(N\frac{\mathrm{d}x}{\mathrm{d}s}\right)=0 \qquad \text{and} \qquad \mathrm{d}\left(N\frac{\mathrm{d}y}{\mathrm{d}s}\right)-f\,\mathrm{d}x=0$$

A first integration gives:

$$N\frac{\mathrm{d}x}{\mathrm{d}s}=N_x \qquad \text{and} \qquad N\frac{\mathrm{d}y}{\mathrm{d}s}=p(x-a)$$

where N_x and a are constants, N_x being the (constant) projection of N on Ox. Eliminating N between the above two equations gives:

$$\frac{\mathrm{d}y}{\mathrm{d}x}=\frac{f}{N_x}(x-a)$$

It immediately follows that

$$y-b=\frac{1}{2}\frac{f}{N_x}(x-a)^2 \qquad \text{and} \qquad N=N_x\sqrt{1+\frac{f^2}{N_x^2}(x-a)^2} \tag{6.62}$$

Equation $(6.62)_1$ shows that the equilibrium shape is a parabola. Integration

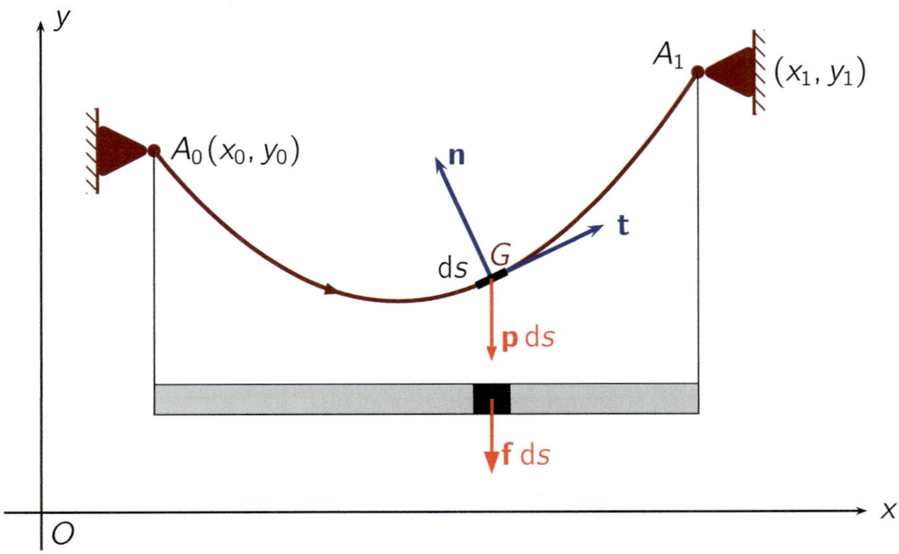

Figure 6.13 A cable under uniform vertical load.

constants N_x, a, and b are determined using the relations:

$$\begin{cases} L = \int_{x_0}^{x_1} \sqrt{1 + \frac{f^2}{N_x^2}(x-a)^2}\,\mathrm{d}x = \frac{N_x}{2f}\left[u\sqrt{1+u^2} + \ln\left(u + \sqrt{1+u^2}\right)\right]_{u_0}^{u_1} \\ y_0 - b = \frac{1}{2}\frac{f}{N_x}(x_0 - a)^2 \\ y_1 - b = \frac{1}{2}\frac{f}{N_x}(x_1 - a)^2 \end{cases}$$

which express that the cable has length L and passes through points A_0 and A_1. Also, the change of variable

$$u = \frac{f}{N_x}(x-a)$$

was used in calculating the integral. Note that a and b are the coordinates of the apex of the parabola (which could eventually fall outside of arc $\widehat{A_0A_1}$) where the radius of curvature takes the value N_x/f. ♡

6.6.3 MIXED LOADING

Cables are often subjected to a combination of distributed and concentrated forces. In such cases, the median line $\widehat{A_0A_n}$ is decomposed into parts denoted $\widehat{A_0A_1}$, $\widehat{A_1A_2}, \ldots, \widehat{A_{n-1}A_n}$ such that on each part taken separately: (i) the distributed force $\mathbf{p}$ (per unit length) is continuous, and (ii) no point forces are applied. Thus, any concentrated forces would be applied at the nodes A_i.

Integration of the equations of equilibrium on each part provides the shape of the curve $\widehat{A_iA_{i+1}}$ and the tension prevailing therein, up to some integration constants.

Then, expressing force equilibrium and the geometric constraints at each node permits to patch together various parts and determine the integration constants.

Example 6.8: The Cable Car

A cable holding a cable car (or teleferique) has one end anchored at point A_0 having coordinates (x_0, y_0). At a given instant, the car of weight P_1 is suspended at point $A_1(x_1, y_1)$. After passing by a pulley, modeled as a point $A_2(x_2, y_2)$, the cable is stretched by a counterweight P_3 fixed at the other end of the cable, A_3. The three parts so defined will be denoted *(1)*, *(2)* and *(3)*, Figure 6.14. The weight of the cable per unit length is denoted p.

From the solution of Example 6.6, the shape of parts *(1)*, *(2)* is hyperbolic. The unit tangent $\mathbf{t}$ at a point G of the median line is given by:

$$\mathbf{t}\begin{cases} \cos\varphi = \frac{\mathrm{d}x}{\mathrm{d}s} = \frac{1}{\cosh\frac{p(x-a)}{N_x}} \\ \sin\varphi = \frac{\mathrm{d}y}{\mathrm{d}s} = \tanh\frac{p(x-a)}{N_x} \end{cases}$$

On part *(1)*, or $\widehat{A_0A_1}$, one has:

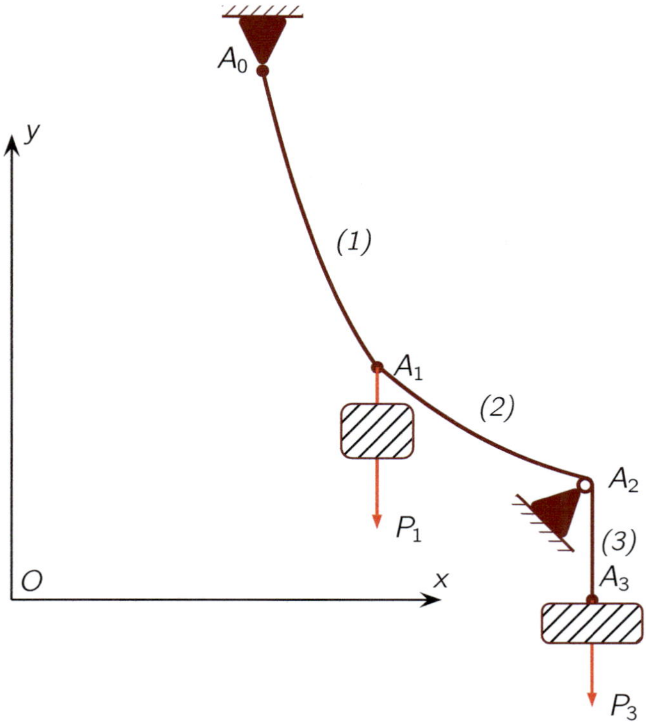

Figure 6.14 A cable under mixed, concentrated and distributed, loads.

- the equation of arc $\widehat{A_0 A_1}$:

$$y - b_1 = \frac{N_{x1}}{p} \cosh \frac{p(x - a_1)}{N_{x1}}$$

- the tension:

$$N_1 = N_{x1} \cosh \frac{p(x - a_1)}{N_{x1}}$$

- The length:

$$L_1 = \frac{N_{x1}}{p} \left[\sinh \frac{p(x - a_1)}{N_{x1}} \right]_{x_0}^{x_1}$$

- The passing through A_0 and A_1:

$$y_0 - b_1 = \frac{N_{x1}}{p} \cosh \frac{p(x_0 - a_1)}{N_{x1}} \qquad \text{and} \qquad y_1 - b_1 = \frac{N_{x1}}{p} \cosh \frac{p(x_1 - a_1)}{N_{x1}}$$

On part *(2)*, analogous formulae apply using integration constants N_{x2}, a_2, and b_2 and coordinates (x_1, y_1) and (x_2, y_2).

On the last part *(3)*, which is straight, one writes:

$$N_3 = P_3 + p(y - y_3), \qquad L_3 = y_2 - y_3, \qquad x = x_2 = x_3$$

Nodal equilibrium writes as follows:

For A_1: $N_{x1} = N_{x2} = N_x$ and $P_2 = N_x \left[\sinh \frac{p(x_1-a_2)}{N_x} - \sinh \frac{p(x_1-a_1)}{N_x}\right]$

For A_2: $N_x \cosh \frac{p(x_2-a_2)}{N_x} = P_3 + p(y_2 - y_3)$

Finally, one writes $L_1 + L_2 + L_3 = L$, the total length of the cable (given).

Solution of the above nonlinear system of equations permits the calculation of:

- integration constants a_1, b_1, a_2, b_2, N_x
- the coordinates (x_1, y_1) of the cable car at A_1
- the equations of catenaries $\widehat{A_0A_1}$ and $\widehat{A_1A_2}$
- the tension N at each point of the cable
- the reactions at the supports in A_0 and A_2

It also allows to calculate the minimum value of counterweight P_3 below which no equilibrium configuration is possible. ♡

PROBLEMS

Problem 6.1 (Elasticity Solution for a Bar) *(i) Verify that the stress state given by Equation* (6.1) *satisfies the boundary conditions on all surfaces (lateral and end cross-sections) and that the strain state, Equation* (6.2)*, is compatible. (ii) Verify the same for the arbitrary beam in Figure 6.6 using Equation* (6.34) *for the stress state and Equation* (6.35) *for the strain.* ⋆

Problem 6.2 (Saint-Venant Principle) *The bar solution presented in Section 6.1 was developed assuming uniformly distributed tractions on end sections. Consider now nonuniform traction distributions on end sections, as would arise in practice due to grip details, etc. State the conditions that must be satisfied by the traction distributions such that the solution given in Section 6.1 remains valid sufficiently far away from end sections. Sketch some traction distributions that meet the stated conditions.* ⋆

Problem 6.3 (Proof) *Show that in a truss each bar is either under tension or compression.* ⋆

Problem 6.4 (Vault under Pressure) *By expressing global equilibrium of the vault in Example 6.4, show that the reactions at the supports are independent of angle α.* ⋆

Problem 6.5 (Suspended Cable under Load) *A cylindrical cable of cross-sectional area A and length ℓ is suspended to the ceiling, as shown in Figure 6.15. It is made of an elastic material with Young's modulus E and mass density ρ. At some distance d from the ceiling, the cable is subjected to a vertical force F. Let* **g** *denote the Earth's gravity field. (i) Determine the normal force N everywhere in the cable. (ii) Find the total elongation of the cable.* ⋆

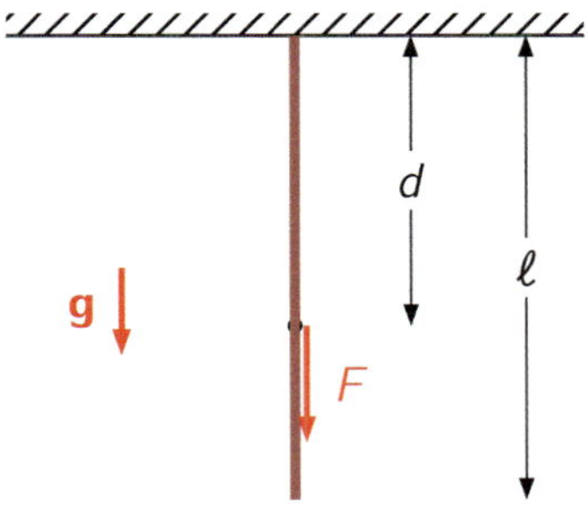

Figure 6.15 Suspended cable under its own weight and a point load.

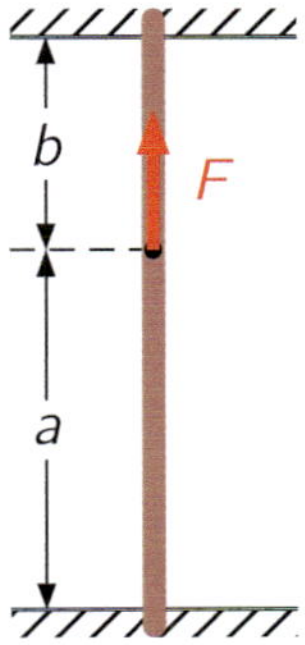

Figure 6.16 Bar fixed at both ends under a point load.

Problem 6.6 (Hyperstatic Straight Bar) *A bar having (per unit length) weight w and fixed at both ends, is subjected to an axial force F at some point along its length, Figure 6.16. Determine the reactions at the links and draw the diagram of the normal force N.* ⋆

Problem 6.7 (Tug of War Game) *A 400 m long rope of constant cross-sectional area S (to be determined) is made of a material with the following characteristics: Young's modulus $E = 60$ GPa and stress at break in tension $\sigma_r = 1$ GPa.*

The rope is pulled by two competing teams of 400 players each, Figure 6.17. The players are equidistant with $L = 50$ cm and each player applies a force $F = 500$ N. The rope is assumed to remain straight. (i) Replace this discontinuous loading with a continuous statically "equivalent" loading and calculate the section area at rupture, S_r. (ii) Find the rope's elongation at rupture. (iii) Find the maximum tension. (iv) Calculate the energy stored in the rope for $S = S_r$. ⋆

Problem 6.8 (Degree of Hyperstatic Character) *Determine whether each of the trusses shown in Figure 6.18 is hyperstatic and if so: (i) state whether the hyperstatic character is internal, external, or both; (ii) give the corresponding values of p and q (see Sections 6.3.1 and 6.4.1 for definitions).* ⋆

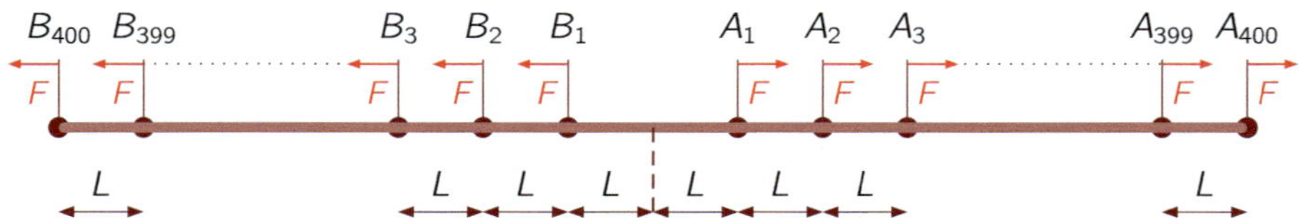

Figure 6.17 Tug of war game.

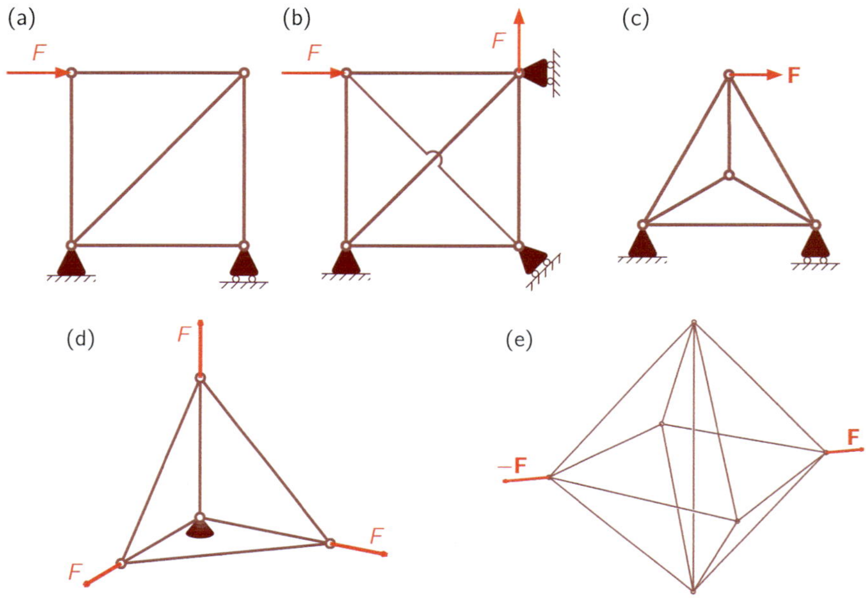

Figure 6.18 Some 2D and 3D trusses.

Problem 6.9 (Energy Stored in a Tapered Bar) *Calculate the potential energy of elastic deformation stored in a tapered solid of revolution when the latter is in pure tension, as shown in Figure 6.19. The net forces* $\mathbf{F}$ *and* $-\mathbf{F}$ *are assumed to be uniformly distributed over cross-sections* S_0 *and* S_1*. The material has Young's modulus* E*.* ⋆

Problem 6.10 (Planar System of Bars) *Three bars are assembled as depicted in Figure 6.20. The angle between each pair of bars is* $120°$*. The system is loaded by an arbitrarily oriented force* $\mathbf{F}$ *as shown. Calculate in terms of the components* X *and* Y *of* $\mathbf{F}$*: (i) the normal forces in the bars and (ii) the displacement of the center node.* ⋆

Problem 6.11 (Isostatic Planar Truss) *Consider the truss in Figure 6.21 made of seven identical bars (same cross-section* S*, same Young's modulus* E*) and loaded*

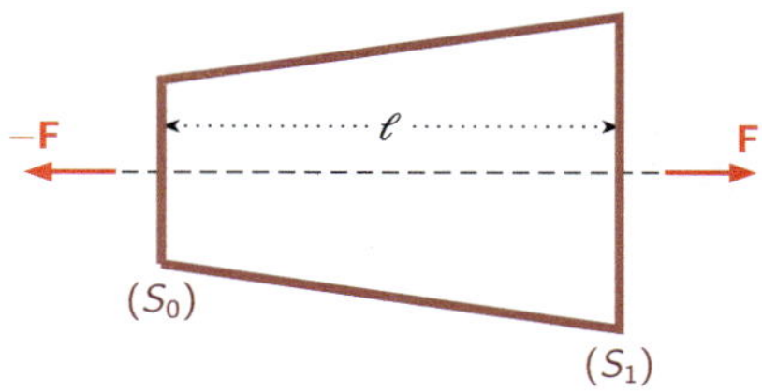

Figure 6.19 Tapered round specimen in tension.

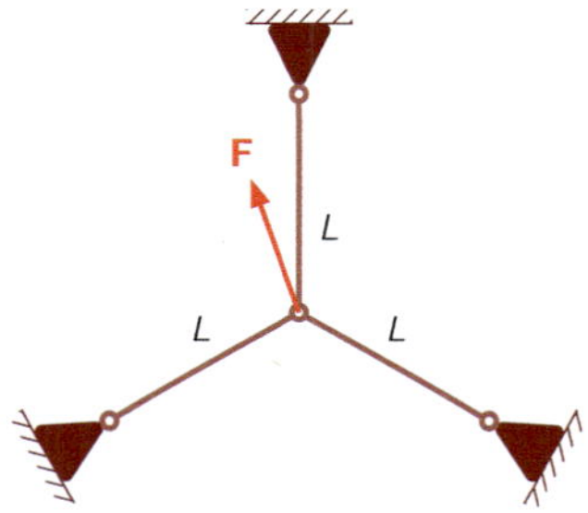

Figure 6.20 Planar system of bars under arbitrary loading.

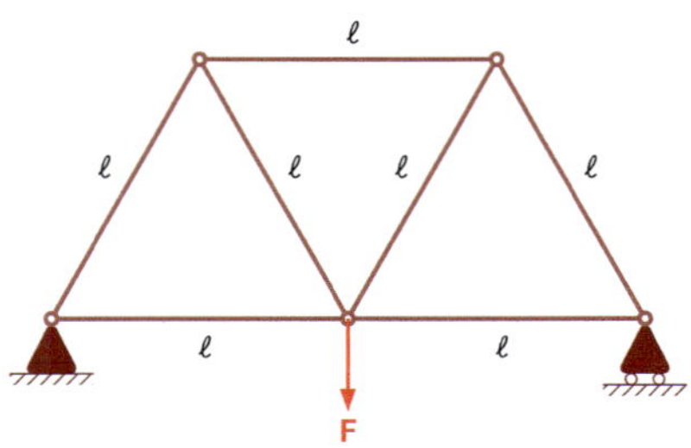

Figure 6.21 Seven bar truss.

as shown. Determine the normal forces in the bars and the displacements of all nodes. ⋆

Problem 6.12 (Five Bar Truss) *Consider the planar truss in Figure 6.22, constrained and loaded as shown. All bars have cross-sectional area S and Young's modulus E. (i) Determine the reaction forces in terms of the applied loads F and P. (ii) Determine the internal forces in the bars. Which bars are under compression? (iii) Determine the displacements of all nodes.* ⋆

Problem 6.13 (Hyperstatic Planar Truss) *The truss shown in Figure 6.23 consists of six bars, four of which are arranged in a square of side length L and two diagonals. The truss is supported as shown and subjected to a force F inclined at 45° downward from the horizontal. All bars have section S and Young's modulus E. Determine the normal forces in the bars.* ⋆

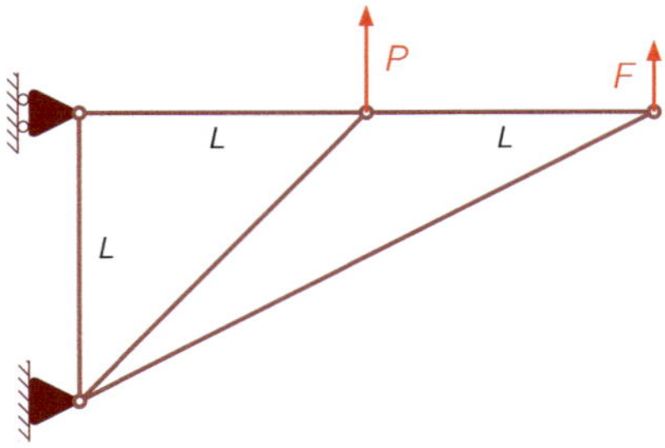

Figure 6.22 Five bar truss.

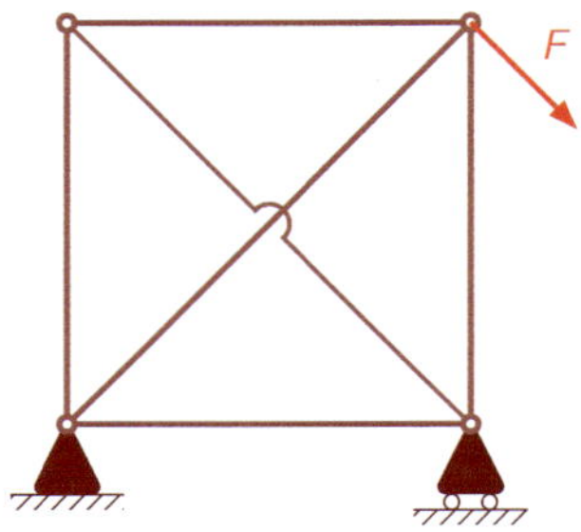

Figure 6.23 Square truss with two diagonals.

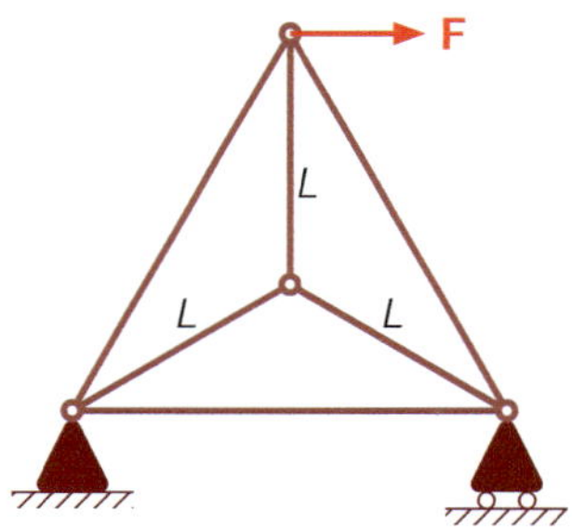

Figure 6.24 Triangular truss under horizontal load.

Problem 6.14 (Planar Hyperstatic Truss) *Consider the truss shown in Figure 6.24. All bars have the same cross-section S and Young's modulus E. (i) Determine the reaction forces at the supports. (ii) Calculate the internal forces in the bars. (iii) Determine the displacements of all the nodes.* ⋆

Problem 6.15 (Analysis of a Planar Truss) *Consider the planar truss depicted in Figure 6.25. Calculate the reactions at the supports, the normal forces in the bars, and the displacement of the point of application of the force.* ⋆

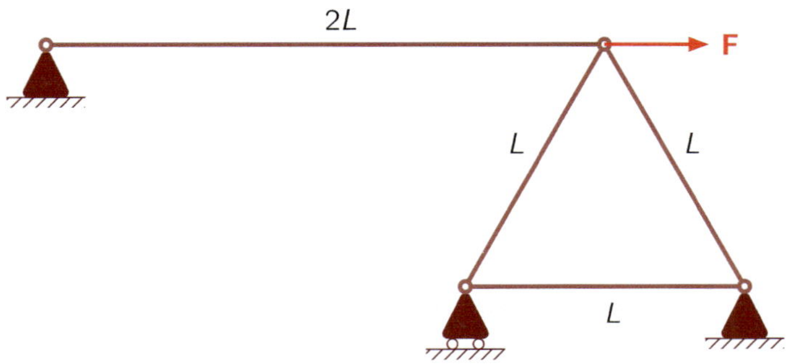

Figure 6.25 Four-bar truss with a redundant support.

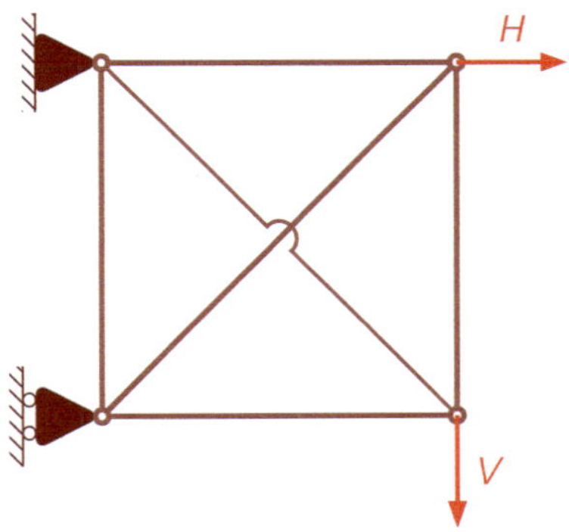

Figure 6.26 Square six-bar truss with two loads.

Problem 6.16 (Six-Bar Truss with Two Loads) *A six-bar square truss is subjected to horizontal and vertical loads, as shown in Figure 6.26. All bars have the same cross-section and Young's modulus with the shortest bars having length L. Calculate the normal forces in the bars and the nodal displacements.* ⋆

Problem 6.17 (Analysis of a Three-Dimensional Truss) *A three-dimensional truss consists of bars arranged along the edges of a trirectangular tetrahedron, as depicted in Figure 6.27. All bars have the same cross-section S and Young's modulus E. The bars oriented along the perpendicular edges have each length ℓ. The truss is fixed at its vertex and loaded as shown. The three forces have the same magnitude and are oriented parallel to the axes of a natural base associated with the truss. Under the action of this loading, determine (i) the normal forces in the bars, (ii) the nodal displacements, and (iii) the elastic potential energy.*

⋆

Problem 6.18 (Three-Dimensional System of Bars) *An undeformable square-shaped plate having weight W is suspended as shown in Figure 6.28 by means of five cables.*

The four lateral cables have each Young's modulus E, section S and length ℓ at rest. For the center cable, these parameters are denoted E', S', ℓ'. Calculate: (i) the tension N in each lateral cable, (ii) the tension N' in the central cable, (iii) the vertical displacement of the plate, and (iv) the elastic potential energy stored in the five cables. ⋆

Problem 6.19 (Hyperstatic Three-Dimensional Truss) *Thirteen bars constitute the edges of a regular octahedron and a diagonal, Figure 6.29. This 3D truss is loaded by two vertical forces of equal magnitude and opposite direction along the diagonal of the octahedron. All bars have the same cross-section S and Young's modulus E. Twelve of the thirteen bars have length L. Determine the normal forces in the bars and the nodal displacements.* ⋆

Problem 6.20 (Rotating Wheel) *Consider a wheel rotating at constant angular velocity ω. The hub and rim are assumed to be rigid (non-deformable). The spokes*

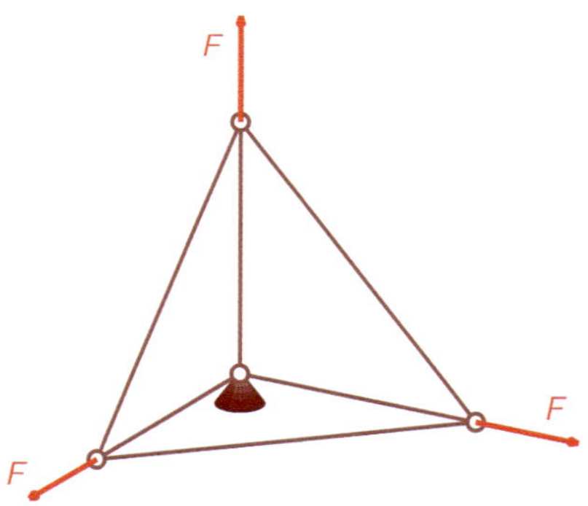

Figure 6.27 Tetrahedral truss under three loads.

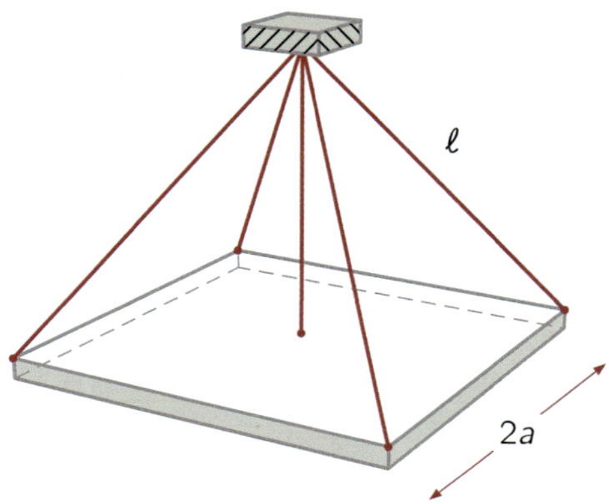

Figure 6.28 Five cables supporting a rigid plate.

are cylinders each of cross-section S, made up of an elastic material (Young's modulus E, mass density ρ). The spokes are assumed to work exclusively in tension or compression. There are n such spokes, regularly spaced, Figure 6.30. Determine the normal force N in each spoke. ⋆

Problem 6.21 (Rotating Beam with Weights) *A straight beam of total length 4L consists of three prismatic parts: a central part of length 2L, section 2S, Young's modulus E, and mass density ρ; and side parts having each length L, section S, Young's modulus E, and mass density ρ. On this beam are placed four point masses M, as shown in Figure 6.31.*

The beam spins around its axis of symmetry with angular velocity ω, which induces a field of centrifugal forces. (i) Calculate the normal force N as a function of

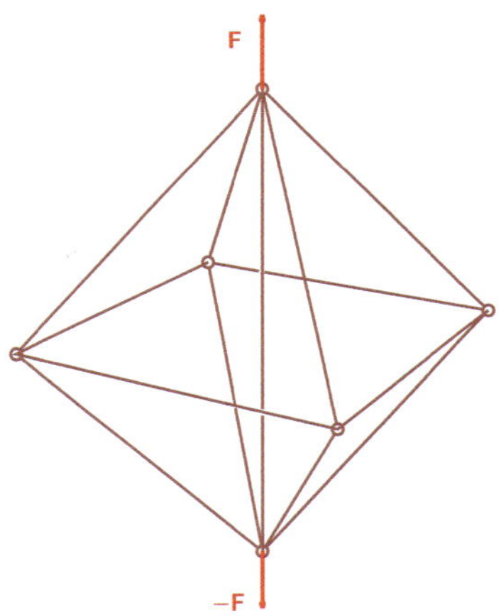

Figure 6.29 Thirteen bars arranged in an octahedral truss.

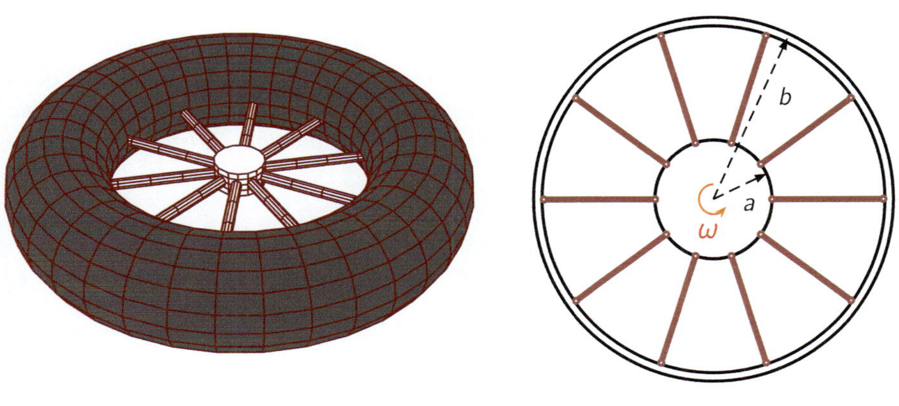

Figure 6.30 Spokes of wheel under centrifugal force.

position along the span. (ii) Draw the diagram of N. (iii) Calculate the elongation of each part of the beam. (iv) Calculate the displacements of the two end points. ⋆

Problem 6.22 (Propeller RPM Limit) *In the 1990s, an ATR-42 propeller blade broke off and went through the cabin, just missing a person. The remaining three blades and hub broke off the gear box and landed on a frozen lake. Consider a 13 ft diameter Hamilton Standard blade (the blade length is then 6.5 ft) weighing 52 lb with a cross-sectional area of 24 sq. in. If the allowable stress is 200 MPa, determine the maximum angular velocity for safe operation of the propeller.* ⋆

Problem 6.23 (Water Tower Sizing) *A water tower of height* $h = 50$ *m supports a water tank weighing* $W = 5000$ *tons. It is made of concrete with effective density accounting for water pressure* $\rho = 5000$ *kg/m*^{3}*. The current cross-section is circular, and the water tower is designed such that the normal stress remains constant along its height. If the allowable stress is 1 MPa, find the minimum base cross-sectional diameter to avoid collapse. Compare with the top section diameter.* ⋆

Problem 6.24 (Vault Design) *A vault has a radius of 10 m and a square cross-section of side length l. If the loading pressure is 1 MPa and the material's Young's modulus is 75 GPa, find the value of l such that the strain does not exceed the failure strain of 0.3%.* ⋆

Problem 6.25 (Underwater Shield Tunnel) *Ninety-eight feet beneath the banks of the Yellow River, a 3.6 mile tunnel consists of 500 pipe rings of underwater shielding. Each ring has a diameter of 17 m and a square cross-section of side length 60 cm. If the pressure on each ring, assumed to be uniformly distributed, is 4 MPa and the material's Young's modulus is 210 GPa, calculate the amount of shrinkage undergone by each ring.* ⋆

Problem 6.26 (Ring under Internal Pressure) *A circular ring of mean radius R and square cross-section of side length* $c \ll R$ *is made up of an elastic material with constants E,* ν*. The median plane is denoted xOy and the principal axes are G,XYZ, Figure 6.32. The ring is subjected to an internal pressure, which manifests as a uniform radial centrifugal load, p, per unit length of the ring. (i) Show that the screwer reduces to the sole normal force N. (ii) Calculate N. (iii) Calculate the variation of mean radius R.* ⋆

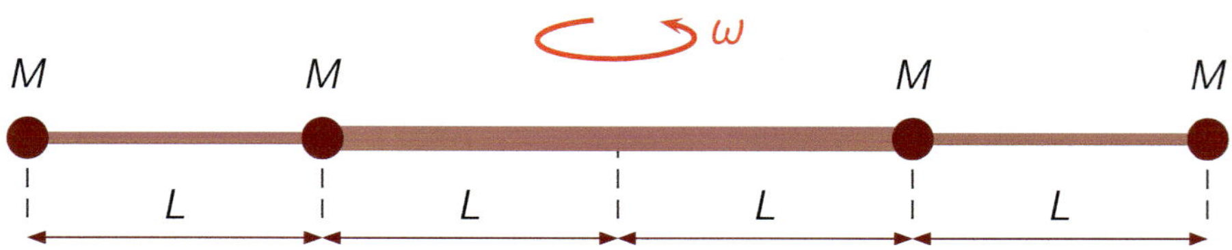

Figure 6.31 Rotating beam with point masses.

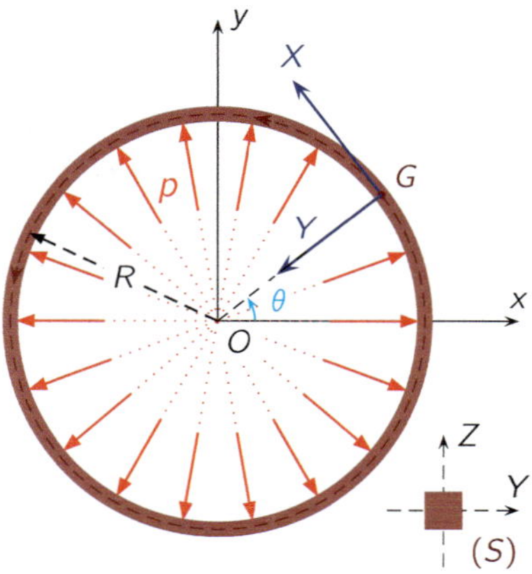

Figure 6.32 Ring under pressure.

Problem 6.27 (Boat Ancre Chain) *A boat is at ancre. There is no current. The wind exerts a force* **F** *on the boat, as sketched in Figure 6.33. The sea floor is at depth h. Let w denote the apparent weight per unit length of the chain. Determine the minimum length ℓ of the chain to be deployed for correct anchorage, i.e., the tangent at the sea floor should be horizontal. Calculate the value of ℓ when $F = 500$ N, $w = 10$ N/m and $h = 3$ m.* ⋆

Problem 6.28 (Cable with Concentrated Loads) *A cable of total length L consists of two parts $A_0A_1A_2$ with length $L_{02} = 7$ m and $A_2A_3A_4$ with length $L_{24} = 7$ m. The cable is made to pass through two frictionless pulleys of fixed locations at A_1 and A_3. The two pulleys are 9 m apart horizontally and 2 m vertically, Figure 6.34. The first cable part is fixed to the ground by means of a spring of stiffness k. A weight P_2 is suspended at A_2. The second cable part is fixed to a cart having weight P_4, which is free to move on a rail inclined at $\gamma = 45°$.*

Find the values of weights P_2 and P_4 in terms of k such that points A_1, A_2, and A_3 are aligned as shown in Figure 6.34. Determine the reactions at the pulleys. ⋆

Problem 6.29 (Rotating Rope) *A rope of length ℓ and mass per unit length ρ is fixed at both ends The rope is modeled as non-stretchable, perfectly flexible and twistable.*

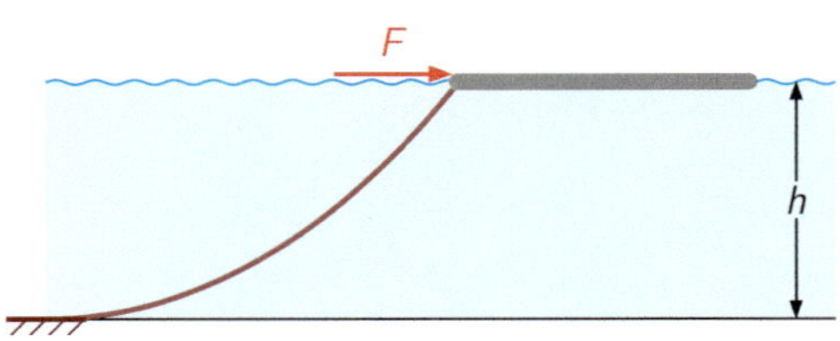

Figure 6.33 Equilibrium of an ancre chain modeled as a catenary.

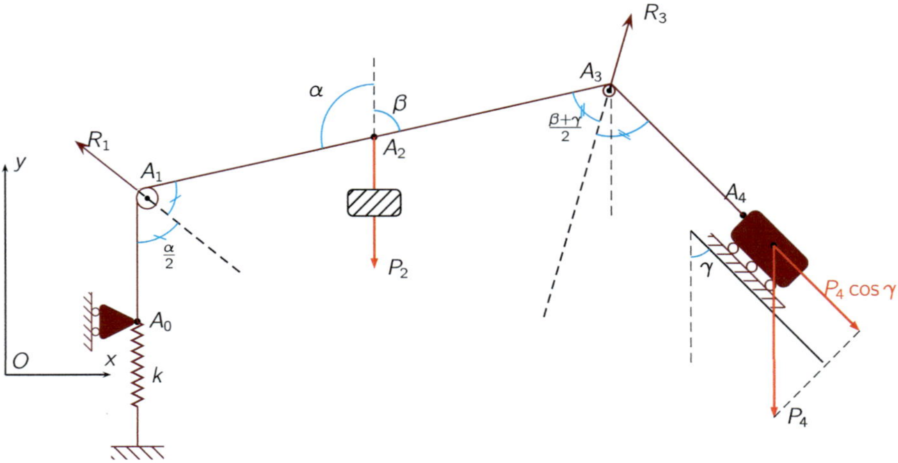

Figure 6.34 Cable subjected to concentrated forces.

Figure 6.35 Rotating rope fixed at both ends.

This rope is constrained to remain in its plane, which rotates at high angular velocity ω, as depicted in Figure 6.35. Calculate the tension in the rope and determine its equilibrium shape. ⋆

7 Bending Moment

7.1 PURE BENDING OF PRISMATIC BEAMS

Consider a prismatic beam, full or hollow, limited by end cross-sections S_0 and S_1 separated by distance L. All analysis is with respect to the *central principal frame of inertia* $\{G_0,xyz\}$ associated with the initial section S_0. Let I_X, I_Y, I_Z be the principal second moments of area of current cross-section S located at x from the origin, Figure 7.1. The current particle (or material point) is denoted $P(x,y,z)$.

On the final section, S_1 is applied a field of normal tractions

$$\sigma_x = -\frac{M}{I_Z}y \tag{7.1}$$

With the above choice of axes and the definition of I_Z, the force field associated with Equation (7.1) has a zero net resultant force and a resultant moment about G_1Z equal to M.

On the initial section, S_0 is applied the opposite loading, that is the couple $-\mathbf{M}$, Figure 7.1.

On each section S, the screwer then reduces to the sole M_Z component of the bending moment with $M_Z = M$. The beam is said to be in a state of *pure bending* in principal plane (xOy).

It is easily verified (see Problem 7.1) that the solution to the so-posed elasticity problem consists of the following fields:

- Stress:

$$[\boldsymbol{\sigma}] = \begin{bmatrix} \sigma_x & 0 & 0 \\ 0 & 0 & 0 \\ 0 & 0 & 0 \end{bmatrix} \quad \text{with} \quad \sigma_x = -\frac{M}{I}y \tag{7.2}$$

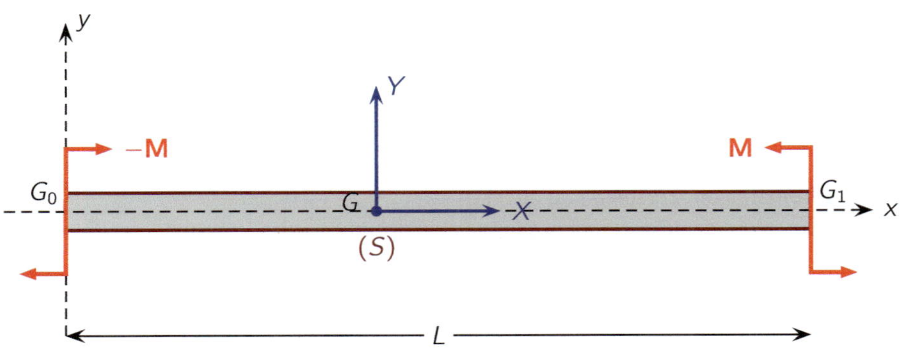

Figure 7.1 Prismatic beam subjected to end couples.

DOI: 10.1201/9781003537229-9

- Strains: using Hooke's law in component form, Equation (3.13f),

$$[\boldsymbol{\varepsilon}] = \begin{bmatrix} \varepsilon_x & 0 & 0 \\ 0 & \varepsilon_y & 0 \\ 0 & 0 & \varepsilon_z \end{bmatrix} \quad \text{with} \quad \begin{cases} \varepsilon_x = \dfrac{\sigma_x}{E} = -\dfrac{M}{EI}y \\ \\ \varepsilon_y = \varepsilon_z = -\nu\varepsilon_x = \nu\dfrac{M}{EI}y \end{cases} \tag{7.3}$$

- Displacements (with respect to a frame tied to S_0):

$$[\mathbf{u}] \begin{cases} u = -\dfrac{M}{EI}xy & +\beta z - \gamma y + a \\ \\ v = \dfrac{M}{2EI}\left[x^2 + \nu(y^2 - z^2)\right] & +\gamma x - \alpha z + b \\ \\ w = \nu\dfrac{M}{EI}zy & +\alpha y - \beta x + c \end{cases} \tag{7.4}$$

- Rotations:

$$[\boldsymbol{\omega}] \begin{cases} \omega_X = \nu\dfrac{M}{EI}z + \alpha \\ \\ \omega_Y = \qquad \beta \\ \\ \omega_Z = \dfrac{M}{EI}x \quad +\gamma \end{cases} \tag{7.5}$$

where E and ν denote Young's modulus and Poisson's ratio.

The terms in gray in Equations (7.4) and (7.5) represent rigid-body motion with integration constants a, b, c being the components of a rigid translation, and α, β, γ those of a rigid rotation. These constants vanish if the reference frame $\{G_0, xyz\}$ is convected with the translation and rotation of a neighborhood of G_0, a situation that will be assumed.

Of particular practical importance are the following conclusions:

1. An elementary domain centered about $G(x,0,0)$ undergoes a translation v_G along Oy and a rotation ω about GZ with

$$v_G = \frac{M}{2EI}x^2, \qquad \omega = \frac{M}{EI}x = \frac{\mathrm{d}v_G}{\mathrm{d}x}$$

   It remains undeformed and stress-free.
2. Any cross-section remains planar and normal to the deformed median line, Figure 7.2.
   Indeed, current cross-section S located at x from the origin undergoes:
   - a translation v_G, named *deflection*;
   - a rotation ω_Z about GZ with

$$\omega_Z = \frac{\mathrm{d}v_G}{\mathrm{d}x} = \omega$$

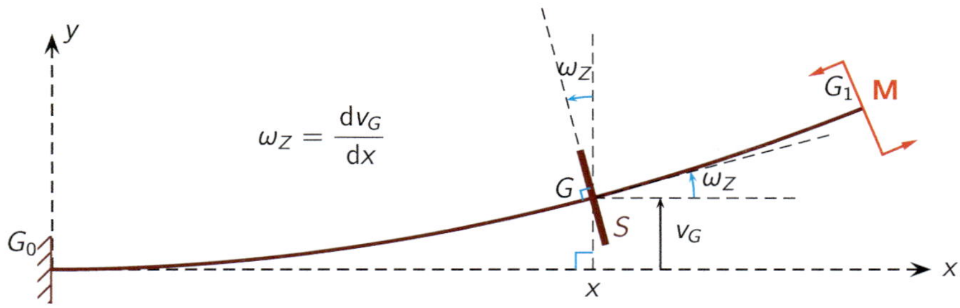

Figure 7.2 Kinematics of pure bending.

- a deformation in its plane.

The section does not warp: any axial displacement of current point P results solely from the rotation ω since by Equation $(7.4)_1$ and Equation $(7.5)_3$ one has $u = -y\,\omega$.

3. The deformed median fiber is a circular arc having curvature:

$$\frac{d^2 v_G}{dx^2} = \frac{M}{EI} = \frac{d\omega}{dx}$$

4. The elastic potential energy stored in an element of a longitudinal fiber centered about $P(x,y,z)$ and having length dx and cross-section dS is

$$\mathcal{U}\, dS\, dx = \frac{1}{2}\sigma_x \varepsilon_x\, dS\, dx = \frac{1}{2}\frac{M^2}{EI^2}\, y^2\, dS\, dx$$

where $\mathcal{U}$ is a specific energy. Integrating over cross-section S, the energy stored in a slice of the beam having thickness dx is then:

$$dW = \frac{1}{2}\frac{M^2}{EI^2}\left(\iint_S y^2\, dS\right) dx = \frac{1}{2}\frac{M^2}{EI}\, dx = \frac{1}{2} M\, d\omega$$

Remark 7.1 *With some restrictions, it is possible to rationalize the second conclusion above, namely that cross-sections remain plane and perpendicular to the deformed median line, without any calculation. This would be the case if the beam's cross-section were symmetrical, say about the y-axis. The configuration (beam and loading) would then be invariant upon rotating the beam about its axis of symmetry by* 180°. *Should the initially plane sections not remain plane they would warp in some way. Since in pure bending all sections undergo the same bending moment, they would all warp in the same manner. Clearly, such warping would not be invariant by the* 180° *rotation about the axis of symmetry. Therefore, symmetry merely forbids warping. The elasticity solution outlined above goes beyond the symmetry argument. Indeed, it clearly shows that warping is absent in pure bending, regardless of whether the cross-section is symmetric.* *

Table 7.1
Principal Formulae for Pure Bending of Prismatic Beams

Case	$\mathbf{M} = \mathbf{M}_Y$	$\mathbf{M} = \mathbf{M}_Z$
Stress	$\sigma_x = \frac{M_Y}{I_Y} z$	$\sigma_x = -\frac{M_Z}{I_Z} y$
Strain	$\varepsilon_x = \frac{\sigma_x}{E}; \quad \varepsilon_y = \varepsilon_z = -\nu\varepsilon_x$	$\varepsilon_x = \frac{\sigma_x}{E}; \quad \varepsilon_y = \varepsilon_z = -\nu\varepsilon_x$
Curvature	$\frac{\mathrm{d}\omega_Y}{\mathrm{d}x} = \frac{M_Y}{EI_Y} = -\frac{\mathrm{d}^2 w_G}{\mathrm{d}x^2}$	$\frac{\mathrm{d}\omega_Z}{\mathrm{d}x} = \frac{M_Z}{EI_Z} = \frac{\mathrm{d}^2 v_G}{\mathrm{d}x^2}$
Stored energy	$\frac{\mathrm{d}W}{\mathrm{d}x} = \frac{1}{2}\frac{{M_Y}^2}{EI_Y}$	$\frac{\mathrm{d}W}{\mathrm{d}x} = \frac{1}{2}\frac{{M_Z}^2}{EI_Z}$
Bending stiffness	EI_Y	EI_Z
Neutral axis	$z = 0$ (y-axis)	$y = 0$ (z-axis)

DEFINITION 7.1 On each cross-section S, the stress-free points are located on a line called the *neutral axis*. ♣

DEFINITION 7.2 The overall bending stiffness of the beam in the (xOy) plane is the ratio:

$$\frac{M}{\omega_1 - \omega_0} = \frac{EI}{L}.$$

A bending stiffness per unit length is defined as follows:

$$\frac{M_Z}{\mathrm{d}\omega_Z/\,\mathrm{d}x} = EI_Z$$

for two cross-sections separated by $\mathrm{d}x$. ♣

The above developments pertain to bending in the (xOy) plane, Figure 7.1. Analogous equations apply under pure bending in the (xOz) plane, as reported in the below summary table.

7.2 SLOUCHY BENDING OF A PRISMATIC BEAM

By definition, this is the case when both M_Y and M_Z are non zero. Displacements, strains, stresses, and energies are then obtained by superposing the two simple cases of Section 7.1, that is either $M_Y = 0$ or $M_Z = 0$; cf. principle of superposition, Theorem 4.2 in Part I.

Lemma 7.1

In slouchy bending of a prismatic beam with distinct second moments of area ($I_Y \neq I_Z$), the neutral axis does not coincide with the axis about which the bending moment is applied. ◇

Proof *By Definition 7.1, the equation of the neutral axis is given by all points satisfying $\sigma_x = 0$, that is (from superposition of the stresses in Table 7.1)*

$$\frac{M_Y}{I_Y}z - \frac{M_Z}{I_Z}y = 0 \tag{7.6}$$

It is clear that, unless $I_Y = I_Z$, this line is different from the axis of bending given by $M_Y z - M_Z y = 0$. ■

Let then α and β denote the polar angles, measured from GY, of the neutral axis Gy^* and bending moment $\mathbf{M}$, respectively (Figure 7.3). One then has

$$\tan\beta = \frac{M_Z}{M_Y}; \quad \tan\alpha = \frac{I_Y}{I_Z}\tan\beta; \quad I_{y^*} = I_Y\cos^2\alpha + I_Z\sin^2\alpha \tag{7.7}$$

Equation $(7.7)_1$ holds by definition, Equation $(7.7)_2$ follows from Equation (7.6), while Equation $(7.7)_3$ gives the second moment of area of the section with respect to the neutral axis by means of transformation law (5.25).

Theorem 7.1

In slouchy bending of a prismatic beam, the normal stress is given by

$$\sigma_x = \frac{M\cos(\beta - \alpha)}{I_{y^*}}z^* \tag{7.8}$$

The beam curvature at abscissa x is given by

$$\frac{d\omega^*}{dx} = \frac{M\cos(\beta - \alpha)}{EI_{y^*}} \tag{7.9}$$

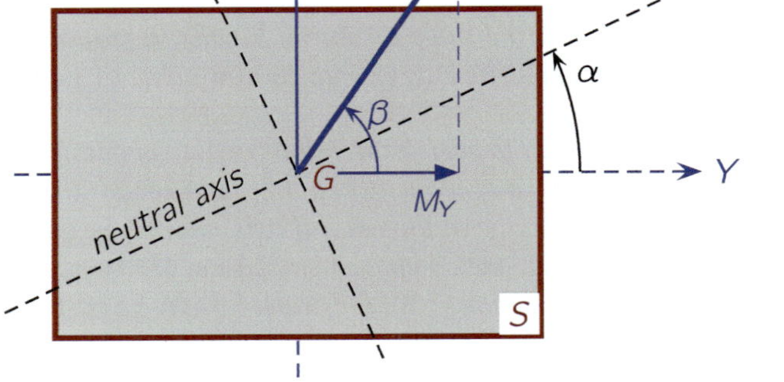

Figure 7.3 Slouchy bending of a symmetric prismatic beam.

The elastic potential energy stored per unit length of the beam is

$$\frac{\mathrm{d}W}{\mathrm{d}x} = \frac{1}{2}\frac{M^2\cos^2(\beta-\alpha)}{EI_{y^*}} \tag{7.10}$$

where M denotes the algebraic measure of the bending moment $\mathbf{M}$ along its axis. ◇

Proof *By Definition 7.1 of the neutral axis, σ_x vanishes on Gy^* and, being a linear function of coordinates on any base, it must then be proportional to z^*. Using Equation (7.6), one gets:*

$$\begin{aligned}
\sigma_x &= \left(\frac{M_Y\cos\alpha}{I_Y} + \frac{M_Z\sin\alpha}{I_Z}\right)z^* \\
&= \frac{1}{I_Z}\left(\frac{M_Y\cos\alpha}{\tan\alpha/\tan\beta} + M_Z\sin\alpha\right)z^* \qquad \textit{using Equation } (7.7)_2 \\
&= \frac{\sin\beta}{I_Z}\left(\frac{M_Y}{\cos\beta}\frac{\cos^2\alpha}{\sin\alpha} + \frac{M_Z}{\sin\beta}\sin\alpha\right)z^* \\
&= \frac{M\sin\beta}{I_Z\sin\alpha}\left(\cos^2\alpha + \sin^2\alpha\right)z^* \qquad \textit{using } M_Y = M\cos\beta \textit{ and } M_Z = M\sin\beta \\
&= \frac{M\sin\beta}{I_Z\sin\alpha}z^*
\end{aligned}$$

While relatively simple, the latter expression mixes starred and unstarred frame quantities (z^* and I_Z); it also does not naturally encompass the trivial case of bending about a principal axis ($\alpha = 0$). Making use of Equation $(7.7)_3$, one obtains the desired expression, Equation (7.8), after simple algebraic manipulations (see Problem 7.5).

For the curvature, write the rotation about Gy^* as $\omega^* = \omega_Y\cos\alpha + \omega_Z\sin\alpha$. Application of the tabulated formulae in Table 7.1 then leads to

$$\frac{\mathrm{d}\omega^*}{\mathrm{d}x} = \frac{1}{E}\left(\frac{M_Y\cos\alpha}{I_Y} + \frac{M_Z\sin\alpha}{I_Z}\right)$$

in which the bracketed term is the same as above. Finally, Equation (7.10) follows directly from the definition of elastic energy with σ_x now given by Equation (7.8). ■

Remark 7.2 *Having mastered the basic expressions of pure bending (see Table 7.1), a mnemonic way to retain those of slouchy bending is to ignore the component of the bending moment perpendicular to the neutral axis. Indeed, in the starred frame associated with the neutral axis, the components of $\mathbf{M}$ are $M_{y^*} = M\cos(\beta-\alpha)$ and $M_{z^*} = M\sin(\beta-\alpha)$. Thus, Equation (7.8), for example, may be cast as follows:*

$$\sigma_x = \frac{M_{y^*}}{I_{y^*}}z^* \tag{7.11}$$

which takes the form of the expression under pure bending (left column of Table 7.1).

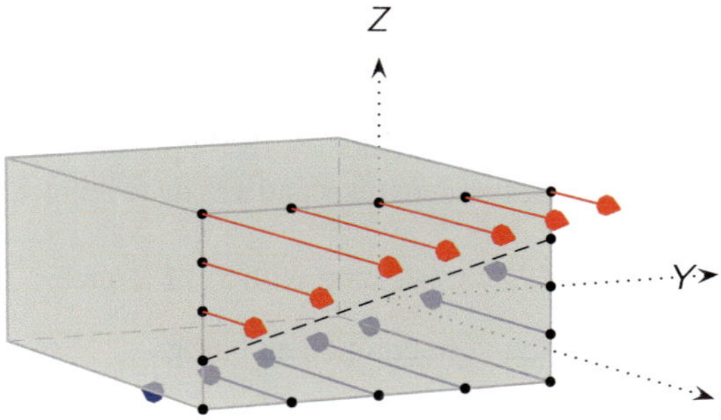

Figure 7.4 Normal tractions under slouchy bending.

Figure 7.4 depicts the distribution of normal tractions acting on the current cross-section. The neutral axis is shown dashed. For clarity, traction vectors are only plotted at surface points. Vectors painted red correspond to tensile states, whereas vectors painted blue are for compression states. The magnitude of each traction only depends on the distance z^* between the point of application of the traction and the neutral axis, as per Equation (7.11). $*$

Remark 7.3 *It is about the neutral axis that the global, bending-induced rotation of current cross-section S takes place, and hence the notation ω^* for that rotation.* $*$

Remark 7.4 *In the literature, one often finds the terms* asymmetric *or* unsymmetrical *bending to refer to what is called* slouchy bending *in this book. In most texts, it is rather explicit that the authors refer to unsymmetric cross-sections when dealing with asymmetric bending. This terminology is purposely avoided here as it is misleading. It is emphasized that slouchy bending applies to both symmetric and unsymmetric cross-sections. Anytime a bending moment is not along a principal axis of inertia, one tackles a situation of "slouchy" bending.*

In the illustration of Figure 7.3, the cross-section admits two axes of symmetry so that the principal axes of inertia are obvious. As indicated in Chapter 5, if the cross-section admits one axis of symmetry, the principal frame of inertia is also determined by the axis of symmetry and the axis perpendicular to it. An example is shown in Figure 7.5. The bending moment being at an angle to the principal axes, a situation of slouchy bending arises. More generally, whether the cross-section admits an axis of symmetry or not is irrelevant to the general problem of bending. One works in the central principal axes of inertia. The symmetry of the cross-section only matters in that it determines the principal frame without any calculation. $*$

7.3 BENDING MOMENT IN ARBITRARY BEAMS

Consider a beam of arbitrary shape and subjected to some arbitrary loading that gives rise on current cross-section S at arclength s to a screwer $\{\mathcal{S}\}$, in particular to a bending moment $\mathbf{M}(s)$, Figure 7.6. Taking into account the basic beam assumptions (the diameter of each cross-section S being small compared with the length of the median line and with the radii of curvature and torsion of that line in G), any thin slice of the beam may be considered prismatic. To a thin slice comprised between cross-sections S and S', located at arclengths s and $s + \mathrm{d}s$ (Figure 7.6), the results of previous sections apply.

In the principal frame $\{G, XYZ\}$ associated with S, $\mathbf{M}$ induces:

- a stress field with the only non-zero component being

$$\sigma_X = \frac{M_Y}{I_Y} Z - \frac{M_Z}{I_Z} Y \tag{7.12}$$

- a corresponding strain field with the following non-zero components:

$$\varepsilon_X = \frac{\sigma_X}{E}; \qquad \varepsilon_Y = \varepsilon_Z = -\nu \varepsilon_X \tag{7.13}$$

a rotation of S' with respect to S about G, defined as $\mathrm{d}\boldsymbol{\omega} = \mathrm{d}\boldsymbol{\omega}_Y + \mathrm{d}\boldsymbol{\omega}_Z$ with

$$\mathrm{d}\boldsymbol{\omega}_Y = \frac{\mathbf{M}_Y}{EI_Y} \mathrm{d}s \qquad \mathrm{d}\boldsymbol{\omega}_Z = \frac{\mathbf{M}_Z}{EI_Z} \mathrm{d}s \tag{7.14}$$

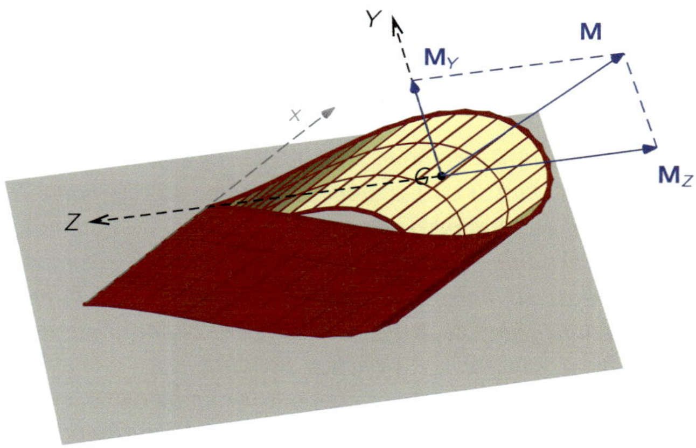

Figure 7.5 Slouchy bending of a prismatic beam with one axis of symmetry.

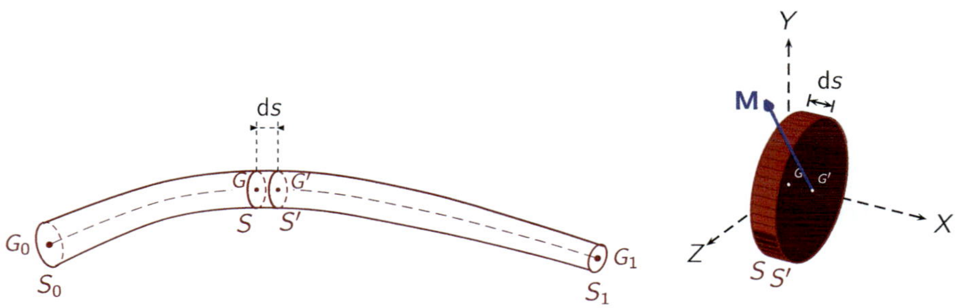

Figure 7.6 Bending moment in an arbitrary beam.

a potential elastic energy stored per unit length given by

$$\frac{dW}{ds} = \frac{1}{2}\frac{M_Y^2}{EI_Y} + \frac{1}{2}\frac{M_Z^2}{EI_Z} \tag{7.15}$$

Remark 7.5 *It is important to keep the indices straight in Equation* (7.14). *For example, writing*

$$d\boldsymbol{\omega} = \frac{\mathbf{M}}{EI} ds$$

would be meaningless because $I_Z \neq I_Y$. *

Remark 7.6 *In Definition 7.1, the neutral axis is a line lying in the plane of the current cross-section. In prismatic beams, the locus of stress-free points is some* neutral *plane, the trace of which in the current cross-section is the neutral axis. In an arbitrary beam, one may eventually refer to a neutral surface. This is the case for a flexible wing with a typical NACA profile, a cross-section which does not admit any axis of symmetry, Figure 7.7.* *

7.4 PLANE BENDING OF A CIRCULAR BEAM

A beam admitting (xOy) as median plane has for its median line $\widehat{G_0G_1}$ a circular arc of radius R and centered at O, Figure 7.8. The beam is loaded in its plane. On current cross-section S, identified by polar angle θ, this loading generally gives rise to a normal (or axial) force N, a transverse shear V_Y and a bending moment M_Z. Let

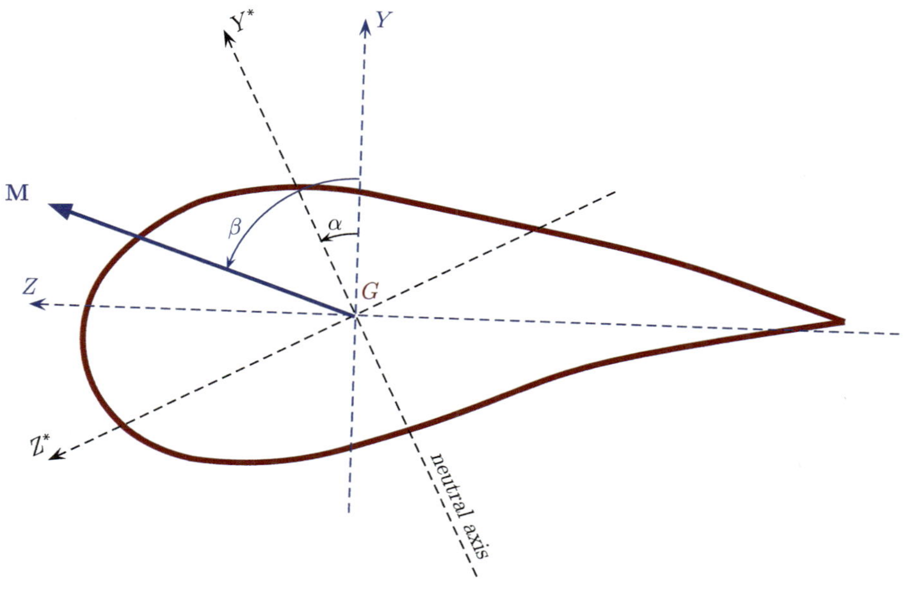

Figure 7.7 Slouchy bending of an asymmetric section in an arbitrary beam.

U and V denote the displacements of centroid G along X and Y, induced by N and M_Z; it will be justified in Chapter 9 that the displacements due to V_Y are negligible. Both U and V are viewed as functions of θ.

In order to obtain the differential equations that govern $U(\theta)$ and $V(\theta)$, the following results from the previous sections will be utilized:

- Specializing Equation (7.14), the cross-sectional rotation ω satisfies

$$\frac{\mathrm{d}\omega}{\mathrm{d}\theta} = \frac{RM}{EI} \tag{7.16}$$

- The displacement along X of a particle P of S having ordinate Y is $U - Y\omega$; see the second conclusion under Section 7.1.
- The axial strain at G is solely due to the normal force N so that

$$\varepsilon_X\Big|_G = \frac{N}{ES} \tag{7.17}$$

Using the relevant expressions giving ε_X and ω in cylindrical coordinates:

$$\varepsilon_X\Big|_G = \varepsilon_\theta\Big|_G = -\frac{V}{R} + \frac{1}{R}\frac{\mathrm{d}U}{\mathrm{d}\theta} \tag{7.18}$$

$$\omega = \omega_Z\Big|_G = \frac{1}{2}\left[\frac{1}{R}\frac{\mathrm{d}V}{\mathrm{d}\theta} - \frac{\partial}{\partial Y}(U - \omega Y) + \frac{U}{R}\right] \tag{7.19}$$

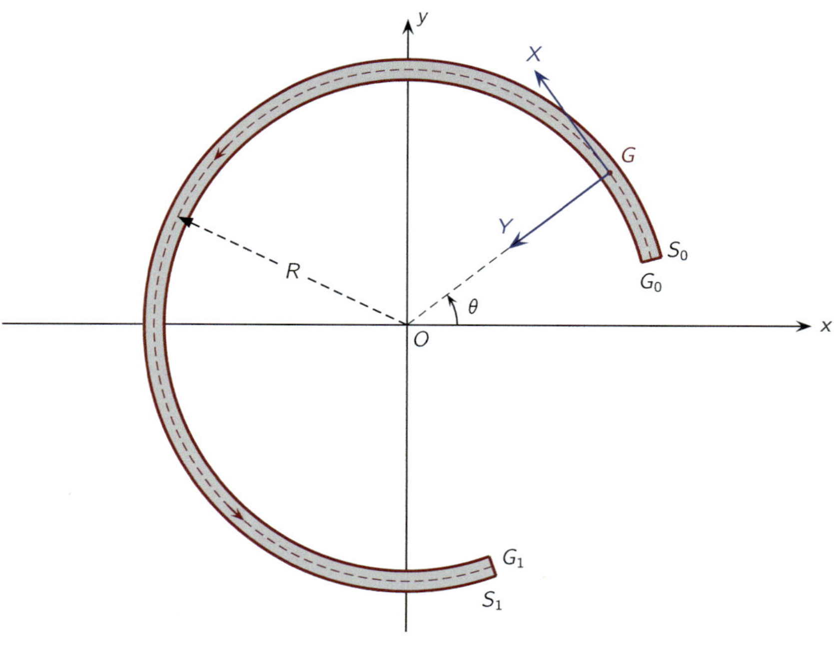

Figure 7.8 Circular beam under plane loading (not shown).

where use has been made of the transformations illustrated in Figure 7.9. In view of Equation (7.17), Equation (7.18) simplifies into

$$\frac{\mathrm{d}U}{\mathrm{d}\theta} = V + \frac{NR}{ES} \tag{7.20}$$

while Equation (7.19) becomes

$$\omega = \frac{1}{R}\left(U + \frac{\mathrm{d}V}{\mathrm{d}\theta}\right) \tag{7.21}$$

The latter, combined with Equation (7.16), allows to eliminate ω such that

$$\frac{RM}{EI} = \frac{1}{R}\left(\frac{\mathrm{d}U}{\mathrm{d}\theta} + \frac{\mathrm{d}^2V}{\mathrm{d}\theta^2}\right) \tag{7.22}$$

from which U may be eliminated using Equation (7.20) to obtain the governing equation for V. In sum, one gets the system of two linear differential equations:

$$\boxed{\begin{cases} \dfrac{\mathrm{d}^2V}{\mathrm{d}\theta^2} + V = \dfrac{R^2M}{EI} - \dfrac{RN}{ES} \\ \dfrac{\mathrm{d}U}{\mathrm{d}\theta} = V + \dfrac{RN}{ES} \end{cases}} \tag{7.23}$$

Integration of Equation (7.23) with known internal resultants $N(\theta)$ and $M(\theta)$ gives the deflection $V(\theta)$ and the axial displacement $U(\theta)$. The calculation of N and M itself requires that the equilibrium equations (5.46) be solved; see Example 5.9 where in the absence of distributed loads the governing equations of N and M reduce to Equation (5.52).

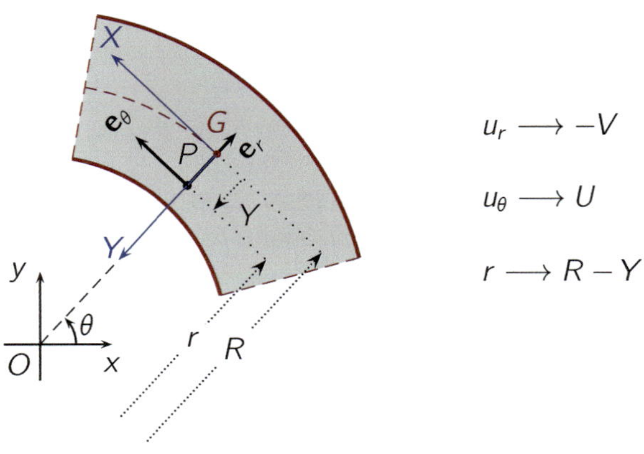

Figure 7.9 Change of axes and variables from the cylindrical frame to the principal frame.

Remark 7.7 *Indices were dropped from I_Z and M_Z in the above for convenience. However, the subscript is kept in V_Y to avoid confusion between the transverse shear and the deflection V measured with respect to the mobile frame. Note that the shear V_Y does not appear in the final equations.* ∗

Remark 7.8 *The axial strain at any point P of S has two components: one that results from N and the other due to M. However, at G the contribution of M vanishes since G belongs to the neutral axis irrespective of the shape of the cross-section.* ∗

Remark 7.9 *A fully circular beam is a special case where end sections S_0 and S_1 are identified with each other. The resulting ring is then internally hyperstatic.* ∗

7.5 PLANE BENDING OF A STRAIGHT BEAM

A straight beam admits (xOy) as a median plane; its median line G_0G_1 lies on the x-axis. Applied loads include forces parallel to Oy and couples parallel to Oz. On current cross-section S, the only non-zero internal resultants are V_Y and M_Z; see Section 5.5.4.

The deflection $v(x)$ of the beam is the vertical displacement of the centroid G of current cross-section S located at x, due to the sole bending moment. It satisfies the fundamental equation:

$$\boxed{\frac{\mathrm{d}^2 v}{\mathrm{d}x^2} = \frac{M_Z}{EI_Z}} \tag{7.24}$$

Proof *The above equation is established by elimination of ω_Z between the two equations:*

$$\omega_Z = \frac{\mathrm{d}v}{\mathrm{d}x}$$

which follows, for example, from specializing Equation (7.21) *to zero curvature, and*

$$\frac{\mathrm{d}\omega_Z}{\mathrm{d}x} = \frac{M_Z}{EI_Z}$$

which specializes Equation $(7.14)_2$ *to a straight beam.* ■

Remark 7.10 *While governing equation (7.24) is formally the same as that established in Section 7.1 for a prismatic beam, it must be recognized that both M_Z and I_Z might be functions of x in Equation* (7.24). ∗

Remark 7.11 *If (xOz) is the median plane then similar expressions hold with $w(x)$ denoting the deflection of G:*

$$\omega_Y = -\frac{\mathrm{d}w}{\mathrm{d}x}, \quad \frac{\mathrm{d}\omega_Y}{\mathrm{d}x} = \frac{M_Y}{EI_Y}, \qquad \textit{hence} \qquad \frac{\mathrm{d}^2 w}{\mathrm{d}x^2} = -\frac{M_Y}{EI_Y} \tag{7.25}$$

∗

Remark 7.12 *Consider a plane curve, say lying in* (xOy)*, defined by the equation* $y = v(x)$*. Its curvature at current point* $P(x,y)$ *is*

$$\frac{1}{R} = \frac{v''}{(1+v'^2)^{\frac{3}{2}}}$$

In bent straight beams, the slope of the tangent to the deformed *median line, i.e.,* $v' = \frac{\mathrm{d}v}{\mathrm{d}x} = \omega_Z$*, is small in magnitude compared to unity. Therefore, the ratio* M_Z/EI_Z *represents the (signed) curvature taken by the median line at current point G.* $*$

Remark 7.13 *It will be shown in Chapter 9 that the deflection due to transverse shearing is negligible in comparison with that due to the bending moment. Equation* (7.24) *is therefore sufficient to determine the deformation of a straight beam in plane bending.* $*$

Remark 7.14 *The governing equation for the deflection, Equation* (7.24)*, is a simple linear* second-*order differential equation, the definition of which is completed by* two *appropriate boundary conditions on either* v *(the deflection) and* $v' \equiv \omega$ *(the rotation), or on* v *alone. Under no circumstances should either* v *or* ω *be discontinuous. Unlike the screwer components, which may undergo a jump in case of applied point forces and couples (see Remark 5.25), deflections and slopes are continuous functions of* x*. However, since the right-hand side of Equation* (7.24) *depends on the bending moment, the differential equation to be solved may vary from one interval to another. In this case, continuity conditions on deflection and slope must be implemented to patch the solutions. In some problems, symmetry arguments may be invoked.* $*$

The plane bending of beams being a very important topic, several examples are provided below which illustrate the main methods used, especially for hyperstatic problems. In many cases, one is only interested in calculating the deflections of particular points. In such cases, some energy theorems (e.g., Castigliano's, Menabrea's) provide the most effective ways to obtain the answers, swiftly and elegantly. Yet, such methods will not be used here.

Example 7.1: Hyperstatic Beam under Its Own Weight

Let $2L$ be the length of the prismatic beam in Figure 7.10, p its weight per unit length, and Ox, Oy, Oz the principal axes of the median cross-section.

Global equilibrium of the beam gives (cf. the free-body diagram of Figure 7.11):

$$R^A = R^B \qquad \text{and} \qquad R^A + R^B + R^O = 2pL$$

The structure is therefore externally hyperstatic of degree 1: there is one redundant support.

Invoking symmetry, it suffices to reason on the part $0 < x < L$, on which one has:

$$V = R^B - p(L-x) \qquad \text{and} \qquad M = R^B(L-x) - \frac{1}{2}p(L-x)^2$$

following the steps of Example 5.7.

The resolution method illustrated here is known as the "double integration" method whereby the governing equation for the deflection, Equation (7.24), is integrated twice on each interval where the bending moment is specified. The deflection differential equation writes then:

$$\frac{\mathrm{d}^2 v}{\mathrm{d}x^2} = \frac{R^B}{EI}(L-x) - \frac{1}{2}\frac{p}{EI}(L-x)^2$$

A first integration gives account being taken of the condition $v'(0) = 0$ (from continuity of the slope at the middle support):

$$\frac{\mathrm{d}v}{\mathrm{d}x} = -\frac{R^B}{2EI}\left[(L-x)^2 - L^2\right] + \frac{p}{6EI}\left[(L-x)^3 - L^3\right]$$

A second integration yields (considering boundary condition $v(0) = 0$):

$$v(x) = -\frac{R^B}{2EI}\left[-\frac{1}{3}(L-x)^3 - L^2 x + \frac{1}{3}L^3\right] + \frac{p}{6EI}\left[-\frac{1}{4}(L-x)^4 - L^3 x + \frac{1}{4}L^4\right]$$

The final boundary condition $v(L) = 0$ allows to determine the hyperstatic unknown R^B so that

$$R^B = R^A = \frac{3}{8}pL; \qquad R^O = \frac{5}{4}pL \tag{7.26}$$

for the global reactions and

$$\begin{cases} V = pL\left(\dfrac{x}{L} - \dfrac{5}{8}\right) \\ M = -\dfrac{1}{8}pL^2\left(4\dfrac{x^2}{L^2} - 5\dfrac{x}{L} + 1\right) \end{cases} \tag{7.27}$$

for the screwer components and

$$v(x) = -\frac{pL^4}{48EI}\left(2\frac{x^4}{L^4} - 5\frac{x^3}{L^3} + 3\frac{x^2}{L^2}\right) \tag{7.28}$$

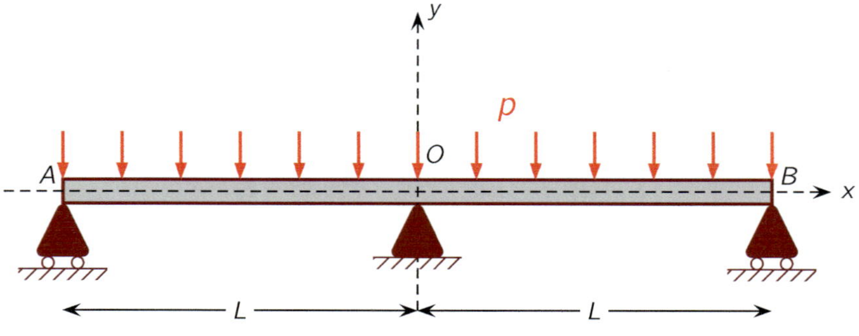

Figure 7.10 Prismatic beam subjected to its own weight on three supports.

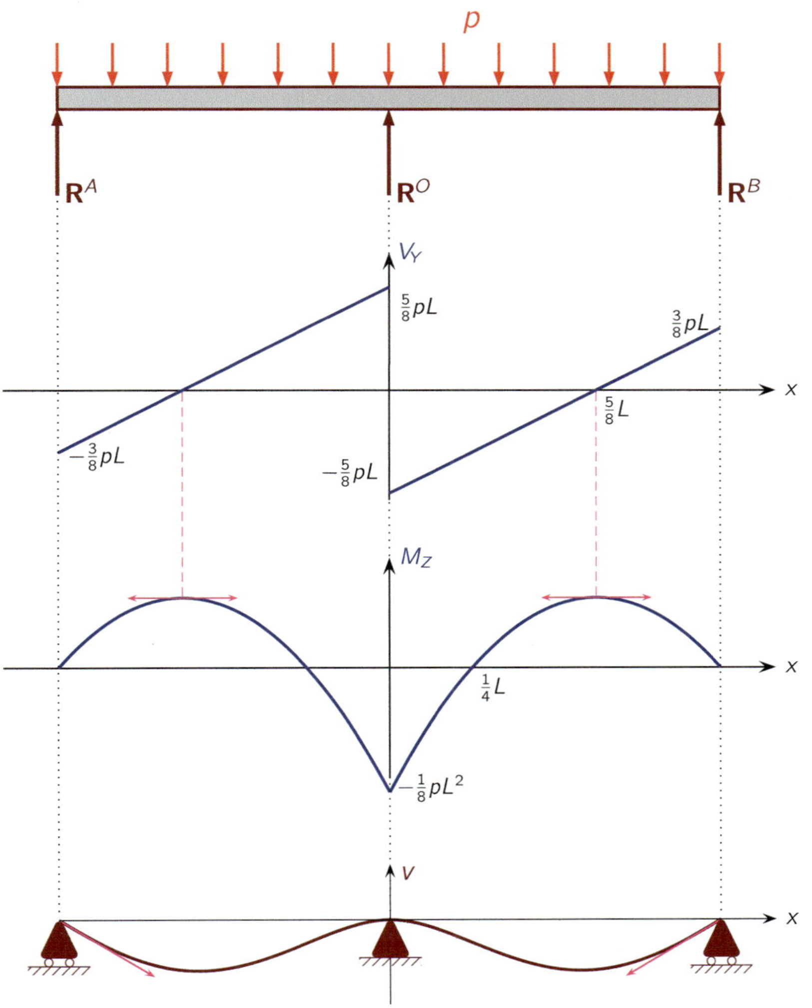

Figure 7.11 Shear, bending moment, and deflection diagrams for the problem of Figure 7.10.

for the deflection.

By symmetry, the solution on part OA of the beam is

$$\begin{cases} V = pL\left(\dfrac{x}{L} + \dfrac{5}{8}\right) \\ M = -\dfrac{1}{8}pL^2\left(4\dfrac{x^2}{L^2} + 5\dfrac{x}{L} + 1\right) \end{cases} \tag{7.29}$$

$$v(x) = -\frac{pL^4}{48EI}\left(2\frac{x^4}{L^4} + 5\frac{x^3}{L^3} + 3\frac{x^2}{L^2}\right) \tag{7.30}$$

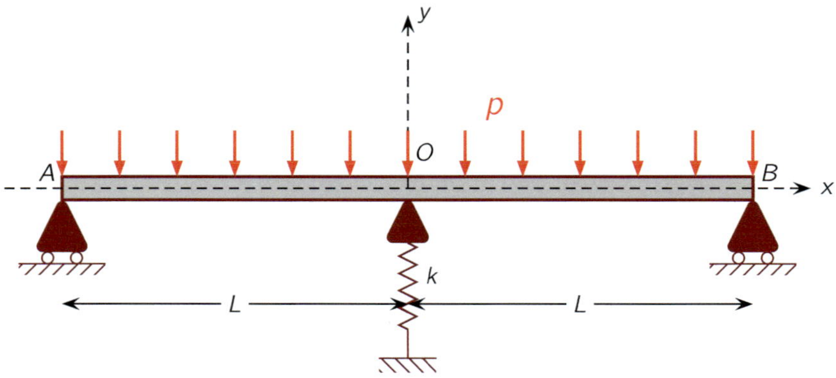

Figure 7.12 Beam subjected to its own weight on two rollers and one elastic support.

A graphical representation of the solution is depicted in Figure 7.11 where the deflections are exaggerated for clarity purposes. ♡

Example 7.2: Hyperstatic Beam on Elastic Support

Consider the prismatic beam of Example 7.1 and replace the middle support with an elastic support having stiffness k, Figure 7.12.

The same global equilibrium equations hold as in Example 7.1 with $R^O = -kv(0)$ here. The structure is externally hyperstatic of degree 1.

Resolution by "double integration" proceeds as in the previous example, the difference being that the condition $v(0) = 0$ is replaced with the condition $v(L) = 0$ after the second integration. Then, the relation $v(0) = -R^O/k$ is used to determine R^O as

$$R^O = \frac{5}{4}\alpha pL \quad \text{with} \qquad \alpha = \frac{1}{1+6\dfrac{EI}{kL^3}} \tag{7.31}$$

and the other reactions as

$$R^B = R^A = \frac{3}{8}\alpha\left(1+16\frac{EI}{kL^3}\right)pL \tag{7.32}$$

Eventually, one gets, first for $0 < x < L$:

$$\begin{cases} V = R^B - p(L-x) \\ M = R^B(L-x) - \dfrac{1}{2}p(L-x)^2 \end{cases} \tag{7.33}$$

for the screwer components and

$$v(x) = -\frac{R^B}{2EI}\left[-\frac{1}{3}(L-x)^3 - L^2x + L^3\right] + \frac{p}{6EI}\left[-\frac{1}{4}(L-x)^4 - L^3x + L^4\right] \tag{7.34}$$

for the deflection.

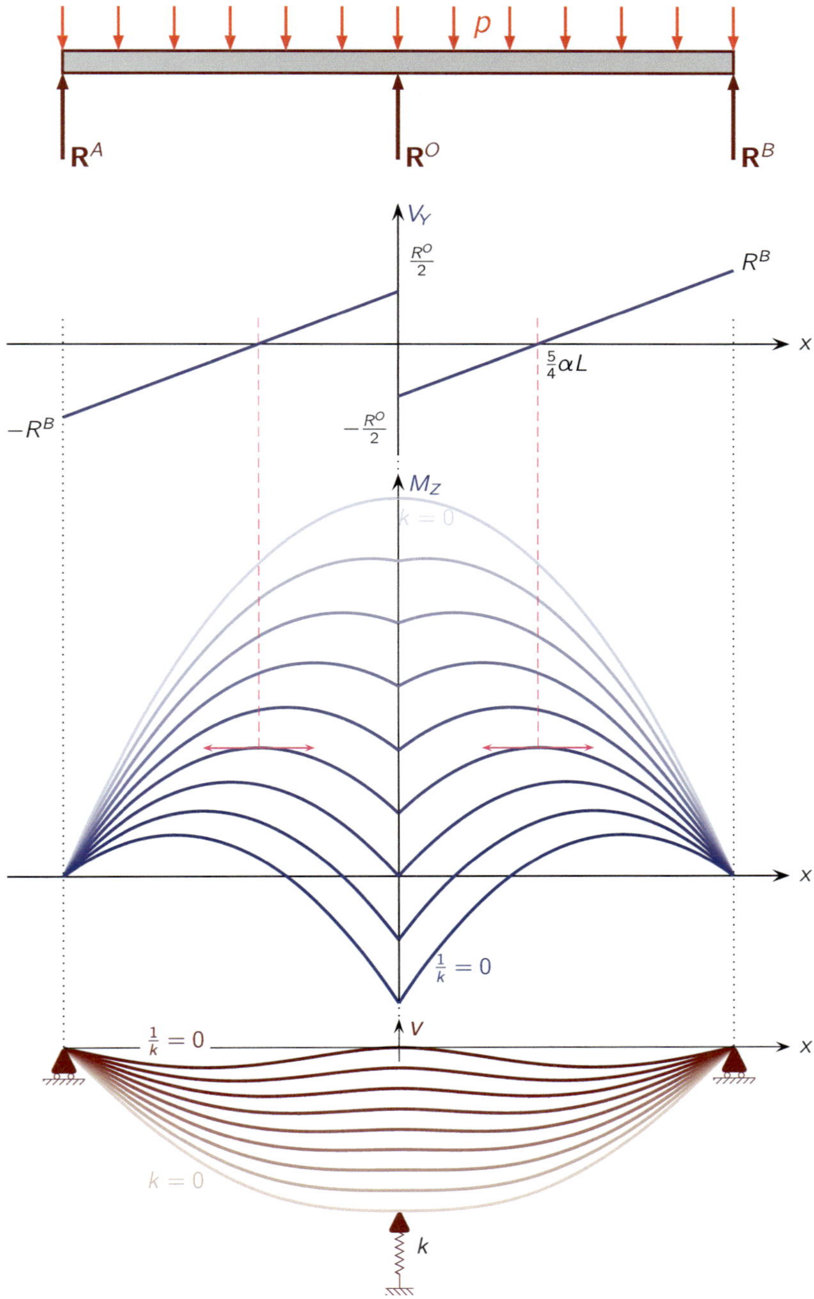

Figure 7.13 Shear, bending moment, and deflection diagrams for the problem of Figure 7.12.

Then, from symmetry (see Remark 7.15) for $-L < x < 0$:

$$\begin{cases} V = -R^B + p(L+x) \\ M = R^B(L+x) - \dfrac{1}{2}p(L+x)^2 \end{cases} \tag{7.35}$$

and

$$v(x) = -\frac{R^B}{2EI}\left[-\frac{1}{3}(L+x)^3 + L^2x + L^3\right] + \frac{p}{6EI}\left[-\frac{1}{4}(L+x)^4 + L^3x + L^4\right] \tag{7.36}$$

For mere convenience, the above expressions are given in terms of R^B, which is determined from Equation (7.32). Particular values include:

$$M(0) = \frac{1}{2}pL^2\left(1 - \frac{5}{4}\alpha\right)$$

A graphical representation of the solution is depicted in Figure 7.13 where, again, the deflections are exaggerated. Various cases are shown, which correspond to a range of values of the spring stiffness k. Note that only one case is shown for the shear diagram for clarity sake. One obviously retrieves the two extreme cases:

- infinitely rigid support $\left(\frac{1}{k} = 0\right)$, which is the case treated in Example 7.1.
- absence of a (middle) support ($k = 0$).

♡

Remark 7.15 *Symmetry is handled above using:*

$$V(-x) = -V(x); M(-x) = M(x); v(-x) = v(x)$$

*

Example 7.3: Prismatic Beam Fixed and Supported

The beam, prismatic of length L, is fixed at one end (section at the origin, S^O) and simply supported at the other end (final section, S^A), as shown in Figure 7.14. It is subjected to a distributed downward vertical loading of variable density: $p(x) = \bar{p}\frac{x}{L}$. The structure is externally hyperstatic of degree 1.

A method based on the principle of superposition, Theorem 4.2 in Part I, is used. The imposed loading may be considered as the superposition of two loadings, denoted (1) and (2) (Figure 7.15) subject to the condition of zero net deflection at A.

In the case (1), isostatic, one finds the deflection at A to be:

$$v^{(1)}(L) = -\frac{11}{120}\frac{\bar{p}L^4}{EI}$$

This case is treated in the last tabulated problem in Section 7.6.1 further below. In the case (2), also isostatic, one finds:

$$v^{(2)}(L) = \frac{R^A L^3}{3EI}$$

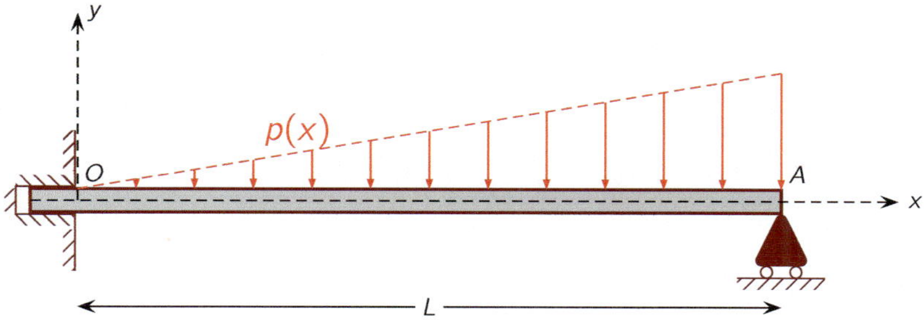

Figure 7.14 Hyperstatic beam subjected to nonuniform loading.

see case 7 in the tabulated results of Section 7.6.1.

Writing the null condition on the net deflection at A, that is,

$$v(L) = v^{(1)}(L) + v^{(2)}(L) = 0$$

yields

$$R^A = \frac{11}{40}\bar{p}L$$

Global equilibrium for the original problem (Figure 7.14) reads:

$$R^O + R^A = \frac{1}{2}\bar{p}L \quad \text{and} \qquad M^O + R^A L = \frac{1}{3}\bar{p}L^2$$

thus leading to all reactions:

$$R^A = \frac{11}{40}\bar{p}L \qquad R^O = \frac{9}{40}\bar{p}L \qquad M^O = \frac{7}{120}\bar{p}L^2 \tag{7.37}$$

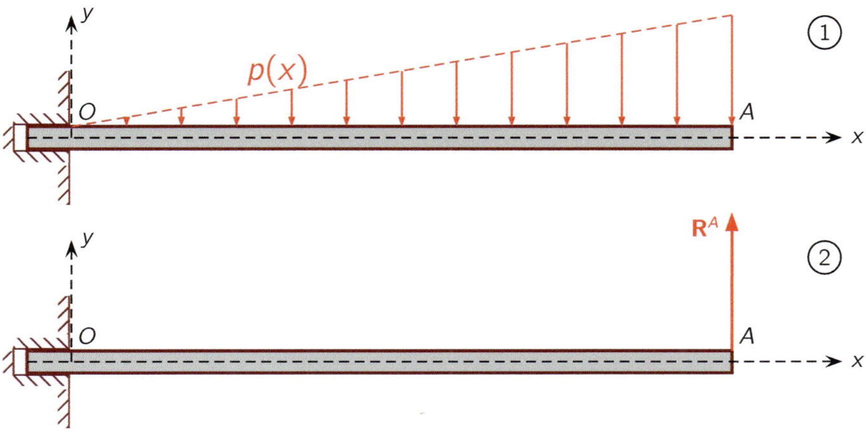

Figure 7.15 Application of superposition principle in solving the problem of Figure 7.14.

On the current cross-section at x, one then obtains (see Problem 7.19)

$$\begin{cases} V = \dfrac{1}{40}\bar{p}L\left(20\dfrac{x^2}{L^2} - 9\right) \\ M = -\dfrac{\bar{p}L^2}{120}\left(20\dfrac{x^3}{L^3} - 27\dfrac{x}{L} + 7\right) \end{cases} \tag{7.38}$$

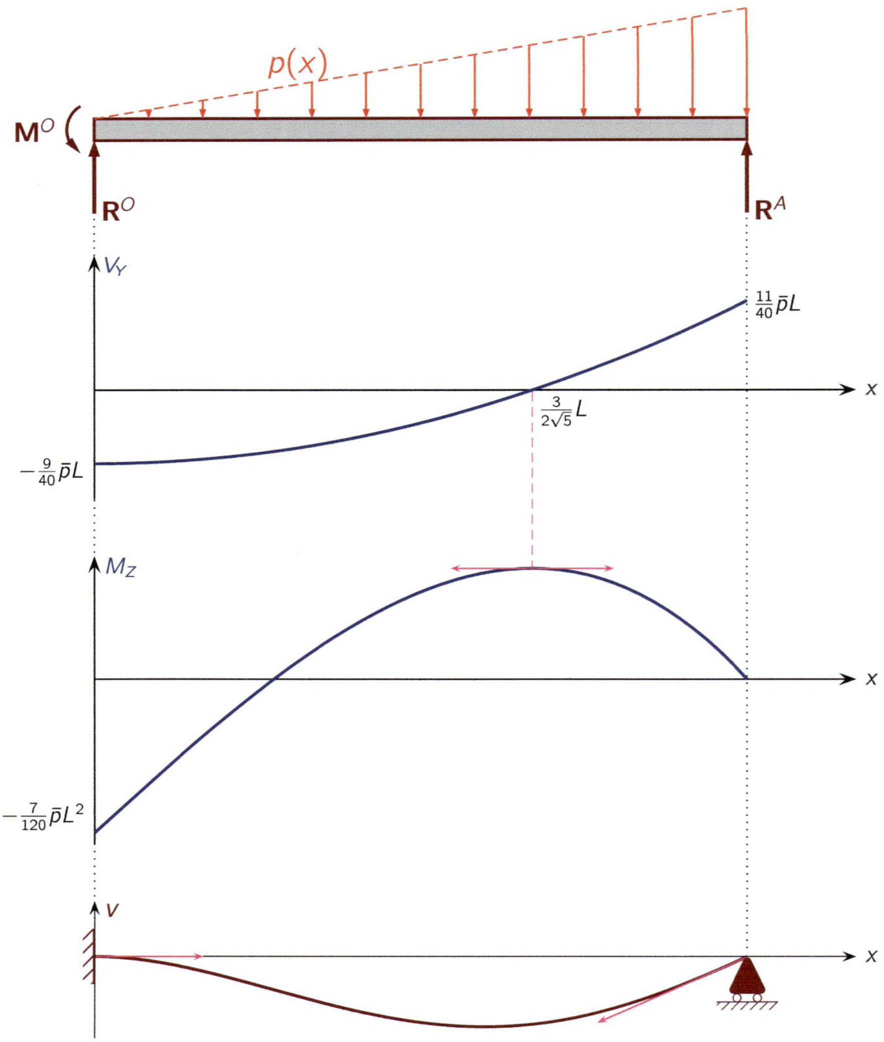

Figure 7.16 Shear, bending moment, and deflection diagrams for the problem of Figure 7.14.

for the screwer components and

$$v(x) = -\frac{\bar{p}L^4}{120EI}\left(\frac{x^5}{L^5} - \frac{9}{2}\frac{x^3}{L^3} + \frac{7}{2}\frac{x^2}{L^2}\right) \tag{7.39}$$

for the deflection.

A graphical representation of the solution is depicted in Figure 7.16 where the deflections are exaggerated for the sake of clarity. ♡

Example 7.4: Bending of a Two-Beam Structure

Consider the structure depicted in Figure 7.17, admitting (xOy) as median plane and consisting of two identical prismatic beams, each of length L, and welded together following section A.

Beam OA is vertical, of axis Oy, fixed at initial section S^O; Beam AB, horizontal and parallel to Ox, is subjected at its tip B to a downward vertical force of intensity P. As usual, $\{G, XYZ\}$ denotes the principal frame of current cross-section S.

Since this two-beam structure is isostatic, global equilibrium fully determines all reactions:

- a vertical reaction $Y^O = P$ (algebraic value along Oy)
- a clamping moment $M^O = PL$ (algebraic value along Oz)

On the current cross-section between O and A, the non-zero components of

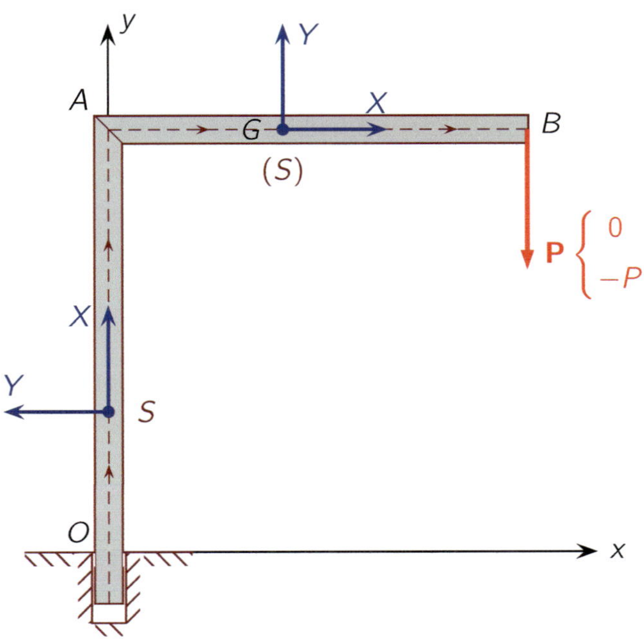

Figure 7.17 A two-beam structure loaded in its plane.

the screwer are: $N = -P$ and $M_Z = -PL$. On AB, on the other hand, the only components are: $V_Y = -P$ and $M_Z = -P(L-x)$.

Neglecting shear deflections, curvatures, rotations, and displacements are calculated on each beam separately.

- On a section S of OA, having coordinates $(0, y)$:
Curvature

$$\frac{1}{R} = -\frac{\mathrm{d}^2u}{\mathrm{d}y^2} = -\frac{PL}{EI} \quad \text{(constant)}$$

Rotation

$$\omega = -\frac{\mathrm{d}u}{\mathrm{d}y} = -\frac{PL}{EI}y; \quad (\omega^A = -\frac{PL^2}{EI})$$

Deflection

$$u = \frac{PL}{2EI}y^2; \quad (u^A = \frac{PL^3}{2EI})$$

Axial displacement

$$v = -\frac{P}{ES}y; \quad (v^A = -\frac{PL}{ES})$$

- On a section S of AB, having coordinates (x, L):
Curvature

$$\frac{1}{R} = \frac{\mathrm{d}^2v}{\mathrm{d}x^2} = -\frac{P}{EI}(L-x)$$

Rotation

$$\omega = \frac{\mathrm{d}v}{\mathrm{d}x} = -\frac{PL^2}{EI}\left(\frac{x}{L} - \frac{1}{2}\frac{x^2}{L^2} + 1\right)$$

Deflection

$$v = -\frac{PL^3}{2EI}\left(\frac{1}{2}\frac{x^2}{L^2} - \frac{1}{6}\frac{x^3}{L^3} + \frac{x}{L}\right) - \frac{PL}{ES}$$

Axial displacement

$$u = u^A = \frac{PL^3}{2EI}$$

Thus, section B undergoes:

- a translation along x: $u^B = \frac{PL^3}{2EI}$
- a translation along y: $v^B = -\frac{4}{3}\frac{PL^3}{EI} - \frac{PL}{ES}$
- a rotation: $\omega^B = -\frac{3}{2}\frac{PL^2}{EI}$

♡

Remark 7.16 *If three straight beams or more are assembled together, the resulting structure becomes more complex to analyze. Examples include gantries and frames. Frames, in particular, are internally hyperstatic, just like rings. Their analysis involves a combination of bending with shear and extension. More generally, the assemblage of arbitrary beams make up an ossature. Methods for analyzing rings, frames, and more generally ossatures are described in Chapter 10.* ∗

7.6 FORMULAE FOR PLANE BENDING

For ease of reference, this section collects solutions to various problems of plane bending for prismatic beams. In all, xOy denotes the median plane, which is also the plane of loading, Ox the median line, Oy the ascending vertical, O the centroid of the initial cross-section, EI the local bending stiffness, V the transverse shear, M the bending moment, v the deflection, and $\omega = \frac{\mathrm{d}v}{\mathrm{d}x}$ the rotation.

7.6.1 ISOSTATIC BEAMS

Loading, Diagrams	Reactions	V and M	v and ω
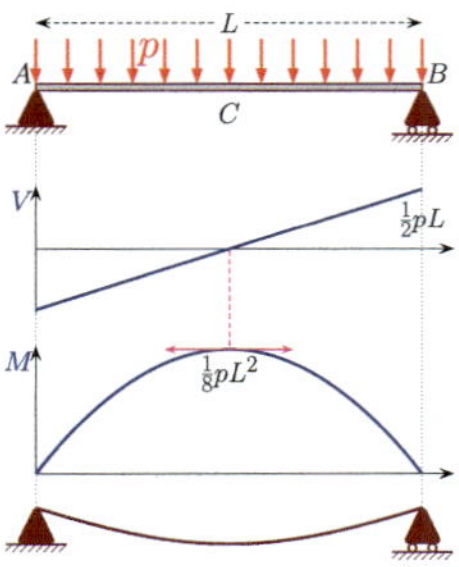	$R^A = \frac{pL}{2}$ $R^B = \frac{pL}{2}$ $AC = \frac{L}{2}$	$V = -pL(\frac{1}{2} - \frac{x}{L})$ $M = \frac{1}{2}pLx(1 - \frac{x}{L})$ $M^C = \frac{1}{8}pL^2$ $V^C = 0$	$v = -\frac{1}{24}\frac{pL^4}{EI}(\frac{x}{L} - 2\frac{x^3}{L^3} + \frac{x^4}{L^4})$ $\omega = -\frac{1}{24}\frac{pL^3}{EI}(1 - 6\frac{x^2}{L^2} + 4\frac{x^3}{L^3})$ $v^C = -\frac{5}{384}\frac{pL^4}{EI}$
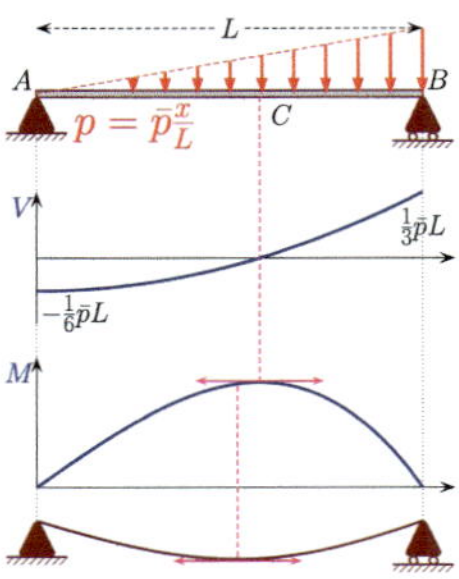	$R^A = \frac{\bar{p}L}{6}$ $R^B = \frac{\bar{p}L}{3}$ $AC = \frac{\sqrt{3}}{3}L$	$V = -\frac{1}{6}\bar{p}L(1 - 3\frac{x^2}{L^2})$ $M = \frac{1}{6}\bar{p}Lx(1 - \frac{x^2}{L^2})$ $M^C = \frac{\sqrt{3}}{27}\bar{p}L^2$ $V^C = 0$	$v = -\frac{\bar{p}L^4}{360EI}(7\frac{x}{L} - 10\frac{x^3}{L^3} + 3\frac{x^5}{L^5})$ $\omega = -\frac{\bar{p}L^3}{360EI}(7 - 30\frac{x^2}{L^2} + 15\frac{x^4}{L^4})$
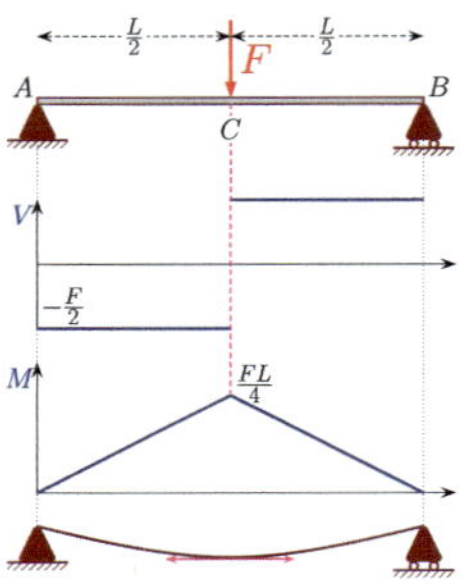	$R^A = \frac{F}{2}$ $R^B = \frac{F}{2}$ $AC = \frac{L}{2}$	$V^{AC} = -\frac{F}{2}$ $V^{CB} = \frac{F}{2}$ $M^{AC} = \frac{1}{2}Fx$ $M^{CB} = \frac{1}{2}F(L-x)$ $M^C = \frac{1}{4}FL$	On AC: $v = -\frac{FL^3}{48EI}(3\frac{x}{L} - 4\frac{x^3}{L^3})$ $\omega = -\frac{FL^2}{16EI}(1 - 4\frac{x^2}{L^2})$ $v^C = -\frac{FL^3}{48EI}$ $\omega^A = -\frac{FL^2}{16EI}$

Loading, Diagrams	Reactions	V and M	v and ω
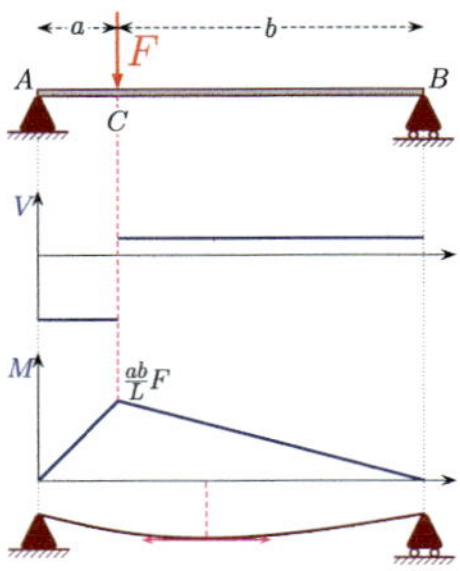	$R^A = -\frac{b}{L}F$ $R^B = -\frac{a}{L}F$	$V^{AC} = -\frac{b}{L}F$ $V^{CB} = \frac{a}{L}F$ $M^{AC} = \frac{b}{L}Fx$ $M^{CB} = \frac{a}{L}F(L-x)$ $M^C = \frac{ab}{L}F$	$v^{AC} = -\frac{F}{6EIL}$ $[ab(2L-a)x - bx^3]$ $v^{CB} = -\frac{F(L-x)}{6EIL}$ $[ab(L+a) - a(L-x)^2]$ $\omega^A = -\frac{Fab}{6EIL}(L+b)$ $\omega^B = \frac{Fab}{6EIL}(L+a)$ $v^C = -\frac{Fa^2b^2}{3EIL}$
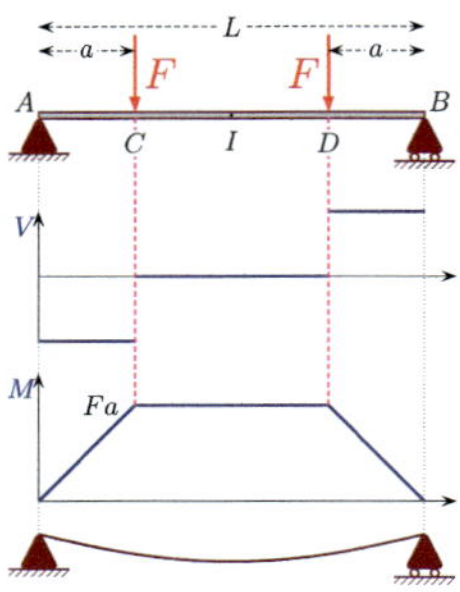	$R^A = F$ $R^B = F$	$V^{AC} = -F$ $V^{CD} = 0$ $V^{DB} = F$ $M^{AC} = Fx$ $M^{CD} = Fa$ $M^{DB} = F(L-x)$	$v^{AC} = -\frac{Fx}{6EI}[3a(L-a) - x^2]$ $v^{CD} = -\frac{Fa}{6EI}[3x(L-x) - a^2]$ $\omega^A = -\frac{F}{2EI}a(L-a)$ $v^C = -\frac{Fa^2}{6EI}(3L-4a)$ $v^I = -\frac{Fa}{24EI}(3L^2 - 4a^2)$
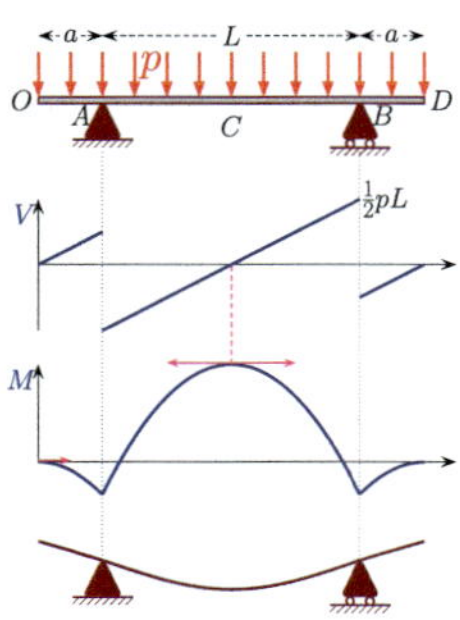	$R^A = \frac{1}{2}p(L+2a)$ $R^B = \frac{1}{2}p(L+2a)$	$V^{OA} = px$ $V^{AB} = p(x - a - \frac{L}{2})$ $V^{BD} = p(x - 2a - L)$ $M^{OA} = -\frac{1}{2}px^2$ $M^{AB} = p[-\frac{x^2}{2}$ $+(a+\frac{L}{2})(x-a)]$ $M^C = \frac{p}{8}(L^2 - 4a^2)$	$v^O = -\frac{pa}{24EI}$ $(3a^3 + 6a^2L - L^3)$ $v^C = -\frac{pL^2}{16EI}(\frac{5L^2}{24} - a^2)$ $\omega^O = \frac{1}{24}\frac{p}{EI}$ $(4a^3 + 6a^2L - L^3)$

Loading, Diagrams	Reactions	V and M	v and ω
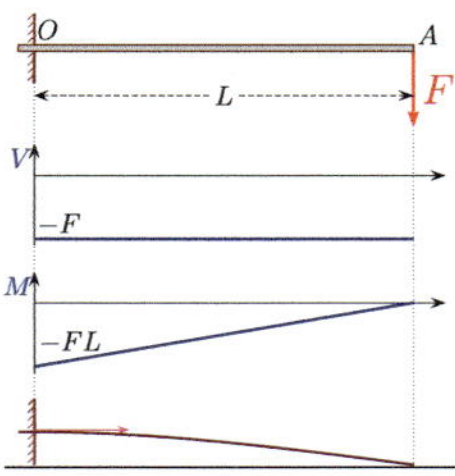	$R^O = F$ $M^O_c = FL$	$V = -F$ $M = -F(L-x)$	$v = -\frac{Fx^2}{6EI}(3L-x)$ $\omega = -\frac{Fx}{2EI}(2L-x)$ $v^A = -\frac{FL^3}{3EI}$ $\omega^A = -\frac{FL^2}{2EI}$
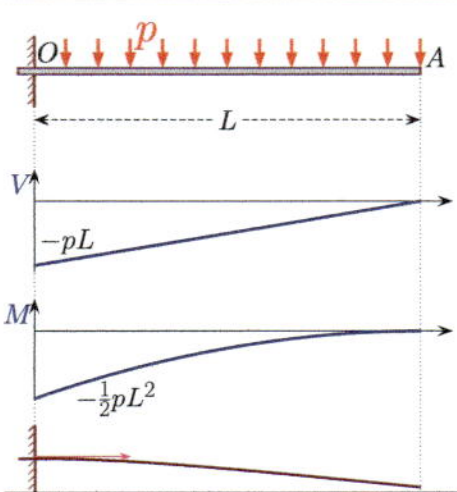	$R^O = pL$ $M^O_c = \frac{1}{2}pL^2$	$V = -p(L-x)$ $M = \frac{1}{2}p(L-x)^2$	$v = -\frac{pL^2}{24EI}$ $x^2\left(\frac{x^2}{L^2} - 4\frac{x}{L} + 6\right)$ $\omega = \frac{p}{6EI}\left[(L-x)^3 - L^3\right]$ $v^A = -\frac{pL^4}{8EI}$
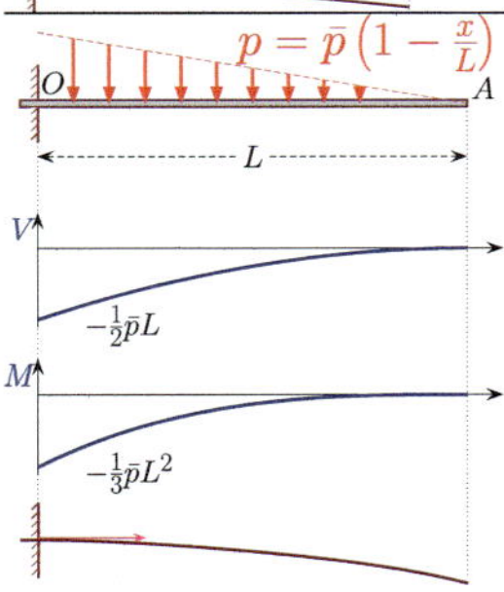	$R^O = \frac{1}{2}\bar{p}L$ $M^O_c = \frac{1}{6}\bar{p}L^2$	$V = -\frac{1}{2}\bar{p}L(1-\frac{x}{L})^2$ $M = -\frac{1}{6}\bar{p}L^2(1-\frac{x}{L})^3$	$v = -\frac{\bar{p}L^2x^2}{120EI}$ $\left(10 - 10\frac{x}{L} + 5\frac{x^2}{L^2} - \frac{x^3}{L^3}\right)$ $\omega = \frac{\bar{p}L^3}{24EI}\left[(1-\frac{x}{L})^4 - 1\right]$ $v^A = -\frac{\bar{p}L^4}{30EI}$ $\omega^A = -\frac{\bar{p}L^3}{24EI}$
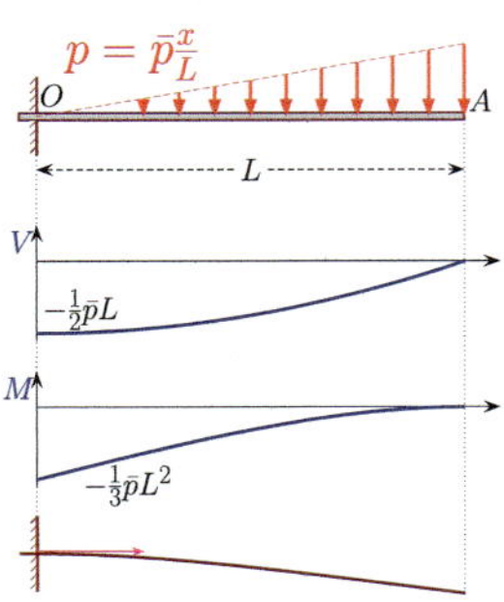	$R^O = \frac{1}{2}\bar{p}L$ $M^O_c = \frac{1}{3}\bar{p}L^2$	$V = -\frac{1}{2}\bar{p}L(1-\frac{x^2}{L^2})$ $M = -\frac{1}{6}\bar{p}L^2\left(2 - 3\frac{x}{L} + \frac{x^3}{L^3}\right)$	$v = -\frac{\bar{p}L^2}{120EI}$ $x^2\left(\frac{x^3}{L^3} - 10\frac{x}{L} + 20\right)$ $\omega = -\frac{\bar{p}L^2}{24EI}$ $x\left(\frac{x^3}{L^3} - 6\frac{x}{L} + 8\right)$ $v^A = -\frac{11\bar{p}L^4}{120EI}$ $\omega^A = -\frac{\bar{p}L^3}{8EI}$

7.6.2 HYPERSTATIC BEAMS (Q=*1*)

As in Section 7.6.1, M_c^X denotes the clamping moment at X. Subscript 'c' is introduced to avoid any confusion between a reaction moment and a particular value of the bending moment at some point X. In the second case, for example, the clamping moment is $M_c^O = \frac{1}{8}pL^2$ but the bending moment at O is $M^O = -M_c^O$.

Loading, Diagrams	Reactions	V and M	v and ω
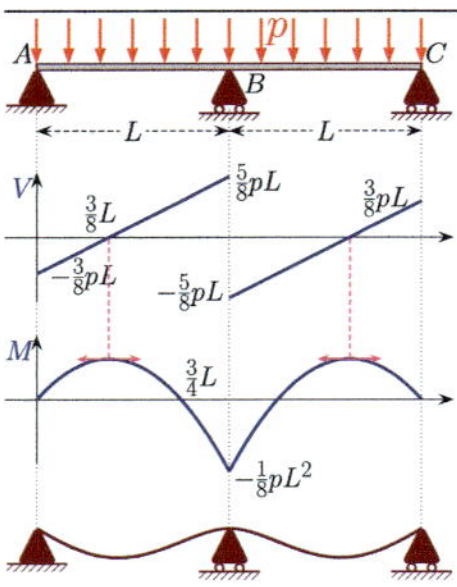	$R^A = \frac{3pL}{8}$ $R^B = \frac{5pL}{4}$ $R^C = \frac{3pL}{8}$	$V^{AB} = pL(\frac{x}{L} - \frac{3}{8})$ $M^{AB} = \frac{3}{8}pLx(1 - \frac{4}{3}\frac{x}{L})$ $M^B = -\frac{1}{8}pL^2$ $M(\frac{3}{8}L) = \frac{9}{128}pL^2$	$v^{AB} = -\frac{1}{48}\frac{pL^4}{EI}$ $(\frac{x}{L} - 3\frac{x^3}{L^3} + 2\frac{x^4}{L^4})$ $\omega^{AB} = -\frac{1}{48}\frac{pL^3}{EI}$ $(1 - 9\frac{x^2}{L^2} + 8\frac{x^3}{L^3})$ $\omega^A = -\frac{1}{48}\frac{pL^3}{EI}$
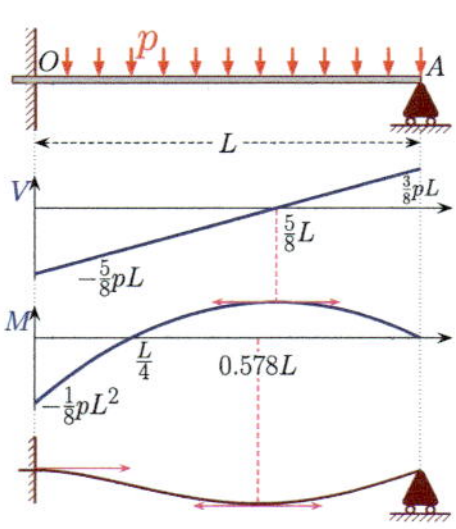	$R^O = \frac{5}{8}pL$ $R^A = \frac{3}{8}pL$ $M_c^O = \frac{1}{8}pL^2$	$V = -pL(\frac{5}{8} - \frac{x}{L})$ $M = -\frac{1}{8}pL^2\ (1 - 5\frac{x}{L} + 4\frac{x^2}{L^2})$ $M(\frac{5}{8}L) = \frac{9}{128}pL^2$	$v = -\frac{pL^4}{48EI}$ $\frac{x^2}{L^2}(3 - 5\frac{x}{L} + 2\frac{x^2}{L^2})$ $\omega = -\frac{pL^3}{48EI}$ $\frac{x}{L}(6 - 15\frac{x}{L} + 8\frac{x^2}{L^2})$ $v^{max} = -0.00542\frac{pL^4}{EI}$ $\omega^A = \frac{pL^3}{48EI}$
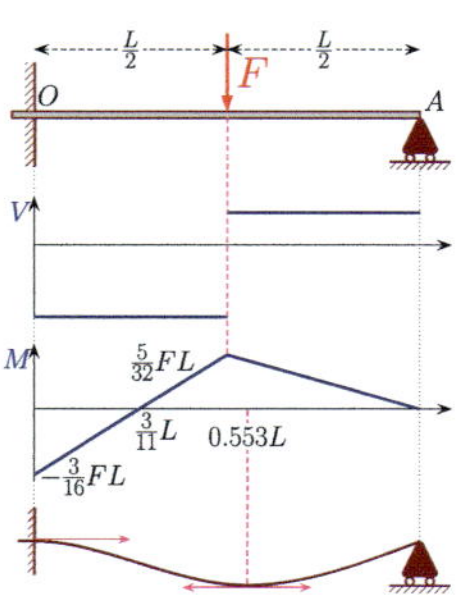	$R^O = \frac{11}{16}F$ $R^A = \frac{5}{16}F$ $M_c^O = \frac{3}{16}FL$	$V^{OI} = -\frac{11}{16}F$ $V^{IA} = \frac{5}{16}F$ $M^{OI} = -\frac{FL}{16}(3 - 11\frac{x}{L})$ $M^{IA} = \frac{5FL}{16}(1 - \frac{x}{L})$	$v^{OI} = -\frac{FL^3}{96EI}\frac{x^2}{L^2}(9 - 11\frac{x}{L})$ $v^{IA} = -\frac{FL^3}{96EI}$ $(-2 + 12\frac{x}{L} - 15\frac{x^2}{L^2} + 5\frac{x^3}{L^3})$ $\omega^{OI} = -\frac{FL^2}{32EI}\frac{x}{L}(6 - 11\frac{x}{L})$ $\omega^{IA} = -\frac{FL^2}{32EI}\ (4 - 10\frac{x}{L} + 5\frac{x^2}{L^2})$ $v^{max} = -0.00932\frac{FL^3}{EI}$

7.6.3 HYPERSTATIC BEAMS (Q=2)

This section contains results on doubly clamped beams so that there are two redundant reactions. As above, the notation M_c^X is used for the clamping moment at X. Hence, at O one has $M^O = -M_c^O$ but at A one writes $M^A = M_c^A$ by virtue of the definition of the screwer and Theorem 5.4.

Loading, Diagrams	Reactions	V and M	v and ω
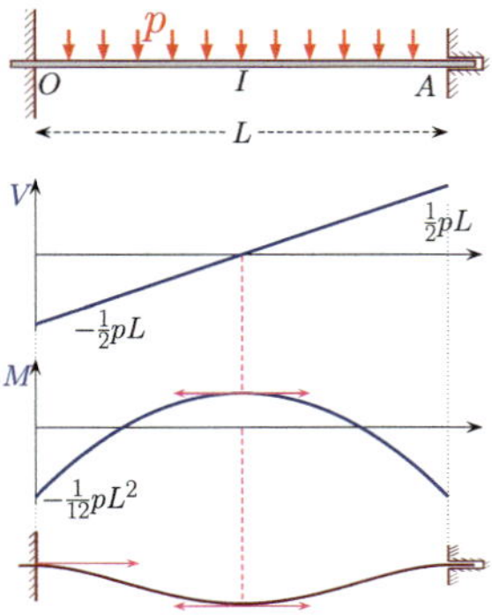	$R^O = \frac{1}{2}pL$ $R^A = \frac{1}{2}pL$ $M_c^O = \frac{1}{12}pL^2$ $M_c^A = -\frac{1}{12}pL^2$	$V = p(x - \frac{L}{2})$ $M = -\frac{1}{12}pL^2$ $(1 - 6\frac{x}{L} + 6\frac{x^2}{L^2})$ $M^I = \frac{1}{24}pL^2$ $M = 0$ for $x = (\frac{1}{2} \pm \frac{\sqrt{3}}{6})L$	$v = -\frac{1}{24}\frac{pL^4}{EI}$ $\frac{x^2}{L^2}(1 - \frac{x}{L})^2$ $\omega = -\frac{1}{12}\frac{pL^3}{EI}$ $\frac{x}{L}(1 - \frac{x}{L})(1 - 2\frac{x}{L})$ $v^{\max} = v^I = -\frac{1}{384}\frac{pL^4}{EI}$
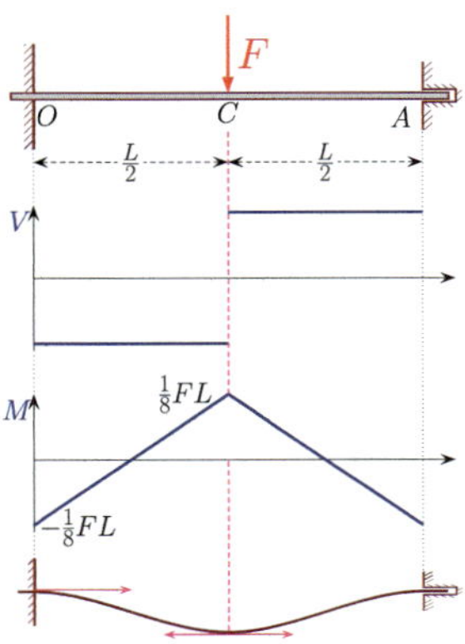	$R^O = \frac{F}{2}$ $R^A = \frac{F}{2}$ $M_c^O = \frac{1}{8}FL$ $M_c^A = -\frac{1}{8}FL$	$V^{OC} = -\frac{F}{2}$ $V^{CA} = \frac{F}{2}$ $M^{OC} = -\frac{FL}{8}$ $(1 - 4\frac{x}{L})$ $M^{CA} = \frac{FL}{8}$ $(3 - 4\frac{x}{L})$	$v^{OC} = -\frac{FL^3}{48EI}$ $(3\frac{x^2}{L^2} - 4\frac{x^3}{L^3})$ $v^{CA} = -\frac{FL^3}{48EI}$ $(4\frac{x^3}{L^3} - 9\frac{x^2}{L^2} + 6\frac{x}{L} - 1)$ $\omega^{OC} = -\frac{FLx}{8EI}$ $(1 - 2\frac{x}{L})$ $\omega^{CA} = -\frac{FL^2}{8EI}$ $(1 - 3\frac{x}{L} + 2\frac{x^2}{L^2})$ $v^C = -\frac{1}{192}\frac{FL^3}{EI}$

Loading, Diagrams	Reactions	V and M	v and ω
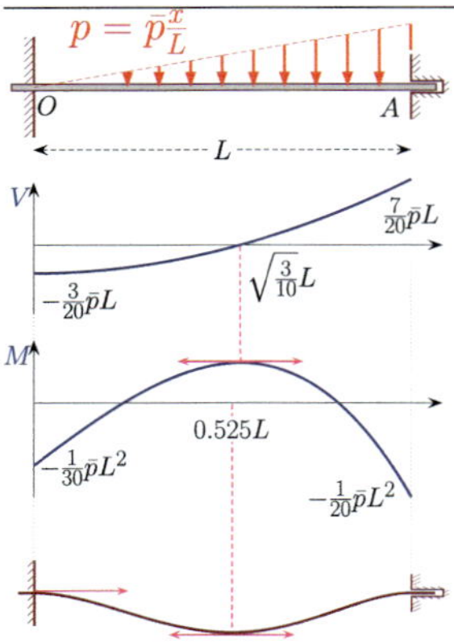	$R^O = \frac{3}{20}\bar{p}L$ $R^A = \frac{7}{20}\bar{p}L$ $M^O_c = \frac{1}{30}\bar{p}L^2$ $M^A_c = -\frac{1}{20}\bar{p}L^2$	$V = \frac{1}{20}\bar{p}L$ $\left(-3+10\frac{x^2}{L^2}\right)$ $M = \frac{1}{60}\bar{p}L^2$ $\left(-2+9\frac{x}{L}-10\frac{x^3}{L^3}\right)$ $M=\frac{3\sqrt{30}-10}{300}\bar{p}L^2$ for $x=\sqrt{\frac{3}{10}}L$	$v = -\frac{1}{120}\frac{\bar{p}L^4}{EI}$ $\frac{x^2}{L^2}\left(2-3\frac{x}{L}+\frac{x^3}{L^3}\right)$ $\omega = \frac{1}{120}\frac{\bar{p}L^3}{EI}$ $\frac{x}{L}\left(-4+9\frac{x}{L}-5\frac{x^3}{L^3}\right)$ $v^{\max}=-0.0013\frac{\bar{p}L^4}{EI}$
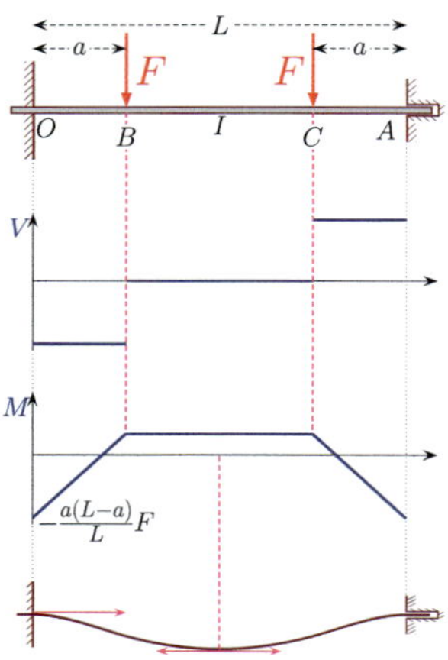	$R^O = F$ $R^A = F$ $M^O_c = FL^*$ with $L^*=\frac{a}{L}\left(1-\frac{a}{L}\right)L$ $M^A_c = -FL^*$	$V^{OB} = -F$ $V^{BC} = 0$ $V^{CA} = F$ $M^{OB} = -FL$ $\left(\frac{a}{L}-\frac{a^2}{L^2}-\frac{x}{L}\right)$ $M^{BC} = \frac{Fa^2}{L}$ $M^{CA} = -FL$ $\left(\frac{a}{L}-\frac{a^2}{L^2}-1+\frac{x}{L}\right)$	$v^{OB} = \frac{FL^3}{6EI}$ $\frac{x^2}{L^2}\left(\frac{x}{L}-3\frac{a}{L}+3\frac{a^2}{L^2}\right)$ $v^{BC} = \frac{Fa^2L}{6EI}$ $\left(3\frac{x^2}{L^2}-3\frac{x}{L}+\frac{a}{L}\right)$ $\omega^{OB} = \frac{FL^2}{2EI}$ $\frac{x}{L}\left(\frac{x}{L}-2\frac{a}{L}+2\frac{a^2}{L^2}\right)$ $\omega^{BC}=\frac{Fa^2}{2EI}\left(2\frac{x}{L}-1\right)$ $v^I=\frac{Fa^2L}{24EI}\left(4\frac{a}{L}-3\right)$

In all drawings of the above $q = 2$ cases, the connection at the right end of the beam is a slider link, which acts like a clamp except that it does not block the horizontal translation. Since the corresponding degree of freedom is not blocked, there is no horizontal reaction at the connection.

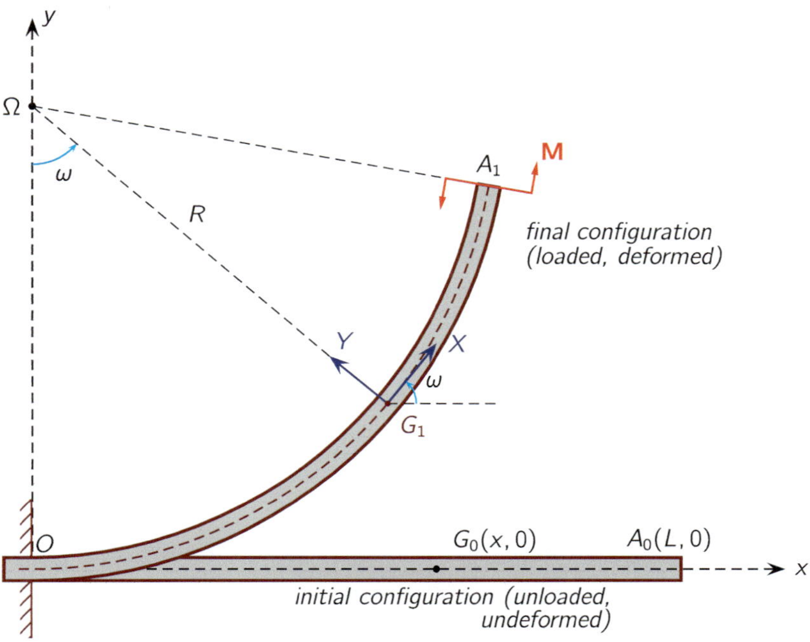

Figure 7.18 Prismatic beam subjected to bending at large displacements and rotations.

7.7 DOMAINS OF VALIDITY

In the majority of beam bending problems, all assumptions of "smallness" are satisfied: small displacements (translations and rotations) and small strains. This notably applies to the Saint-Venant problem discussed in Section 7.1.

All expressions developed in this chapter, except those involving the components u, v, w of the displacement vector, remain valid at finite displacements (large translations and rotations) with small strains. In such cases, the equilibrium equations must be written in the deformed (or final) configuration of the structure, the latter no longer being conflated with the initial (undeformed and stress-free) configuration. A simple example of such a bending problem is given below in small strains but large displacements.

Example 7.5: Bending with Large Displacements and Rotations

Consider a prismatic beam that is very long and very flexible in its median plane (xOy). The beam is clamped following its initial section S^O and subjected to a bending couple $\mathbf{M}$ on its final section S^A, Figure 7.18. All fibers undergo small strains, all elongations $\delta l/l$ being small in magnitude in comparison with unity. However, the translations of the centroids G are not small in comparison with the length L of the beam and the rotations of the cross-sections are not small in comparison with a radian.

On the current cross-section, in its deformed configuration, the screwer reduces to the bending moment $M_Z = M$, which gives at current point $P(Y,Z)$ within that cross-section a stress $\sigma_X = -\frac{M}{I}Y$.

The cross-sections remain normal to the deformed median fiber; the latter is a circular arc having radius R such that

$$\frac{1}{R} = \frac{M}{EI} = \frac{\mathrm{d}\omega}{\mathrm{d}s}.$$

The current cross-section S, of initial abscissa x and of final arclength (or curvilinear abscissa) $s = x$, undergoes a rotation:

$$\omega = \frac{M}{EI}x = \frac{M}{EI}s$$

Its centroid G undergoes a translation $\overrightarrow{G_0G_1}$, the coordinates of G_0 being $(x,0)$, while those of G_1 are

$$x_1 = R\sin\omega, \quad \text{and} \quad y_1 = R - R\cos\omega$$

with

$$\omega = \frac{M}{EI}s \quad \text{and} \quad R = \frac{EI}{M}.$$

In addition, the global rigidity of the beam, in bending, is

$$\frac{M}{\omega_1} = \frac{EI}{L}.$$

This closes the example and, with it, the chapter! ♡

PROBLEMS

Problem 7.1 (Elasticity Solution for Beam Bending) *Show that Equations* (7.2) *to* (7.4) *are indeed the solution to the elasticity problem defined in Section 7.1 and Figure 7.1.* ⋆

Problem 7.2 (Saint-Venant Principle) *The pure bending solution presented in Section 7.1 was developed assuming linearly distributed normal tractions on end sections, Equation* (7.1). *Consider now alternative traction distributions on end sections, as would arise in practice due to fixture and set-up variations. State the conditions that must be satisfied by the traction distributions such that the solution given in Section 7.1 remains valid sufficiently far away from end sections. Sketch some traction distributions that meet the stated conditions.* ⋆

Problem 7.3 (Euler–Bernoulli Theory) *As an alternative to the theory of elasticity approach to beam bending, an older theory proceeds from the following three assumptions regarding the current cross-section: ($\mathcal{A}_1$) it is infinitely rigid in its own plane; ($\mathcal{A}_2$) it remains planar after deformation; ($\mathcal{A}_3$) it remains normal to the deformed axis of the beam. Under the action of a bending moment M about the z-axis, show the following propositions in succession.*

1. *The axial displacement is given by*

$$u(x,y,z) = -y\omega_z = -y\frac{\mathrm{d}v_G}{\mathrm{d}x} \tag{7.40}$$

2. *The axial stress is given by*

$$\sigma_x = -\frac{M}{I}y \tag{7.41}$$

3. *The curvature is given by*

$$\frac{\mathrm{d}^2 v_G}{\mathrm{d}x^2} = \frac{M}{EI} \tag{7.42}$$

Above, $\omega_z(x)$ denotes the overall rotation of the section, assumed to be small, $v_G(x)$ is the y displacement of the centroid, E is Young's modulus and $I \equiv I_z$ is the second moment of area. Discuss the necessity of assumptions $(\mathcal{A}_1)$–$(\mathcal{A}_3)$ to the result of Equation (7.42). ⋆

Problem 7.4 (Limitation of Euler–Bernoulli Theory) *1. Building on the results of Problem 7.3, show that*

$$\frac{\partial \sigma_x}{\partial x} = -Ey\frac{\mathrm{d}^2\omega_z}{\mathrm{d}x^2} \tag{7.43}$$

Conclude.

2. *By starting from kinematic assumptions, the Euler–Bernoulli (EB) theory has a broader scope than the theory of elasticity solution for pure bending in Section 7.1, which will be extended to general bending in Section 9.1 of Chapter 9. Does the EB theory violate any fundamental principle other than local axial equilibrium?*

⋆

Problem 7.5 (Slouchy Bending Formulae) *Referring to Figure 7.3, the normal stress due to bending may be expressed in the starred frame as $\sigma_x = ay^* + bz^*$. 1. Verify that $a = 0$. 2. Work out the details to show that b is given through Equation* (7.8). ⋆

Problem 7.6 (Beam Assembly in Series) *Two beams are assembled in series, as shown in Figure 7.19; they are welded together end to end. The Ox-axis corresponds to the neutral axis while Oy, Oz are the central principal axes of inertia of S^O. If E_iI_i is the (per unit length) bending stiffness of beam i ($i = 1,2$), what is the bending stiffness EI/L of the assembly with $L = L_1 + L_2$?* ⋆

Problem 7.7 (Beam Assembly in Parallel) *Two beams are assembled in parallel, as shown in Figure 7.20; beam (i) occupies the entire cavity of beam (ii). The Ox, Oy, and Oz axes are the central principal axes of both beams. If E_iI_i/L is the bending stiffness of beam i ($i = 1,2$), what is the bending stiffness EI/L of the assembly?* ⋆

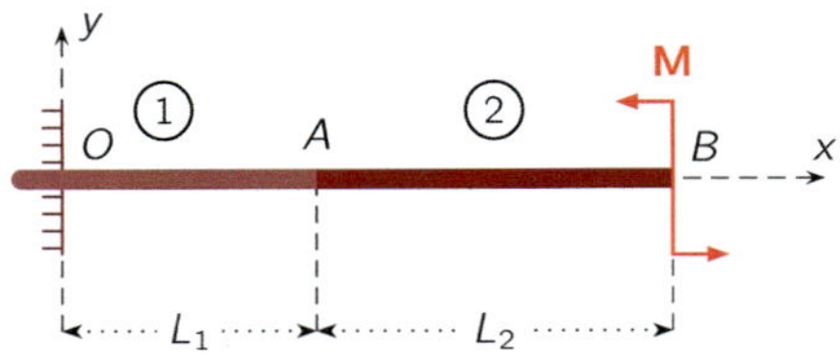

Figure 7.19 Two beams in series.

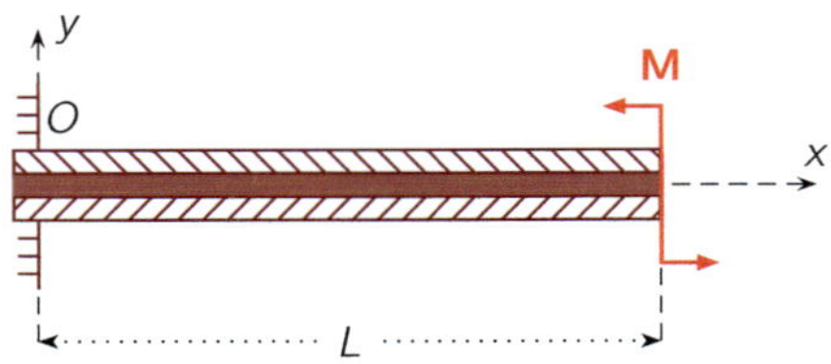

Figure 7.20 Two beams in parallel.

Problem 7.8 (Assembly of Two Identical Beams) *Consider the structure shown in Figure 7.21 and composed of: two identical prismatic beams AA′ and BB′ admitting Oxy as median plane and having characteristics L (length), S (cross-sectional area), E (Young's modulus); a bar A′B′ of length h, connecting the two beams by means of two articulations. The two beams are clamped as shown in the figure and a "bending" couple M is applied to the bar A′B′. Calculate the internal resultants in the two beams and the "bending stiffness" of their assembly.* ⋆

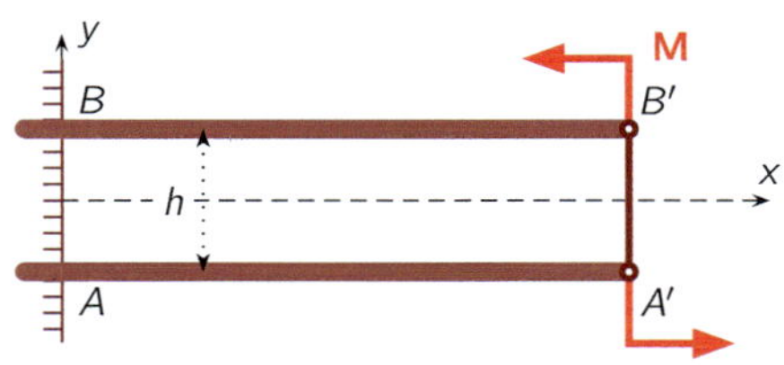

Figure 7.21 Two identical beams connected by a bar.

Figure 7.22 Clamped beam.

Problem 7.9 (Stiffness Matrix of Clamped Beam) *The clamped beam of Figure 7.22 is under the action of a tip couple Γ and a tip load F. The beam has characteristics L (length), S (cross-sectional area), $I = I_z$ (second moment of area) and E (Young's modulus). Find the stiffness matrix $\mathbb{K}$ such that:*

$$\begin{Bmatrix} F \\ \Gamma \end{Bmatrix} = [\mathbb{K}] \begin{Bmatrix} v_A \\ \omega_A \end{Bmatrix}$$

and its inverse $\mathbb{S} = \mathbb{K}^{-1}$. ⋆

Problem 7.10 (Asymmetric Three-Point Bending) *Consider the simply supported beam with a downward force applied at a distance from the support, Figure 7.23. Study the variation of the shear and bending moment diagrams with the point of application of the force. Determine the corresponding deflections. Conclude.* ⋆

Problem 7.11 (Asymmetric Four Point Bending) *Consider Case 5 in the table of plane bending, Section 7.6. Place the origin O at the center. Establish expressions of shear, bending moment, and deflection as functions of x. Can the magnitude of the load be large enough for the midpoint O to deflect less than the points of application of the loads? Justify your answer without any calculation. Consider now two loads F (at C) and P (at D), as shown in Figure 7.24. What is the minimum load imbalance (measured by the ratio F/P) such that the midpoint would deflect less than either of C or D?* ⋆

Problem 7.12 (Comparing Deflections in Bending) *In the first five tabulated cases of simply supported beams, Section 7.6.1, all deflections look alike as they*

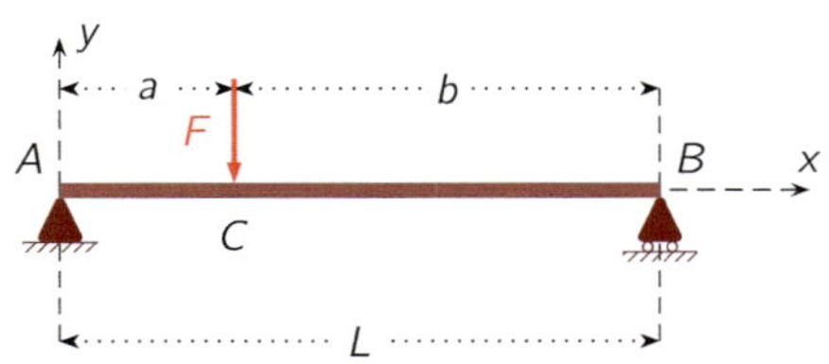

Figure 7.23 Isostatic beam under asymmetric 3-point bending.

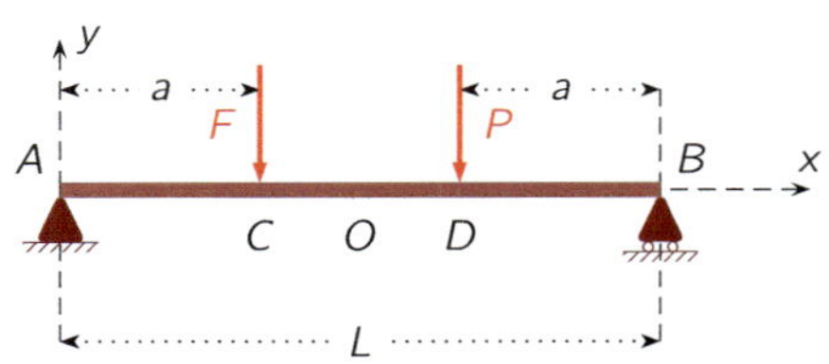

Figure 7.24 Isostatic beam under asymmetric 4-point bending.

were plotted using arbitrary magnitudes. Analyze what type of loading leads to the largest maximum deflection. Comparison needs to be made using appropriately normalized loads. ⋆

Problem 7.13 (Beam on Two Supports under Its Own Weight) *Consider Case 6 in the table of plane bending, Section 7.6. This case is re-analyzed in more detail here using the notations of Figure 7.25. Establish expressions of shear, bending moment and deflection as functions of x and parameterized by d and L. Does the beam always deflect as shown in the table? If not, what other shapes are possible? Quantify your answer based on the ratio d/L.* ⋆

Problem 7.14 (Variant of Four-Point Bending) *Consider the beam shown in Figure 7.26. Establish expressions of shear, bending moment and deflection as functions of x and parameterized by a and L. Compare the shape of the deflected beam with that of Problem 7.11 when $P = F$.* ⋆

Problem 7.15 (Hyperstatic Beam under Tip Load) *Consider the straight beam (OAB), admitting a median plane, defined in Figure 7.27, and subjected to a tip load* **F** *at B. The bending stiffness is EI. Determine: (i) the components of the screwer on each cross-section of the beam; draw the shear and bending moment diagrams;*

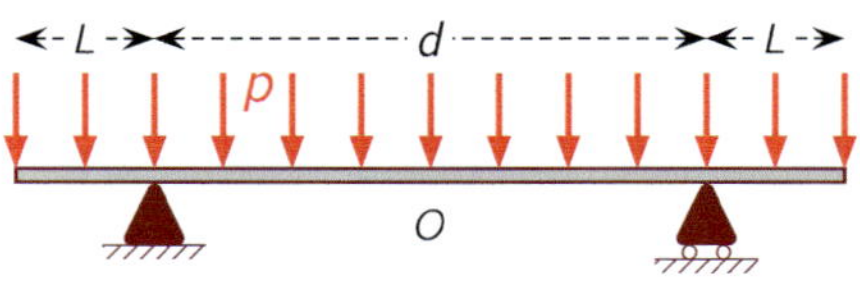

Figure 7.25 Analysis of case 6 in Section 7.6.

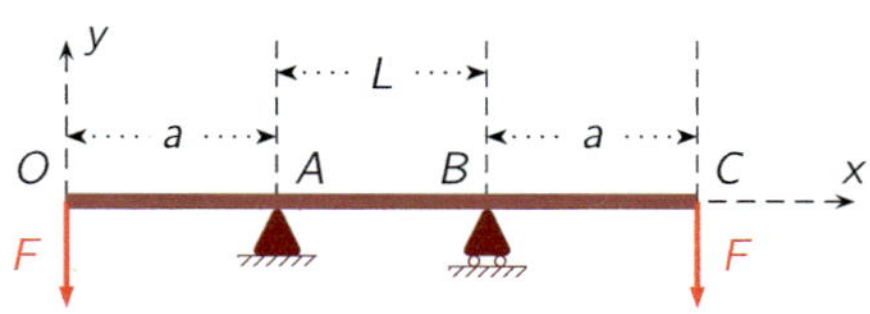

Figure 7.26 Simply supported beam under two tip loads.

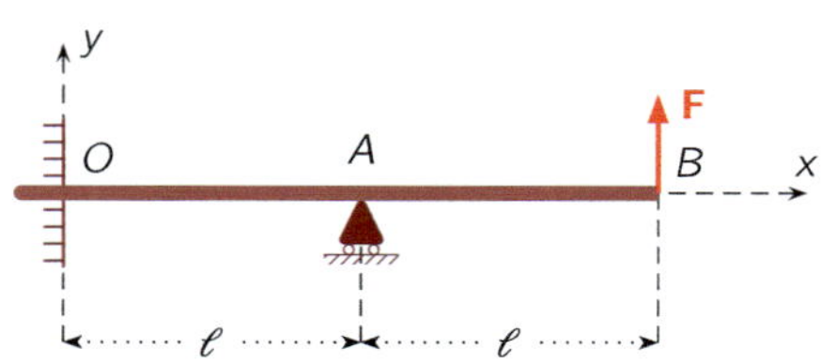

Figure 7.27 Plane bending by tip load.

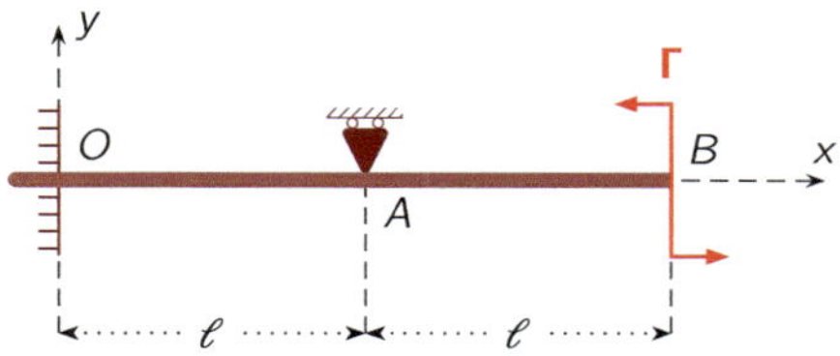

Figure 7.28 Plane bending by tip couple.

(ii) the reactions at O and A; (iii) the maximum tensile stress assuming that the cross-section is constant and rectangular in shape with height h and width w. ⋆

Problem 7.16 (Hyperstatic Beam under Tip Couple) *A prismatic beam of length 2ℓ and bending stiffness EI is clamped at O, simply supported at A, and subjected to a bending couple Γ at B, Figure 7.28. Determine: (i) the reactions at O and A; (ii) the deflection and rotation at B.* ⋆

Problem 7.17 (Plane Bending of Hyperstatic Beam) *A straight beam admitting xOy as median plane is clamped at one end and subjected to a slider link at the other, Figure 7.29. The beam is loaded by two pure opposite couples, as shown. The beam's lineal stiffness is EI. Determine the reactions at both ends and the screwer components on each cross-section. Draw their diagrams.* ⋆

Problem 7.18 (Plane Bending of Hyperstatic Straight Beam) *A prismatic beam with principal axes Oxyz and bending stiffness EI is loaded as depicted in Figure 7.30. (i) Determine the reactions at the supports. (ii) Calculate the rotation ω^A of the cross-section at A. (iii) Draw the shear and bending moment diagrams.* ⋆

Problem 7.19 (Nonuniform Loading of a Hyperstatic Beam) *Work out the details of Example 7.3.* ⋆

Problem 7.20 (Role of Redundant Supports) *Evaluate the decrease in maximum shear V_Y when a middle support is introduced in Example 7.3.* ⋆

Problem 7.21 (Hyperstatic Beam under an Arbitrary Point Load) *For the beam constrained and loaded as shown in Figure 7.31, establish expressions of the shear force, bending moment and deflection as functions of x and parameterized by a and b.* ⋆

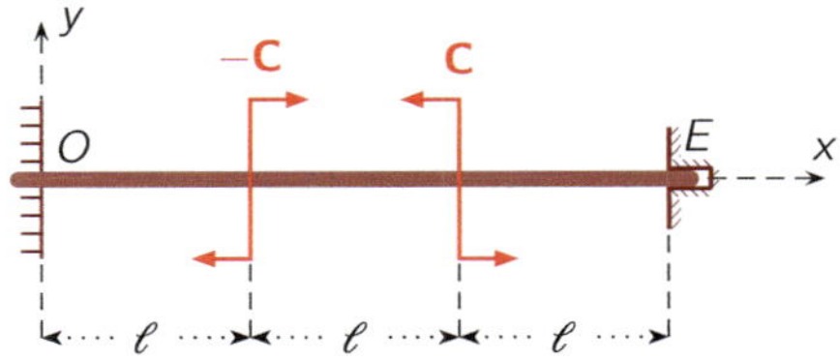

Figure 7.29 Beam under two point couples.

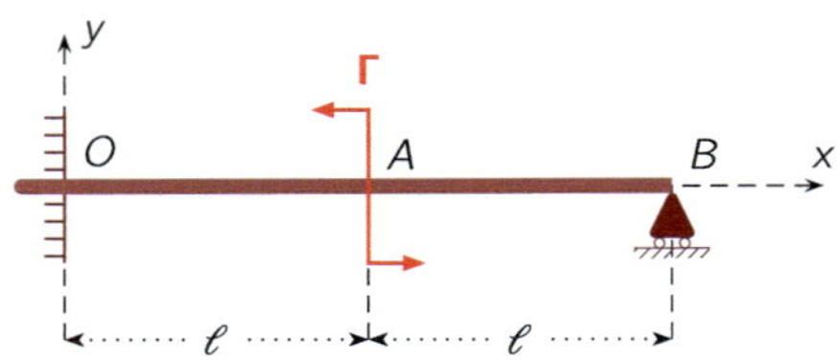

Figure 7.30 Beam under point couple at mid-span.

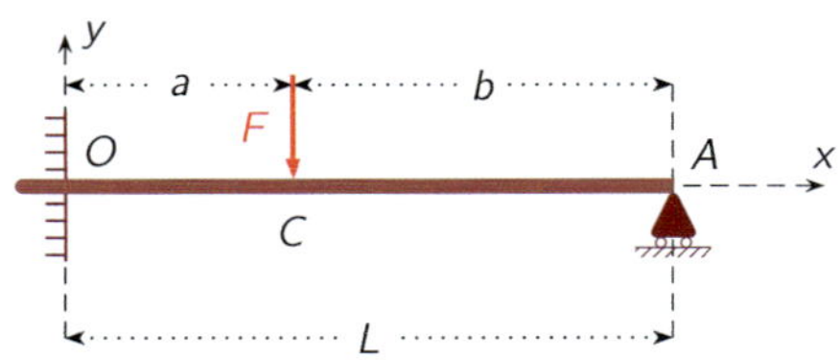

Figure 7.31 Generalization of case 3 in Section 7.6.2.

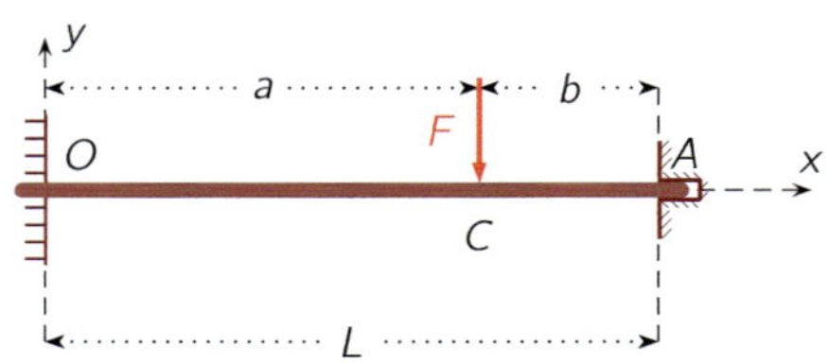

Figure 7.32 Generalization of case 2 in Section 7.6.3.

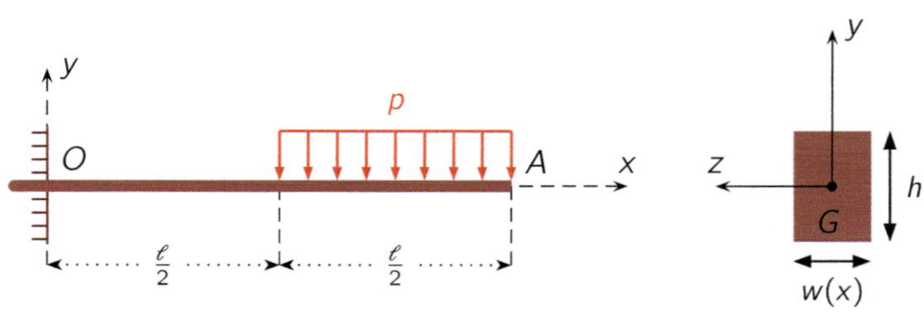

Figure 7.33 Beam under a distributed load and its rectangular cross-section.

Problem 7.22 (Doubly Clamped Beam under an Arbitrary Point Load) *For the beam constrained and loaded as shown in Figure 7.32, establish expressions of the shear force, bending moment and deflection as functions of x and parameterized by a and b.* ⋆

Problem 7.23 (Plane Bending of Beam with Variable Width) *Consider the straight beam with median plane (xOy), of length ℓ, clamped at the left end O and free at the other end, defined in Figure 7.33.*

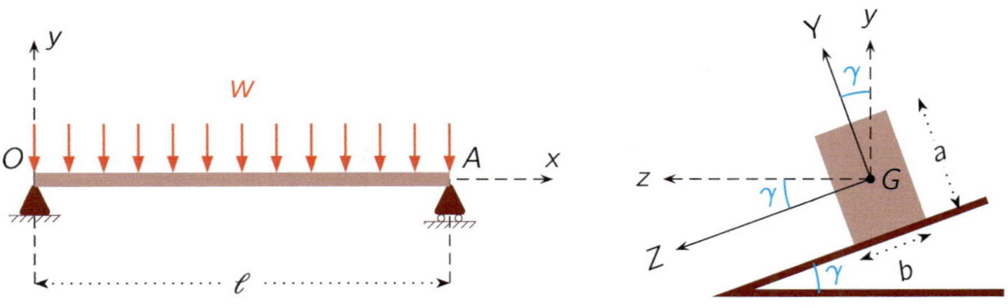

Figure 7.34 Beam under its own weight on inclined supports.

Its cross-section is rectangular of height h and variable width $w(x)$.

The beam is subjected on half of its span, toward tip A, to a downward, uniformly distributed load of intensity p per unit length of the beam.

Determine how the width $w(x)$ should vary span-wise such that the so-defined beam possesses the same bending resistance (the maximum normal stress $\sigma^{\max}$ is then constant span-wise.) ⋆

Problem 7.24 (Slouchy Bending of Straight Beam) *A prismatic beam having length ℓ and a rectangular cross-section with dimensions a, b rests on two supports at its ends O and A, Figure 7.34. The edges of the supports are inclined at an angle γ from the horizontal, as sketched in the figure. The beam is subjected to its own weight $\mathcal{W} = w\ell$. Let E denote Young's modulus. (i) Determine the components of the screwer on section S located at x. (ii) Calculate the normal stress σ_X at current point $P(X, Y, Z)$ and give the position of the neutral axis, e.g. through its polar angle β from GY. (iii) Where does σ_X reach its maximum? (iv) Calculate the deformed shape of the neural fiber.* ⋆

Problem 7.25 (Plane Bending of Straight Beam with Varying Inertia) *A straight beam of negligible weight and with median plane xOy is composed of two hollow circular cylindrical parts of equal lengths, constant wall thickness t, and mean radii R_1 and R_2, Figure 7.35. The beam is connected to the exterior by an articulation joint at A and a simple support at C. It is subjected to a uniformly distributed load of intensity*

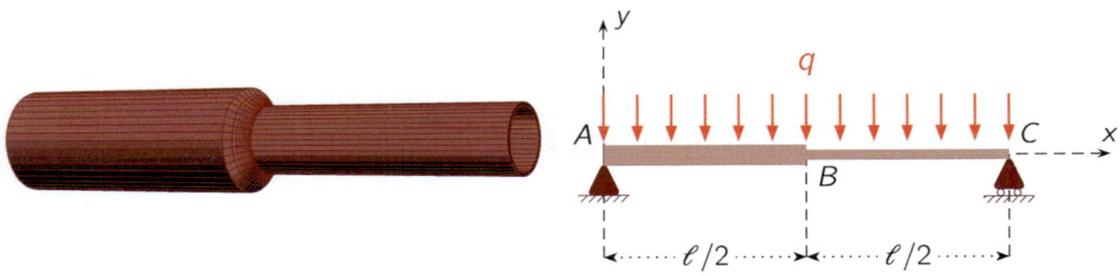

Figure 7.35 Beam with varying inertia under uniform load.

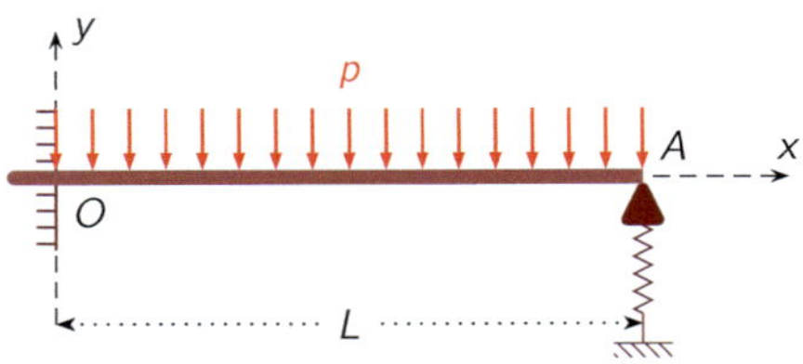

Figure 7.36 Beam with a tip spring under its own weight.

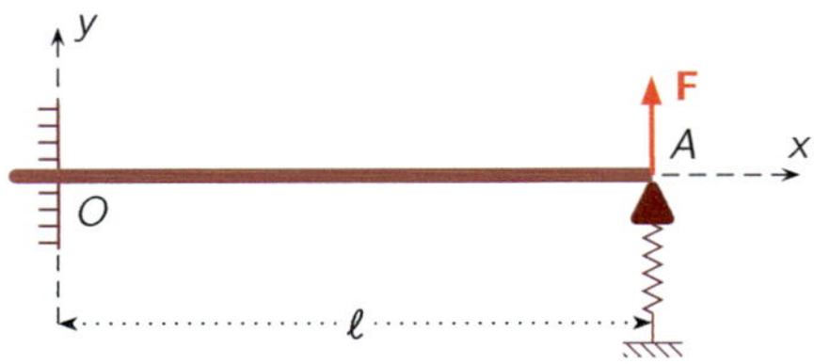

Figure 7.37 Hyperstatic beam with a tip spring under a tip force.

q. (i) Determine the components of the screwer on the current cross-section. (ii) Give the expression of the deflection of the mean fiber. (iii) Calculate the maximum tensile stress. (iv) Calculate the elastic energy stored during deformation. ⋆

Problem 7.26 (Plane Bending of Straight Beam on Elastic Support) *The beam shown in Figure 7.36 admits (xOy) as plane of symmetry. It is characterized with length L, second moment of area $I = I_Z$ and Young's modulus E. It is clamped at one end and stands on an elastic support (having stiffness k) at the other end. Calculate the reactions and the deflection when the beam is subjected to its own weight $P = pL$.* ⋆

Problem 7.27 (Straight Beam on Elastic Support) *The prismatic beam in Figure 7.37 is clamped at O and connected to the ground at A via a spring of stiffness k. Applied at A is an ascending vertical force* **F**. *The beam has characteristics E (Young's modulus) and I (second moment of area). Determine the tip deflection at A.* ⋆

Problem 7.28 (Beam on Three Supports) *A beam of length 3ℓ sits on two elastic supports and a pin while supporting its own weight (intensity per unit length w), Figure 7.38. The spring at A has stiffness k and the spring at B double stiffness, 2k. Calculate the reaction forces.* ⋆

Problem 7.29 (Hyperstatic Beam under Nonuniform Load) *The straight beam in Figure 7.39 is pinned at O and semi-clamped (slider link) at A. The beam is subjected to a distributed vertical load that increases linearly from the left to the right*

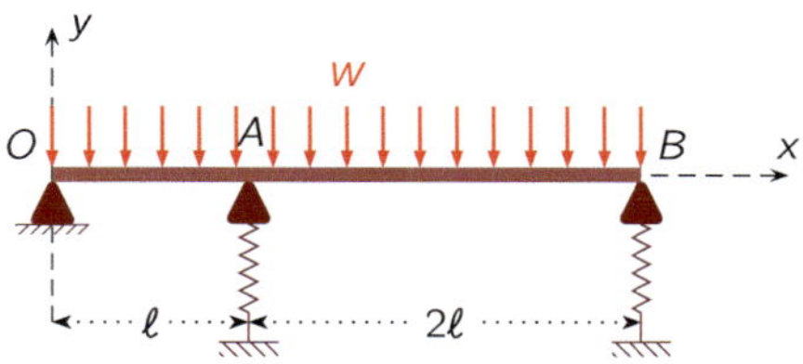

Figure 7.38 Hyperstatic beam on two springs.

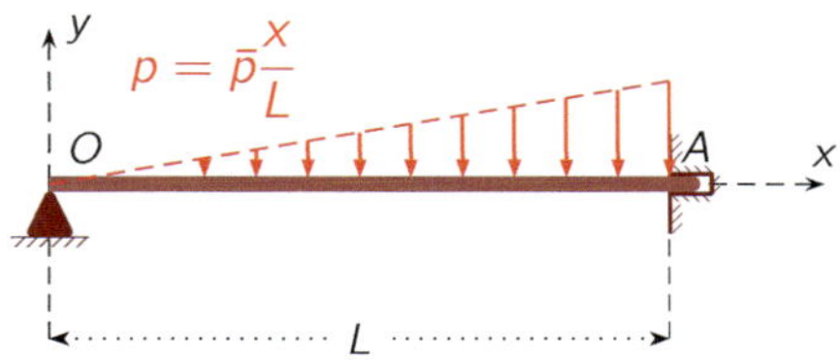

Figure 7.39 Pinned-clamped beam under non-uniformly distributed load.

end. Here, $\bar{p}$ denotes the maximum intensity at A. The beam has characteristics L (length), S (cross-sectional area), E (Young's modulus), and $I = I_z$ (second moment of area). (i) Calculate the reactions at O and A. (ii) Find the value and location of maximum normal stress. (iii) Give the expression of the deflection. (iv) Draw the shear and bending moment diagrams. ⋆

Problem 7.30 (Fourth-Order Deflection Equation) *Show that the deflection of an initially straight beam subjected to a distributed vertical load of intensity $p(x)$ per unit length of the median line obeys the following differential equation:*

$$v'''' \equiv \frac{\mathrm{d}^4 v}{\mathrm{d}x^4} = \frac{p}{EI}$$

where EI is the flexural rigidity, which is taken to be constant. Discuss the type and order of boundary conditions needed to integrate the above differential equation. ⋆

Problem 7.31 (Isostatic Beam Using the Fourth-Order ODE) *Consider a clamped-free beam under a triangular load, which is the last case tabulated in Section 7.6.1. Determine the tip deflection and rotation by using the fourth-order differential equation derived under Problem 7.30.* ⋆

Problem 7.32 (Hyperstatic Beam Using the Fourth-Order ODE) *For the beam shown in Figure 7.30, determine the deflection and rotation by integrating the fourth-order differential equation derived under Problem 7.30.* ⋆

Problem 7.33 (Plane Bending of Curved Beam) *A quarter-circular beam admitting a median plane is clamped at one end and loaded by an arbitrary plane force*

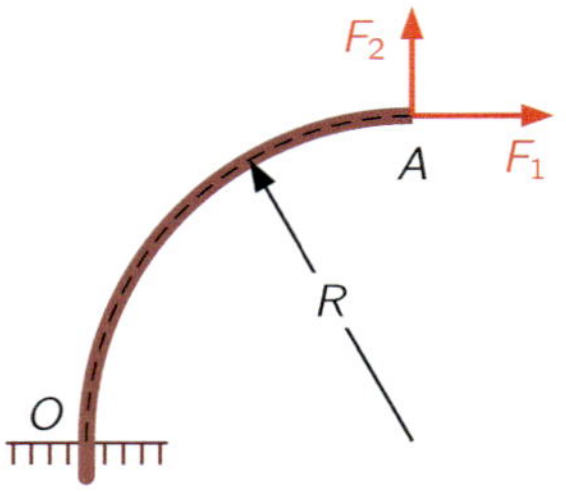

Figure 7.40 Curved beam under arbitrary plane tip load.

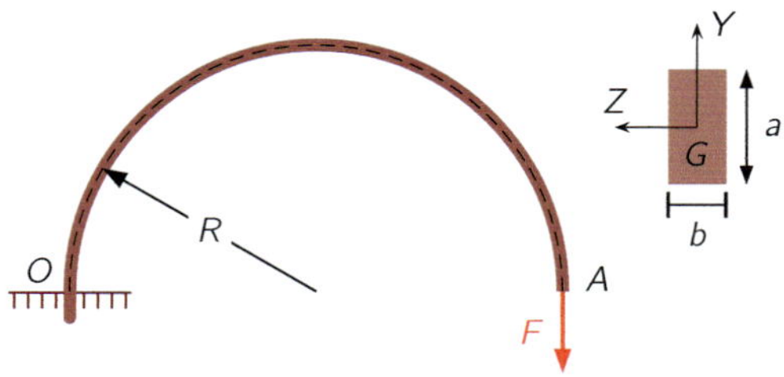

Figure 7.41 Semi-circular beam under vertical tip load.

with components F_1, F_2, as shown in Figure 7.40. (i) Determine the components of the screwer acting on the current cross-section. (ii) Determine the components u_A, v_A of the tip displacement as well as the rotation ω_A of the cross-section at A. Contributions due to transverse shear and normal force can be neglected. ⋆

Problem 7.34 (Plane Bending of Semi-Circular Beam) *A semi-circular beam admitting a median plane is clamped at one end and loaded by a vertical force at the other end, Figure 7.41. The cross-section is rectangular with dimensions a and b. (i) Determine the components of the screwer acting on the current cross-section S. (ii) Calculate the normal stress σ_X and find the neutral axis in S. 1. Determine the components U and V of the displacement of G center of S as well as the rotation ω of S. The contribution of the transverse shear to deflections and rotation can be neglected in comparison with those due to the normal force and the bending moment. (iii) Give the values of U, V, ω in A.* ⋆

Problem 7.35 (Beam Lying on Elastic Foundation) *A prismatic beam of length 2L is simply supported at mid-span and lies on a thick rubber mat having stiffness k, Figure 7.42. The beam is under the action of its own weight $P = 2pL$. (i) Calculate the deflection of the beam. (ii) Infer the tip deflection. (iii) Study the limit case of an infinitely long beam ($L \to \infty$).* ⋆

Problem 7.36 (Plane Bending of Two-Beam System) *A beam OA of length 2L and a beam BC of length L are pin-jointed at B, as shown in Figure 7.43. Beam OA is clamped at one end and loaded by a vertical force of magnitude F at mid-span. Beam*

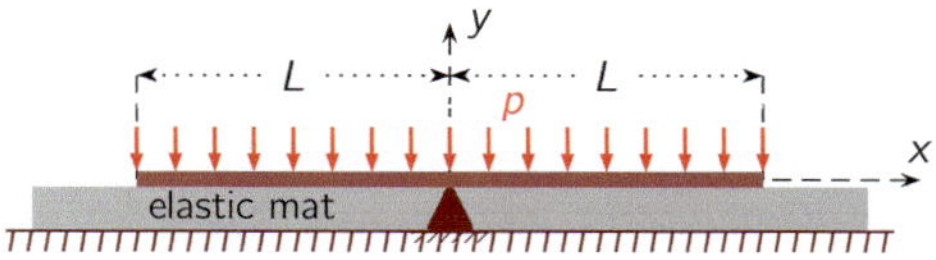

Figure 7.42 Beam on elastic floor.

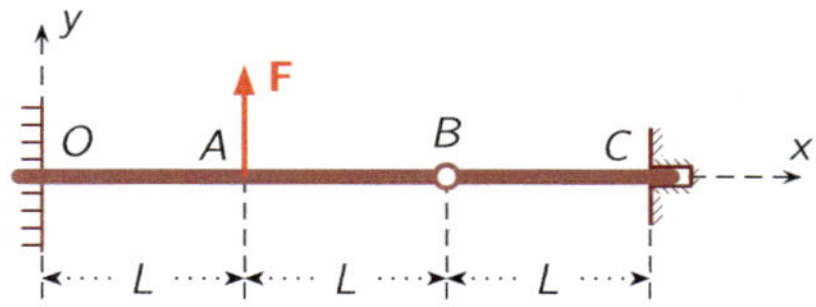

Figure 7.43 Two-beam system with a hinge.

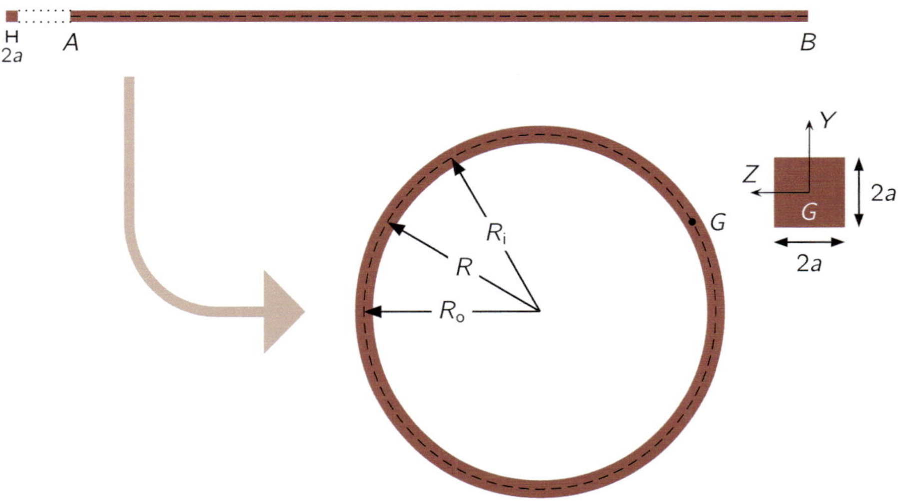

Figure 7.44 Making a ring from a long beam.

BC has a sliding link at C. The two beams are otherwise identical (same Young's modulus E, second moment of area I and cross-section S). The plane of the figure is a plane of symmetry for the structure. (i) Determine the reactions. (ii) Calculate the deflection. (iii) Give in particular the expressions of the deflections at A and B, as well as the rotations at B for both beams. ⋆

Problem 7.37 (Bending of a Long Beam) *A straight beam is bent to make a ring, as depicted in Figure 7.44.The beam has length $L = 20$ m, a square cross-section of side length $2a = 1$ cm and is made of plexiglas ($E = 2.90\,10^9$ MPa, $\nu = 0.4$). The beam is deformed in its median plane (of symmetry) until the two ends A and B are welded together. In the current cross-section S of the ring, centered about G:*

(i) Show that among all six components of the screwer, only one is non-zero; which one? (ii) Calculate the strain $\varepsilon_X(Y,Z)$ at the current point P having coordinates Y, Z. (iii) Calculate the corresponding stress field $\sigma_X(Y,Z)$. (iv) Calculate the bending moment. (v) State if the problem satisfies the assumption of small strains or small displacements. (vi) Determine numerically the values of maximum strain and maximum stress. ⋆

8 Torsion of Beams

8.1 PURE TORSION OF A CYLINDRICAL BEAM

In many mechanical systems, power transmission is effected by cylindrical shafts, which may be full or hollow. Examples include turbomachine shafts, ship and plane propeller shafts, transmission shafts in ground vehicles, and coupling shafts.

If M^t denotes the magnitude of the torsion couple (or simply the torque) in such a shaft rotating at angular velocity $\frac{\mathrm{d}\varphi}{\mathrm{d}t}$ (rad/s), then the transmitted power is

$$\mathcal{P} = M^t \frac{\mathrm{d}\varphi}{\mathrm{d}t}$$

The analysis of axisymmetric torsion of a cylindrical shaft was carried out in Part I; see Example 4.2. The results are recalled below, using a cylindrical coordinate system about the x-axis, Figure 8.1.

Let L be the length of the shaft along its x axis, R_i and R_o the inner and outer radii, S_0 and S_1 the initial and final cross-sections, centered about G_0 and G_1.

Applied on section S_1 is a field of circumferential (hoop) tractions $\vec{\tau}$ satisfying

$$\tau = \frac{M^t}{I_x} r \tag{8.1}$$

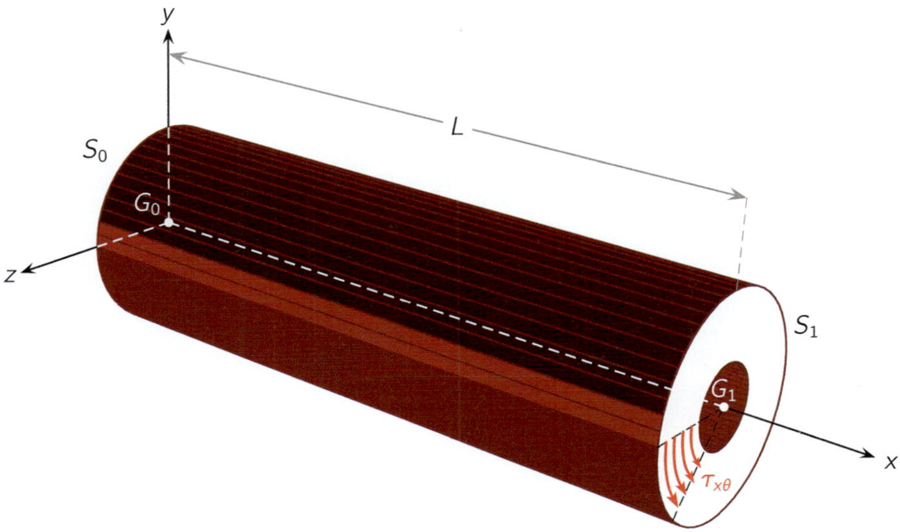

Figure 8.1 Cylindrical beam subjected to end torsion couples.

DOI: 10.1201/9781003537229-10

with

$$\begin{cases} \tau = \tau_{x\theta} : \text{signed magnitude of hoop traction } \overrightarrow{\tau} \\ M^t : \text{applied torque about the } x \text{ axis} \\ I_x = \dfrac{\pi}{2}(R_\text{o}^4 - R_\text{i}^4) : \text{polar moment of area} \\ r : \text{distance between the axis and current point } P \text{ where } \overrightarrow{\tau} \text{ acts} \end{cases}$$

In section S_0, the opposite loading is applied.

By way of consequence, the hollow beam is in a state of *pure torsion*; the torsion moment M_X is the same on all cross-sections and equal to M^t. At current point $P(r, \theta, x)$, one has:

- Stress:

$$[\boldsymbol{\sigma}] = \begin{bmatrix} 0 & 0 & 0 \\ 0 & 0 & \tau_{x\theta} \\ 0 & \tau_{x\theta} & 0 \end{bmatrix} \qquad \text{with} \quad \tau_{x\theta} = \frac{M^t}{I_x} r \tag{8.2}$$

- Strains: using Hooke's law in component form, Equation (3.13f) or Equation (4.18f),

$$[\boldsymbol{\varepsilon}] = \begin{bmatrix} 0 & 0 & 0 \\ 0 & 0 & \gamma_{x\theta} \\ 0 & \gamma_{x\theta} & 0 \end{bmatrix} \qquad \text{with} \quad \gamma_{x\theta} = \frac{1+\nu}{E} \tau_{x\theta} = \frac{M^t}{2GI_x} r \tag{8.3}$$

  where $G = \mu = \frac{E}{2(1+\nu)}$ is the elastic shear modulus; see Equation (3.17).

- Rotations: up to a rigid-body rotation,

$$\{\boldsymbol{\omega}\} \begin{cases} \omega_r = 0 \\ \omega_\theta = 0 \\ \omega_x = \dfrac{M^t}{GI_x} x \end{cases} \tag{8.4}$$

- Displacements: up to a rigid-body displacement,

$$\{\mathbf{u}\} \begin{cases} u_r = 0 \\ u_\theta = v = \dfrac{M^t}{GI_x} x r \\ u_x = 0 \end{cases} \tag{8.5}$$

The following observations are in order:

1. The cross-sections rotate about the x axis with respect to each other, without warping or deforming, Figure 8.2.

   To fix ideas, take the initial section S_0 as a solid of reference, then ω_x given by Equation (8.4) is the rotation about the x axis of current section S located at height x (Figure 8.2). A cross-section S' located at $x + \mathrm{d}x$ turns with respect to a neighboring section S located at x by an angle:

$$\mathrm{d}\omega_x = \frac{M^t}{GI_x}\,\mathrm{d}x \tag{8.6}$$

   In particular, the final section S_1 located at height L turns with respect to S_0 by an angle:

$$\omega_1 = \frac{M^t L}{GI_x} \tag{8.7}$$

   ω_1 represents the twist angle of the beam and the quantity $\frac{\mathrm{d}\omega_x}{\mathrm{d}x}$ is the twist per unit length.

   Not only does the current cross-section not warp ($u_x = 0$) but also it does not deform since the purely circumferential displacement of particle P given by Equation (8.5) may be written as $v = \omega_x r$. In other words, this displacement is consistent with the overall rotation of S about the x axis.
2. The elastic potential energy stored in an element of a longitudinal fiber centered about $P(r,\theta,x)$ and having length $\mathrm{d}x$ and cross-section $\mathrm{d}S$ is:

$$\mathcal{U}\,\mathrm{d}S\,\mathrm{d}x = \frac{1}{2}2\tau_{x\theta}\gamma_{x\theta}\,\mathrm{d}S\,\mathrm{d}x = \frac{1}{2}\frac{{M^t}^2}{GI_x^2}r^2\,\mathrm{d}S\,\mathrm{d}x$$

   where $\mathcal{U}$ is a specific energy; see Equation (3.31). Integrating over cross-

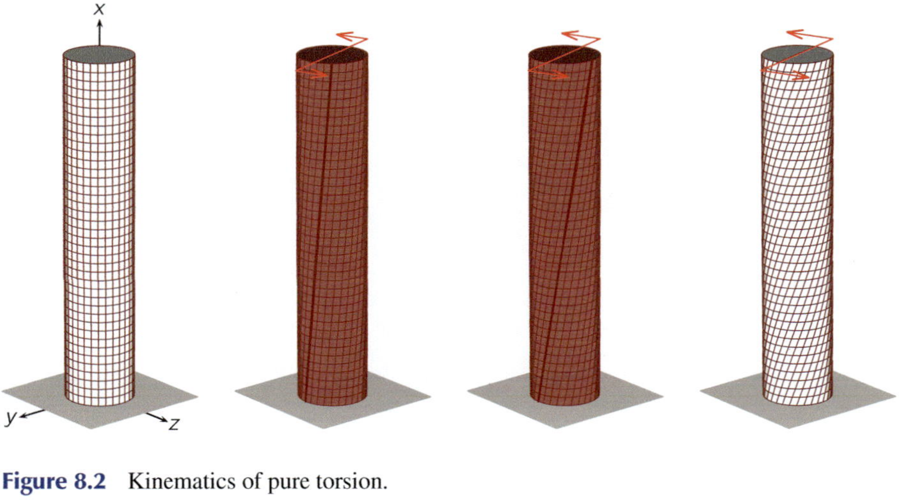

Figure 8.2 Kinematics of pure torsion.

section S, the energy stored in a slice of thickness $\mathrm{d}x$ is then:

$$\mathrm{d}W = \frac{1}{2}\frac{M^{t2}}{GI_x^2}\left(\iint_S r^2\,\mathrm{d}S\right)\mathrm{d}x = \frac{1}{2}\frac{M^{t2}}{GI_x}\mathrm{d}x = \frac{1}{2}M^t\,\mathrm{d}\omega_x$$

DEFINITION 8.1 The overall torsional stiffness of the shaft is the ratio:

$$\frac{GI_x}{L} = \frac{M^t}{\omega_1 - \omega_0}$$

For two cross-sections separated by $\mathrm{d}x$, a torsion stiffness per unit length is defined as

$$\frac{M^t}{\mathrm{d}\omega_x/\,\mathrm{d}x} = GI_x$$

♣

Remark 8.1 *The above results established for a hollow shaft apply to a solid shaft of radius R by posing $R_\mathrm{i} = 0$ and $R_\mathrm{o} = R$, Figure 8.3.* *

Remark 8.2 *The shaft is said to be thin-walled if its thickness $t = R_\mathrm{o} - R_\mathrm{i}$ is small compared with the mean radius $R = \frac{R_\mathrm{i}+R_\mathrm{o}}{2}$, Figure 8.4. In this case, the polar moment of inertia I_x becomes*

$$I_x = 2\pi R^3 t \tag{8.8}$$

and the (uniform) shear stress $\tau_{x\theta}$ is given by

$$\tau_{x\theta} = \tau = \frac{M^t}{2\pi R^2 t} \tag{8.9}$$

*

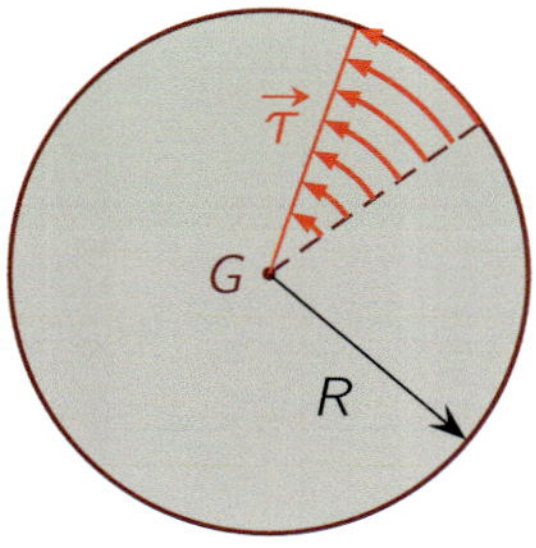

Figure 8.3 Solid shaft.

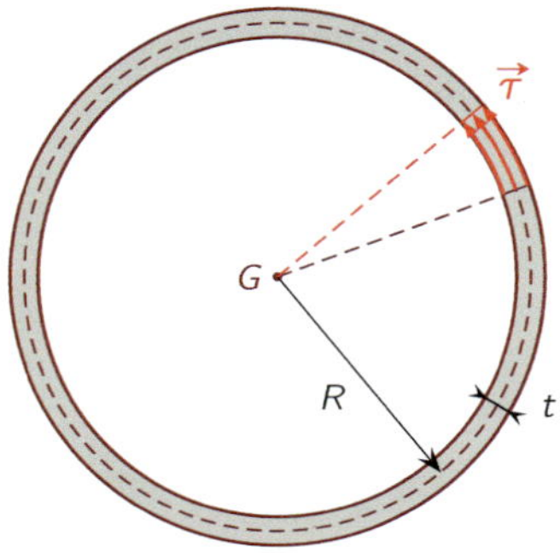

Figure 8.4 Thin-walled shaft.

8.2 TORSION OF SOLID PRISMATIC BEAMS

Consider a full prismatic (or cylindrical) beam with an arbitrary cross-section, Figure 8.5. As usual, the initial and final sections are denoted by S_0 and S_1 while the current cross-section located at abscissa x is denoted S. Also, L is the length of the beam and $-M^t$ and M^t are the torques applied to S_0 and S_1, respectively.

The axes are Ox, Oy and Oz where O is a particle of S_0 and Ox is parallel to the generatrices.

As in any problem of elastostatics, the solution must satisfy pointwise the following conditions:

1. the local equilibrium, that is $\mathbf{div}\,\boldsymbol{\sigma} + \mathbf{b} = \mathbf{0}$;
2. Hooke's law;
3. the boundary conditions; and
4. the strain-displacement (and rotation-displacement) relations.

These conditions are here listed in the order they are considered in a stress-based method; see Section 4.2.2 in Part I. In this case, the strain-displacement relations are equivalent to the compatibility equations; see Section 1.2 in Part I.

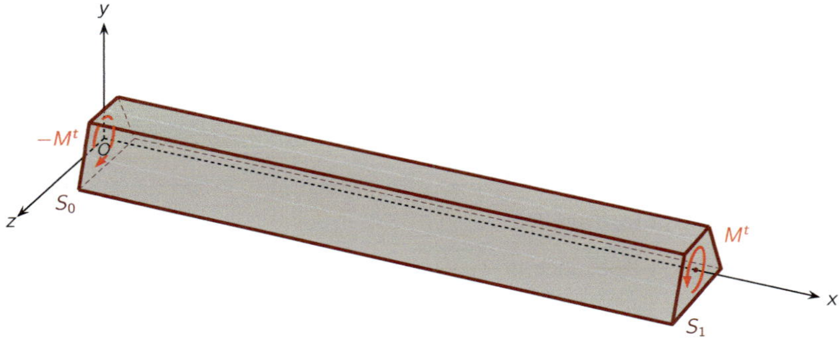

Figure 8.5 Prismatic beam subjected to end torsion couples.

The method of Saint-Venant proceeds so as to determine in the beam the relevant fields of strain, stress, rotations, and translations, by assuming that the stress matrix anywhere in the beam takes the form:

$$[\boldsymbol{\sigma}] = \begin{bmatrix} 0 & \tau_{xy} & \tau_{xz} \\ \tau_{xy} & 0 & 0 \\ \tau_{xz} & 0 & 0 \end{bmatrix}$$

8.2.1 LOCAL EQUILIBRIUM

In the absence of body forces, equilibrium is specialized using (see Equations (2.8) or Equations (4.9))

$$\frac{\partial \tau_{xy}}{\partial y} + \frac{\partial \tau_{xz}}{\partial z} = 0; \qquad \frac{\partial \tau_{xy}}{\partial x} = 0; \qquad \frac{\partial \tau_{xz}}{\partial x} = 0 \tag{8.10}$$

The last two equations indicate that the shear stresses cannot depend on x. A sufficient condition for the first equilibrium equation is that the components τ_{xy} and τ_{xz} derive from a stress function $\varphi(y,z)$, to be determined, such that

$$\vec{\tau} \quad \begin{cases} \tau_{xy} = -\dfrac{\partial \varphi}{\partial z} \\ \tau_{xz} = \dfrac{\partial \varphi}{\partial y} \end{cases} \tag{8.11}$$

The function φ is assumed to be endowed with third-order partial derivatives that are continuous on S. As in any method based on stress functions, there is an issue of completeness. However, if a solution having the above form were to hold, then it is *the* solution to the so-posed problem; recall the theorem of uniqueness of solutions in linear elasticity, Section 4.4.1 in Part I.

8.2.2 HOOKE'S LAW

It implies that the strain matrix takes the following form:

$$[\boldsymbol{\varepsilon}] = \begin{bmatrix} 0 & \gamma_{xy} & \gamma_{xz} \\ \gamma_{xy} & 0 & 0 \\ \gamma_{xz} & 0 & 0 \end{bmatrix} \quad \text{with} \quad \begin{cases} \gamma_{xy} = \dfrac{1}{2G}\tau_{xy} = -\dfrac{1}{2G}\dfrac{\partial \varphi}{\partial z} \\ \gamma_{xz} = \dfrac{1}{2G}\tau_{xz} = \dfrac{1}{2G}\dfrac{\partial \varphi}{\partial y} \end{cases} \tag{8.12}$$

where $G = \mu = \frac{E}{2(1+\nu)}$ is the shear modulus, as above.

8.2.3 BOUNDARY CONDITIONS

8.2.3.1 On Lateral Surface S_2

Surface S_2 is traction-free since no surface forces are applied on it. This condition writes

$$-\tau_{xy}\,\mathrm{d}z + \tau_{xz}\,\mathrm{d}y = 0 \qquad \text{on } \Gamma, \text{boundary of } S \tag{8.13}$$

where Γ denotes the closed curve, intersection of lateral surface S_2, and current cross-section S, Figure 8.6.

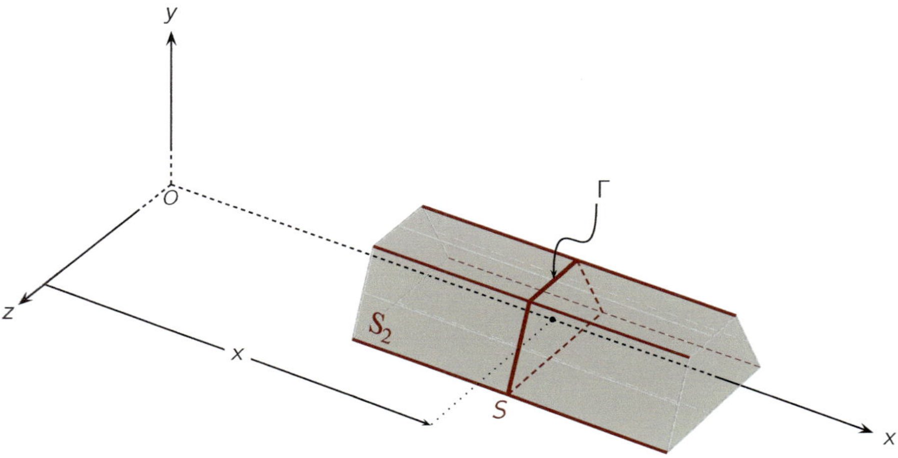

Figure 8.6 Extruding boundary Γ of section S along x gives lateral surface S_2.

The traction-free boundary condition, Equation (8.13), may now be expressed in terms of the stress function $\varphi(y,z)$ using Equation (8.11):

$$\frac{\partial \varphi}{\partial y}\,\mathrm{d}y + \frac{\partial \varphi}{\partial z}\,\mathrm{d}z = \mathrm{d}\varphi = 0$$

The function $\varphi(y,z)$ is therefore constant on the contour Γ. Since this function is defined up to an additive constant by Equation (8.11), one may take

$$\varphi(y,z) = 0 \qquad \text{on } \Gamma, \text{boundary of } S \tag{8.14}$$

with no loss of generality.

8.2.3.2 On End Sections S_0 and S_1

The way in which the net torques are applied to S_0 and S_1 (Figure 8.1) is not specified. On the other hand, whatever shear traction distributions may prevail on these end sections, they must satisfy the following conditions:

$$\left.\begin{aligned} \iint_S \tau_{xy}\,\mathrm{d}S = V_y = 0 \\ \iint_S \tau_{xz}\,\mathrm{d}S = V_z = 0 \end{aligned}\right\} \quad \text{no transverse shear} \tag{8.15}$$

$$\iint_S (y\,\tau_{xz} - z\,\tau_{xy})\,\mathrm{d}S = M_x = M^t \tag{8.16}$$

where the traction-screwer relations, Equation (5.35), have been invoked. The first two integrals can be transformed into contour integrals using Riemann's formula:

$$\left.\begin{aligned} &\iint_S -\frac{\partial\varphi}{\partial z}\,\mathrm{d}y\,\mathrm{d}z = \oint_\Gamma \varphi\,\mathrm{d}y = 0 \\ &\iint_S \frac{\partial\varphi}{\partial y}\,\mathrm{d}y\,\mathrm{d}z = \oint_\Gamma \varphi\,\mathrm{d}z = 0 \end{aligned}\right\} \quad \text{since } \varphi = 0 \text{ on } \Gamma \text{ as per Equation (8.14)}$$

The third one can be integrated by parts to yield

$$M^t = -2\iint_S \varphi(y,z)\,\mathrm{d}y\,\mathrm{d}z \tag{8.17}$$

8.2.4 COMPATIBILITY EQUATIONS

The strain–displacement relations and the (infinitesimal) rotation–displacement relations read (cf. Equations (1.3) and (1.4) in Chapter 1, Part I)

$$2\boldsymbol{\varepsilon} = \nabla\mathbf{u} + \nabla^T\mathbf{u} \qquad \text{and} \qquad 2\boldsymbol{\Omega} = \nabla\mathbf{u} - \nabla^T\mathbf{u}$$

These kinematic relations do not directly bring any information on stress or, by way of implication, on φ. Recall, however, that the strain–displacement relations are equivalent to the compatibility equations (cf. (1.16a)–(1.16f) in Chapter 1) which here reduce to (see Problem 8.18)

$$\frac{\partial}{\partial y}\left(\frac{\partial\tau_{xz}}{\partial y} - \frac{\partial\tau_{xy}}{\partial z}\right) = 0 \qquad \text{and} \qquad \frac{\partial}{\partial z}\left(\frac{\partial\tau_{xy}}{\partial z} - \frac{\partial\tau_{xz}}{\partial y}\right) = 0 \tag{8.18}$$

or, in terms of the stress function $\varphi(y,z)$:

$$\frac{\partial}{\partial y}(\Delta\varphi) = 0 \qquad \text{and} \qquad \frac{\partial}{\partial z}(\Delta\varphi) = 0$$

where Δ stands for the Laplacian.

The function $\varphi(y,z)$ is thus solution to a Poisson-like equation:

$$\Delta\varphi = B$$

where the constant B is determined, in principle, using Equation (8.17). This constant is therefore proportional to the torque M^t and depends on the geometry of the cross-sections S.

Dimensional analysis shows that B scales with the ratio of M^t by some quantity J that has dimension L^4 of a second moment of area. The above equation may thus be written with $B = 2M^t/J$ where the factor of 2 is for convenience, as will be evident below. The quantity J is called the torsion constant.

Integration of the rotation–displacement relations gives

$$\boldsymbol{\omega}\quad\begin{cases}\omega_x = \dfrac{M^t}{GJ}x & +\alpha\\[2ex] \omega_y = \dfrac{1}{2G}\left(\dfrac{\partial\varphi}{\partial y} - 2\dfrac{M^t}{J}y\right) & +\beta\\[2ex] \omega_z = \dfrac{1}{2G}\left(\dfrac{\partial\varphi}{\partial z} - 2\dfrac{M^t}{J}z\right) & +\gamma\end{cases} \tag{8.19}$$

where integration constants α, β, and γ are the components of an overall rotation of the beam about point O.

Integration of the strain–displacement relations finally gives

$$\mathbf{u}\quad\begin{cases}u = \dfrac{1}{G}\chi(y,z) + \beta z - \gamma y + a\\[2ex] v = -\dfrac{M^t}{GJ}xz \ + \gamma x - \alpha z + b\\[2ex] w = \dfrac{M^t}{GJ}xy \ + \alpha y - \beta x + c\end{cases} \tag{8.20}$$

where integration constants a, b, and c are the components of an overall translation.

The function $\chi(y,z)$ is defined up to an additive constant by the relations:

$$\frac{\partial\chi}{\partial y} = -\frac{\partial\varphi}{\partial z} + \frac{M^t}{J}z \tag{8.21}$$

$$\frac{\partial\chi}{\partial z} = \frac{\partial\varphi}{\partial y} - \frac{M^t}{J}y \tag{8.22}$$

It is completely determined by adding the condition:

$$\chi(0,0) = 0$$

It is called the *warping* function and determines the out-of-plane displacement $u(y,z)$ responsible for cross-sectional warping.

8.2.5 PRACTICAL RESOLUTION OF SAINT-VENANT'S PROBLEM

Given the torque (M^t) and the geometry (shape of S) the solution proceeds in three steps:

1. Determine the stress function $\varphi(y,z)$ by solving the Dirichlet-type problem:

$$\begin{aligned}&\Delta\varphi = 2\frac{M^t}{J} && \text{on } S\\ &\varphi = 0 && \text{on } \Gamma, \text{boundary of } S\end{aligned} \tag{8.23}$$

2. Determine the constant J by applying Equation (8.17) and then the warping function $\chi(y,z)$ using Equations (8.21) and (8.22).
3. Infer:
   - the stresses using Equation (8.11)
   - the strains using Hooke's law; cf. Equation (8.12)
   - the rotations using Equation (8.19)
   - the translations using Equation (8.20).

8.2.6 CONCLUSIONS: WARPING AND CENTER OF TORSION

Two important results follow from the above treatment of the Saint-Venant torsion problem.

First Conclusion: Warping – The cross-sections warp because the axial displacement $u(y,z)$ is a nonlinear function of y and z.

In addition, u being independent of x, the warping is the same for all cross-sections.

The only exception pertains to beams of revolution studied in Section 8.1 for which there is no warping.

Second Conclusion: Axis of torsion and center of torsion The cross-sections rotate with respect to each other about an axis, which is parallel to the generators. It is called the *axis of torsion* of the beam and is denoted the $\bar{x}$-axis. This axis intersects current cross-section S at a point C, termed *center of torsion*.

Suppose, to set the ideas, that the neighborhood of particle O does not undergo any displacement or rotation; one then has $\alpha = a = b = c = 0$ as well as

$$\beta = -\frac{1}{2G}\left(\frac{\partial \varphi}{\partial y}\right)_O = -(\gamma_{xz})_O \qquad \text{and} \qquad \gamma = -\frac{1}{2G}\left(\frac{\partial \varphi}{\partial z}\right)_O = (\gamma_{xy})_O$$

where use has been made of the second and third of Equations (8.19) and subscript O means that the quantities are evaluated at point O, that is for $y = z = 0$.

It follows that the transverse displacement of current particle P within S has components:

$$\begin{cases} v = \left(-\dfrac{M^t}{GJ}z + \gamma\right)x & \text{along } Oy \\[2ex] w = \left(\dfrac{M^t}{GJ}y - \beta\right)x & \text{along } Oz \end{cases} \tag{8.24}$$

By way of consequence, the point C of coordinates

$$y_C = \frac{GJ}{M^t}\beta \qquad \text{and} \qquad z_C = \frac{GJ}{M^t}\gamma$$

has zero displacement ($u^C = v^C = w^C = 0$).

In the frame $\{C,\bar{x}\bar{y}\bar{z}\}$ parallel to $\{O,xyz\}$ but having C as origin, the transverse displacement of P has components:

$$\begin{cases} \bar{v} = -\dfrac{M^t}{GJ}\bar{x}\bar{z} = -\omega_x\bar{z} \\ \bar{w} = \dfrac{M^t}{GJ}\bar{x}\bar{y} = \omega_x\bar{y} \end{cases} \tag{8.25}$$

with $\bar{x} = x$ and ω_x the rotation from Equation (8.19). As illustrated in Figure 8.7, such a displacement field in the plane of S corresponds to a rotation by angle ω_x (small) about point C.

In other words, the displacement of P in the barred frame takes the form: $\bar{\mathbf{u}} = \bar{\boldsymbol{\omega}} \times \overrightarrow{CP}$ where $[\bar{\boldsymbol{\omega}}]^T = [\omega_x \; 0 \; 0]$. Alternatively, in polar coordinates with C as pole, the radial displacement of P is nil, its orthoradial displacement being $r\,\omega_x = \frac{M^t}{GJ}rx$.

In sum, each cross-section S undergoes, in addition to the warping $u(y,z)$, an overall rotation about the axis of torsion $C\bar{x}$ known as the *angle of twist*, given by

$$\omega_x = \frac{M^t}{GJ}x$$

Likewise, the rotation $\mathrm{d}\omega_x$ of cross-section S' located at $x + \mathrm{d}x$ with respect to S at x satisfies, based on Equation (8.19):

$$\boxed{\frac{\mathrm{d}\omega_x}{\mathrm{d}x} = \frac{M^t}{GJ}} \tag{8.26}$$

The above two formulas are similar to those obtained for a beam of revolution, Equations (8.4) and (8.6), respectively, with the polar moment of inertia I_x replaced with J.

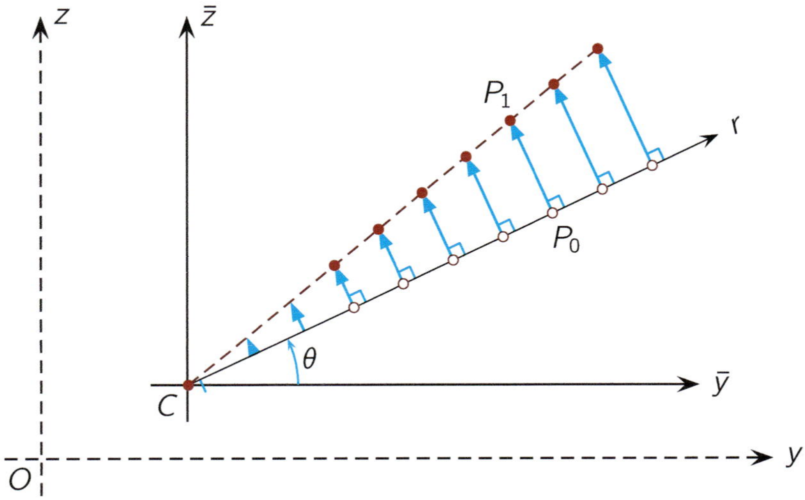

Figure 8.7 Center of torsion C: a particle initially at P_0 occupies location P_1 after torsion.

DEFINITION 8.2 The torsional stiffness per unit length is defined as

$$\frac{M^t}{\mathrm{d}\omega_x/\,\mathrm{d}x} = GJ$$

Its inverse $1/GJ$ is the torsibility (per unit length).

The overall torsional stiffness of the beam is the ratio:

$$\frac{GJ}{L} = \frac{M^t}{\omega_1 - \omega_0}$$

Its inverse is the global torsibility. ♣

Remark 8.3 *If the cross-section S admits an axis of symmetry, then the latter necessarily contains point C. If S admits two axes of symmetry, their intersection G is also center of torsion.* ∗

Remark 8.4 *The expression obtained in Section 8.1 for the stored energy remains formally valid provided that I_x is replaced with J. Indeed, if a slice of the beam is considered, which has thickness* dx *and is bounded by sections S and S', then when the torque increases from 0 to its final value M^t, S' rotates with respect to S about the x axis by an angle* $\mathrm{d}\omega_x$. *The corresponding work done is thus* $\mathrm{d}W = \frac{1}{2}M^t\,\mathrm{d}\omega_x$. *The strain energy stored per unit length is therefore*

$$\frac{\mathrm{d}W}{\mathrm{d}x} = \frac{1}{2}\frac{{M^t}^2}{GJ} \tag{8.27}$$

since the twist per unit length is given by M^t/GJ. ∗

8.2.7 APPLICATION EXAMPLES

Example 8.1: Torsion of Elliptic Cross-Section

Consider the beam in Figure 8.8 under torsional loading. Each cross-section is subjected to the same resultant torque M^t.

Following the steps of Section 8.2.5, the obtained solution consists of (see Problem 8.17)

$$\varphi(y,z) = -\frac{M^t}{S}\left(1 - \frac{y^2}{a^2} - \frac{z^2}{b^2}\right) \tag{8.28}$$

$$J = \frac{\pi a^3 b^3}{a^2 + b^2} \tag{8.29}$$

$$\vec{\tau}\quad \begin{cases} \tau_{xy} = -\dfrac{2M^t}{Sb^2}z \\ \tau_{xz} = \dfrac{2M^t}{Sa^2}y \end{cases} \tag{8.30}$$

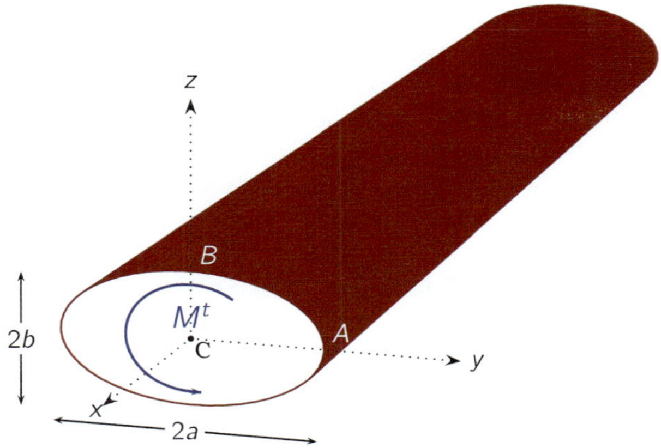

Figure 8.8 Beam with an elliptic cross-section under a torque.

$$u(y,z) = -\frac{a^2 - b^2}{\pi a^3 b^3}\frac{M^t}{G}yz \tag{8.31}$$

$$\boldsymbol{\omega} \begin{cases} \omega_x = \dfrac{M^t}{GJ}x = \dfrac{a^2 + b^2}{\pi a^3 b^3}\dfrac{M^t}{G}x \\ \omega_y = -\dfrac{1}{\pi a b^3}\dfrac{M^t}{G}y \\ \omega_z = -\dfrac{1}{\pi a^3 b}\dfrac{M^t}{G}z \end{cases} \tag{8.32}$$

Contour lines of constant shear stress are ellipses, as depicted in Figure 8.9a. The shear traction $\vec{\tau}$ varies linearly along any radial line emanating from the center C, Figure 8.9b. The maximum shear stress occurs at the ends of the minor axis, Figure 8.9b. Its magnitude is:

$$\tau^{\max} = \frac{2M^t}{bS}$$

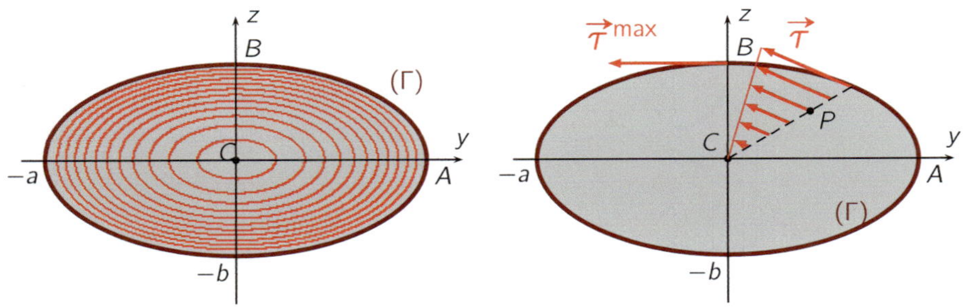

Figure 8.9 Shear stress distribution in an elliptic section.

The expression of the axial displacement u shows that it scales with yz. This means that the contour lines for the warped section are hyperbolas admitting the axes of the ellipse as asymptotes. For a positive torque, as shown in Figure 8.8, particles belonging to the upper right or lower left quadrants rise above the section while those in the other two quadrants sink beneath it. The principal axes of the ellipse remain unaffected by warping; in general, they only rotate due to the twist.

By posing $a = b = R$ one retrieves the formulas for a solid circular beam, Section 8.1. ♡

Example 8.2: Equilateral Triangular Cross-Section

For a beam with a cross-section in the shape of an equilateral triangle, the center of gravity is evidently center of torsion, Figure 8.10.

Taking $3a$ as the height of the triangle, the three sides belong to lines having equations

$$\left.\begin{aligned} & z+\tfrac{\sqrt{3}}{3}y-2\tfrac{\sqrt{3}}{3}a=0 && \text{for } (AB) \\ & y+a=0 && \text{for } (BC) \\ & z-\tfrac{\sqrt{3}}{3}y+2\tfrac{\sqrt{3}}{3}a=0 && \text{for } (CA) \end{aligned}\right\} \tag{8.33}$$

The simplest stress function that would vanish on each of the above three lines,

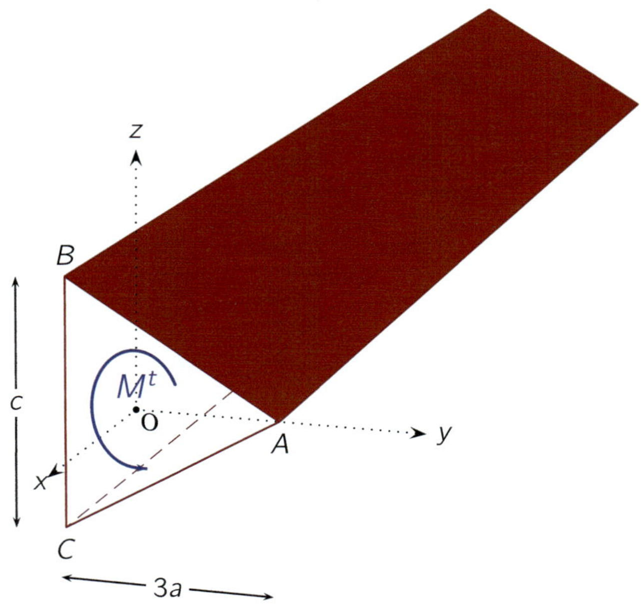

Figure 8.10 Beam with a triangular cross-section under a torque.

by virtue of boundary condition $(8.23)_2$, is

$$\varphi(y,z) = \frac{5\sqrt{3}}{54}\frac{M^t}{a^5}\left(z+\frac{\sqrt{3}}{3}y-2\frac{\sqrt{3}}{3}a\right)(y+a)\left(z-\frac{\sqrt{3}}{3}y+2\frac{\sqrt{3}}{3}a\right) \tag{8.34}$$

where the prefactor has been determined by setting the Laplacian equal to $2M^t/J$, Equation (8.23), and evaluating the integrals in Equation (8.17); see Section 8.2.5. One finds

$$J = \frac{9\sqrt{3}}{5}a^4 \tag{8.35}$$

The shear traction is obtained using Equation (8.11) as

$$\vec{\tau}\quad \begin{cases} \tau_{xy} = -\dfrac{10\sqrt{3}}{54}\dfrac{M^t}{a^5}(y+a)z \\[2ex] \tau_{xz} = -\dfrac{5\sqrt{3}}{54}\dfrac{M^t}{a^5}(z^2-y^2+2ay) \end{cases} \tag{8.36}$$

It is maximum at each side's midpoint (Figure 8.11) where its magnitude is given by

$$\tau^{\max} = \frac{5\sqrt{3}}{18}\frac{M^t}{a^3} = 20\frac{M^t}{c^3} \tag{8.37}$$

with $c = 2a\sqrt{3}$ denoting the side length. The shear stress vanishes at the vertices as well as the center.

Integrating Equations (8.21) and (8.22) to obtain the warping function, the axial displacement is found to be

$$u(y,z) = -\frac{1}{2}\frac{M^t}{GJ}\frac{z}{a}\left(y^2-\frac{1}{3}z^2\right) = \frac{1}{G}\chi(y,z) \tag{8.38}$$

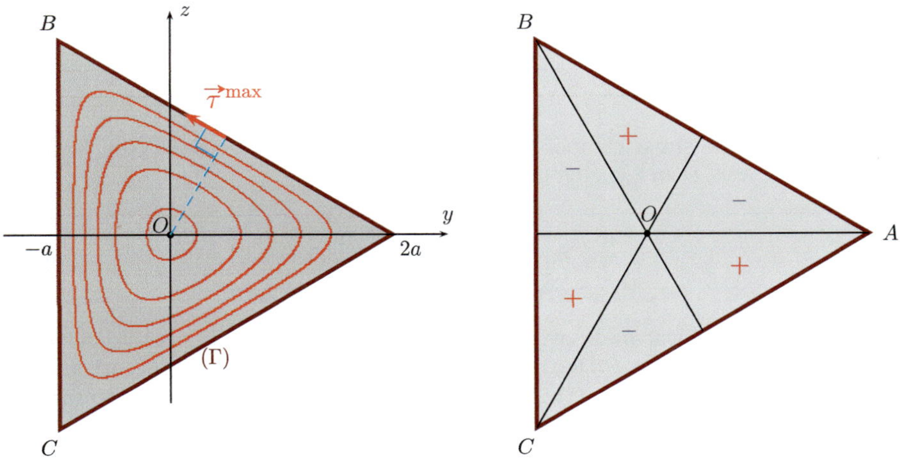

Figure 8.11 Shear stress distribution in a triangular section and axial displacement sectors.

and the rotations from Equation (8.19):

$$\boldsymbol{\omega} \quad \begin{cases} \omega_x = \dfrac{M^t}{GJ} x \\[2ex] \omega_y = \dfrac{M^t}{4GJa} (z^2 - y^2 - 2ay) \\[2ex] \omega_z = \dfrac{M^t}{2GJa} (y - a) z \end{cases} \tag{8.39}$$

The axial displacement in Equation (8.38) vanishes when $z = 0$ or $z = \pm\sqrt{3}y$. In other words, the three medians of the triangle have zero axial displacement. Each median rotates in its plane due to twist. The three medians divide the triangle into six sectors. As depicted in Figure 8.11, sectors with a $+$ sign move above the plane of the undeformed cross-section whereas those with a $-$ sign move below it, resulting in warping. ♡

Remark 8.5 *For an elliptical cross-section (Example 8.1) as well as the triangular one (Example 8.2), the stress function was simply obtained from an equation defining the contour of the cross-section so as to satisfy the traction-free boundary condition on the lateral surface, Equation* (8.14). *Unfortunately, this method is not general. The case of a rectangular section illustrates that; also see Problem 8.19.* $*$

Example 8.3: Rectangular Cross-Section

Consider the rectangular prism of Figure 8.12 under pure torsion. Following the previous two examples, it is tempting to adopt as a trial stress function one that would vanish on each of the four lines defining the boundary of the rectangle. Such a choice would lead to a gross approximation (see Problem 8.19).

The exact solution for the stress function is given in terms of the series

$$\varphi(y,z) = \frac{32}{\pi^3}\frac{a^2 M^t}{J} \sum_{n=1,3,5}^{\infty} \frac{(-1)^{\frac{n-1}{2}}}{n^3} \left[1 - \frac{\cosh \dfrac{n\pi z}{2a}}{\cosh \frac{n\pi b}{2a}} \right] \cos \frac{n\pi y}{2a} \tag{8.40}$$

from which the stress components τ_{xy} and τ_{xz} are readily obtained (see Problem 8.20). Also, from Equation (8.17) one gets

$$J = \frac{16}{3} a^3 b \left[1 - \frac{192}{\pi^5} \frac{a}{b} \sum_{n=1,3,5}^{\infty} \frac{1}{n^5} \tanh \frac{n\pi b}{2a} \right] \tag{8.41}$$

Contours of equal shear are shown in Figure 8.13a. The stress is zero at the vertices and at the center of the rectangle. It is maximum at the middle of each long side, as illustrated in Figure 8.13b. For the selected b/a ratio, which is greater than unity, the stress variation at mid-height is quasi-linear, although the detailed expressions are quite complex (see Problem 8.20). In making the plots of Figure 8.13, the first five terms of the open series are enough for convergence.

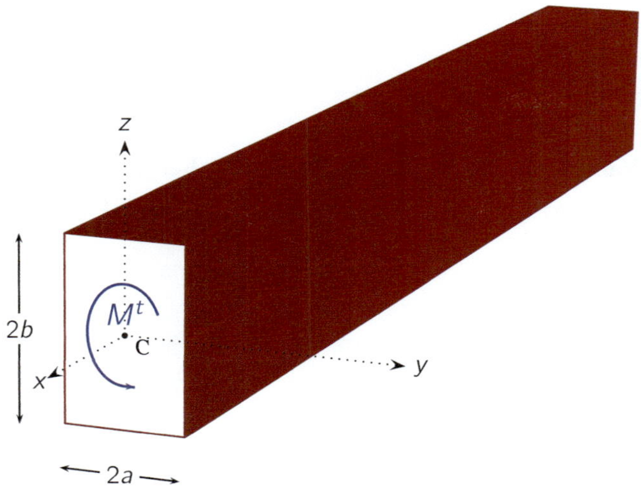

Figure 8.12 Beam with a rectangular cross-section under a torque.

For slender sections satisfying $b \gg a$, the maximum shear stress is given by the expression

$$\tau^{\max} = 2\frac{aM^t}{J}\left[1 - \frac{8}{\pi^2}\sum_{n=1,3,5}^{\infty}\frac{1}{n^2\cosh\frac{n\pi b}{2a}}\right] \quad (8.42)$$

In practice, the two important coefficients are J and $\tau^{\max}$, given by Equa-

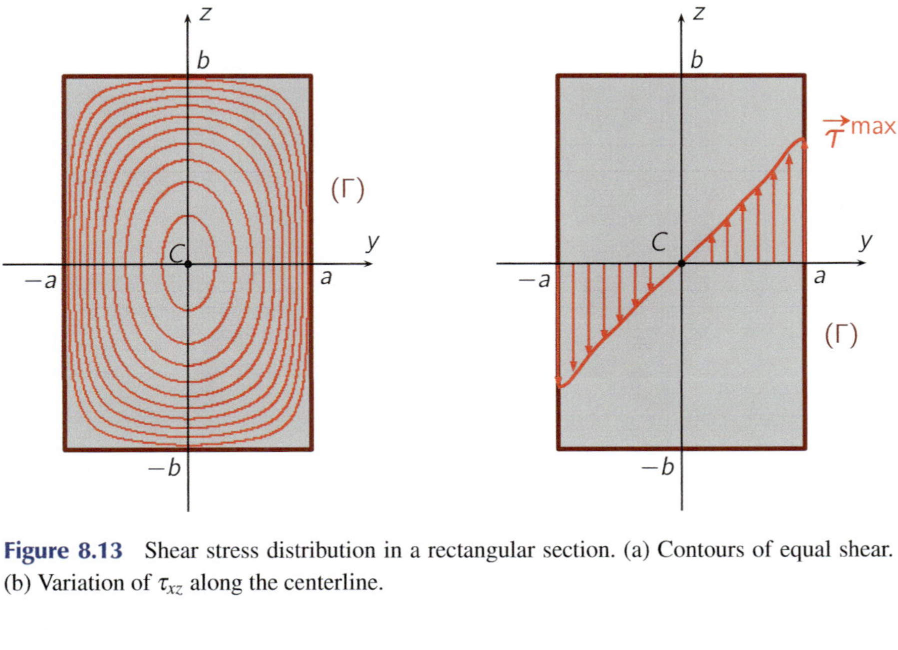

Figure 8.13 Shear stress distribution in a rectangular section. (a) Contours of equal shear. (b) Variation of τ_{xz} along the centerline.

tions (8.41) and (8.42), respectively. Table 8.1 gives some of their values for various b/a ratios.

Table 8.1

Numerical Values of J and τ^{max} in a Rectangular Bar for Various b/a Ratios

$\frac{b}{a}$	1	1.5	2	2.5	3	4	5	10	100
$\frac{J}{(2a)^2(2b)^2}$	0.141	0.131	0.115	0.099	0.088	0.070	0.058	0.031	0.003
$\tau^{max}\frac{(2a)^2(2b)}{M^t}$	4.801	4.326	4.061	3.888	3.745	3.548	3.433	3.205	3

In particular, for a square-prismatic beam ($b/a = 1$) one has:

$$J = 0.141\,S^2 \quad \text{and } \tau^{max} = 4.801\,\frac{M^t}{2aS}$$

with $S = 4a^2$ the surface area. For a thin rectangular strip ($b \gg a$), one has

$$J = \frac{1}{3}\frac{a}{b}S^2 \quad \text{and } \tau^{max} = 3\,\frac{M^t}{2aS}$$

where $S = 4ab$.

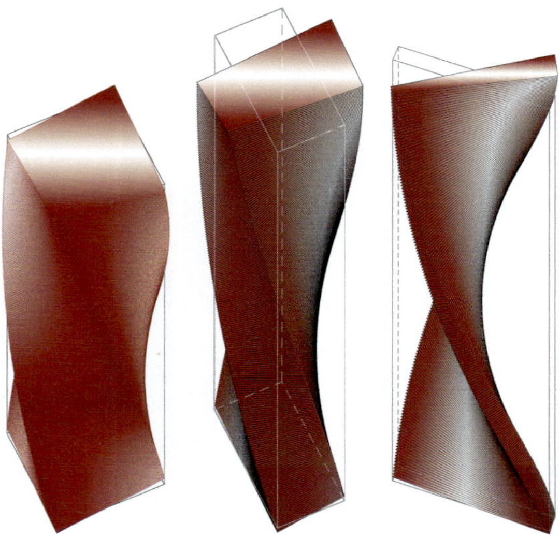

Figure 8.14 Twisting and warping of a square, rectangular ($b/a = 2$) and thin ($b/a = 10$) beams.

Integrating Equations (8.21) and (8.22), the warping function is obtained as (see Problem 8.21)

$$\chi(y,z) = \frac{M^t}{J}\left[yz + \frac{32}{\pi^3}a^2 \sum_{n=1,3,5}^{\infty} \frac{(-1)^{\frac{n-1}{2}}}{n^3} \frac{\sinh\frac{n\pi z}{2a}\sin\frac{n\pi y}{2a}}{\cosh\frac{n\pi b}{2a}}\right] \tag{8.43}$$

Here, the first three terms are sufficient because of the $1/n^3$ scaling. Figure 8.14 depicts the warping of some beams, including a square-prismatic one. The drawings were made using the above warping function and the twist angle. ♡

Remark 8.6 *The constant J that enters the general formulas of beam torsion is not to be confused with the polar moment of inertia I_x. Only for a circular beam would the equality $J = I_x$ hold.* ∗

8.3 TORSION OF HOLLOW PRISMATIC BEAMS

In Section 8.2, the otherwise arbitrary prismatic beam was assumed to be solid, that is with no holes. This assumption was made to simplify the analysis.

In the case of a hollow beam that may contain one or more holes, the cross-section S constitutes a multiply connected domain; its boundary Γ is the union of several closed and disjoint curves: one external contour, say Γ_0, and n internal contours $\Gamma_1, \Gamma_2, \ldots, \Gamma_n$, Figure 8.15.

All results in Section 8.2 remain valid with due account of the following:

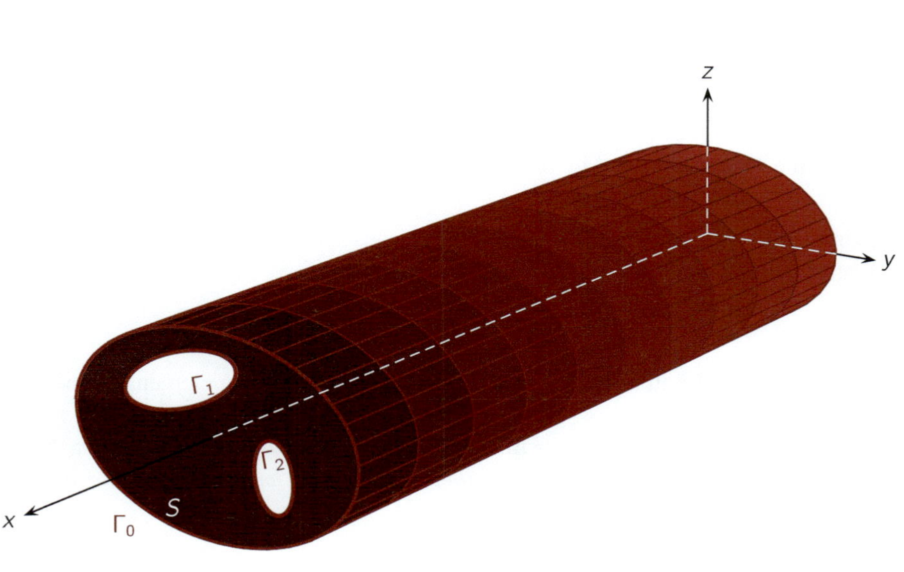

Figure 8.15 Hollow prismatic beam under torsion.

- The value of the stress function $\varphi(y,z)$ is taken to be zero on Γ_0 in order to satisfy the traction-free condition on the lateral surface. Then, the (constant) values $\varphi_1, \varphi_2, \ldots, \varphi_n$ taken by function φ on $\Gamma_1, \Gamma_2, \ldots, \Gamma_n$ are inferred. These values are generally different from one contour to the other.
- The formula that permits the calculation of J, Equation (8.17), is amended as follows:

$$M^t = -2\left[(\varphi_1 \mathcal{A}_1 + \cdots + \varphi_n \mathcal{A}_n) + \iint_S \varphi(y,z)\,\mathrm{d}y\,\mathrm{d}z\right] \tag{8.44}$$

where $\mathcal{A}_1, \mathcal{A}_2, \ldots, \mathcal{A}_n$ denote the areas enclosed by $\Gamma_1, \Gamma_2, \ldots, \Gamma_n$.
- The formulas giving the angle of twist, Equation (8.26), and the energy of twist, Equation (8.27), are applicable with the appropriate value of J.

8.4 FLUX AND CIRCULATION OF SHEAR TRACTION

The analysis carried out in Section 8.2, and extended in Section 8.3, calls for the following observations:

- Relations (8.11), which derive the traction vector $\vec{\tau}$ from the torsion function $\varphi(y,z)$, show that this vector is obtained from the vector $\nabla\varphi$ by a rotation of $\frac{\pi}{2}$ in the plane of the cross-section S; the curves $\varphi = \text{const}$, known as shear lines, are the force lines of field $\vec{\tau}$.
- The divergence of vector $\vec{\tau}$ is nil ($\mathbf{div}\,\vec{\tau} = 0$); the traction vector is therefore a curl; it derives from a vector potential.
- The algebraic value, curl $\vec{\tau}$, of the pseudo-vector $\mathbf{curl}\,\vec{\tau}$ on the x axis equals the Laplacian of the torsion function:

$$\text{curl}\,\vec{\tau} = \Delta\varphi = 2\frac{M^t}{J} \tag{8.45}$$

Consider first solid sections with no holes.

Theorem 8.1: Shear Flux Theorem

The outward flux of shear traction vector $\vec{\tau}$ vanishes across any closed curve $\mathcal{C}$ drawn on current cross-section S:

$$\forall \mathcal{C}, \quad \Phi_{\mathcal{C}} = \oint_{\mathcal{C}} \vec{\tau} \cdot \mathbf{n}\,\mathrm{d}l = 0 \tag{8.46}$$

with $\mathbf{n}$ the outward unit normal at current point $P \in \mathcal{C}$ and $\mathrm{d}l$ an element of line. ♢

Proof *Let $\mathcal{C}$ be an arbitrary closed curve drawn on current cross-section S, as illustrated in Figure 8.16. $\mathcal{C}$ is also called a contour. The vector $\vec{\tau}$ being a curl, Riemann's formula permits to write*

$$\oint_{\mathcal{C}} \vec{\tau} \cdot \mathbf{n}\,\mathrm{d}l = \iint_{\mathcal{D}} \mathbf{div}\,\vec{\tau}\,\mathrm{d}S$$

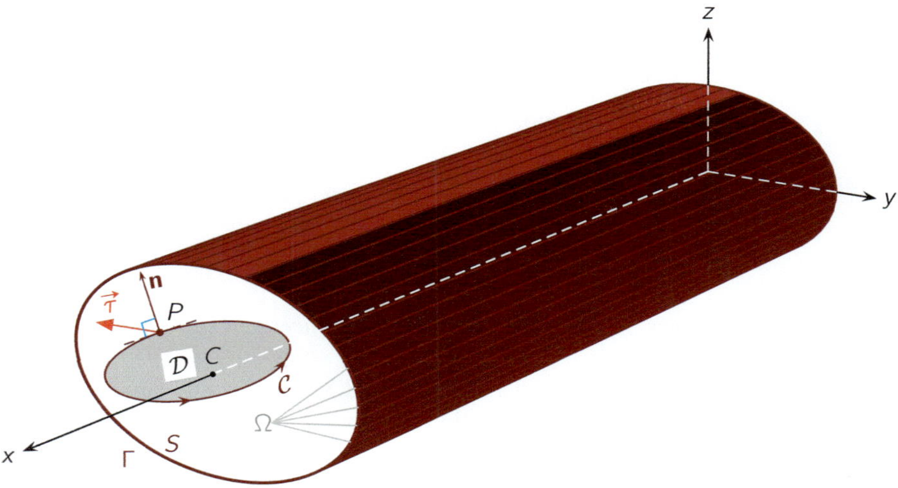

Figure 8.16 Flux and circulation of shear traction.

where $\mathcal{D}$ denotes the domain enclosed by $\mathcal{C}$. The equilibrium statement implies that $\vec{\tau}$ is pointwise divergence-free; therefore the integrand in the above equation is zero. ■

It follows from Theorem 8.1 that the same flux traverses all semi-lines emanating from a given point Ω on S, Figure 8.16.

Theorem 8.2: Shear Circulation Theorem

The circulation of shear traction vector $\vec{\tau}$ along any closed curve $\mathcal{C}$ drawn in the positive sense on current cross-section S is given by

$$\forall \mathcal{C}, \quad \oint_{\mathcal{C}} \vec{\tau} \cdot \mathbf{dl} = 2\frac{M^t}{J}\mathcal{A}(\mathcal{D}) \tag{8.47}$$

where $\mathbf{dl}$ denotes an (oriented) tangential line element and $\mathcal{A}(\mathcal{D})$ the area of domain $\mathcal{D}$ bounded by $\mathcal{C}$. In terms of the angle of twist,

$$\forall \mathcal{C}, \quad \oint_{\mathcal{C}} \vec{\tau} \cdot \mathbf{dl} = 2G\mathcal{A}\frac{\mathrm{d}\omega_x}{\mathrm{d}x} \tag{8.48}$$

◇

Proof *Consider the same closed curve in Figure 8.16 enclosing domain $\mathcal{D}$. Stokes'* *formula shows that the circulation of $\vec{\tau}$ is actually the flux of its curl across $\mathcal{D}$:*

$$\begin{aligned}
\oint_{\mathcal{C}} \vec{\tau}\cdot \mathbf{dl} &= \iint_{\mathcal{D}} \mathbf{curl}\,\vec{\tau}\cdot \mathbf{dS} && \textit{Stokes' theorem}\\
&= \iint_{\mathcal{D}} \Delta\varphi \,\mathrm{d}y\,\mathrm{d}z && \textit{using Equation (8.45)}\\
&= \iint_{\mathcal{D}} 2\frac{M^t}{J}\,\mathrm{d}y\,\mathrm{d}z && \textit{by virtue of Equation (8.23)}\\
&= 2\frac{M^t}{J}\iint_{\mathcal{D}} \mathrm{d}S && J \textit{ and } M^t \textit{ constant}
\end{aligned}$$

Equation (8.47) *follows. To prove Equation* (8.48) *it suffices to recall Equation* (8.26) *and use the previous result.* ■

For a hollow beam (see Figure 8.15 in Section 8.3), the flux and circulation formulas, Equations (8.46) and (8.47), remain valid even if the curve $\mathcal{C}$ encloses one or more holes. This follows from a closer examination of the above proofs.

Theorems 8.1 and 8.2 giving the shear flux and circulation are essential to simplifying the equations of torsion of arbitrary prismatic beams in the case of thin-walled sections.

8.5 TORSION OF THIN-WALLED TUBES

As seen in Section 8.2 and illustrated in Examples 8.1–8.3, determining the shear stresses in a cross-section constitutes a relatively complex problem.

This problem becomes much simpler in thin-walled tubes, i.e., beams in which the cross-section is made up of one or more bands having each a thickness that is much smaller than the diameter of the cross-section. The thickness t is generally uniform, the tube being obtained by folding of a rectangular sheet, but it can also vary, in a continuous and progressive fashion, Figure 8.17. Since the wall thickness is small, a point is used to refer to a location along the wall and it is tacitly understood that current point P is associated with some wall thickness t.

DEFINITION 8.3 A line segment within S that has length t (thickness of the section at current point P) and is normal in P to the median line $\mathscr{C}$ is called a cut (Figure 8.17). ♣

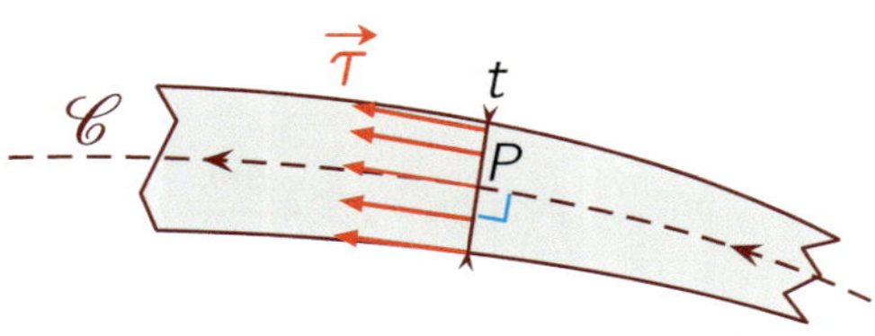

Figure 8.17 Two consequences of the thin wall assumption.

At both ends of the cut, the shear traction vector $\vec{\tau}$ is tangent to the contour of S; this strictly follows from the traction-free condition on either lateral surface (inner and outer). Because the thickness t is small, $\vec{\tau}$ may be considered as uniform across the cut and oriented tangentially to $\mathscr{C}$ at P, Figure 8.17.

DEFINITION 8.4 The flux of shear traction $\vec{\tau}$ across a cut of thickness t is simply called *shear flow* and is denoted by $\phi = \tau t$. ♣

Three cases are to be considered with the degree of intrinsic complexity going *crescendo* in the order they are listed: simple closed tube; multi-cell closed tube; and open simple tube.

8.5.1 SIMPLE CLOSED TUBE

The cross-section S is formed by a single band, closing on itself and admitting median line $\mathscr{C}$. The median line is oriented counterclockwise about the axis of torsion, Figure 8.18.

If the median line is parameterized by arclength s, then it must be recognized that, in general, $\phi(s) = \tau(s)t(s)$ where both ϕ and τ are counted positively in the direction of oriented line $\mathscr{C}$.

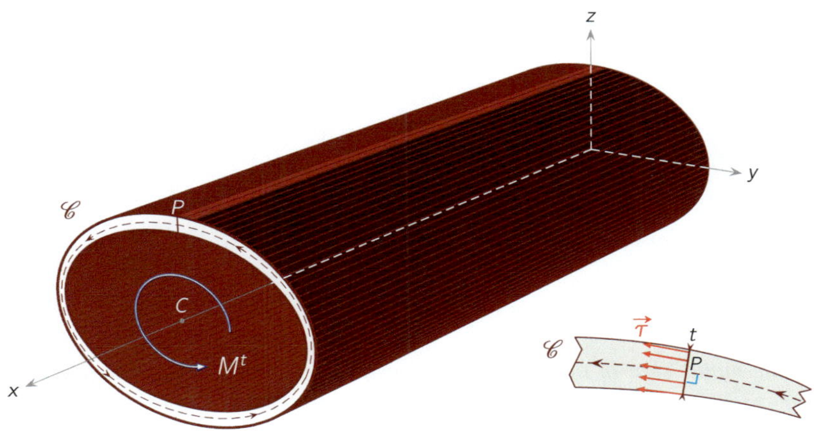

Figure 8.18 One-cell closed section under torque M^t.

Theorem 8.3

In a simple closed tube, the shear flow is constant. ◇

Proof *Apply Theorem 8.1 to the particular contour C shown in Figure 8.19:*

$$\oint_C \vec{\tau} \cdot \mathbf{n}\, \mathrm{d}l = 0$$

This contour is the union of four open (oriented) curves, denoted $C_1, \ldots, C_4$, not to be confused with $\mathscr{C}$, the median line. C_1 and C_3 are cuts, as per Definition 8.3, whereas C_2 and C_4 are (traction-free) parts of the outer and inner walls of cross-section S.

Since $C = C_1 \cup C_2 \cup C_3 \cup C_4$, the above contour integral may be written as the sum of four integrals:

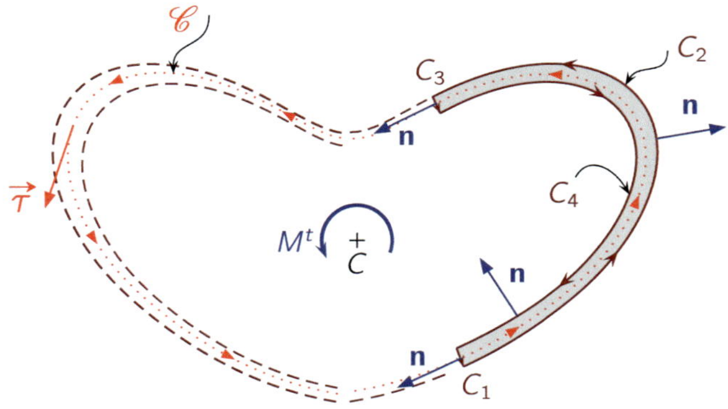

Figure 8.19 Shear flux across contour C bounding grayed area.

$$\int_{C_1} \vec{\tau} \cdot \mathbf{n}\, \mathrm{d}l + \int_{C_2} \vec{\tau} \cdot \mathbf{n}\, \mathrm{d}l + \int_{C_3} \vec{\tau} \cdot \mathbf{n}\, \mathrm{d}l + \int_{C_4} \vec{\tau} \cdot \mathbf{n}\, \mathrm{d}l = 0$$

On C_2 and C_4, $\vec{\tau} \perp \mathbf{n}$ so that the integrand vanishes pointwise. On C_1 and C_3, $\vec{\tau} \parallel \mathbf{n}$ and since $\vec{\tau}$ is uniform the above equation reduces to

$$-(\tau t)\big|_1 + (\tau t)\big|_3 = 0$$

where the minus sign in the first term arises from the outer normal definition of $\mathbf{n}$ and the orientation of shear flow (shown with red arrows). Given that cuts C_1 and C_3 are arbitrary, the product τt, defined above as the shear flow ϕ, is indeed constant along $\mathscr{C}$. ■

It follows from Theorem 8.3 that if the wall thickness is uniform, then the shear stress itself is constant in the tube. In general, however, that need not be the case. It also follows that the shear stress in the wall scales as the inverse of the wall thickness. The thinner the wall the larger the shear stress.

Theorem 8.4

In a simple closed tube, the shear flow is given by Bredt's formula:

$$\boxed{\phi = \frac{M^t}{2\mathcal{A}}} \tag{8.49}$$

where $\mathcal{A}$ is the area of the cavity. ◇

Proof *The screwer–traction relations allow to relate the resultant torque to the shear traction distribution. Specifically, a variant of Equation (5.35)$_1$ reads*

$$\mathbf{M}^t = \oint_{\mathscr{C}} \overrightarrow{CP} \times \vec{\tau}\, t\, \mathrm{d}l \tag{8.50}$$

where a first integration through the thickness was carried out on account of $\vec{\tau}$ being uniform on any cut.

In polar coordinates (r,θ) of pole C, one has

$$\overrightarrow{CP} \begin{cases} r \\ 0 \end{cases} \qquad \textit{and} \quad \vec{\tau} \begin{cases} \tau \dfrac{\mathrm{d}r}{\mathrm{d}l} \\ \tau\, r \dfrac{\mathrm{d}\theta}{\mathrm{d}l} \end{cases}$$

Hence,

$$\begin{aligned} M^t &= \oint_{\mathscr{C}} \tau\, t\, r^2 \frac{\mathrm{d}\theta}{\mathrm{d}l}\, \mathrm{d}l \\ &= \phi \oint_{\mathscr{C}} r^2\, \mathrm{d}\theta \qquad \textit{since } \phi = \textit{const. by Theorem 8.3} \\ &= 2\phi\mathcal{A} \qquad \textit{enclosed area formula in polar coordinates} \end{aligned}$$

Here, $\mathcal{A}$ is the area bounded by median line $\mathscr{C}$ or, roughly speaking, the area of the cavity since the wall thickness is small. ■

Finally, to complete the solution one needs to determine the constant J in order to compute the twist from Equation (8.26). To this end, the shear circulation formula, Equation (8.47), is invoked using the median line $\mathscr{C}$ for the closed curve in Theorem 8.2, that is:

$$\begin{aligned} \oint_{\mathscr{C}} \vec{\tau} \cdot \mathbf{dl} &= \oint_{\mathscr{C}} \frac{\phi}{t}\, \mathrm{d}l \qquad \vec{\tau} \text{ tangent to } \mathscr{C} \text{ and definition of } \phi \\ &= \phi \oint_{\mathscr{C}} \frac{\mathrm{d}l}{t} \qquad \text{since } \phi = \text{const. by Theorem 8.3} \\ &= 2\mathcal{A}\frac{M^t}{J} \qquad \text{circulation theorem, Equation (8.47)} \end{aligned}$$

Combining the right-hand sides of the last two equations leads to the desired relationship between J and ϕ. Elimination of the latter by way of Bredt's formula, Equation (8.49), gives an explicit expression for J in terms of geometrical parameters alone:

$$\boxed{J = \frac{4\mathcal{A}^2}{\oint_{\mathscr{C}} \frac{\mathrm{d}l}{t}}} \tag{8.51}$$

If the wall thickness is uniform, then

$$J = \frac{4t\,\mathcal{A}^2}{\mathcal{L}} \tag{8.52}$$

with $\mathcal{L}$ the perimeter of $\mathscr{C}$, the median line. Also, in that case the (constant) value of the shear stress is given by

$$\tau = \frac{M^t}{2t\mathcal{A}} \tag{8.53}$$

8.5.1.1 Summary

Thanks to the flux and circulation theorems, solution to the torsion problem in terms of a stress function defined by a partial differential equation and a surface integral condition, becomes a solution in terms of shear flow defined by a simple algebraic equation and J given by a simple integral in case of nonuniform wall thickness.

Example 8.4: Thin Rectangular Tube

Consider a box-like beam having a rectangular cross-section, Figure 8.20, and let a and b be, respectively, the half-width and half-height of the section, and t the wall thickness of the box.

The area enclosed by the median line (or simply area of the cavity) is:

$$\mathcal{A} = 4ab$$

Using Equation (8.53), obtain the (uniform) shear stress τ as:

$$\tau = \frac{M^t}{8tab}$$

Using Equation (8.52), the torsional stiffness is expressed as:

$$GJ = Gt\frac{16a^2b^2}{a+b}$$

This completes the solution. ♡

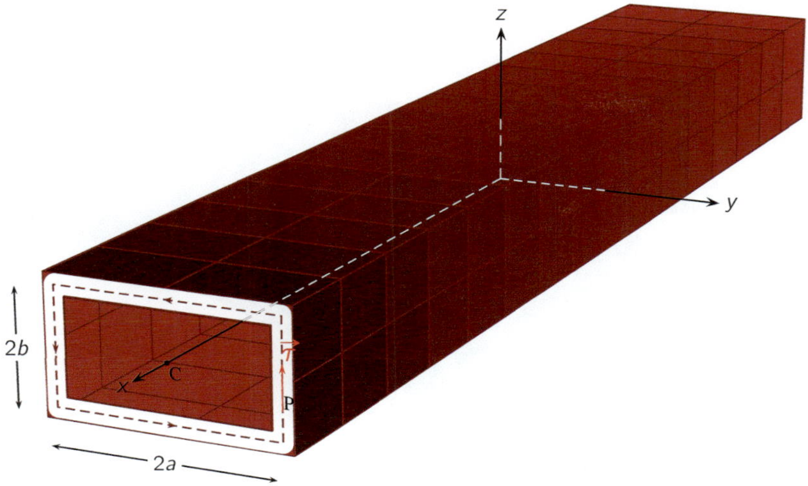

Figure 8.20 Rectangular box section.

8.5.2 MULTI-CELL CLOSED TUBE

The beam is made up of a cylindrical (or prismatic) external wall and one or more internal partitions (or webs), the whole so delimiting n cells (or cavities), Figure 8.21. The cross-section S consists of multiple branches meeting at junctions (or nodes). On S, with each cell i ($i = 1, 2, \ldots, n$) is associated an oriented closed curve $\mathscr{C}^i$ that encloses it.

As in the simple tube case, the shear stress $\vec{\tau}$ is taken to be uniform through the thickness so that the shear flow $\phi = \tau t$ may be defined on each cut.

Theorem 8.5: Branch Rule

In a multi-cell closed tube, the shear flow is constant per branch. ◇

Proof *It follows the same steps as for Theorem 8.3 but the two limiting cuts C_1 and C_3 must belong to the same branch, i.e., no junction should fall between them.* ■

Theorem 8.6: Node Rule

At a node of a multi-cell closed tube, the shear-flows toward the node (or away from it) add up to zero. ◇

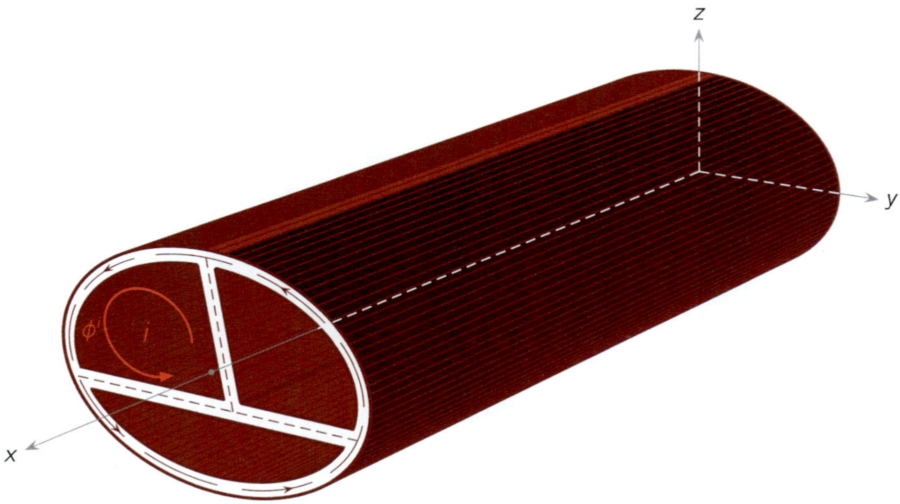

Figure 8.21 Closed section with multiple walls.

Proof *Any given node is common to at least three branches. Apply Theorem 8.1 to the particular contour $\mathcal{C}$ shown in Figure 8.22:*

$$\oint_{C} \vec{\tau} \cdot \mathbf{n} \, \mathrm{d}l = 0$$

This contour is the union of multiple open (oriented) curves, among which only a few (three in Figure 8.22) have non-zero flux. Those are cuts, as per Definition 8.3, and are denoted C_1, C_2, C_3, having thicknesses t_1, t_2, t_3. On all other parts, $\vec{\tau} \perp \mathbf{n}$ so that the flux vanishes. The above contour integral thus reduces to three integrals:

$$\int_{C_1} \vec{\tau} \cdot \mathbf{n} \, \mathrm{d}l + \int_{C_2} \vec{\tau} \cdot \mathbf{n} \, \mathrm{d}l + \int_{C_3} \vec{\tau} \cdot \mathbf{n} \, \mathrm{d}l = 0$$

On each cut, the shear is uniform and tangent to the median line so that

$$(\tau t)\big|_1 + (\tau t)\big|_2 + (\tau t)\big|_3 = 0$$

where, for the sake of this proof, the median lines on all three branches are oriented the same way (either toward the node or away from it).

More generally, for a node shared by b branches the node rule reads

$$\phi\big|_1 + \cdots + \phi\big|_b = 0 \tag{8.54}$$

where it is emphasized that all median lines are oriented either towards or away from the node. ■

In general, a multi-cell section would have n cells, b branches and N nodes where n, b and N are topologically related. In order to keep the notation economical, a

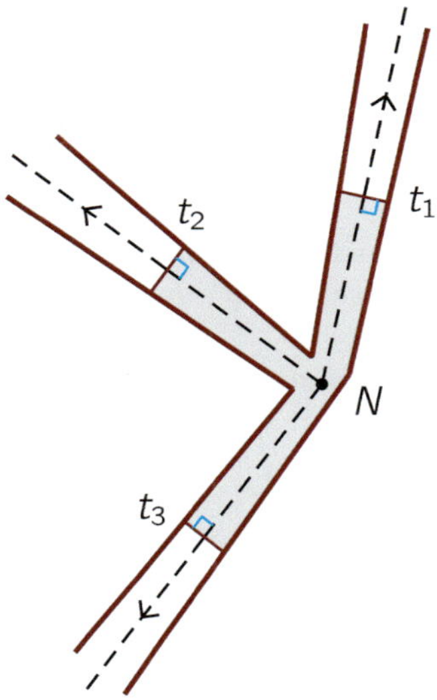

Figure 8.22 Junction of three walls at node N.

shear flow ϕ^i is associated with cell i and is counted positively consistent with the orientation of its median line. The following is adopted:

DEFINITION 8.5 A shear flow ϕ^i associated with cell i having median line $\mathscr{C}^i$ is defined such that, on each branch limiting the cell, the *actual* shear flow (evaluated algebraically on $\mathscr{C}^i$) is constant and equal to:

ϕ^i if the branch separates the exterior from cell i;
$\phi^i - \phi^j$ if the branch separates cells i and j. ♣

The above definition automatically satisfies the node rule (Theorem 8.6) and the branch rule (Theorem 8.5). The practical problem reduces therefore to determining the n unknown shear-flows ϕ^i ($i = 1, \ldots, n$) and the coefficient J.

Theorem 8.7: Cell Rule

In a multi-cell closed tube consisting of n cells, the n shear-flows ϕ^i and coefficient J are determined from the $n+1$ equations:

$$\frac{1}{2\mathcal{A}^i}\oint_{\mathscr{C}^i}\frac{\phi}{t}\,\mathrm{d}l = \frac{M^t}{J} \qquad \text{for } i = 1,2,\ldots,n \tag{8.55}$$

$$2(\phi^1\mathcal{A}^1 + \phi^2\mathcal{A}^2 + \cdots + \phi^n\mathcal{A}^n) = M^t \tag{8.56}$$

where $\mathcal{A}^i$ denotes the area of cell i, i.e., the area enclosed by $\mathscr{C}^i$. ◇

Proof *Equation* (8.55) *follows from applying the circulation formula, Equation* (8.47)*, to the median line of cell i, that is by applying Theorem 8.2 taking* $C = \mathscr{C}^i$ *for each cell separately.*

To establish Equation (8.56)*, the screwer–traction relation, Equation* 5.35_1*, reads*

$$\mathbf{M}^t = \iint_S \overrightarrow{CP} \times \vec{\tau}\,\mathrm{d}S$$

The median line $\mathscr{C}^i$ *of the* i^{th} *cell is further divided into the external wall* C^i *and the union of internal ones partitioning cell i from neighboring cells. Formally,* C^{ij} *is defined as the putative partition between cells i and j. By way of extension, if such a partition does not exist then* C^{ij} *is taken to degenerate to a point so that* $\int_{C^{ij}}(\cdot) = 0$. *Then,*

$$\mathbf{M}^t = \sum_{i=1}^{n}\int_{C^i}\overrightarrow{CP}\times\vec{\tau}\,t\,\mathrm{d}l + \sum_{i=1}^{n}\sum_{j>i}^{n}\int_{C^{ij}}\overrightarrow{CP}\times\vec{\tau}\,t\,\mathrm{d}l$$

where a first integration through the thickness has been carried out since $\vec{\tau}$ *is uniform on any cut. In writing the above sum, one must recognize that any branch common to two neighboring cells would be counted twice without the prescription* $j > i$ *in the second sum.*

Next, introducing polar coordinates, as in the proof of Theorem 8.4, and replacing the product τt *with the shear flow, the above equation, projected on the torsion axis, becomes*

$$M^t = \sum_{i=1}^{n}\int_{C^i}\phi^i r^2\,\mathrm{d}\theta + \sum_{i=1}^{n}\sum_{j>i}^{n}\int_{C^{ij}}(\phi^i - \phi^j)r^2\,\mathrm{d}\theta$$

with due account of Definition 8.5. The constant shear flows are kept inside the integrals for the purpose of the following manipulations, which are needed to rigorously obtain Equation (8.56)*:*

$$\begin{aligned} M^t &= \sum_{i=1}^{n}\int_{C^i}\phi^i r^2\,\mathrm{d}\theta + \sum_{i=1}^{n}\sum_{j\neq i}^{n}\int_{C^{ij}}(\phi^i - \phi^j)r^2\,\mathrm{d}\theta \\ &\quad - \sum_{i=1}^{n}\sum_{j<i}^{n}\int_{C^{ij}}(\phi^i - \phi^j)r^2\,\mathrm{d}\theta \qquad \textit{breaking the second term into two sums} \\ &= \sum_{i=1}^{n}\left\{\int_{C^i}\phi^i r^2\,\mathrm{d}\theta + \sum_{j\neq i}^{n}\int_{C^{ij}}\phi^i r^2\,\mathrm{d}\theta\right\} \\ &\quad - \sum_{i=1}^{n}\sum_{j\neq i}^{n}\int_{C^{ij}}\phi^j r^2\,\mathrm{d}\theta - \sum_{i=1}^{n}\sum_{j<i}^{n}\int_{C^{ij}}(\phi^i - \phi^j)r^2\,\mathrm{d}\theta \qquad \textit{rearranging} \end{aligned}$$

$$= \sum_{i=1}^{n} \oint_{\mathscr{C}^i} \phi^i r^2 \, d\theta - \sum_{i=1}^{n}\sum_{j<i}^{n} \int_{C^{ij}} \phi^j r^2 \, d\theta - \sum_{i=1}^{n}\sum_{j>i}^{n} \int_{C^{ij}} \phi^j r^2 \, d\theta$$

$$- \sum_{i=1}^{n}\sum_{j<i}^{n} \int_{C^{ij}} (\phi^i - \phi^j) r^2 \, d\theta \qquad \textit{using } \mathscr{C}^i = C^i \cup \left(\bigcup_{j\neq i} C^{ij} \right)$$

and breaking one sum

$$= \sum_{i=1}^{n} \oint_{\mathscr{C}^i} \phi^i r^2 \, d\theta - \sum_{i=1}^{n}\sum_{j>i}^{n} \int_{C^{ij}} \phi^j r^2 \, d\theta - \sum_{i=1}^{n}\sum_{j<i}^{n} \int_{C^{ij}} \phi^i r^2 \, d\theta$$

$$= \sum_{i=1}^{n} \oint_{\mathscr{C}^i} \phi^i r^2 \, d\theta$$

The last two double sums cancel out. To see this recall that, by virtue of Definition 8.5,

$$\int_{C^{ij}} \phi^i r^2 \, d\theta = - \int_{C^{ji}} \phi^i r^2 \, d\theta$$

so that the last double sum may be written as follows:

$$- \sum_{i=1}^{n}\sum_{j<i}^{n} \int_{C^{ij}} \phi^i r^2 \, d\theta = \sum_{i=1}^{n}\sum_{j<i}^{n} \int_{C^{ji}} \phi^i r^2 \, d\theta = \sum_{i=1}^{n}\sum_{i<j}^{n} \int_{C^{ij}} \phi^j r^2 \, d\theta$$

because i and j commute in the embedded sum.

The above result is explicitly evaluated to deliver in fine Equation (8.56)*:*

$$M^t = \sum_{i=1}^{n} \oint_{\mathscr{C}^i} \phi^i r^2 \, d\theta = \sum_{i=1}^{n} \phi^i \oint_{\mathscr{C}^i} r^2 \, d\theta = \sum_{i=1}^{n} 2\phi^i \mathcal{A}^i$$

■

Remark 8.7 *The branch rule (Theorem 8.5) implies that, in any given cell, the shear flow is piece-wise constant. It is therefore emphasized that Equation* (8.56) *is not as trivial as it may appear since only the values of shear flow in external walls appear.* *

Remark 8.8 *The total moment carried by the multi-cell cross-section is not equal to the sum of the moments carried by individual cells (see Problem 8.23). Interestingly, what Equation* (8.56) *expresses is that the total torque* M^t *results from the n moments* $M_i^t = 2\phi^i \mathcal{A}^i$ *associated with the n shear-flows* ϕ^i*. In other words,* M_i^t *does not represent the full torque carried by cell i.* *

Remark 8.9 *The beginner may find the above proof of Theorem 8.7 too abstract. A simple way to access its essence is to consider a few cells, say three as in Problem 8.24, and calculate the total torque.* *

Example 8.5: Rectangular Box with Two Partitions

Consider the three-cell box shown in Figure 8.23. Skin and partitions have the same thickness t.

Equation (8.55), spelled for each cell, reads:

$$\frac{1}{2\cdot 2a^2}\left[4\frac{\phi^1}{t}a+2\frac{\phi^1-\phi^2}{t}a\right]=\frac{M^t}{J}$$

$$\frac{1}{2\cdot 2a^2}\left[2\frac{\phi^2-\phi^1}{t}a+2\frac{\phi^2-\phi^3}{t}a+2\frac{\phi^2}{t}a\right]=\frac{M^t}{J}$$

$$\frac{1}{2\cdot 2a^2}\left[4\frac{\phi^3}{t}a+2\frac{\phi^3-\phi^2}{t}a\right]=\frac{M^t}{J}$$

In applying Equation (8.55) to each cell, it is emphasized that the shear flow ϕ is piece-wise constant per branch. The constant values ϕ^1, ϕ^2, ϕ^3 are defined consistent with Definition 8.5. Equation (8.56) completes the set:

$$2(\phi^1+\phi^2+\phi^3)(2a^2)=M^t$$

Solving the above system of four equations in the four unknowns ϕ^1, ϕ^2, ϕ^3 and J gives:

$$\phi^1=\phi^3=\frac{1}{13}\frac{M^t}{a^2} \qquad \text{and} \qquad \phi^2=\frac{5}{52}\frac{M^t}{a^2}$$

then

$$J=\frac{104}{7}a^3t$$

The shear flows in other inter-cell branches may be inferred from the above expressions. ♡

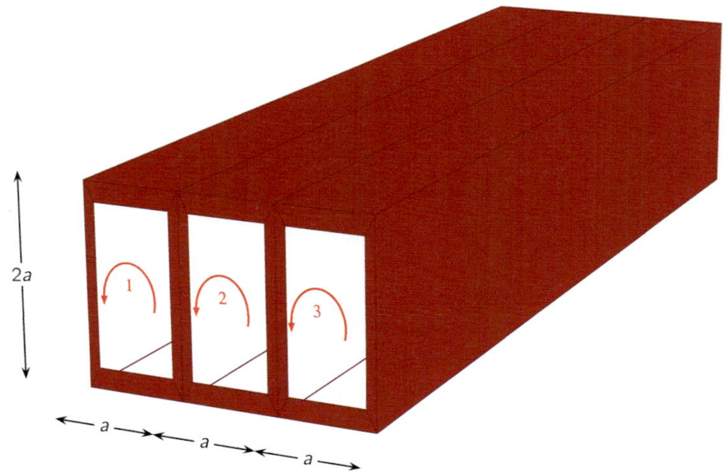

Figure 8.23 Three-box beam.

8.5.3 SIMPLE OPEN TUBE

The cross-section S consists of a single, open band admitting $\mathscr{C}$ as an oriented median line, of length $\mathcal{L}$, Figure 8.24.

The shear flow being the same on all cuts but nil on the end cuts must therefore be nil on every cut. It follows that one can no longer assume a uniform distribution of $\vec{\tau}$ across the thickness. The variation of shear stress must be symmetric with respect to the center P of the cut (inset in Figure 8.24). The magnitude τ^{ex} of τ on the cut extremes varies with the position of the cut and vanishes on the initial and final cuts.

A classical approximation is to consider that the stress field and the torsional stiffness remain unchanged upon developing the beam, by unrolling it, so as to make it into a rectangular planar plate of length L and width $\mathcal{L}$.

If the wall thickness is uniform, the resulting beam of length L has a thin rectangular cross-section (height $\mathcal{L}$, thickness t), as analyzed in Example 8.3. It was found there in the limit $t \ll \mathcal{L}$:

$$\boxed{J = \frac{1}{3}\mathcal{L}t^3 \qquad \text{and} \qquad \tau^{\max} = 3\frac{M^t}{t^2\mathcal{L}}} \tag{8.57}$$

the maximum stress being attained at mid-height.

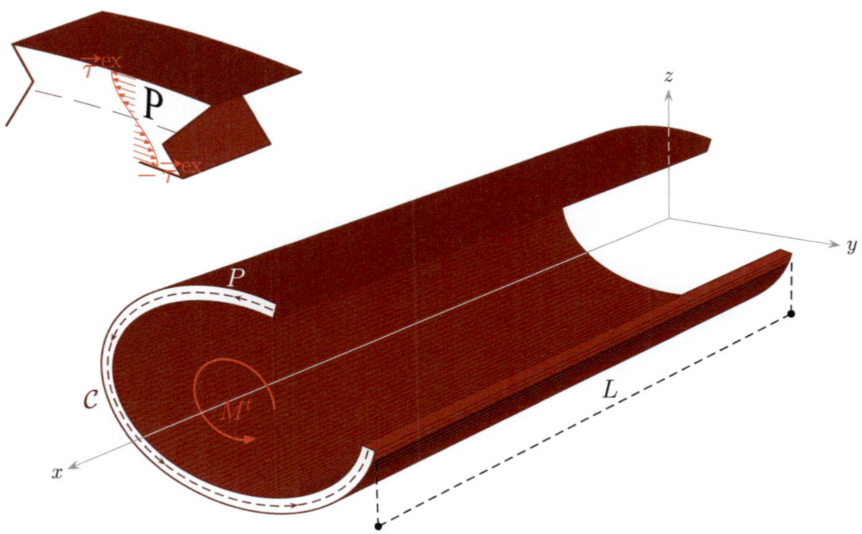

Figure 8.24 Simple open tube under torsion.

Example 8.6: Slit-Open Circular Tube

A circular tube having thickness t and radius R ($t \ll R$) is slit open along one generatrix.

Equation (8.57) gives:

$$\tau^{\max} = \frac{3}{2\pi}\frac{M^t}{Rt^2} \qquad \text{and} \quad J = \frac{2\pi}{3}Rt^3$$

Compare with

$$\tau^{\max} = \frac{1}{2\pi}\frac{M^t}{R^2 t} \qquad \text{and} \quad J = 2\pi R^3 t$$

for a closed thin-walled circular tube, Equations (8.52) and (8.53).

The ratio of maximum shears is: $3\frac{R}{t} \gg 1$

The ratio of torsional stiffnesses is: $\frac{1}{3}\left(\frac{t}{R}\right)^2 \ll 1$

The open tube is much less stiff than the closed tube; the same torque produces stresses (and stress gradients) that are much greater than in the closed tube. ♡

8.6 NON-PRISMATIC BEAMS

The results of all previous sections were obtained for prismatic beams. For an arbitrary beam (with curvature, torsion, washout) and subjected to some loading giving rise, among other things, to a torque $\mathbf{M}^X$, consider a thin slice of the beam comprised between cross-sections S and S' located at arclengths s and $s + \mathrm{d}s$, Figure 8.25.

From the assumptions of slender beams:

- The diameter of S is small in comparison with the length of the median line, the radius of curvature and the radius of torsion in G, centroid of S.
- The shape, dimensions and washout of S vary slowly and continuously with s (or do not vary at all).

Such a slice may then be considered as prismatic so that all previous results can be applied to the slice, that is in differential form. Let us recall the principal results, with $\mathbf{M}^t$ denoting the resultant torque on S:

- Obtain on S the stress field $\vec{\tau}$, the center of torsion C, and the torsional stiffness GJ using Saint-Venant's method.

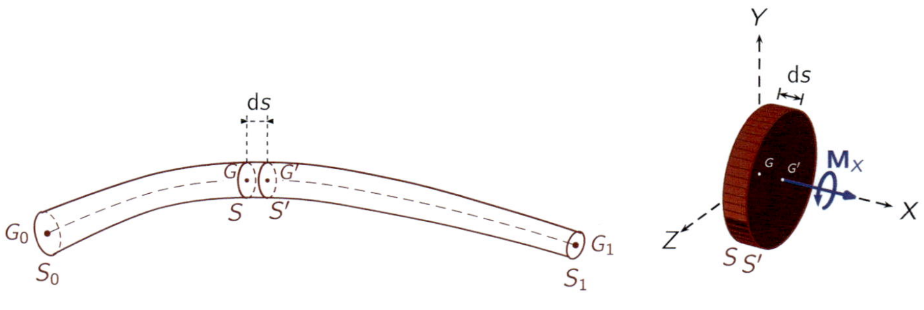

Figure 8.25 Twisting of an arbitrary beam.

- The flux and circulation formulas for the traction vector $\vec{\tau}$ on any closed curve $\mathcal{C}$ drawn on S.
- Cross-section S' rotates globally with respect to S about the axis of torsion parallel to the longitudinal axis CX, the angle of twist being

$$\mathrm{d}\boldsymbol{\omega} = \frac{1}{GJ}\mathbf{M}^t\,\mathrm{d}s \tag{8.58}$$

- The elastic potential energy of torsion in the slice of thickness $\mathrm{d}s$ is

$$\mathrm{d}W = \frac{1}{2}\frac{M^{t2}}{GJ}\,\mathrm{d}s \tag{8.59}$$

- The formulas obtained in the case of thin-walled tubes (shear-flow, stress and stiffness) remain valid.

8.7 DOMAINS OF VALIDITY

1. The theory of torsion developed in this chapter assumes that all cross-sections are free to warp. This condition may not be realized on certain cross-sections, for instance on fixed end sections.

   The results of the theory are therefore not valid in the neighborhood of such sections; in particular, the torsional stiffness of these sections is greater; see Problem 8.16.
2. In the outlined theory, the torques applied to end sections were assumed to result from a specific distribution of shear tractions. In practice, torques may be applied in various ways and on more than just two sections. Near such loaded sections, the results of the general theory are not exact.
3. If on a cross-section prevails a transverse shear $\mathbf{V}$, in addition to a longitudinal moment $\mathbf{M}_X$, then the shear-traction field $\vec{\tau}$ on S admits $\mathbf{V}$ as a general resultant and $\mathbf{M}_X$ as a resultant in G. The torque $\mathbf{M}^t$ is the resultant moment of the same traction loads $\vec{\tau}\,\mathrm{d}S$, but in C, center of torsion of S.
   The torque $\mathbf{M}^t$ is thus related to the longitudinal moment $\mathbf{M}_X$ by the couple transport equation:

$$\mathbf{M}^t = \mathbf{M}_X + \overrightarrow{CG} \times \mathbf{V} \tag{8.60}$$

   In particular, $\mathbf{M}^t$ and $\mathbf{M}_X$ are identical when the vector product $\overrightarrow{CG} \times \mathbf{V}$ vanishes, that is in the following cases:
   - C and G are the same, e.g., for sections admitting two axes of symmetry;
   - $\mathbf{V}$ is zero (as in the present chapter);
   - $\overrightarrow{CG}$ and $\mathbf{V}$ are parallel.
4. For very long prismatic beams (e.g., wire of a torsional pendulum) the twist ω_X and the transverse displacements v and w do not necessarily remain small. All formulas remain, however, valid except those concerning the transverse displacements v and w, namely Equations $(8.5)_2$, $(8.20)_{2,3}$, as well as Equations (8.24) and (8.25).

PROBLEMS

Problem 8.1 (Elasticity Solution for Cylindrical Beam) *Show that Equations* (8.2), (8.3) *and* (8.5) *are indeed the solution to the elasticity problem defined in Section 8.1 for the torsion of an eventually hollow circular cylindrical bar, Figure 8.1.* ⋆

Problem 8.2 (Saint-Venant Principle) *The pure torsion solution presented in Section 8.1 was developed assuming linearly distributed shear tractions on end sections, Equation* (8.1)*. Consider now alternative traction distributions on end sections, as would arise in practice due to fixture and set-up variations. State the conditions that must be satisfied by the traction distributions such that the solution given in Section 8.1 remains valid sufficiently far away from end sections. Sketch some traction distributions that meet the stated conditions.* ⋆

Problem 8.3 (Solid vs. Hollow Shafts) *Compare the transmitted powers of two shafts, one solid and one hollow, having the same weight.* ⋆

Problem 8.4 (Solid vs. Hollow Shaft) *Compare the weights of two (circular) shafts, one hollow the other solid, having the same length L and made up of the same material: (i) when the maximum stress is the same in both shafts; (ii) when the deformation is the same (same twist or relative rotation of extreme sections). Let R denote the radius of the solid shaft, P^{s} its weight, R^{i} and R^{o} the inner and outer radii of the hollow shaft, with $\rho = R^{\mathrm{i}}/R^{\mathrm{o}}$, and P^{h} its weight. Provide numerical results for $\rho = 0.6$ and make your observations.* ⋆

Problem 8.5 (Dimensioning a Transmission Shaft) *A solid cylindrical shaft of revolution having length L and shear modulus G transmits a maximum power $\mathcal{P}$ at angular velocity of n in rotations per minute (RPM). If the shear stress at break is τ_{R} and if k denotes the safety factor, determine the radius R of the shaft and the relative rotation of end sections. Carry out the numerical application in the following two cases: (i) $L = 3$ m, $G = 90$ GPa, $\mathcal{P} = 300$ hp, $n = 3000$ RPM, $\tau_{\mathrm{R}} = 250$ MPa, $k = 5$. (ii) $L = 3$ m, $G = 90$ GPa, $\mathcal{P} = 500$ hp, $n = 600$ RPM, $\tau_{\mathrm{R}} = 250$ MPa, $k = 5$.* ⋆

Problem 8.6 (Staged Transmission Shaft) *A transmission shaft having length 4L is controlled by three stages, as shown in Figure 8.26. It is subjected to a motor torque Γ. Due to friction, the stages apply counter-torques $-C_1\dot{\varphi}$, $-C_2\dot{\varphi}$, $-C_3\dot{\varphi}$ where $\dot{\varphi}$ is the angular velocity of the shaft. Draw the diagram of the internal torque M_x and determine the available output couple Γ'. What is the shaft efficiency, i.e., the ratio of output power to input power?* ⋆

Problem 8.7 (Measurement of Shear Modulus) *A torsion pendulum consists of a cylindrical wire and a disk, Figure 8.27. The wire is made of a homogeneous elastic material of unknown shear modulus G and has radius r and length ℓ. The disk, which is suspended by the wire to a fixed body, has radius R and mass m. Derive the expression giving the shear modulus G of the material of the wire in terms of the period P*

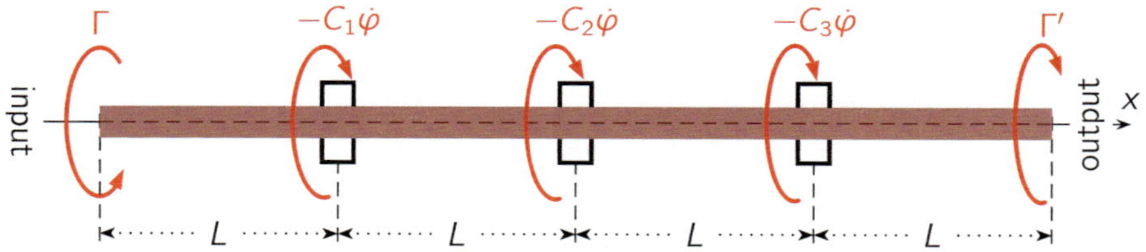

Figure 8.26 A three-stage controlled shaft.

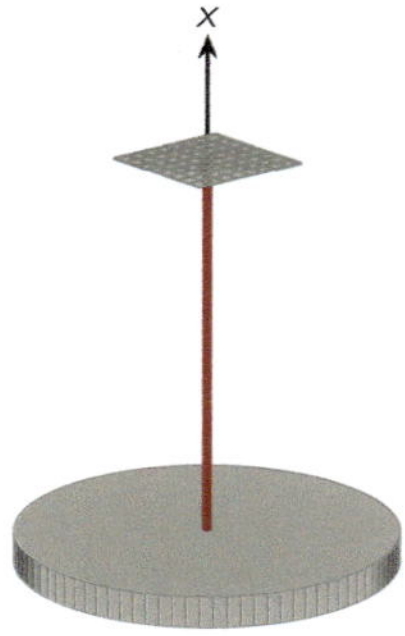

Figure 8.27 Torsion pendulum.

of the pendulum. We shall neglect the inertia of the wire in comparison with that of the disk. ⋆

Problem 8.8 (Torsion of a Doubly-Clamped Beam) *A prismatic beam admitting the median plane of Figure 8.28 is clamped at both ends and subjected to a torque* Γ *along its span, as shown. Determine: (i) the reactions at the supports. (ii) the rotation of the section of application of the torque* Γ*. We shall take GJ for the torsion stiffness.* ⋆

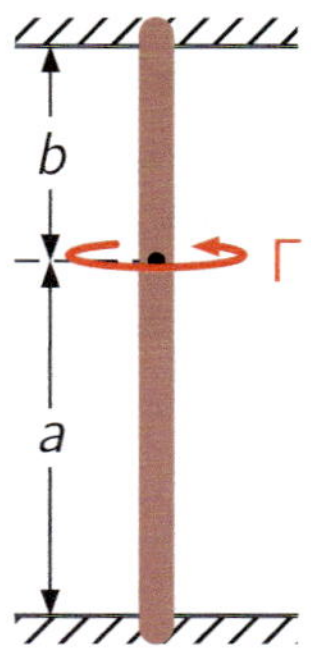

Figure 8.28 Hyperstatic bar under torsion.

Problem 8.9 (Shaft Assembly in Parallel) *A beam with a circular cross-section is made up of two elastic materials, as depicted in Figure 8.29. The materials are assumed to be perfectly bonded together. Material 1 has Young's modulus E_1 and Poisson's ratio ν_1. Correspondingly, material 2 has elastic constants E_2 and ν_2. Determine the torsional stiffness of the assembly.* ⋆

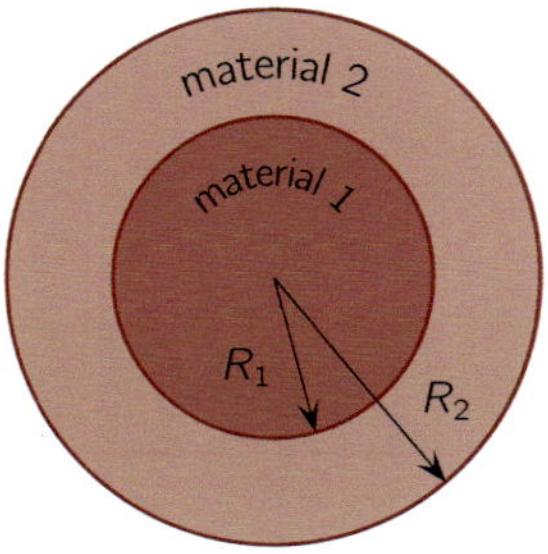

Figure 8.29 Two shafts in parallel.

Problem 8.10 (Shaft Assembly in Series) *Two beams of torsion stiffnesses G_1J_1, G_2J_2 and respective lengths L_1, L_2 are assembled end to end, as shown in Figure 8.30. Determine the coefficient of torsional stiffness K_T of the assembly, defined as the ratio of the applied torque to the end twist.* ⋆

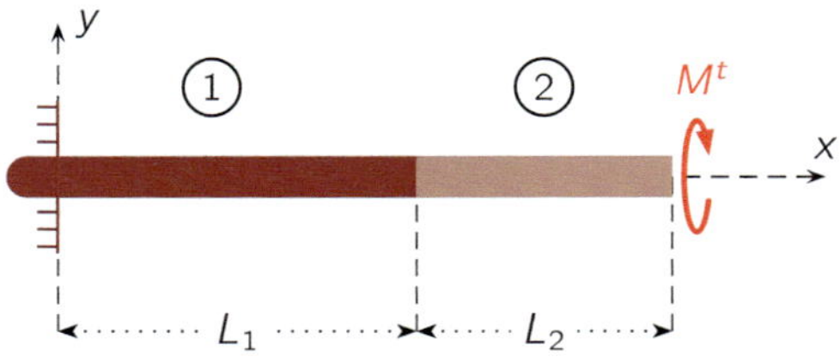

Figure 8.30 Two shafts in series.

Problem 8.11 (Torsional Stiffness of Various Thin Profiles) *(i) Determine the torsional stiffnesses of the beams whose cross-sections are shown in Figure 8.31. All sections have uniform thickness t. (ii) Rank-order the stiffnesses assuming that the beams are made of the same material and have the same wall thickness. (iii) Calculate the values of J for $R = 30$ mm and $t = 1$ mm.*

⋆

Problem 8.12 (Torsion of a Two-Box Section) *Consider the beam of Figure 8.32. Skin and web have the same thickness t, small in comparison with the dimensions of the section. (i) Calculate the shear stress in every point of the section when a torque M^t is applied. (ii) Calculate the torsional stiffness GJ. (iii) What is the inner web useful for?* ⋆

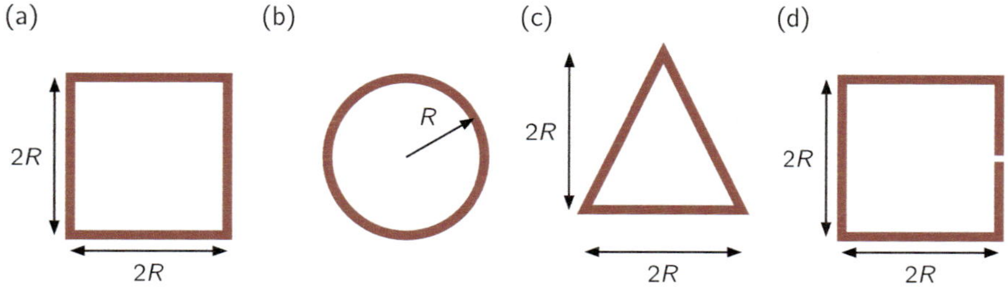

Figure 8.31 Various thin-walled sections. The profile in (d) is slit open at right.

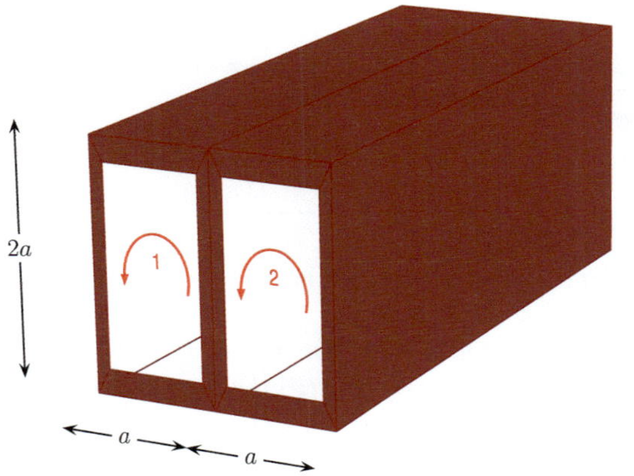

Figure 8.32 Twin-cell box.

Problem 8.13 (Torsional Stiffness of a Fuselage) *A portion of fuselage is modeled as a prismatic beam with thin walls, (Figure 8.33). The material is isotropic linear elastic with shear modulus G. The fuselage is made of a cylindrical skin of radius R and uniform thickness t and a floor of thickness t welded to the skin as shown in*

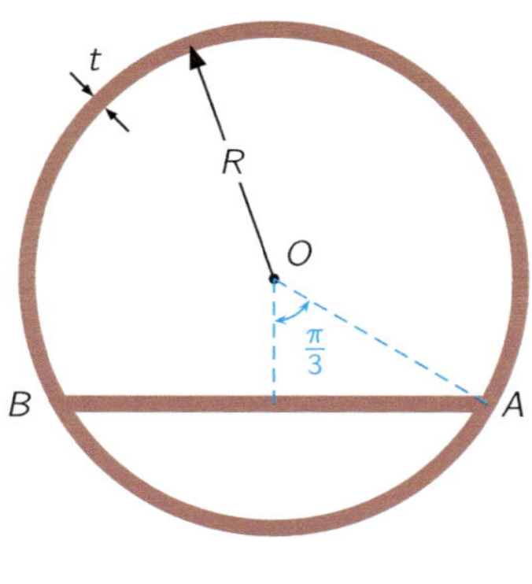

Figure 8.33 Idealization of a fuselage section.

the figure. During a maneuver, the cross-section is subjected to a torque M^t. (i) Determine the shear flow in the cross-section. (ii) Calculate the torsion constant J. (iii) Examine the influence of adding the floor to the overall stiffness. ⋆

Problem 8.14 (Analysis of an Elevator under Torsion) *The elevator depicted in Figure 8.34 in a planar section, is subjected to a torque M^t. The shear flows in the walls DAC, $\widehat{CBD}$ and DC are denoted ϕ_1, ϕ_2, ϕ_3, respectively. All dimensions are*

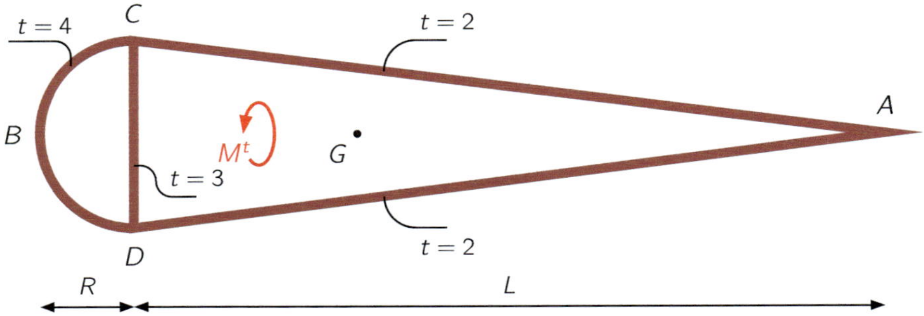

Figure 8.34 Elevator idealized as a two-cell box.

given in mm. Also, L = 600 and R = 75. The material has shear modulus G = 26 GPa. Calculate: (i) the shear flows ϕ_1, ϕ_2, ϕ_3; (ii) the torsional stiffness (per unit length) GJ. (iii) the relative rotation of two sections distant by 1 m for an applied couple of 10^4N · m. ⋆

Problem 8.15 (Torsion by Antisymmetric Bending of Two Beams) *Two identical parallel beams are clamped at one end and joined together by a bar, as shown in Figure 8.35. Each beam has Young's modulus E, cross-section S, length L, and second moment of the area $I \equiv I_Y$. The principal axes of both beams are parallel to the x, y, and z axes depicted in Figure 8.35. The distance between the neutral fibers of the two beams is denoted by h. Also, the bar is modeled as rigid and the connections to the beams are simple pin joints. Each beam is loaded in bending, as shown in the figure. Calculate the torsion stiffness of the assembly.* ⋆

Problem 8.16 (Vlasov's Nonuniform Torsion) *The Saint-Venant theory of torsion is formulated for traction boundary conditions. As a result, the solution is invariant by any arbitrary translation. In particular, all sections are free to warp, including end sections. Consider an I beam clamped at one end and subjected to a net torque M^t on the other. The dimensions for the skin and web are shown in Figure 8.36. The aim is to estimate the increase in torsional stiffness of the beam due to constrained warping. To this end, use the result from Problem 8.15.* ⋆

Problem 8.17 (Torsion of a Solid Elliptical Section) *Consider a prismatic beam having an elliptical cross-section with semi-axes a and b ($a > b$); see (Figure 8.37).*

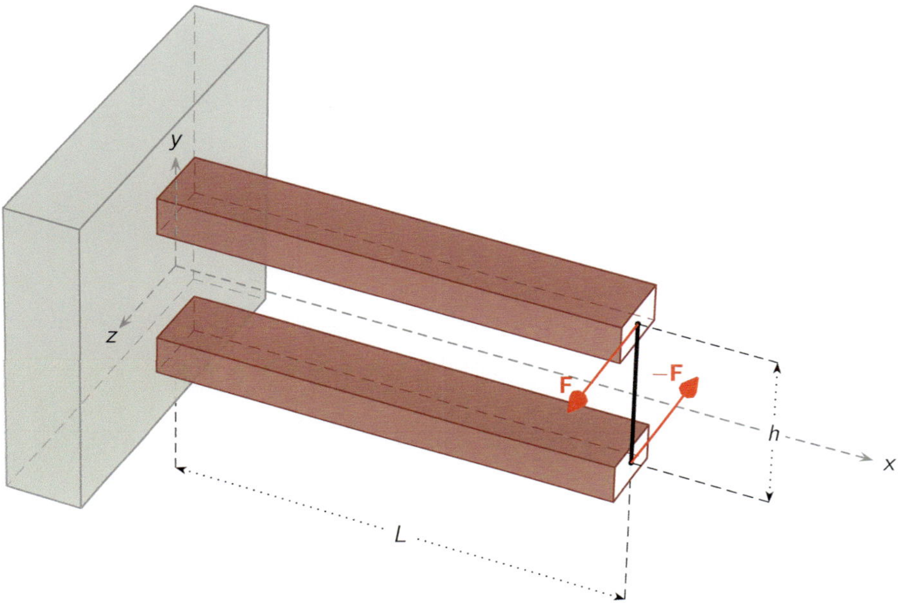

Figure 8.35 Beam pair under antisymmetric bending.

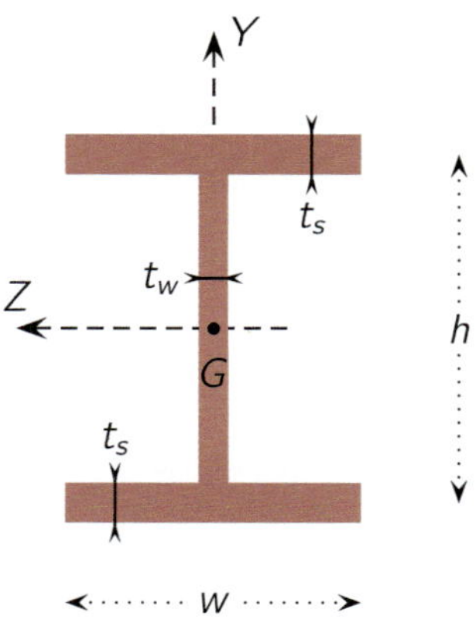

Figure 8.36 I-beam section.

Within the framework of Saint-Venant's theory, determining the solution amounts to finding the stress function $\varphi(y,z)$*, which obeys a Dirichlet boundary-value problem, Equation* (8.23)*, with* M^t *the torque, and J obtained from Equation* (8.17)*; see Section 8.2.5.*

1. *Write the equation defining contour* Γ.

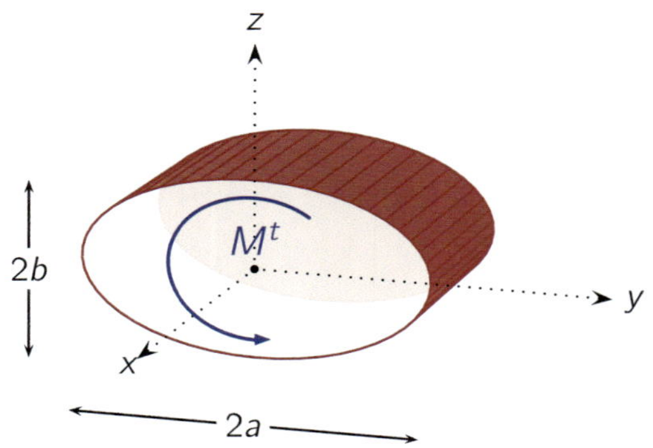

Figure 8.37 Solid elliptical section.

2. *Show that the following stress function verifies boundary condition* $(8.23)_2$:

$$\varphi(y,z) = \left(1 - \frac{y^2}{a^2} - \frac{z^2}{b^2}\right) f(y,z)$$

where $f(y,z)$ *is a priori unknown.*
3. *Use the simplest trial function* $f(y,z)$ *to find a relationship between* f *and* J.
4. *Determine* J. *Compare its expression with that giving the polar moment of inertia* I_x.
5. *Determine the warping function* $\chi(y,z)$.
6. *Complete the resolution to find stresses, strains, rotations and translations.*

⋆

Problem 8.18 (Compatibility Equations in Torsion) *(i) Which of the six compatibility equations given in Chapter 1, (1.16a)–(1.16f), lead to the two equations used in Section 8.2.4? (ii) Use the mnemonic way of Equation* (1.18) *to obtain the two useful equations. Hint: try* $i = x,\ j = z,\ k = z,\ l = y$ *for one of them.* ⋆

Problem 8.19 (Polynomial Stress Function for a Rectangular Section) *Using a polynomial trial stress function that vanishes on each of the two vertical and each of the two horizontal lines containing the sides of the rectangle, obtain the prefactor constant that satisfies the equations of the problem.* ⋆

Problem 8.20 (Torsion of a Rectangular Section) *Using the stress function given in Example 8.3 determine the expressions of the shear stresses* τ_{xy} *and* τ_{xz}. *Give the expression of maximum shear and simplify it in the case of* $b \gg a$. ⋆

Problem 8.21 (Warping of a Rectangular Section) *Determine the warping function for a rectangular section with dimensions* $2a$ *and* $2b$. *One may start from the corresponding stress function in Example 8.3 and integrate Equations* (8.21) *and* (8.22) *to obtain the result. Plot the warped section. Plot the deformed configuration of a beam after a twist of your choosing showing the warping.* ⋆

Problem 8.22 (On Warping of Ellipses versus Rectangles) *The warping of an elliptical section scales as yz whereas that of a rectangular section has an additional open series term. Which section warps more for a given applied torque and a set b/a ratio?* ⋆

Problem 8.23 (Moment Carried by a Thin-Walled Cell) *Consider a multi-cell tube as in Figure 8.21. Show that the moment carried by the i*[th] *cell is given by*

$$\iint_{S^i} \overrightarrow{CP} \times \vec{\tau}\, \mathrm{d}S = \oint_{\mathscr{C}^i} \phi^i r^2\, \mathrm{d}\theta - \sum_{j \neq i}^{n} \int_{C^{ij}} \phi^j r^2\, \mathrm{d}\theta$$

where S^i denotes the cross-section of the cell, $\mathscr{C}^i$ its median line, and C^{ij} the oriented median line of the branch common to cells i and j. Conclude. ⋆

Problem 8.24 (Quasi-Proof of Theorem 8.7) *Consider an arbitrary three-cell thin-walled tube. In that simple context, show that Equation* (8.56) *holds by explicitly expanding the expression of the moment.* ⋆

9 Shear Force

Each of the previous three chapters started with analyzing a state where one component of the screwer acted alone on all cross-sections of a prismatic beam. This led us to study for instance pure extension, Section 6.1, pure bending, Section 7.1, or pure torsion, Section 8.1. The same is not possible for studying the shear force. A state of pure shearing acts on a given beam cross-section S if the axial force and moments of bending and torsion are all nil on that section. Such a state can only occur in some particular sections. Consider, for instance, a straight beam admitting XOY and XOZ as median planes. The last two moment-equilibrium equations write (see Section 5.5.4)

$$m_Y + \frac{\mathrm{d}M_Y}{\mathrm{d}x} - V_Z = 0, \qquad m_Z + \frac{\mathrm{d}M_Z}{\mathrm{d}x} + V_Y = 0 \tag{9.1}$$

and show that the shearing forces V_Y and V_Z are related to the bending moments M_Z and M_Y, respectively.

Figure 9.1 provides two examples of cross-sections S_0 loaded in pure shear.

This chapter contains the analysis of stresses, strains, and displacements produced by a shearing force $\mathbf{V}$. It suffices to study the effects of the sole principal component V_Y, the general case being treated by decomposing $\mathbf{V}$ into $\mathbf{V}_Y$ and $\mathbf{V}_Z$, then superposing the results.

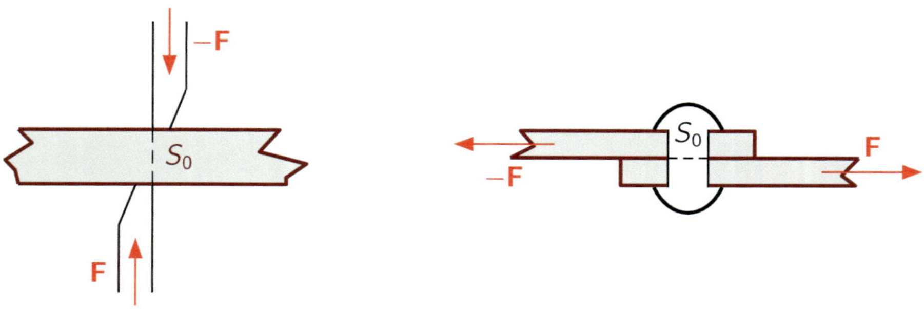

Figure 9.1 (a) Beam taken between shear jaws. (b) Assembly rivet on two sheets loaded in-plane.

9.1 SAINT-VENANT'S THEORY

Consider a full prismatic (or cylindrical) beam of length L and arbitrary cross-section, limited by end sections S_0 and S_1, and reported to the principal frame $\{G_0, xyz\}$ of S_0, with S denoting the current cross-section located at x, Figure 9.2.

The beam is loaded as follows:

DOI: 10.1201/9781003537229-11

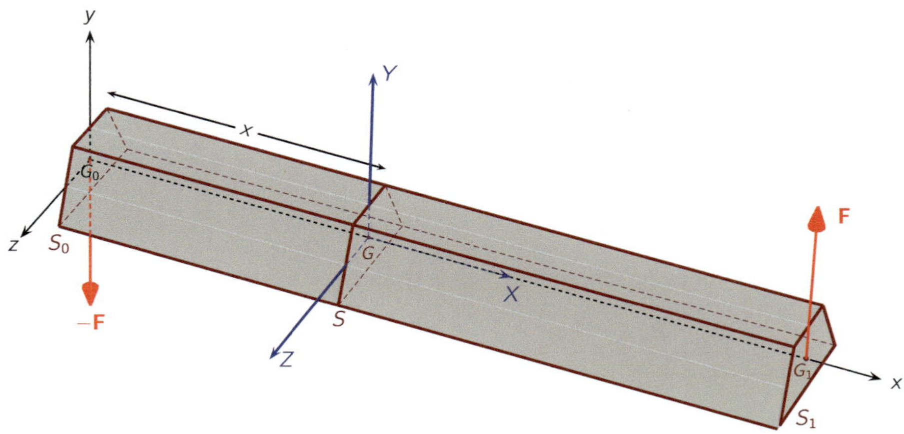

Figure 9.2 Prismatic beam under opposite end forces and a distribution of normal stresses on S_0 (not shown).

- On S_1, a distribution of shear tractions $\vec{\tau}$ (a field of tangential stresses that will be specified later) is applied which admits $\mathbf{F} = F\mathbf{J}$, parallel to G_0y as a resultant force, and a zero resultant moment (of torsion) M^t about C_1 (center of torsion of S_1).
- On S_0, the opposite distribution of shear tractions $-\vec{\tau}$ *and* a distribution of normal stresses:
$$\sigma_x(0,y) = -\frac{FL}{I_Z}y$$
that ensures the global equilibrium of moments, with I_Z denoting the principal second moment of area of S about GZ and y the ordinate of current point P.

As a result, the screwer on S has two non-zero components:

- the shearing force: $V_Y = F$
- the bending moment: $M_Z = V_Y(L-x) = F(L-x)$

The method of Saint-Venant proceeds so as to determine the strains, stresses, rotations, and translations, by assuming that the stress matrix anywhere in the beam takes the form:

$$[\boldsymbol{\sigma}] = \begin{bmatrix} \sigma_x & \tau_{xy} & \tau_{xz} \\ \tau_{xy} & 0 & 0 \\ \tau_{xz} & 0 & 0 \end{bmatrix}$$

where the normal stress σ_x results from the bending component of the loading.

The sought fields constitute *the* solution of the problem if and only if they satisfy pointwise the following conditions:

1. the local equilibrium, that is $\mathbf{div}\,\boldsymbol{\sigma} + \mathbf{b} = \mathbf{0}$;
2. Hooke's law;

3. the strain-displacement and rotation-displacement relations; and
4. the boundary conditions.

These conditions are listed in the order they will be considered below following a stress-based method, in which case the strain-displacement relations are equivalent to the compatibility equations; see Section 1.2 in Part I.

9.1.1 LOCAL EQUILIBRIUM

In the absence of body forces, equilibrium is specialized using

$$\frac{\partial \sigma_x}{\partial x} + \frac{\partial \tau_{xy}}{\partial y} + \frac{\partial \tau_{xz}}{\partial z} = 0; \qquad \frac{\partial \tau_{xy}}{\partial x} = 0; \qquad \frac{\partial \tau_{xz}}{\partial x} = 0 \tag{9.2}$$

Since the normal stress σ_x is solely due to bending and the bending moment is fully determined, one must have

$$\sigma_x(x,y) = -\frac{M_Z}{I_Z} y = -\frac{V_Y}{I_Z}(L-x)y \tag{9.3}$$

The last two relations in Equation (9.2) indicate that the shear stresses cannot depend on x. A sufficient condition for the first equilibrium equation is that the components τ_{xy} and τ_{xz} derive from a stress function $\varphi(y,z)$, to be determined, such that the shear traction vector $\vec{\tau}$ is given by

$$\begin{cases} \tau_{xy} = \dfrac{\partial \varphi}{\partial z} + f(y,z) \\ \tau_{xz} = -\dfrac{\partial \varphi}{\partial y} \end{cases} \tag{9.4}$$

The function $f(y,z)$ is needed to ensure axial equilibrium. It is assumed to be endowed with second-order partial derivatives that are continuous on S (i.e. of class $\mathscr{C}^2$). Likewise, the function φ is assumed to be of class $\mathscr{C}^3$ on S. The uniqueness of solutions in elasticity (see Section 4.4.1 in Part I) is guaranteed if a solution having the above form were to hold, in spite of completeness issues associated with the pair φ, f.

Inserting Equation (9.4) into Equation (9.2) gives:

$$\frac{\partial \sigma_x}{\partial x} + \frac{\partial f}{\partial y} = 0$$

which on account of Equation (9.3) leads to the following form of function f:

$$f(y,z) = \frac{V_Y}{2I_Z}\left(g(z) - y^2\right) \tag{9.5}$$

where $g(z)$ is a function to be determined from the compatibility conditions.

9.1.2 HOOKE'S LAW

It implies that the strain matrix takes the following form:

$$[\boldsymbol{\varepsilon}] = \begin{bmatrix} \varepsilon_x & \gamma_{xy} & \gamma_{xz} \\ \gamma_{xy} & \varepsilon_y & 0 \\ \gamma_{xz} & 0 & \varepsilon_z \end{bmatrix} \qquad \text{with:} \begin{cases} \varepsilon_x = \dfrac{\sigma_x}{E} = -\dfrac{V_Y}{EI_Z}(L-x)y \\ \varepsilon_y = \varepsilon_z = -\nu\varepsilon_x \\ \gamma_{xy} = \dfrac{1}{2G}\tau_{xy} \\ \gamma_{xz} = \dfrac{1}{2G}\tau_{xz} \end{cases} \tag{9.6}$$

where $G = \mu = \frac{E}{2(1+\nu)}$ is the elastic shear modulus; see Equation (3.17).

9.1.3 COMPATIBILITY EQUATIONS

Strain compatibility is invoked in lieu of the strain–displacement relations; it is expressed through the following two equations:

$$\frac{\partial}{\partial y}\left(\frac{\partial \gamma_{xy}}{\partial z} - \frac{\partial \gamma_{xz}}{\partial y}\right) = 0 \qquad \text{and} \qquad \frac{\partial^2 \varepsilon_z}{\partial x \partial y} + \frac{\partial}{\partial z}\left(\frac{\partial \gamma_{xy}}{\partial z} - \frac{\partial \gamma_{xz}}{\partial y}\right) = 0 \tag{9.7}$$

Indeed, because $\gamma_{yz} = 0$ and the other two shear strains are only functions of y and z while the normal strains ε_x, ε_y, ε_z are linear functions of x and y, the first four compatibility relations in Equation (1.16f) are identically satisfied. Then, Equation (1.16e) reduces to Equation $(9.7)_1$ and Equation (1.16f) reduces to Equation $(9.7)_2$. On using the mnemonic way underlying Equation (1.18) see Problem 9.20.

Using Hooke's law, Equation (9.6), then equilibrium in the form of Equation (9.4), the above compatibility statements can be expressed in terms of the stress function $\varphi(y,z)$ as follows:

$$\frac{\partial}{\partial y}\left(\Delta\varphi + \frac{\partial f}{\partial z}\right) = 0 \qquad \text{and} \qquad \frac{\partial}{\partial z}(\Delta\varphi) + \frac{V_Y}{I_Z}\left\{-\frac{\nu}{1+\nu} + \frac{1}{2}g''(z)\right\} = 0 \tag{9.8}$$

where Δ stands for the Laplacian; see Problem 9.20.

The first compatibility condition, Equation $(9.8)_1$, simplifies since $\frac{\partial f}{\partial z} \propto g'(z)$ is independent of y, see Equation (9.5).

The first term in the second compatibility equation is a function of both y and z whereas the second term (between braces) is only a function of z. A simple choice that satisfies all governing equations is one that would make the braced term identically vanish, i.e. choose the function $g(z)$ such that:

$$g''(z) = 2\frac{\nu}{1+\nu}$$

which leads to

$$g(z) = \frac{\nu}{1+\nu}z^2$$

having taken $g(0) = 0$ and $g'(0) = 0$ to eliminate rigid-body motions.

Two important consequences follow. First, the function $f(y,z)$ in Equation (9.5) is now fully determined and is independent of the shape of the cross-section:

$$f(y,z) = \frac{V_Y}{2I_Z}\left(\frac{\nu}{1+\nu}z^2 - y^2\right) \tag{9.9}$$

Second, the Laplacian of φ must be a constant. Furthermore, this constant is necessarily equal to 0. Indeed, it can be shown that $\Delta\varphi$ is proportional to a torsion-induced twist. As long as the shear traction distribution is as specified above, Figure 9.2, the net torque about the center of torsion is zero and there is no twist.

The function $\varphi(y,z)$ is therefore harmonic:

$$\Delta\varphi = 0 \tag{9.10}$$

Along with Equation (9.9), Equation (9.10) is the main outcome of strain compatibility.

Remark 9.1 *While the final solution to the Saint-Venant problem posed above is unique, there are many different combinations of functions $\varphi(y,z)$ and $g(z)$ that would deliver that solution. The advanced student may refer to Section 9.9 for further in-depth discussion of this point.* $\ast$

9.1.4 STRAIN–DISPLACEMENT AND ROTATION–DISPLACEMENT RELATIONS

Integration of the rotation–displacement relations gives

$$\boldsymbol{\omega} \quad \begin{cases} \omega_x = \nu\dfrac{V_Y}{EI_Z}(L-x)z & +\alpha \\[2ex] \omega_y = -\dfrac{1+\nu}{E}\dfrac{\partial\varphi}{\partial y} + \nu\dfrac{V_Y}{EI_Z}yz & +\beta \\[2ex] \omega_z = -\dfrac{1+\nu}{E}\dfrac{\partial\varphi}{\partial z} + \dfrac{V_Y}{2EI_Z}\left[y^2 - (L-x)^2\right] & +\gamma \end{cases} \tag{9.11}$$

where α, β, and γ denote the components of a solid rotation.

Integration of the strain–displacement relations gives

$$\mathbf{u} \quad \begin{cases} u = 2\dfrac{1+\nu}{E}\chi(y,z) + \dfrac{V_Y}{2EI_Z}\left[(L-x)^2 y + \nu y z^2 - \dfrac{2+\nu}{3}y^3\right] & +\beta z - \gamma y + a \\[2ex] v = \dfrac{\nu V_Y}{2EI_Z}(L-x)(y^2 - z^2) + \dfrac{V_Y}{6EI_Z}(L-x)^3 & +\gamma x - \alpha z + b \\[2ex] w = \dfrac{\nu V_Y}{EI_Z}(L-x)yz & +\alpha y - \beta x + c \end{cases} \tag{9.12}$$

where a, b, and c are the components of a solid translation.

The function $\chi(y,z)$ is defined up to an additive constant by the relations:

$$\frac{\partial \chi}{\partial y} = \frac{\partial \varphi}{\partial z} \tag{9.13}$$

$$\frac{\partial \chi}{\partial z} = -\frac{\partial \varphi}{\partial y} \tag{9.14}$$

One implication of the above expression for the axial displacement $u(x,y,z)$ is that cross-sections warp. Put otherwise, initially plane sections do not remain plane. There is no exception to this warping.

Remark 9.2 *Unlike in the torsion of prisms, the function $\chi(y,z)$ does not embody all information about cross-sectional warping. Indeed, the bracketed term in the expression of u contains some nonlinear terms that contribute to warping.* ∗

Remark 9.3 *Function $\chi(y,z)$ is also harmonic. Equations* (9.13) *and* (9.14) *represent Cauchy's conditions for conjugate functions.* ∗

9.1.5 BOUNDARY CONDITIONS

9.1.5.1 On Lateral Surface S_2

Surface S_2 is free of any tractions. Thus,

$$\tau_{xy}\,\mathrm{d}z - \tau_{xz}\,\mathrm{d}y = 0 \qquad \text{on } \Gamma, \text{boundary of } S, \tag{9.15}$$

see Figure 9.3. In terms of the stress function $\varphi(y,z)$, Equation (9.15) may be rewritten using Equation (9.4) as

$$\frac{\partial \varphi}{\partial z}\,\mathrm{d}z + f(y,z)\,\mathrm{d}z + \frac{\partial \varphi}{\partial y}\,\mathrm{d}y = 0$$

Hence, on the contour of S, the unknown function $\varphi(y,z)$ must satisfy

$$\mathrm{d}\varphi = \frac{V_Y}{2I_Z}\left(y^2 - \frac{\nu}{1+\nu}z^2\right)\mathrm{d}z \qquad \text{on } \Gamma, \text{boundary of } S \tag{9.16}$$

Integration of Equation (9.16) on Γ provides the values of φ therein, up to some additive constant.

9.1.5.2 On End Sections S_0 and S_1

The boundary conditions on these extreme sections (see Figure 9.2) are satisfied. In fact, invoking the traction-screwer relations, Equation (5.35), one verifies the follow-

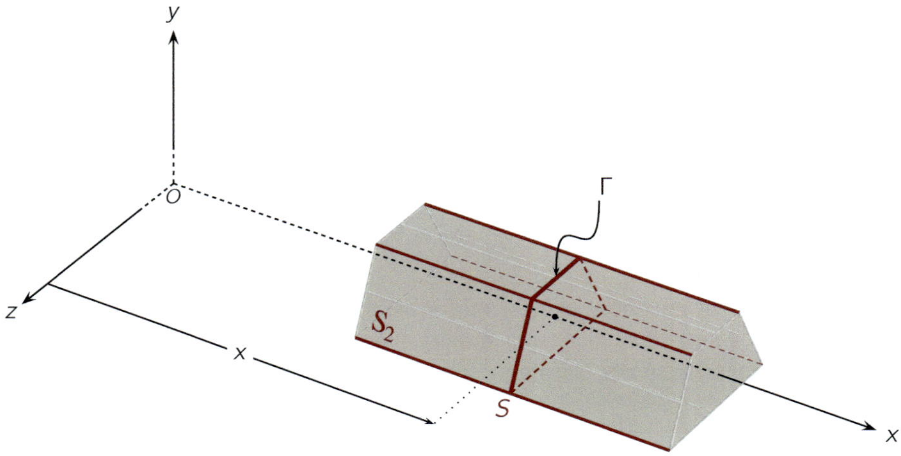

Figure 9.3 Γ denotes the closed curve, intersection of lateral surface S_2 and current cross-section S.

ing on every cross-section:

$$\iint_S \tau_{xy}\,\mathrm{d}S = V_Y = F \tag{9.17}$$

$$\iint_S \tau_{xz}\,\mathrm{d}S = V_Z = 0 \tag{9.18}$$

$$\iint_S \overrightarrow{CP} \times \vec{\tau}\,\mathrm{d}S = \mathbf{M}^t = \mathbf{0} \tag{9.19}$$

The last two equations signify no transverse shear along Z and no net torque about the center of twist C.

Proof *Let us show the first of these, Equation* (9.17)*:*

$$\begin{aligned}
\iint_S \tau_{xy}\,\mathrm{d}S &= \iint_S \left[\frac{V_Y}{2I_Z}\left(\frac{\nu}{1+\nu}z^2 - y^2\right) + \frac{\partial\varphi}{\partial z}\right]\mathrm{d}y\,\mathrm{d}z \qquad \textit{using (9.4) and (9.9)}\\
&= \frac{V_Y}{2I_Z}\left(\frac{\nu}{1+\nu}I_Y - I_Z\right) + \oint_\Gamma -\varphi\,\mathrm{d}y \qquad \textit{(Riemann's formula)}\\
&= \frac{V_Y}{2I_Z}\left(\frac{\nu}{1+\nu}I_Y - I_Z\right) + \oint_\Gamma y\,\mathrm{d}\varphi \qquad \textit{(integration by parts)}\\
&= \frac{V_Y}{2I_Z}\left(\frac{\nu}{1+\nu}I_Y - I_Z\right) + \oint_\Gamma y\frac{V_Y}{2I_Z}\left(y^2 - \frac{\nu}{1+\nu}z^2\right)\mathrm{d}z \quad \textit{using (9.16)}\\
&= \frac{V_Y}{2I_Z}\left(\frac{\nu}{1+\nu}I_Y - I_Z\right) + \frac{V_Y}{2I_Z}\iint_S \left(3y^2 - \frac{\nu}{1+\nu}z^2\right)\mathrm{d}y\,\mathrm{d}z\\
&\qquad \textit{(Riemann's formula)}\\
&= \frac{V_Y}{2I_Z}\left(\frac{\nu}{1+\nu}I_Y - I_Z + 3I_Z - \frac{\nu}{1+\nu}I_Y\right) = V_Y
\end{aligned}$$

Equation (9.18) *is verified in a similar way. The third boundary condition, Equation* (9.19) *is also verified because of the absence of twist. Further details are discussed in Section 9.9.* ■

Finally, on section S_0, the axial stresses σ_x do result in a bending moment equal to FL, according to Equation (9.3).

9.1.6 PRACTICAL RESOLUTION OF SAINT-VENANT'S PROBLEM

Given the shearing force (V_Y), the geometry (shape of S) and the absence of torque ($M^t = 0$) the solution proceeds in three steps:

1. Determine the stress function $\varphi(y,z)$ by solving the Dirichlet-type problem:

$$\begin{aligned} \Delta\varphi &= 0 && \text{in } S \\ \mathrm{d}\varphi &= \frac{V_Y}{2I_Z}\left(y^2 - \frac{\nu}{1+\nu}z^2\right)\mathrm{d}z && \text{on } \Gamma, \text{boundary of } S \end{aligned} \tag{9.20}$$

2. Determine the conjugate harmonic function $\chi(y,z)$ using Equations (9.13) and (9.14).
3. Infer:
   - the stresses using Equation (9.4), along with Equation (9.3)
   - the strains using Hooke's law; cf. Equation (9.6)
   - the rotations using Equation (9.11)
   - the translations using Equation (9.12).

9.1.7 CONCLUSIONS

Three key results follow from the Saint-Venant bending problem.

First Conclusion: Warping – The cross-sections warp because, for a given cross-section (fixed x), the axial displacement $u(y,z)$ is a nonlinear function of y and z. Unlike in torsion, the warping is not the same for all cross-sections; the coupling between shear and bending makes u dependent upon x. This appears by inspecting Equation $(9.12)_1$.

Second Conclusion: Rotation of Cross-Sections – The cross-sections rotate with respect to each other about the Z-axis. Equation $(9.11)_3$ implies:

$$\frac{\partial \omega_z}{\partial x} = \frac{V_Y(L-x)}{EI_Z} = \frac{M_Z}{EI_Z}$$

Cross-section S' rotates with respect to S by the angle

$$\mathrm{d}\omega_z = \frac{M_Z}{EI_Z}\,\mathrm{d}x$$

The rotation is about axis GZ and is exclusively due to the bending moment M_Z.

Third Conclusion: Curvature of Deformed Axial Fibers – Equation $(9.12)_2$ gives:

$$\frac{d^2 v}{dx^2} = \frac{V_Y(L-x)}{EI_Z} = \frac{M_Z}{EI_Z} = \frac{\partial \omega_z}{\partial x}$$

Therefore, every axial fiber takes, at abscissa x, the curvature:

$$\frac{1}{R} = \frac{d^2 v}{dx^2} = \frac{M_Z}{EI_Z}$$

Remark 9.4 *Recall from the study of bending that the above equations hold irrespective of whether the cross-sections remain plane or not. In fact, in the vast majority of cases, plane sections do not remain plane!* $*$

9.1.8 APPLICATION EXAMPLES

Example 9.1: Shearing of Circular Solid Section

Consider the beam in Figure 9.4 under a shear load V_Y. The load shown is the resultant shear on the cross-section, which in general must be concurrent with some bending moment (not shown).

Following the steps of Section 9.1.6, the stress function is (see Problem 9.21)

$$\varphi(Y,Z) = \frac{V_Y}{8(1+\nu)I_Z}\left[(1+2\nu)\left(Y^2Z - \frac{1}{3}Z^3\right) + (3+2\nu)R^2Z\right] \tag{9.21}$$

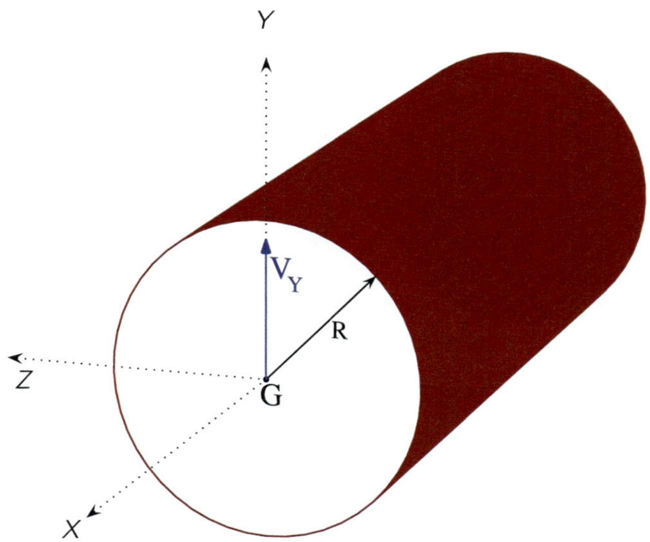

Figure 9.4 Beam with a circular cross-section under shear.

The shear traction vector is thus

$$\vec{\tau}\quad\begin{cases}\tau_{XY}=\dfrac{1}{8}\dfrac{3+2\nu}{1+\nu}\dfrac{V_Y}{I_Z}\left(R^2-Y^2-\dfrac{1-2\nu}{3+2\nu}Z^2\right)\\[2mm] \tau_{XZ}=-\dfrac{1}{4}\dfrac{1+2\nu}{1+\nu}\dfrac{V_Y}{I_Z}\quad YZ\end{cases}\tag{9.22}$$

where $I_Z=\frac{\pi}{4}R^4$.

The shear field is depicted in Figure 9.5 laying emphasis on key features. On the external surface, the shear is tangential to the circular boundary. On either the Y or Z axes, the shear is vertical (along Y). Material lines along which the shear traction vector $\vec{\tau}$ is tangential are known as *shear lines*. They are depicted as white in the figure. The external boundary as well as the vertical diameter also constitute shear lines.

The shear stress attains its maximum at the center G, Figure 9.5. Its magnitude there is given by the following expression:

$$\tau^{\max}=\tau^G=\frac{1}{8}\frac{3+2\nu}{1+\nu}\frac{V_Y}{I_Z}R^2\tag{9.23}$$

and thus depends on Poisson's ratio.

On the Z-axis the shear variation is quadratic:

$$\tau(Y=0,Z)=\frac{1}{8}\frac{3+2\nu}{1+\nu}\frac{V_Y}{I_Z}R^2\left(1-\frac{1-2\nu}{3+2\nu}\frac{Z^2}{R^2}\right)\tag{9.24}$$

The shear decreases from its maximum at G to reach the following value at the

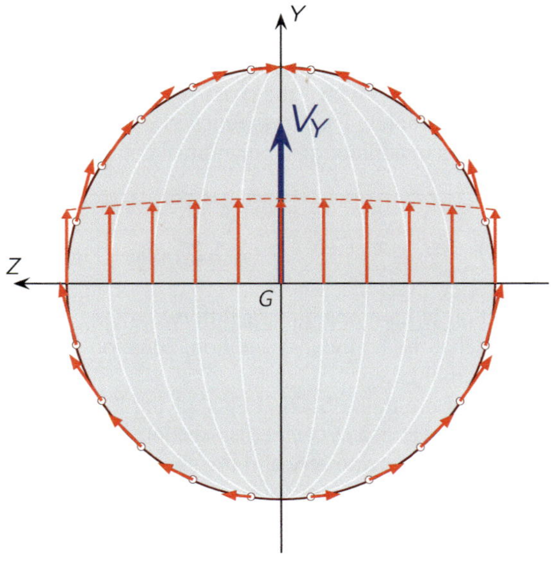

Figure 9.5 Distributions of shear tractions (red) and shear lines (white) in a circular section.

ends of the horizontal diameter:

$$\tau(0, \pm R) = \frac{1}{4}\frac{1+2\nu}{1+\nu}\frac{V_Y}{I_Z}R^2 \tag{9.25}$$

with the rate of decrease being modulated by Poisson's ratio.

On the Y-axis the shear also varies in a quadratic fashion:

$$\tau(Y, 0) = \frac{1}{8}\frac{3+2\nu}{1+\nu}\frac{V_Y}{I_Z}R^2\left(1 - \frac{Y^2}{R^2}\right) \tag{9.26}$$

The shear decreases from its maximum at G to zero at the ends of the vertical diameter. While $\tau^{\max}$ depends on Poisson's ratio the rate of decrease (along the vertical diameter) does not.

The conjugate function $\chi(Y,Z)$ is given by (posing $\chi(0,0) = 0$)

$$\chi(Y,Z) = \frac{1}{8}\frac{3+2\nu}{1+\nu}\frac{V_Y}{I_Z}\left[R^2 + \frac{1+2\nu}{3+2\nu}\left(\frac{1}{3}Y^2 - Z^2\right)\right]Y \tag{9.27}$$

With φ and χ so specified, the rotations and displacements are readily calculated using Equations (9.11) and (9.12).

The shearing of circular cross-sections is very important as it corresponds to rivets and bolts, among others. ♡

Remark 9.5 *The "engineering formula" for the shear stress provides a good approximation of the maximum shear stress, but assumes that the shear lines are vertical, i.e.. along the net shear force. This violates the traction-free condition on the curved surfaces.* *

Example 9.2: Rectangular Solid Section

Let $2a$ and $2b$ denote the height and width of the rectangular section, Figure 9.6; one then has: $I_Z = \frac{4}{3}a^3b = \frac{1}{3}a^2S$ with S the cross-sectional area.

The solution is quite complex. The stress function $\varphi(Y,Z)$ is presented in terms of a double Fourier series in addition to a polynomial component:

$$\varphi = \frac{V_Y}{2I_Z}\left(a^2Z - \frac{\nu}{1+\nu}\frac{Z^3}{3}\right) + \frac{8}{\pi^4}\frac{\nu}{1+\nu}\frac{b^3V_Y}{I_Z}\sum_{m=0}^{\infty}\sum_{n=1}^{\infty}\frac{(-1)^{m+n}\cos\frac{(2m+1)\pi Y}{2a}\sin\frac{n\pi Z}{b}}{(2m+1)n\left[(2m+1)^2\frac{b^2}{4a^2} + n^2\right]}$$

That the above function is harmonic is straightforward (see Problem 9.23). Two methods for showing that the boundary condition, Equation (9.16), is satisfied are implemented in Problem 9.23.

The components of $\vec{\tau}$ are then obtained by applying the formulae in (9.4) (see Problem 9.23). As in the case of a circular cross-section, the horizontal component τ_{XZ} vanishes on both axes of symmetry and the shear there is vertical. The shear vanishes on the top and bottom points lying on the Y-axis $\tau^T = \tau^B = 0$. On the external boundary, the shear is parallel to the contour, Figure 9.7.

In general, the relevant shear distribution is at mid-height ($Y = 0$), as illustrated in the figure for $a/b = 1.46$ and $a/b = 0.5$ using $\nu = 0.25$. The plate limit ($a/b \ll 1$)

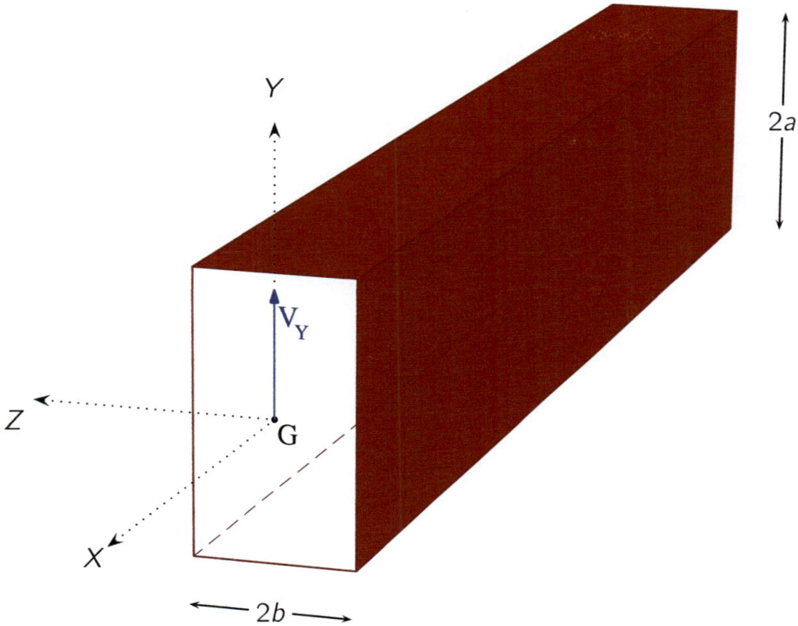

Figure 9.6 Beam with a rectangular cross-section under shear.

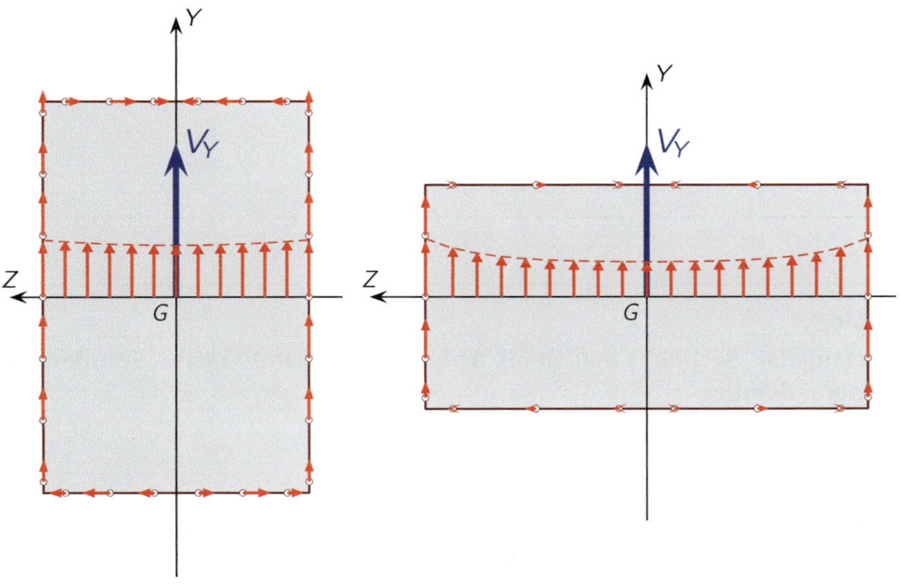

Figure 9.7 Some aspects of shear flow for $a/b > 1$ (a) and $a/b < 1$ (b).

will be discussed below. Particular values of the shear stress magnitude are at the center G:

$$\tau^G = \frac{3}{2}\frac{V_Y}{S}\left[1 - \frac{\nu}{1+\nu}\frac{b^2}{a^2}\left(\frac{1}{3} + \frac{4}{\pi^2}\sum_{n=1}^{\infty}\frac{(-1)^n}{n^2\cosh\frac{n\pi a}{b}}\right)\right]$$

and at the midpoints of the vertical sides (Figure 9.6):

$$\tau^R = \tau^L = \frac{3}{2}\frac{V_Y}{S}\left[1 + \frac{\nu}{1+\nu}\frac{b^2}{a^2}\left(\frac{2}{3} - \frac{4}{\pi^2}\sum_{n=1}^{\infty}\frac{1}{n^2\cosh\frac{n\pi a}{b}}\right)\right]$$

Letting

$$s_1 = \sum_{n=1}^{\infty}\frac{(-1)^{n+1}}{n^2\cosh\frac{n\pi a}{b}} \quad \text{and} \quad s_2 = \sum_{n=1}^{\infty}\frac{1}{n^2\cosh\frac{n\pi a}{b}}$$

one may write

$$\begin{cases} \tau^G = \frac{3}{2}\frac{V_Y}{S}\left[1 - \frac{\nu}{1+\nu}\frac{b^2}{a^2}\left(\frac{1}{3} - \frac{4}{\pi^2}s_1\right)\right] \\ \tau^R = \frac{3}{2}\frac{V_Y}{S}\left[1 + \frac{\nu}{1+\nu}\frac{b^2}{a^2}\left(\frac{2}{3} - \frac{4}{\pi^2}s_2\right)\right] \end{cases}$$

Table 9.1 provides some values of s_1 and s_2 for various ratios a/b. The series converges fast.

When a/b is greater than 2.5, s_1 and s_2 are negligible and practically equal to the first term in the series, that is $\left(\cosh\frac{\pi a}{b}\right)^{-1}$. To an excellent approximation, the shear values at the center and at the midpoints of the vertical sides are then

$$\tau^G = \frac{3}{2}\frac{V_Y}{S}\left(1 - \frac{1}{3}\frac{\nu}{1+\nu}\frac{b^2}{a^2}\right); \qquad \tau^R = \frac{3}{2}\frac{V_Y}{S}\left(1 + \frac{2}{3}\frac{\nu}{1+\nu}\frac{b^2}{a^2}\right)$$

The maximum shear is attained at the midpoints of the vertical sides, but the variation across the beam's width is very small. The larger the a/b ratio, the smaller the variation.

For a slender cross-section ($a/b \gg 1$), one gets

$$\tau^G = \tau^R = \tau^L = \frac{3}{2}\frac{V_Y}{S}$$

Table 9.1

Rectangular Section Coefficients Entering the Saint-Venant Solution of General Bending

a/b	0.1	0.5	1	1.5	2	2.5
s_1	0.7981	0.3787	8.535×10^{-2}	1.793×10^{-2}	3.733×10^{-3}	7.764×10^{-4}
s_2	1.3018	0.4223	8.721×10^{-2}	1.801×10^{-2}	3.737×10^{-3}	7.764×10^{-4}

For wide beams having flat cross-sections ($a/b < 1$), larger variations arise between points G and R, as illustrated in Figure 9.7. For $a/b = 0.5$ and $\nu = 0.25$ one finds $\tau^G = 0.55\tau^R$. As the ratio a/b becomes even smaller, the gradient in shear along the horizontal axis intensifies, and the above approximate formulae are not applicable.

Interestingly, when a/b exceeds about 15 (for usual values of ν) the location of the maximum shear stress shifts from the midpoints on the vertical sides (points R and L) to points on the top and bottom faces near the corners. Not only does the location of shear maxima change but also their orientation. This behavior is depicted in Figure 9.8.

This type of stress concentrationis peculiar to the shearing of thin-walled rectangular sections. ♡

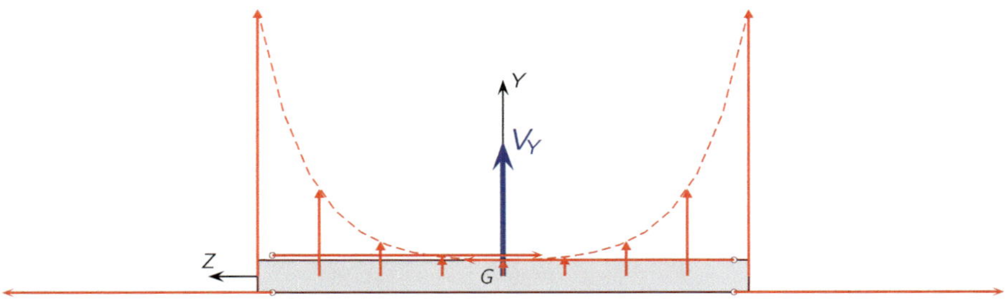

Figure 9.8 Singularity in shear in the plate limit.

9.2 ENERGY AND DEFLECTION DUE TO SHEAR FORCE

The potential elastic energy stored per unit volume in the neighborhood of material point P is

$$\mathcal{U} = \frac{1}{2}\mathrm{tr}\,(\boldsymbol{\sigma}\,\boldsymbol{\varepsilon}) = \frac{1}{2}\sigma_x\varepsilon_x + \tau_{xy}\gamma_{xy} + \tau_{xz}\gamma_{xz} \tag{9.28}$$

or, by virtue of Hooke's law

$$\mathcal{U} = \frac{1}{2}\frac{\sigma_x^2}{E} + \frac{1}{2}\frac{\tau^2}{G} \tag{9.29}$$

with $\tau^2 = \tau_{xy}^2 + \tau_{xz}^2$, as in the previous examples. This energy density is therefore the sum of two contributions:

- $\frac{1}{2}\frac{\sigma_x^2}{E}$, induced by the bending moment M_Z;
- $\frac{1}{2}\frac{\tau^2}{G}$, induced by the transverse shear V_Y.

Consider now a slice of the beam between cross-sections S and S', located at x and $x + \mathrm{d}x$, respectively, as shown in Figure 9.9. The energy stored in a fiber element having length $\mathrm{d}x$ and cross-sectional area $\mathrm{d}S$ is thus:

$$\mathcal{U}\,\mathrm{d}x\mathrm{d}S = \left(\frac{1}{2}\frac{\sigma_x^2}{E}\mathrm{d}S + \frac{1}{2}\frac{\tau^2}{G}\mathrm{d}S\right)\mathrm{d}x$$

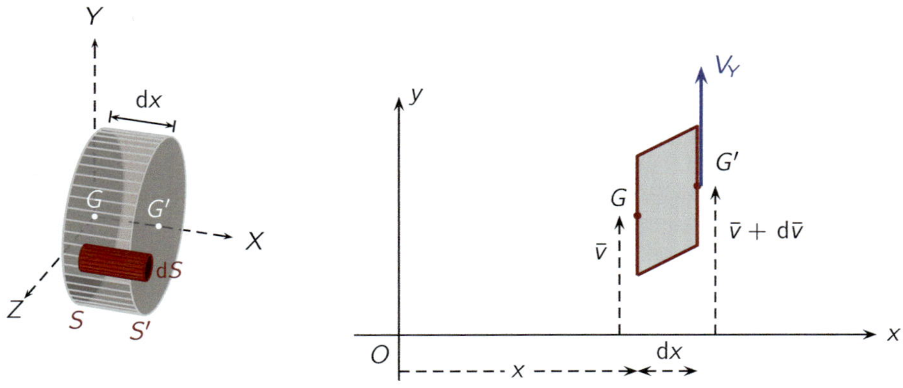

Figure 9.9 Deflection due to shear.

For the entire slice, this energy amounts to:

$$\mathrm{d}W = \iint_S \mathcal{U} \,\mathrm{d}x\mathrm{d}S = \left(\frac{1}{2} \iint_S \frac{\sigma_x^2}{E} \,\mathrm{d}S + \frac{1}{2} \iint_S \frac{\tau^2}{G} \,\mathrm{d}S \right) \mathrm{d}x$$

Since $\sigma_x = -\frac{M_Z}{I_Z} y$ from Equation (9.3), the elastic energy stored per unit length of the beam is thus

$$\frac{\mathrm{d}W}{\mathrm{d}x} = \frac{1}{2} \frac{M_Z^2}{EI_Z} + \frac{1}{2} \iint_S \frac{\tau^2}{G} \,\mathrm{d}S = \left(\frac{\mathrm{d}W}{\mathrm{d}x} \right)_M + \left(\frac{\mathrm{d}W}{\mathrm{d}x} \right)_V \tag{9.30}$$

Each cross-section being under the action of a bending moment M_Z and a transverse shear V_Y, one associates to these a bending energy, W^M, and a shear energy, W^V. The latter does not admit a closed-form expression because the shear stress τ is, in general, the solution of a complex problem. However, at the current material point, the shear stress τ must be proportional to the transverse shear V_Y, and so one may write

$$\frac{\mathrm{d}W^V}{\mathrm{d}x} = \frac{1}{2} k_Y \frac{V_Y^2}{GS} \tag{9.31}$$

where the non-dimensional factor k_Y is the *reduced section coefficient*. It depends on the geometry of the cross-section and Poisson's ratio. The ratio S/k_Y is the reduced section. To calculate k_Y, one must integrate the square of the shear stress over the cross-section.

Example 9.3: Reduced Section Coefficient for Circular Solid Section

Refer back to Figure 9.4. Equating the two expressions of the shear energy (per unit length), Equations (9.30) and (9.31), one has

$$k_Y \frac{V_Y^2}{S} = \iint_S \tau^2 \,\mathrm{d}S$$

The shear stress distribution was determined in Example 9.1. Operating the natural change of variable $Y = r\cos\theta, Z = r\sin\theta$ then $\bar{r} = r/R$ and noting that $I_Z = SR^2/4$:

$$k_Y \frac{V_Y^2}{\pi R^2} = \int_{r=0}^{R} \int_{\theta=0}^{2\pi} \tau^2(r,\theta)\,\mathrm{d}S$$

$$k_Y = \frac{1}{4\pi(1+\nu)^2} \int_0^1 \int_0^{2\pi} \Big[(3+2\nu)^2(1-\bar{r}^2\cos^2\theta)^2 + (1-2\nu)^2\bar{r}^4\sin^4\theta$$
$$-2(3+2\nu)(1-2\nu)\bar{r}^2\sin^2\theta(1-\bar{r}^2\cos^2\theta) + (1+2\nu)^2\bar{r}^4\sin^2 2\theta\Big] R^2\bar{r}\,\mathrm{d}\bar{r}\,\mathrm{d}\theta$$

This involves a collection of elementary integrals of powers of trigonometric functions that are straightforward to calculate using the identities

$$\cos^2\theta = \frac{1}{2} + \frac{1}{2}\cos 2\theta$$
$$\sin^2\theta = \frac{1}{2} - \frac{1}{2}\cos 2\theta$$

and noting that all integrals of $\cos n\theta$ or $\sin n\theta$ ($n \in \mathbb{N}$) over a period are nil. For example, $\int_0^{2\pi} \cos^2\theta\,\mathrm{d}\theta = \frac{1}{2}2\pi = \pi$.

All calculations done, one finds

$$k_Y = \frac{8\nu^2 + 14\nu + 7}{6(1+\nu)^2} \tag{9.32}$$

For common values of Poisson's ratio, one has

$$k_Y = \begin{cases} 1.173 & \text{for } \nu = 0.25 \\ 1.176 & \text{for } \nu = 0.3 \end{cases}$$

The reduced section coefficient is actually not so different from unity and rather constant over the entire range $0 \le \nu \le 0.5$. For auxetic materials, values as large as $k_Y = 4$ may be attained (see Problem 9.22). ♡

Likewise, the total deflection, $\bar{v}$, taken by the center G may be decomposed into two parts:

$$\bar{v} = \bar{v}^M + \bar{v}^V$$

Theorem 9.1

If the bending portion $\bar{v}^M$ is defined as that which obeys the usual differential equation:

$$\frac{\mathrm{d}^2\bar{v}^M}{\mathrm{d}x^2} = \frac{M_Z}{EI_Z}$$

then the deflection due to shear alone satisfies

$$\frac{\mathrm{d}\bar{v}^V}{\mathrm{d}x} = k_Y \frac{V_Y}{GS} \tag{9.33}$$

◇

Proof *It suffices to apply Castigliano's theorem to the energy* $\mathrm{d}W^V$ *stored in the slice S–S′ (Figure 9.9) working in the frame* $\{G, XYZ\}$, *assumed to follow the overall motion of cross-section S; one then gets:*

$$\frac{\partial}{\partial V_Y}(\mathrm{d}W^V) = \frac{\mathrm{d}\bar{v}^V}{\mathrm{d}x}\,\mathrm{d}x$$

which then implies, given Equation (9.31)*:*

$$k_Y \frac{V_Y}{GS}\,\mathrm{d}x = \frac{\mathrm{d}\bar{v}^V}{\mathrm{d}x}\,\mathrm{d}x$$

■

9.3 BENDING WITH SHEAR: GENERAL CASE

Pure bending of prismatic beams was analyzed in Chapter 7. So far in this chapter, a special case of bending with shearing of prisms was tackled by means of the Saint-Venant method (Section 9.1), namely that of a constant shear force along the beam.

In both cases, the same formulae were obtained for the bending moment; let us recall the main results:

- induced normal stress: $\sigma_x = -\dfrac{M}{I}Y$
- overall rotation of S' relative to S: $\mathrm{d}\omega_Z = \dfrac{M}{EI}\mathrm{d}X$
- strain energy (per unit length): $\dfrac{\mathrm{d}W^M}{\mathrm{d}X} = \dfrac{1}{2}\dfrac{M^2}{EI}$
- curvature of axial fibers: $\dfrac{\mathrm{d}^2 v}{\mathrm{d}X^2} = \dfrac{M}{EI}$

There are, of course, other loading cases where each cross-section of the prism undergoes a transverse shear force V associated with a bending moment M via the equilibrium relation:

$$m + \frac{\mathrm{d}M}{\mathrm{d}X} + V = 0$$

with m a distributed couple; see Chapter 5.

By way of extension, the results of Saint-Venant's theory apply. In particular,

- M induces on section S the normal stresses:

$$\sigma_X = -\frac{M}{I}Y \qquad \text{(from which } \varepsilon_X = \frac{\sigma_X}{E} \text{ and } \varepsilon_Y = \varepsilon_Z = -\nu\varepsilon_X)$$

- V induces shears τ_{XY} and τ_{XZ} that are determined by means of the stress function $\varphi(Y,Z)$. One also gets $\gamma_{XY} = \frac{1}{2G}\tau_{XY}$ and $\gamma_{XZ} = \frac{1}{2G}\tau_{XZ}$.

- The deflection $v(X)$ taken by the centroid G on the median line reads $v(X) = v^M(X) + v^V(X)$ with

$$v^M(X) \text{ the deflection due to } M \text{ satisfying: } \frac{\mathrm{d}^2 v^M}{\mathrm{d}X^2} = \frac{M(X)}{EI}$$

$$v^V(X) \text{ the deflection due to } V \text{ satisfying: } \frac{\mathrm{d}v^V}{\mathrm{d}X} = k\frac{V(X)}{GS}$$

- M produces a rotation of section S' located at $X + \mathrm{d}X$ relative to S located at X, about GZ, the angle of rotation being (in radian)

$$\mathrm{d}\omega = \frac{M}{EI}\,\mathrm{d}X$$

- The elastic potential energies of bending and shear, per unit length, are given by

$$\frac{\mathrm{d}W^M}{\mathrm{d}X} = \frac{1}{2}\frac{M^2}{EI} \quad \text{and} \quad \frac{\mathrm{d}W^V}{\mathrm{d}X} = \frac{1}{2}k\frac{V^2}{GS}$$

The latter formula allows to calculate the reduced section coefficient k.

Thus, most results established in the case of constant V remain valid when V is function of X. To see this, note that the axial stress is now given by

$$\sigma_X = \frac{Y}{I}\int V(X)\,\mathrm{d}X - \frac{C}{I}Y$$

where C is a constant having units of a couple. Equilibrium then only involves the X derivative of the axial stress so that Equation (9.5) remains valid. Subsequently, the treatment using the stress function remains essentially the same. Obviously, some specific formulae do change. For example, terms not related to φ or χ in Equations (9.11) and (9.12) are bending-specific. Determining the full displacement and rotation fields for general loadings is of little interest in the spirit of beam theory. What matters are the above stated relations pertaining to the deflection of the centroid and the rotation of axial fibers. The student has learned in Chapter 7 how to determine these for general bending with shear.

Example 9.4: Simply Supported Beam under Its Own Weight

The beam shown in Figure 9.10 has length L and is assumed to have a circular cross-section of radius R; it is supported at its two ends and subjected only to its own weight $P = pL$.

On cross-section S at X act a shear $V = V_Y = pX - pL/2$ and a bending moment $M = M_Z = \frac{1}{2}pX(L - X)$ such that

$$\frac{\mathrm{d}^2 v^M}{\mathrm{d}X^2} = \frac{M}{EI} = \frac{p}{2EI}X(L-X)$$

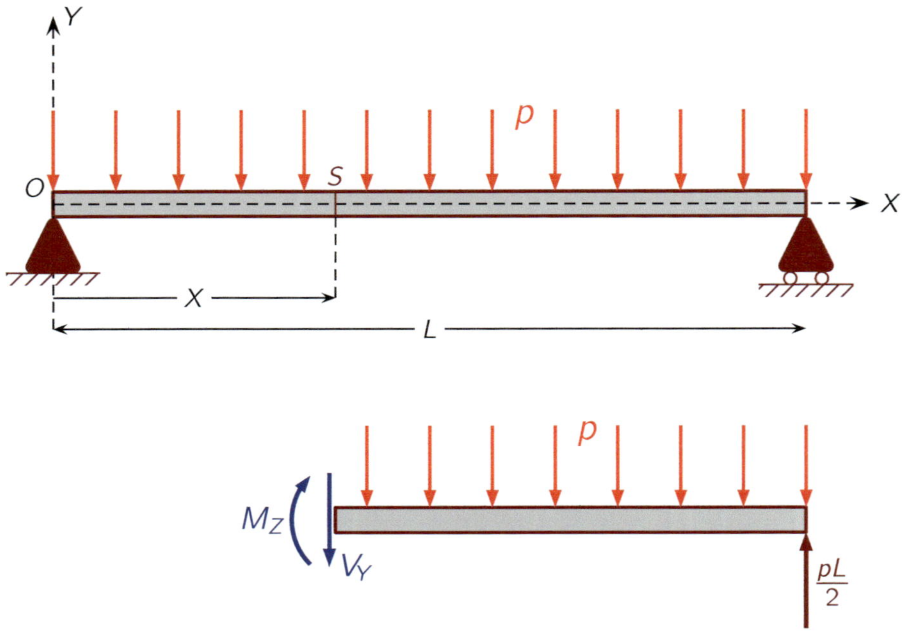

Figure 9.10 Simply supported beam under its own weight.

$$\frac{dv^V}{dX} = k\frac{V}{GS} = k\frac{p}{GS}\left(X - \frac{L}{2}\right)$$

Integrating gives $v = v^M + v^V$ with

$$v = \frac{p}{2}\left(-\frac{X^4}{12EI} + \frac{LX^3}{6EI} + k\frac{X^2}{GS} - k\frac{LX}{GS} + AX + B\right)$$

where constants A and B are determined by the conditions at the supports

$$v(0) = 0 = v(L) \quad \Longrightarrow \quad B = 0 \text{ and } A = -\frac{L^3}{12EI}$$

Finally,

$$v(X) = \underbrace{\frac{-p}{24EI}X(L-X)(-X^2+LX+L^2)}_{v^M} - \underbrace{\frac{pk}{2GS}X(L-X)}_{v^V}$$

The deflection is maximum at mid-span (see Problem 9.26) where it takes the following value:

$$v^{\max} = \underbrace{-\frac{5}{384}\frac{pL^4}{EI}}_{v^M} \underbrace{-\frac{1}{8}k\frac{pL^2}{GS}}_{v^V}$$

Numerical Application: The beam is made of steel ($\rho = 7.8 \times 10^{-3}$ kg/m^3, $E =$ 200 GPa, $\nu = 0.29$), has length $L = 4$ m and radius $R = 1$ cm.

It follows that $p = \pi R^2 \rho g$ with $g = 9.81$ m/s^2. The reduced section coefficient was calculated in Example 9.1. Also, $G = E/2/(1+\nu)$, $S = \pi R^2$ and $I_Z = \pi R^4/4$.

One infers

$$v^{\max} = -\frac{5}{96}\frac{\rho g L^4}{E R^2}\left(1 + \frac{4}{5}\frac{8\nu^2 + 14\nu + 7}{1+\nu}\frac{R^2}{L^2}\right)$$

Numerically,

$$v^{\max} = -0.051(1 + 0.000045)\text{ m} = 5.10\text{ cm}$$

In conclusion, as discussed in Chapter 7 on bending, the deflection caused by the transverse shearing is negligible for very long beams in comparison with the deflection due to bending. ♡

9.4 FLUX AND CIRCULATION OF SHEAR TRACTION

This section establishes two key formulae, as in the case of torsion. Because of their fundamental nature and generality they are stated below as theorems.

The two theorems will be essential to establishing the Bredt equation, as well as enabling key simplifications of the elasticity problem for thin-walled sections. The shear flux theorem will be used more frequently. The circulation theorem will be key for general multi-cell thin-walled sections.

9.4.1 FUNDAMENTAL THEOREMS

The current cross-section S is supposed to be under the action of a transverse shear V_Y associated with a shear field $\vec{\tau}$.

Let $\mathcal{C}$ be an arbitrary closed curve drawn on S, oriented in the positive (anticlockwise) sense about the X-axis, Figure 9.11. The curve bounds a domain $\mathcal{D}$ with area A. At current point $P \in \mathcal{C}$, $\mathbf{n}$ denotes the outward unit normal, $\mathrm{d}\mathbf{l}$ an (oriented) tangential line element and $\mathrm{d}l$ an element of line.

Theorem 9.2: Shear Flux Theorem

The outward flux, $\Phi^{\mathcal{C}}$, of vector $\vec{\tau}$ across $\mathcal{C}$ drawn on current cross-section S is given by

$$\forall \mathcal{C}, \quad \Phi^{\mathcal{C}} = \oint_{\mathcal{C}} \vec{\tau}\cdot\mathbf{n}\,\mathrm{d}l = -\frac{V_Y}{I_Z}\mathcal{Q}_Z(\mathcal{D}) \tag{9.34}$$

where $\mathcal{Q}_Z(\mathcal{D})$ is the static moment of inertia of domain $\mathcal{D}$ about axis GZ:

$$\boxed{\mathcal{Q}_Z(\mathcal{D}) = \iint_{\mathcal{D}} y\,\mathrm{d}S} \tag{9.35}$$

◇

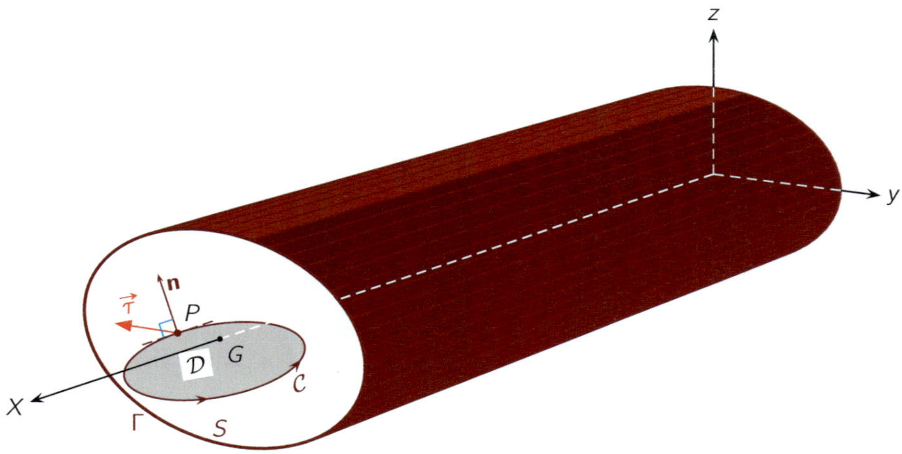

Figure 9.11 Flux and circulation of shear traction.

The above formula, Equation (9.34), is referred to as the Bredt equation. The static moment of inertia is also known as the first moment of area.

Proof *The outward flux can be calculated using Ostrogradski's formula (a variant of the divergence theorem):*

$$\oint_{\mathcal{C}} \vec{\tau} \cdot \mathbf{n}\, \mathrm{d}l = \iint_{\mathcal{D}} \mathbf{div}\, \vec{\tau}\, \mathrm{d}S = \iint_{\mathcal{D}} \left(\frac{\partial \tau_{xy}}{\partial y} + \frac{\partial \tau_{xz}}{\partial z} \right) \mathrm{d}S$$

where $\mathcal{D}$ denotes the domain enclosed by $\mathcal{C}$. The equilibrium statement constrains the divergence of the shear traction to be balanced by the axial variation of the normal stress, see Equation (9.2)*. Hence,*

$$\Phi^{\mathcal{C}} = \iint_{\mathcal{D}} -\frac{V_Y}{I_Z}\, y\, \mathrm{d}S = -\frac{V_Y}{I_Z} \iint_{\mathcal{D}} y\, \mathrm{d}S$$

which delivers the Bredt equation. ■

Remark 9.6 *The Bredt equation is general and is not restricted to the case of constant V. This amounts to considering the y-derivative of function f.* ∗

Theorem 9.3: Shear Circulation Theorem

The circulation of shear traction vector $\vec{\tau}$ along any closed curve $\mathcal{C}$ drawn in the positive sense on current cross-section S is given by

$$\forall \mathcal{C}, \quad \oint_{\mathcal{C}} \vec{\tau} \cdot \mathbf{dl} = -\frac{\nu}{\nu+1} \frac{V_Y}{I_Z} \mathcal{Q}_Y(\mathcal{D}) \qquad (9.36)$$

where $\mathcal{Q}_Y(\mathcal{D})$ is the static moment of domain $\mathcal{D}$ about axis GY:

$$\boxed{\mathcal{Q}_Y(\mathcal{D}) = \iint_{\mathcal{D}} z\,\mathrm{d}S} \tag{9.37}$$

◇

Proof *Consider the same closed curve in Figure 9.11 enclosing domain $\mathcal{D}$. Stokes' formula shows that the circulation of $\vec{\tau}$ is actually the flux of its curl across $\mathcal{D}$:*

$$\begin{aligned}
\oint_C \vec{\tau}\cdot \mathbf{dl} &= \iint_{\mathcal{D}} \mathbf{curl}\,\vec{\tau}\cdot \mathbf{dS} && \textit{Stokes' theorem}\\
&= -\iint_{\mathcal{D}} \left(\Delta\varphi + \frac{\partial f}{\partial z}\right)\mathrm{d}y\,\mathrm{d}z && \textit{using Equation (9.4)}\\
&= -\iint_{\mathcal{D}} \frac{V_Y}{I_Z}\frac{\nu}{\nu+1} z\,\mathrm{d}y\,\mathrm{d}z && \textit{by virtue of Equations (9.9) and (9.10)}\\
&= -\frac{\nu}{\nu+1}\frac{V_Y}{I_Z}\iint_{\mathcal{D}} z\,\mathrm{d}S && I_Z \textit{ and } V_Y \textit{ constant on } S
\end{aligned}$$

which leads to Equation (9.36) having adopted definition (9.37). ■

Remark 9.7 *The above proof amounts to the z-derivative of function f, which implies that it is not general, as it depends on the choice of $g(z)$!* ∗

Remark 9.8 *Static moments $\mathcal{Q}_Y(\mathcal{D})$ and $\mathcal{Q}_Z(\mathcal{D})$ of domain $\mathcal{D}$ about axes GY and GZ, respectively, have the dimensions of a volume (units of L^3). They can be positive, negative or nil. For instance,*

- *$\mathcal{Q}_Y(\mathcal{D})$ is zero when the center of $\mathcal{D}$ lies on axis GY;*
- *$\mathcal{Q}_Z(\mathcal{D})$ is zero when the center of $\mathcal{D}$ lies on axis GZ.*

As a corollary, flux and circulation vanish if the contour $\mathcal{C}$ is identified with the boundary Γ of cross-section S. The physical meaning of this result is two-fold:

1. *The vanishing of the net flux of $\vec{\tau}$ across Γ is an expression of overall equilibrium.*
2. *The vanishing of the net circulation of $\vec{\tau}$ along Γ appears to be an expression of geometric compatibility.*

Further details are appended to Section 9.9 ∗

Remark 9.9 *The static moments can be calculated by direct integration, Equations (9.35) and (9.37). Alternatively, one may use the following formulae:*

$$\mathcal{Q}_Y(\mathcal{D}) = A\,Z_\Omega \qquad \textit{and} \qquad \mathcal{Q}_Z(\mathcal{D}) = A\,Y_\Omega \tag{9.38}$$

where Y_Ω and Z_Ω are the coordinates of the center Ω of $\mathcal{D}$, and A is the area of $\mathcal{D}$. ∗

9.4.2 APPLICATIONS OF THE BREDT EQUATION

Determining the shear stress field by the Saint-Venant method of Section 9.1 is in general a complex problem that may require the implementation of numerical techniques.

An approximate method, based on the Bredt equation established above, permits to determine the shear field, under certain conditions.

Suppose that the *shear lines* are known on cross-section S, exactly or approximately, Figure 9.12. Shear lines are defined as those lines which are tangent, in each of their points, to the shear traction vector $\vec{\tau}$ prevailing at that point.

Consider two such neighboring shear lines and a common perpendicular of length l. The three lines define a domain $\mathcal{D}$, as illustrated in Figure 9.12. Applying the Bredt formula, Equation (9.34), to $\mathcal{D}$ writes

$$\phi = \tau l = -\frac{V_Y}{I_Z} \mathcal{Q}_Z(\mathcal{D}) \tag{9.39}$$

because, by definition, the shear flux across shear lines vanishes, as $\vec{\tau} \perp \mathbf{n}$ there.

Example 9.5: Rectangular Cross-Section

For a rectangle of length $2a$ and width $2b$ with $a > b$, it is a good approximation to (i) consider the lines parallel to GY as shear lines, and (ii) that the shears $\vec{\tau}$ are uniform on a horizontal section line, i.e. on a parallel to GZ, Figure 9.12.

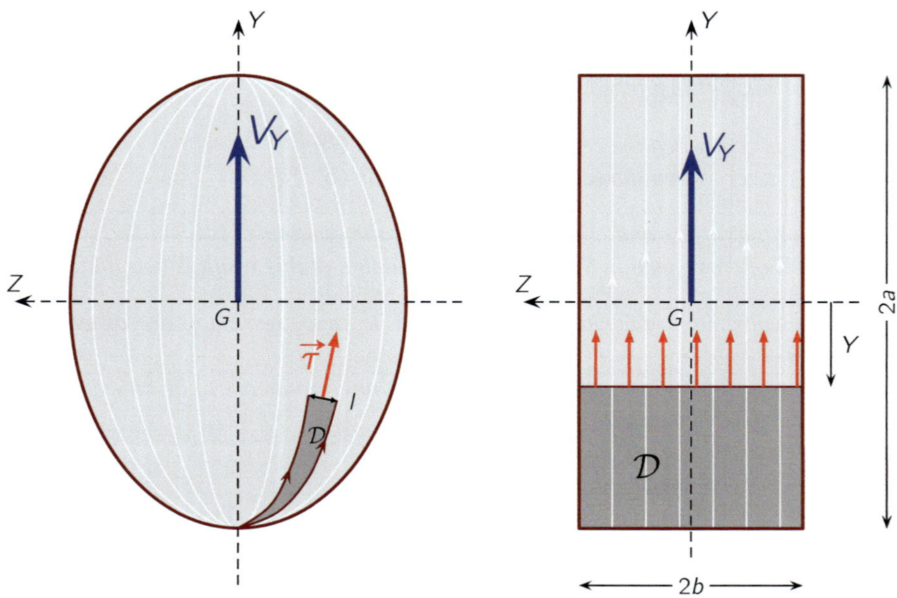

Figure 9.12 Shear lines and the Bredt approximation.

In this case, one may write

$$\tau 2b = \tau_{XY} 2b = -\frac{V_Y}{I_Z} 2b(Y+a)\frac{Y-a}{2}$$

where the second format of the Bredt equation has been used. Hence,

$$\tau = \tau_{XY} = \frac{V_Y}{2I_Z}(a^2 - Y^2) = \frac{3}{2}\frac{V_Y}{S}\left(1 - \frac{Y^2}{a^2}\right)$$

By way of consequence, the reduced section coefficient is found to be independent of Poisson's ratio:

$$k_Y = \frac{6}{5} = 1.2$$

The greater the ratio a/b, the more accurate are the above results. ♡

9.5 HOLLOW PRISMATIC BEAM

In what precedes, the prismatic beam was assumed to be solid in order to simplify the analysis. Here, the previous results are generalized to hollow sections.

In a hollow beam that contains one or more cavities, the cross-section S constitutes a multiply connected planar domain. Its boundary Γ is the union of a number of disjoint closed curves: an external contour Γ^0 separating S from the exterior and n internal contours $\Gamma^1, \Gamma^2, \ldots, \Gamma^n$, Figure 9.13.

All results obtained previously for solid sections are extended to hollow sections as follows:

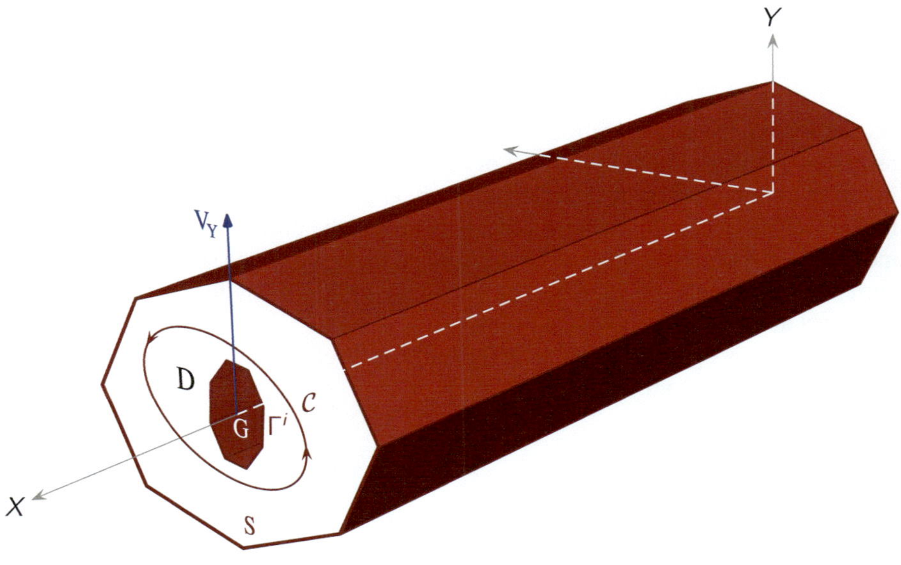

Figure 9.13 Hollow prismatic beam under shear.

1. The stress function $\varphi(y,z)$ is determined by solving the Dirichlet problem:

$$\begin{aligned} &\Delta\varphi = 0 && \text{in } S \\ &\mathrm{d}\varphi = \frac{V_Y}{2I_Z}\left(Y^2 - \frac{\nu}{1+\nu}Z^2\right)\mathrm{d}Z && \text{on } \Gamma = \bigcup_{i=0}^{n}\Gamma^i, \text{boundary of } S \end{aligned} \tag{9.40}$$

which supplants Equation (9.20).

2. The flux formula, hence the Bredt equation, remain valid:

$$\Phi^{\mathcal{C}} = -\frac{V_Y}{I_Z}\mathcal{Q}_Z(\mathcal{D}) \tag{9.41}$$

If the closed curve $\mathcal{C}$ drawn on S encloses one or more cavities (Figure 9.13) then the domain $\mathcal{D}$ is the intersection of S with the domain bounded by $\mathcal{C}$. In other words, cavities are excluded.

3. If the closed curve $\mathcal{C}$ does not enclose a cavity then the circulation formula, Equation (9.36), applies as is.
If it encloses one or more cavities, say cavities $1,2,\ldots,p$ having boundaries $\Gamma^1,\Gamma^2,\ldots,\Gamma^p$, then the formula reads

$$\oint_{\mathcal{C}}\vec{\tau}\cdot\mathbf{dl} = \oint_{\Gamma^1}\vec{\tau}\cdot\mathbf{dl} + \cdots \oint_{\Gamma^p}\vec{\tau}\cdot\mathbf{dl} - \frac{\nu}{\nu+1}\frac{V_Y}{I_Z}\mathcal{Q}_Y(\mathcal{D}) \tag{9.42}$$

where $\mathcal{D}$ denotes the planar domain, multiply connected of order p, limited by $\mathcal{C}, \Gamma^1,\ldots,\Gamma^p$, as for the extended flux theorem, Equation (9.41).

4. Equations (9.31) and (9.33) giving the energy density and deflection due to a shear force remain unchanged.

9.6 SHEARING OF THIN-WALLED SECTIONS

As for torsion, the determination of the shear field $\vec{\tau}$ due to a shear force becomes a relatively simple problem in the case of thin-walled beams. The same notation used in the chapter on torsion will be adopted here and it will be assumed that the shear traction vector $\vec{\tau}$ is uniform across the thickness of the profile and tangential to the median line, Figure 9.14. One is then compelled to reason on the shear flow $\phi = \tau t$ at current point P on the oriented median line of the profile, with t the current thickness of the cut at P.

There are three generic cases, as in torsion. However, the order of increasing complexity is different from Chapter 8.

9.6.1 OPEN SECTION

Orienting the median line $\mathscr{C}$ of the profile, as depicted in Figure 9.15, let $\mathcal{D}$ be the domain comprised between the cut at the origin and the current cut, centered about P, having thickness t.

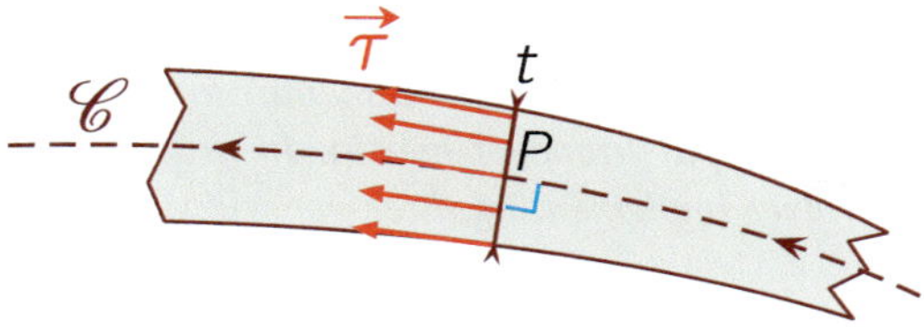

Figure 9.14 Implications of the thin wall approximation.

Theorem 9.4

In a simple open tube, the shear flow is given by

$$\boxed{\phi = \tau t = -\frac{V_Y}{I_Z}\,\mathcal{Q}_Z(\mathcal{D})} \tag{9.43}$$

◇

Proof *It suffices to apply the shear flux formula of Theorem 9.2 to the domain $\mathcal{D}$ defined above. The arguments used in torsion apply here in that the shear flux vanishes on the skins and the open cut. The only term left in the flux formula is that which traverses the cut at P, called shear flow ϕ.*

Alternatively, the simplified Bredt equation, Equation (9.39), *may be invoked,*

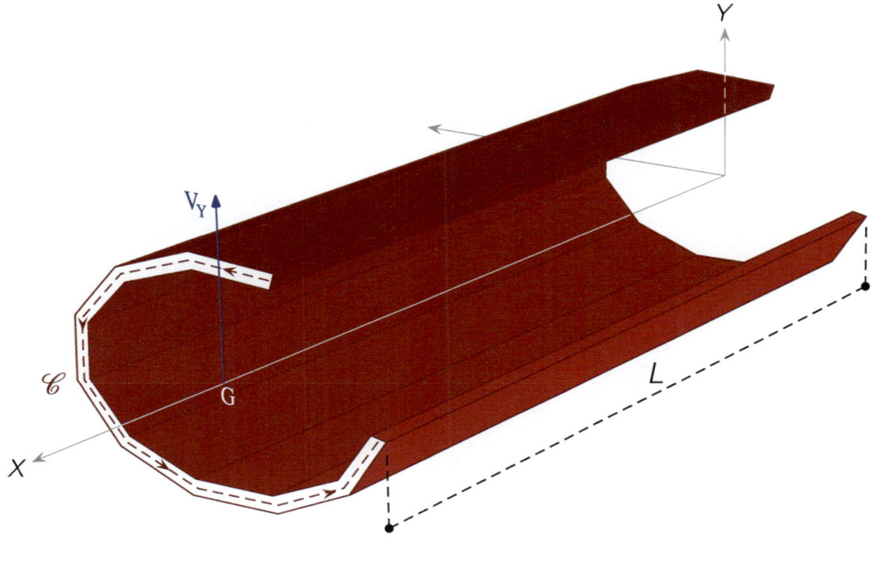

Figure 9.15 Simple open tube under shear.

since the outer and inner walls are shear lines. ■

Remark 9.10 *It is emphasized that the fundamental formulae are expressed with respect to the central principal frame of inertia. In practice, therefore, solving a shear flow problem involves implementing three generic steps:*

1. **Geometry:** *determine key geometrical parameters: (i) cross-sectional area S; (ii) position of the centroid G; (iii) second moment of area I_Z.*
2. **Parameterization:** *decide how to parameterize the median line $\mathscr{C}$. In general, the arclength on $\mathscr{C}$ is to be used. If the arc length is denoted by l then the shear flow $\phi(l)$ and shear stress $\tau(l)$ may be viewed as functions of l.*
3. **Integration:** *Apply the Bredt Equation* (9.43) *by noting that*

$$\mathcal{Q}_Z(\mathcal{D}) = \int_{l_0}^{l} Y\,t\,\mathrm{d}l \tag{9.44}$$

where a first integration has been carried out through the thickness. In addition, if the thickness is uniform, then t comes out of the integral. *

Example 9.6: Semi-Circular Thin Profile

Let R be the radius and t the uniform thickness of the semi-circular beam shown in Figure 9.16.

The aim is to determine the shear flow in the walls. To do so, implement the three steps listed above.

1. **Geometry:** The geometrical parameters, area S, position y^G of the centroid, and second moment of the area I_Z, were determined in Chapter 5; see Example 5.2 and Table 5.2:

$$S = \pi R t, \qquad y^G = \frac{2}{\pi} R, \qquad I_Z = \frac{\pi^2 - 8}{2\pi} R^3 t$$

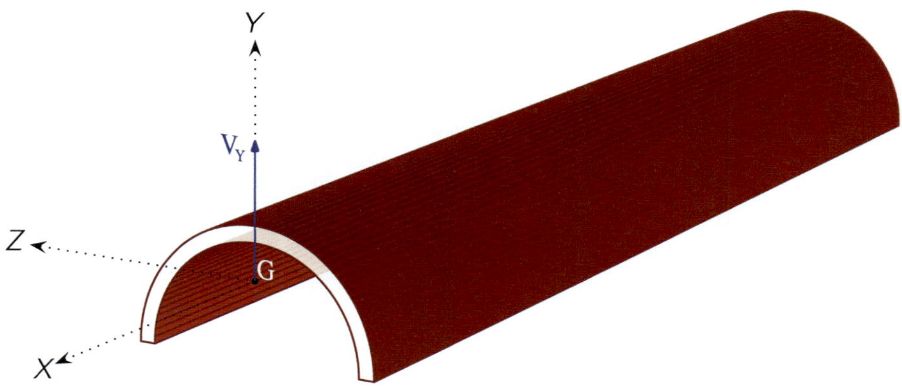

Figure 9.16 Semi-circular thin-walled beam sheared along the symmetry line.

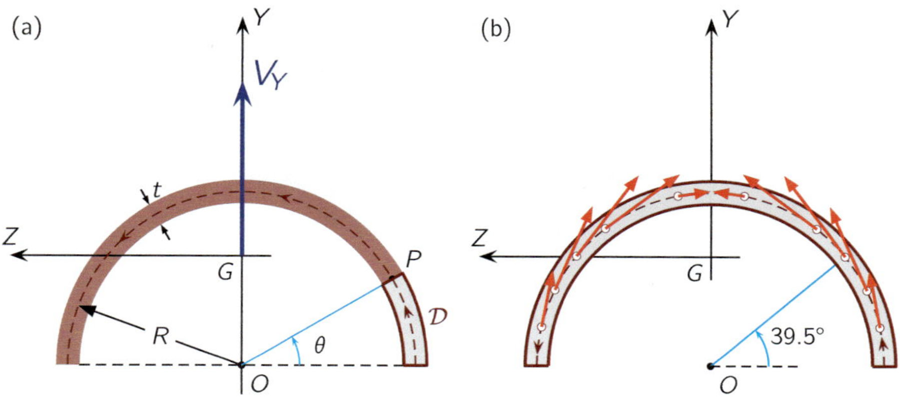

Figure 9.17 Solution for the semi-circular section. (a) Application of the Bredt equation. (b) Shear flow.

The expression of I_Z is complex because it is the moment about GZ and thus requires application of Huygens' (parallel axis) theorem, Equation (5.21). Also, Guldin's first theorem, Equation (5.12), may be invoked to obtain y^G without any integration.

2. **Parameterization:** The natural parameter along the median line is the angle θ defined in Figure 9.17a. The shear flow will therefore be given in terms of θ.
3. **Integration:** Applying the Bredt equation, Equation (9.43), involves calculating the first moment $\mathcal{Q}_Z(\mathcal{D})$ of domain $\mathcal{D}$ shown gray in Figure 9.17a. Domain $\mathcal{D}$ is comprised between the shear-free cut at $\theta = 0$ and the current cut P at θ. A formula suitable to thin-walled sections, as given under Remark 9.10, is used, noting that the element of line $\mathrm{d}l$ is here $R\mathrm{d}\theta$:

$$\mathcal{Q}_Z(\mathcal{D}) = t\int_0^{\theta} YR\mathrm{d}\theta$$

where an "abuse of notation" is made for θ as the integrand and bound on the integral (location of current point P) for mere convenience. It is emphasized that Y denotes here the location of material point P using the centroid G as the origin. Translating the origin of coordinates to point O (see Figure 9.17a) with $Y = y - y^G$ gives:

$$\begin{aligned}\mathcal{Q}_Z(\mathcal{D}) =& Rt\int_0^{\theta}\left(y - y^G\right)\mathrm{d}\theta = Rt\left\{\left(\int_0^{\theta} R\sin\theta\,\mathrm{d}\theta\right) - y^G\theta\right\}\\ =& R^2t\left\{\left[-\cos\theta\right]_0^{\theta} - \frac{2}{\pi}\theta\right\} = R^2t\left[1 - \cos\theta - \frac{2}{\pi}\theta\right]\end{aligned}$$

where the expression of y^G from step 1 above has been used. Finally, one gets the shear flow ϕ using Equation (9.43) then the shear stress $\tau(\theta) = \phi/t$

$$\tau = -\frac{2\pi^2}{\pi^2 - 8}\frac{V_Y}{S}\left[1 - \cos\theta - \frac{2}{\pi}\theta\right]$$

The shear flow is depicted in Figure 9.17b. The shear is maximum at $\theta = 39.5°$. Symmetry of the flow with respect to the y-axis is verified due to $\tau(\pi - \theta) = -\tau(\theta)$,

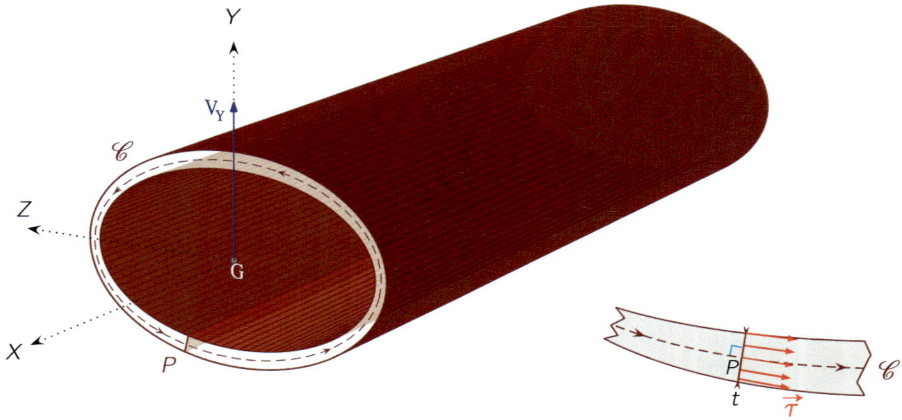

Figure 9.18 Simple closed tube under shear.

which may be checked using the above expression. The shear must vanish on the axis of symmetry, which is verified since $\tau(\pi/2) = 0$. ♡

9.6.2 SIMPLE CLOSED SECTION

Let the median line $\mathscr{C}$ of the profile be oriented anticlockwise, as depicted in Figure 9.18. Some arbitrary cut centered at P_0 is chosen as the origin of cuts. Let ϕ_0 denote the shear flow at P_0. The problem is hyperstatic of degree 1. Denoting by $\mathcal{D}$ the domain comprised between the P_0 cut and the current cut at P, the Bredt equation allows to relate ϕ, ϕ_0 and the static moment $\mathcal{Q}_Z(\mathcal{D})$. To resolve the indeterminacy, one would invoke the circulation formula. The extension of the latter to hollow sections, see Equation (9.42) of Section 9.5, leads however to an identically verified equation. Thus, before proceeding with a method of calculation of ϕ, a key result is first established.

Lemma 9.1: Circulation in a Closed Thin-Walled Section

The shear flow circulation vanishes in a closed thin-walled section subjected to transverse shear:

$$\oint_{\mathscr{C}} \frac{\phi}{t}\, \mathrm{d}l = 0 \tag{9.45}$$

◇

Proof *An energy argument is used. Consider a slice of the beam having thickness* dX*, Figure 9.19. The slice is cut open parallel to the* X*-axis following the original cut* P_0 *so as to expose the forces acting on the two lips of the cut, that is* $\pm\phi_0\,\mathrm{d}X$*,* *parallel to the X-axis. The relative displacement of the two lips is zero, which may be expressed using Castigliano's theorem as follows:*

$$\frac{\partial}{\partial \phi_0}\left(\frac{\mathrm{d}W^V}{\mathrm{d}X}\right) = 0$$

where W^V is the shear energy with

$$\frac{\mathrm{d}W^V}{\mathrm{d}X} = \frac{1}{2G}\iint_S \tau^2 \,\mathrm{d}S = \frac{1}{2G}\oint_{\mathscr{C}} \tau^2 t \,\mathrm{d}l$$

and

$$\tau = \frac{1}{t}\left[\phi_0 - \frac{V_Y}{I_Z}\mathcal{Q}_Z(\mathcal{D})\right]$$

from the Bredt equation. The lemma follows. ■

Remark 9.11 *Formally, Equation* (9.45) *would follow from applying the shear circulation formula, Equation* (9.36) *of Theorem 9.3, to the median line $\mathscr{C}$; having used the fact that the static moment of the cross-section is nil, $\mathcal{A}_Z(S) = 0$; see Remark 9.8. One would then obtain Equation* (9.45).

While having a certain mnemonic value, this reasoning is flawed because the closed thin-walled section may be viewed as a special case of a hollow section. The extension of Theorem 9.3 to hollow beams (see Section 9.5) involves the format of Equation (9.42) *for the circulation equation, which is identically verified and thus does not produce the desired result.* *

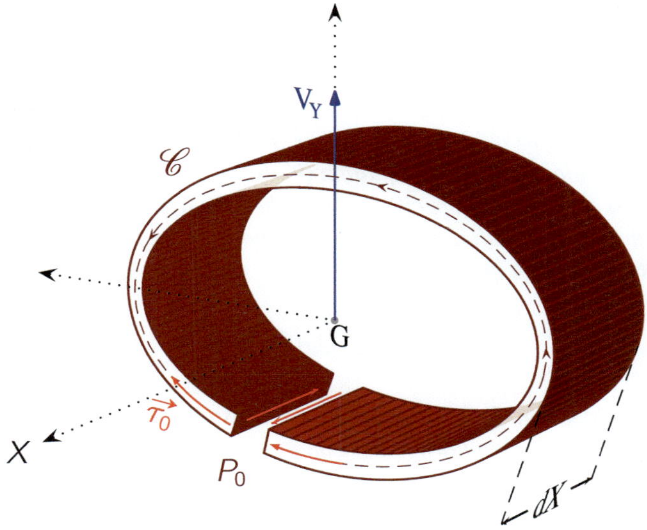

Figure 9.19 Proof of Equation (9.45) using an energy method.

Next, the following theorem permits the determination of the shear field.

Theorem 9.5

In a simple closed section, the shear flow due to V_Y is given by

$$\boxed{\phi - \phi_0 = -\frac{V_Y}{I_Z}\,\mathcal{Q}_Z(\mathcal{D})} \qquad (9.46)$$

where the hyperstatic unknown ϕ_0 is given by:

1. $\phi_0 = 0$ if
   a. the Y-axis (direction of shearing) is an axis of symmetry for cross-section S; and
   b. the origin of all cuts, P_0, is chosen on the axis of symmetry.
2. Otherwise,

$$\phi_0 \oint_{\mathscr{C}} \frac{\mathrm{d}l}{t} = \frac{V_Y}{I_Z} \oint_{\mathscr{C}} \frac{\mathcal{Q}_Z(\mathcal{D})}{t}\,\mathrm{d}l \qquad (9.47)$$

which, should the thickness be uniform, simplifies into

$$\phi_0 = \frac{V_Y}{\mathcal{L} I_Z} \oint_{\mathscr{C}} \mathcal{Q}_Z(\mathcal{D})\,\mathrm{d}l \qquad (9.48)$$

where $\mathcal{L}$ denotes the total length of median line $\mathscr{C}$.

◇

Proof *Equation* (9.46) *results from direct application of the Bredt equation to domain* $\mathcal{D}$.

If the cross-section is symmetric about the axis of shear, then the shear flow must vanish on the symmetry axis. It then suffices to choose P_0 *on the symmetry axis to proceed with calculating* ϕ *using Equation* (9.46).

If there is no symmetry, as specified above, then the hyperstatic character is lifted by invoking the shear circulation formula of Lemma 9.1. Substituting for ϕ *in Equation* (9.45) *its expression in Equation* (9.46), *one gets*

$$\oint_{\mathscr{C}} \left[\frac{\phi_0}{t} - \frac{V_Y}{t I_Z}\mathcal{Q}_Z(\mathcal{D})\right] \mathrm{d}l = 0$$

which delivers Equation (9.47) *after taking the constants out of the integrals.*

If the wall thickness is uniform, then Equation (9.48) *follows directly by denoting*

$$\mathcal{L} = \oint_{\mathscr{C}} \mathrm{d}l$$

the perimeter of $\mathscr{C}$. ■

An example for each case is given below.

Example 9.7: Thin Circular Profile

Let R and t respectively denote the mean radius and wall thickness of the beam shown in Figure 9.20 with $t \ll R$. There is symmetry about the Y-axis in this case.

Taking the origin of cuts P_0 to lie on the Y-axis, as shown in the figure, one gets from Equation (9.46):

$$\phi = -\frac{V_Y}{I_Z} Q_Z(\mathcal{D}) = -\frac{V_Y t}{I_Z} \int_0^\theta (-R\cos\theta) R \, d\theta = \frac{V_Y t}{I_Z} R^2 \sin\theta$$

with $\phi_0 = 0$ by symmetry. The domain $\mathcal{D}$ is illustrated in Figure 9.21a.

Having $S = 2\pi R t$ and $I_Z = \pi R^3 t$ (see Table 5.1), one finds

$$\phi = 2t \frac{V_Y}{S} \sin\theta$$

Hence, dividing by the thickness, the shear stress is

$$\tau = 2 \frac{V_Y}{S} \sin\theta$$

In particular, the maximum shear stress is found at the horizontal cuts. This was expected since the shear stress must vanish on vertical cuts lying on the symmetry axis (see Figure 9.21b). The expression for the maximum shear stress is given by

$$\tau^{\max} = \tau(\pi/2) = 2 \frac{V_Y}{S}$$

The reduced section coefficient is found to be $k_Y = 2$. ♡

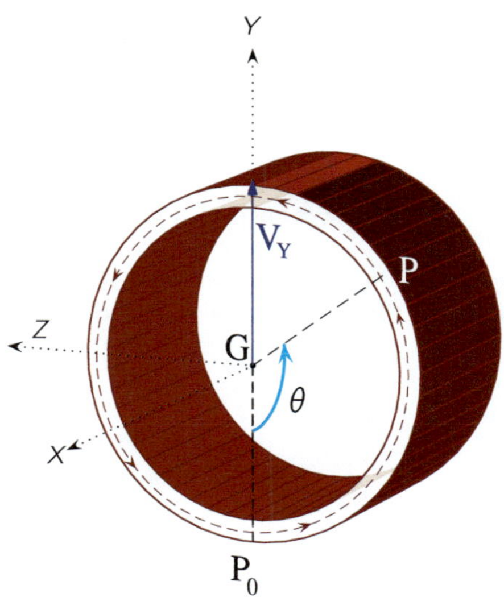

Figure 9.20 Thin-walled circular section under shear.

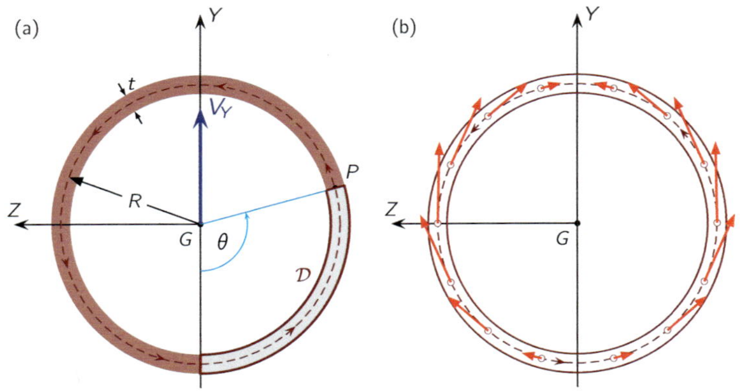

Figure 9.21 Solution for the circular section. (a) Application of the Bredt equation. (b) Shear flow.

The following example illustrates a more general procedure where symmetry breaks down, and the hyperstatic unknown ϕ_0 must be determined using the circulation formula.

Example 9.8: D-Shaped Thin Profile

Let R and t respectively denote the mean radius and wall thickness of the beam shown in Figure 9.22 with $t \ll R$.

The geometry of the problem is dealt with by considering the inverted D section as made up of two thin-walled parts: a rectangular part and a semi-circular part, which were both tackled in Chapter 5; see Tables 5.1 and 5.2. The key results are: (i) The cross-sectional area is $S = (2+\pi)Rt \simeq 5.142Rt$. (ii) The centroid G lies on the symmetry axis Oz and is obtained using Equation (5.16), which leads to

$$z^G = \frac{2}{2+\pi}R \simeq 0.389R$$

(iii) second moment of area

$$I_Z = \frac{3\pi+4}{6}R^3 t \simeq 2.237R^3 t$$

The shear force V_Y is not along an axis of symmetry. This is a type-2 problem with uniform wall thickness (see Theorem 9.5). Taking the origin at cut A (Figure 9.23), one gets:

On Circular Part $\widehat{AB}$: see Figure 9.23a

$$\phi = \phi^A - \frac{V_Y}{I_Z} t R^2 \sin\theta$$

On Straight Part $[AB]$: see Figure 9.23b

$$\phi = \phi^A + \frac{V_Y}{I_Z}\frac{t}{2}(R^2 - Y^2)$$

where the first integral on $\widehat{AB}$ is formally calculated as in Example 9.7. Note that the orientation of the median line must be respected in applying the Bredt equation,

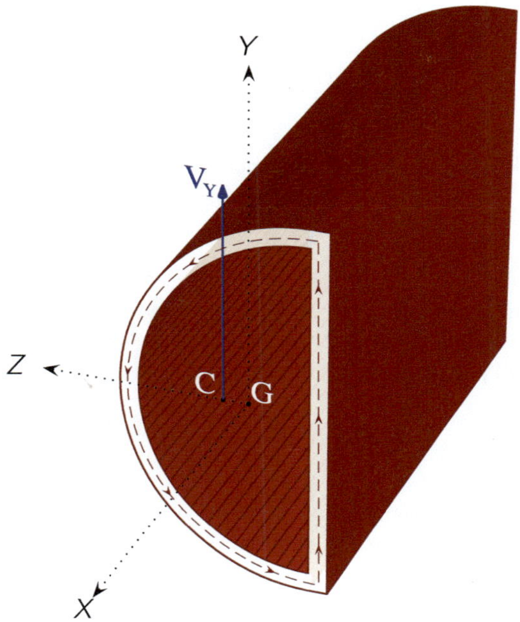

Figure 9.22 Inverted D-section subjected to vertical shear.

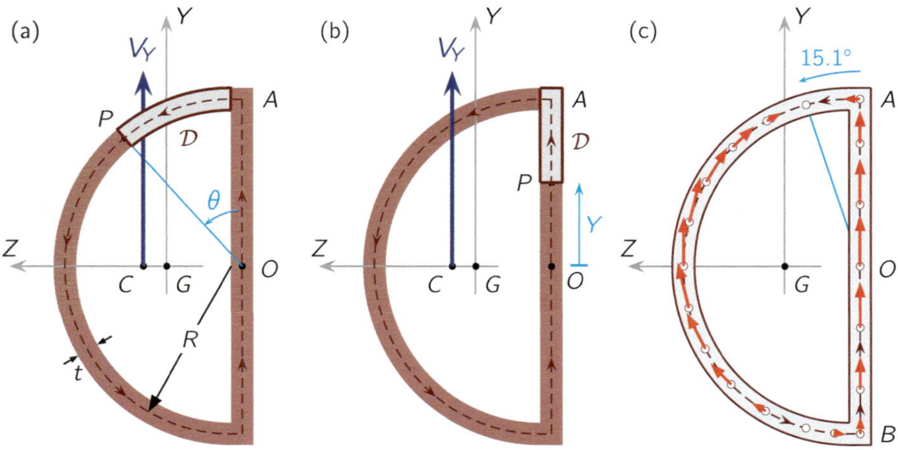

Figure 9.23 Two-step process for determining shear flow in the (a) semi-circular portion and (b) straight portion with the solution sketched in (c).

Equation (9.46), to the straight part $[AB]$, namely

$$\phi^A - \phi = -\frac{V_Y}{I_Z}\int_Y^R Y\,t\,\mathrm{d}Y$$

The particular shear flow value ϕ^A is still unknown. It is determined using the circulation formula, Equation (9.48), with $\mathcal{L} = (2+\pi)R$ the perimeter of the profile:

$$\phi^A = \frac{4}{3(\pi+2)}tR^2\frac{V_Y}{I_Z} \simeq 0.259\,tR^2\frac{V_Y}{I_Z} \simeq 0.596\frac{V_Y}{S}t$$

On the circular part, the shear is maximum at D:

$$\tau^D \simeq -0.74\frac{V_Y}{I_Z}R^2$$

It decreases until it becomes zero at $\theta = 15.1°$; see Figure 9.23c. On the small portion between A and the cut at $\theta = 15.1°$ the shear flow is reversed.

On the straight part, the shear is maximum at O:

$$\tau^O \simeq 0.76\frac{V_Y}{I_Z}R^2 = \tau^{\max}$$

It decreases steadily in magnitude toward cuts A and B while keeping the same orientation as that of the applied shear force V_Y.

The shear stresses at D and O are quite close, but the shear maximum is at mid-height of the straight portion. ♡

9.6.3 MULTI-CELL CLOSED SECTION

Shear flow calculations are most complicated in the transverse shearing of multi-cell closed sections. This is so because, unlike in torsion, (i) the shear flow varies in any given branch; and (ii) in general, the problem is internally hyperstatic.

DEFINITION 9.1 A multi-cell closed section consists of two or more closed *cells*, each being associated with its median line $\mathscr{C}^i$. Any two neighboring cells share a *branch*. Junctions where multiple branches meet are called *nodes*. ♣

To determine the shear flow, and hence the shear stresses acting in the walls of a multi-cell closed section, three rules are generally needed.

9.6.3.1 Branch Rule

Consider two cuts t_0 and t on the same branch (between two nodes) in such a way that the positive sense on the median line goes from t_0 to t, Figure 9.24.

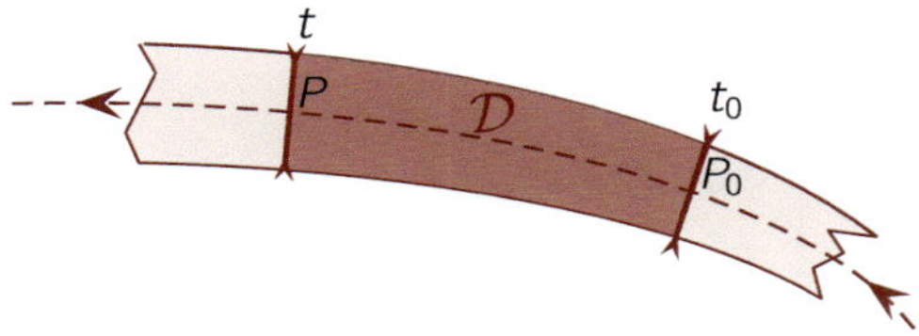

Figure 9.24 Branch rule.

Theorem 9.6: Branch Rule

In a branch, the shear flow is given by

$$\phi = \phi_0 - \frac{V_Y}{I_Z} \mathcal{Q}_Z(\mathcal{D}) \tag{9.49}$$

◇

Proof *The result follows directly from applying the Bredt equation to the domain $\mathcal{D}$ highlighted in Figure 9.24 and using the thin-wall assumptions.* ■

9.6.3.2 Node Rule

Consider a node N shared by n branches (four in the illustration of Figure 9.25). Orient the median lines in the same way, either into or away from node N. The shear flow in each of the branches varies, as per Equation (9.49). However, the following must hold at nodes.

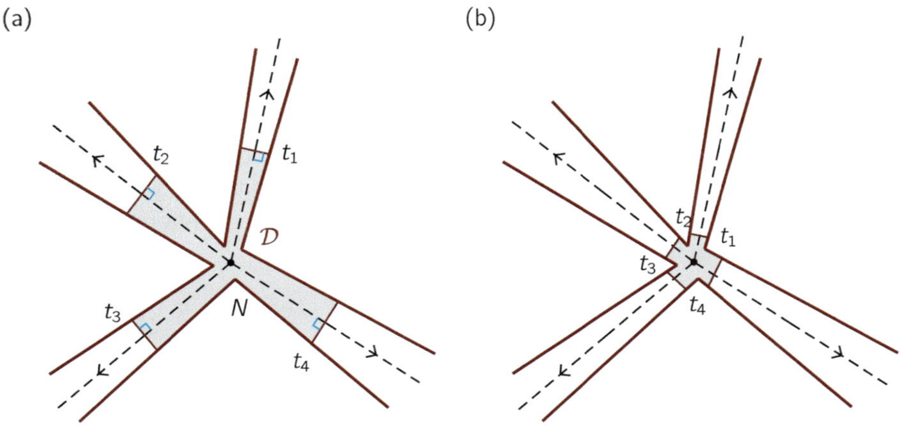

Figure 9.25 Node rule.

Theorem 9.7: Node Rule

The algebraic sum of the shear flows away (or toward) any node N vanishes. Formally,

$$\phi_1^N + \phi_2^N + \cdots + \phi_n^N = 0 \tag{9.50}$$

where n is the number of branches sharing node N and the superscript indicates that all shears are evaluated at the node. ◇

Proof *Consider the domain $\mathcal{D}$ comprised between current cuts $t_1, t_2, \ldots, t_n$ and the inner and outer walls, Figure 9.25a. Application of the Bredt equation to that domain yields*

$$\phi_1 + \phi_2 + \cdots + \phi_n = -\frac{V_Y}{I_Z} \mathcal{Q}_Z(\mathcal{D})$$

Let now the size of $\mathcal{D}$ go to zero (Figure 9.25b); in other words, the cuts tend toward node N. The static moment of a domain vanishing in size must vanish. Hence, Equation (9.50) is obtained in the limit, where it is understood that superscript N implies that each shear flow is evaluated at the cut N viewed in its respective branch. ■

9.6.3.3 Cell Rule

Formally, let C, B and N be integers representing the number of cells, branches and nodes, respectively. On each branch, one may write the equivalent of Equation (9.49), which would introduce B unknowns since, on each branch, the origin is arbitrarily chosen. These B unknowns are constrained by the equivalent of Equation (9.50) applied to each of the N nodes. In general, there are more branches than there are nodes (at least three branches are needed to make up a node). Therefore, the B unknown ϕ_0^b $(b = 1, \ldots, B)$ are so far governed by $N < B$ equations. The problem is thus hyperstatic. The larger the difference between the number of branches and the number of nodes, the higher the degree of indeterminacy.

The branch and the node rules both follow from the Bredt equation, i.e., the shear flux theorem. Thus, they are expressions of equilibrium. To lift the indeterminacy, the conditions of "geometric fit" must be invoked in that the cells must deform in concert without causing gaps or interpenetrations. As mentioned in Remark 9.8 in Section 9.4.1, the circulation equation provides just this type of information about the compatibility of deformations.

Theorem 9.8: Cell Rule

The circulation of shear flow vanishes separately in each cell:

$$\forall i, \quad \oint_{\mathscr{C}^i} \frac{\phi}{t} \, \mathrm{d}l = 0 \tag{9.51}$$

where $\mathscr{C}^i$ denotes the median line of cell i. ◇

Proof *The energy argument developed in the proof of Equation* (9.45) *is insensitive to the cell having some walls branching out at some locations. It therefore delivers Equation* (9.51) *in the same way when applied to an isolated cell.* ■

By writing down the three rules for all branches, nodes and cells of the cross-section, one obtains a linear system of equations, the resolution of which gives the shear flow everywhere.

Remark 9.12 *The C equations (9.51) provide more than what is needed to lift the hyperstatic character of the problem. In fact, $C-1$ equations are needed if there are at least two cells (see Problem 9.28).* *

Example 9.9: Empennage under Symmetric Shear

The cross-section of an aircraft empennage is idealized as shown in Figure 9.26. The wall thickness t is uniform. The empennage is under the action of a shear force V_Y along the horizontal axis. Calculate the shear flow in the walls.

Solution:

The geometry of the problem is worked out under Problem 9.17. Here, it suffices to denote the position of the centroid $y_G = \overline{OG}$ where O is the center of spar CD. Also, the second moment of area is denoted $I \equiv I_Z$, and let $V = V_Y$.

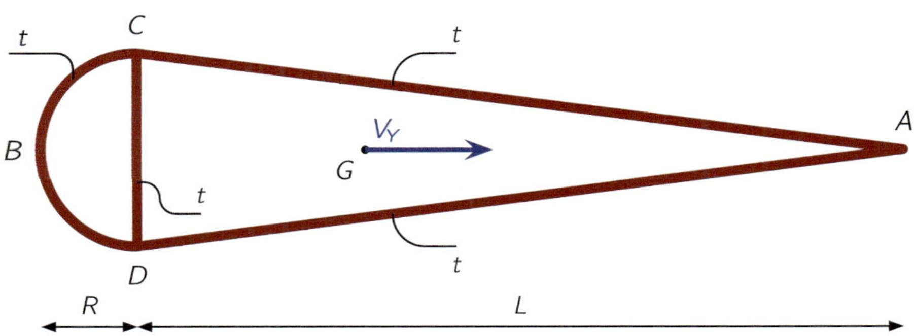

Figure 9.26 Idealized empennage under shear: symmetric case.

The empennage is symmetric about the y-axis and the applied shear does not break symmetry. There is no shear on any cut that lies on the axis of symmetry; see Theorem 9.5 in Section 9.6.2. Thus, the shear flow is zero on cuts A, O and B.

On the Spar CD: Consider a cut located at some height $z = Z$; see Figure 9.27a. The Bredt equation gives

$$\phi(z) - \phi^O = -\frac{V}{I}\int_0^z Y\,t\,\mathrm{d}z^* \tag{9.52}$$

where the simple formula, Equation (9.44) under Remark 9.10, has been used. All points on the spar have the same $Y = -y^G$ to first order (thin walls). Therefore,

$$\phi(z) = -\frac{V}{I}(-y^G tz) = \frac{Vt}{I}y^G z \tag{9.53}$$

On Curved Part $\widehat{CBD}$: Consider a cut at some angle θ measured from the vertical Oz-axis, Figure 9.27b. Using cut B as the origin, obtain

$$\phi(\theta) - \phi^B = -\frac{V}{I}\int_{\frac{\pi}{2}}^{\theta}(y^* - y^G)tR\,\mathrm{d}\varphi \tag{9.54}$$

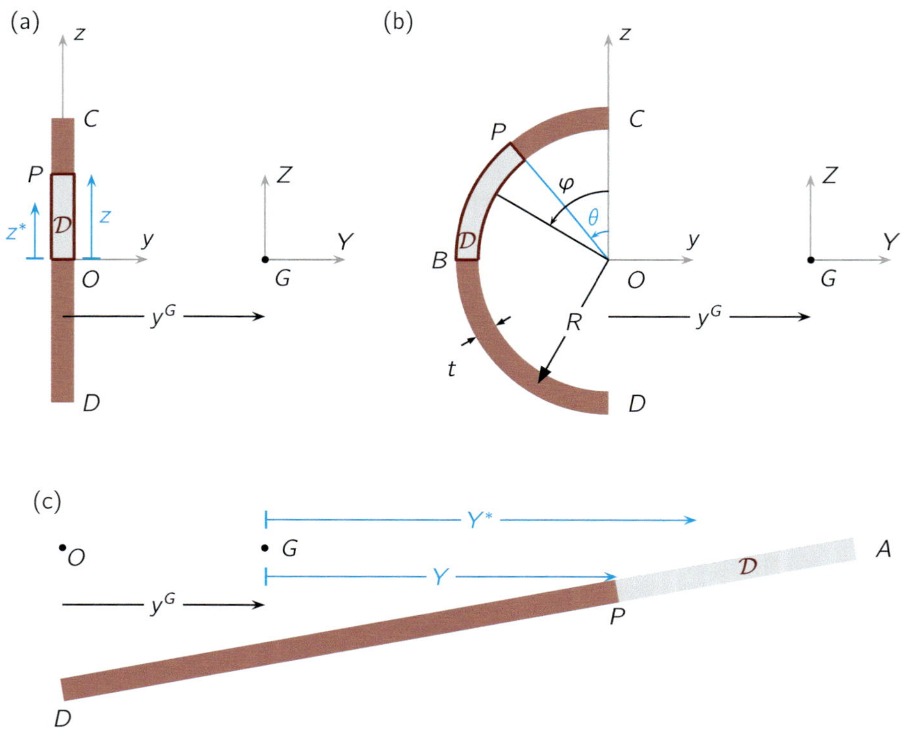

Figure 9.27 Domains for calculating the first moment of area.

Hence,

$$\phi(\theta) = -\frac{V}{I}\int_{\frac{\pi}{2}}^{\theta} Rt(-R\sin\varphi - y^G)\,\mathrm{d}\varphi \tag{9.55}$$

Integrate to get

$$\phi(\theta) = \frac{V}{I}Rt\left[-R\cos\theta + y^G\left(\theta - \frac{\pi}{2}\right)\right] \tag{9.56}$$

On the Straight Part DA: Consider a cut located between D and A at some Y:

$$\phi(Y) - \phi^A = -\frac{V}{I}\int_{Y^A}^{Y} Y^* t\frac{h}{L}\,\mathrm{d}Y^* \tag{9.57}$$

where $Y^A = L - y^G$ and

$$h = \sqrt{L^2 + R^2}$$

denotes the length of the inclined part DA. Then,

$$\phi(Y) = -\frac{Vth}{2IL}\left(Y^2 - (L - y^G)^2\right) \tag{9.58}$$

Finally, on AC obtain by symmetry:

$$\phi(Y) = \frac{Vth}{2IL}\left(Y^2 - (L - y^G)^2\right) \tag{9.59}$$

One may check that the node rule is verified at both C and D. The result comes from the formula giving the position y^G of the centroid. Also, in general empennage parts may have different thicknesses depending on design constraints. This situation of varying thickness is considered under Problem 9.17 where additional details may be worked out along with a numerical application. ♡

Example 9.10: Circular Section with a Central Web

A thin-walled circular beam has a central web and thus constitutes a two-cell closed section, Figure 9.28. R and t denote the radius and wall thickness. The two cells are referred to as L and R.

The geometry was worked out in Chapter 5 with

$$S = 2(\pi + 1)Rt \simeq 8.283Rt, \qquad I_Z = \left(\frac{2}{3} + \pi\right)R^3 t \simeq= 3.808R^3 t$$

The natural parameterization is to use Y for the web and the angle θ for the circular walls.

In the web (segment $[AB]$), the branch rule, Equation (9.49), is expressed as follows:

$$\phi_s^B - \phi = -\frac{V_Y}{I_Z}\int_Y^R Yt\,\mathrm{d}Y \quad \Longrightarrow \quad \phi(Y) = \phi_s^B + \frac{V_Y t}{2I_Z}\left(R^2 - Y^2\right)$$

where ϕ_s^B stands for the shear flow at node B viewed from the straight branch $[AB]$.

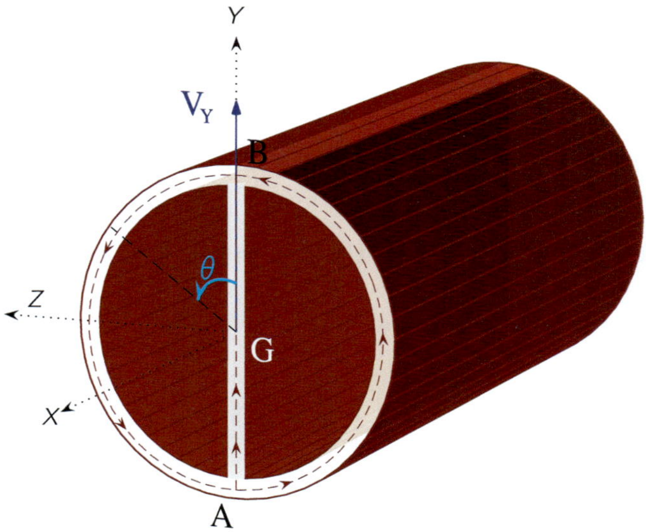

Figure 9.28 Beam with a circular section and a central web under shear.

On the left semi-circular branch $\widehat{BA}$, Figure 9.29a, the branch rule is written as follows:

$$\phi - \phi_L^B = -\frac{V_Y}{I_Z}\int_0^\theta (R\cos\theta)tR\,\mathrm{d}\theta \quad \Longrightarrow \quad \phi(\theta) = \phi_L^B - \frac{V_Y}{I_Z}R^2 t\sin\theta$$

where ϕ_L^B is the shear flow at node B viewed from the left branch $\widehat{BA}$.

The shear flow on the right branch $\widehat{AB}$ can be determined by symmetry. In particular, $\phi_R^B = -\phi_L^B$.

The node rule at B writes, see Equation (9.50):

$$\phi_s^B + \phi_R^B - \phi_L^B = \phi_s^B - 2\phi_L^B = 0 \tag{9.60}$$

and account has been taken of the median line orientations on the three branches sharing B.

The problem is hyperstatic of degree 1. To lift the indeterminacy, applying the cell rule, Equation (9.51), to cell L gives (t uniform)

$$\int_{\widehat{BA}} \phi(\theta)R\,\mathrm{d}\theta + \int_{[AB]} \phi(Y)\,\mathrm{d}Y = 0$$

Substituting for $\phi(\theta)$ and $\phi(Y)$ their expressions on the respective branches above:

$$\int_0^\pi \left[\phi_L^B - \frac{V_Y}{I_Z}R^2 t\sin\theta\right] R\,\mathrm{d}\theta + \int_{-R}^{R}\left[\phi_s^B + \frac{V_Y t}{2I_Z}\left(R^2 - Y^2\right)\right]\mathrm{d}Y = 0$$

All calculations done, obtain the following relationship between the unknowns ϕ_s^B and ϕ_L^B:

$$\pi\phi_L^B + 2\phi_s^B = \frac{4}{3}\frac{V_Y}{I_Z}R^2 t \tag{9.61}$$

Resolution of the system of equations {(9.60),(9.61)} leads to

$$\phi_s^B = \frac{8}{3(\pi+4)}\frac{V_Y}{I_Z}R^2 t, \qquad \phi_L^B = \frac{4}{3(\pi+4)}\frac{V_Y}{I_Z}R^2 t = -\phi_R^B$$

Finally, after rearranging,

- On the straight part $[AB]$:

$$\tau = \frac{\phi}{t} = \frac{V_Y}{S}\frac{3(\pi+1)}{2+3\pi}\left[\frac{28+3\pi}{3(\pi+4)} - \frac{Y^2}{R^2}\right] = \frac{V_Y}{S}\left(1.900 - 1.088\frac{Y^2}{R^2}\right)$$

- On the left circular part $\widehat{BA}$,

$$\tau = \frac{\phi}{t} = \frac{V_Y}{S}\frac{3(\pi+1)}{2+3\pi}\left[\frac{8}{3(\pi+4)} - 2\sin\theta\right] = \frac{V_Y}{S}\,(0.406 - 2.175\sin\theta)$$

- On the right circular part $\widehat{AB}$,

$$\tau = \frac{\phi}{t} = \frac{V_Y}{S}\frac{3(\pi+1)}{2+3\pi}\left[-\frac{8}{3(\pi+4)} - 2\sin\theta\right] = \frac{V_Y}{S}\,(-0.406 - 2.175\sin\theta)$$

The resulting shear flow distribution is depicted in Figure 9.29b.
On the circular parts, the shear is maximum at $\theta = \pm\pi/2$:

$$\tau\left(\pm\frac{\pi}{2}\right) \simeq \mp 1.769\,\frac{V_Y}{S}$$

It decreases until it becomes zero at $\theta = 10.8°$. On the small portion between B and the cut at $\theta = 10.8°$ the shear flow is reversed.

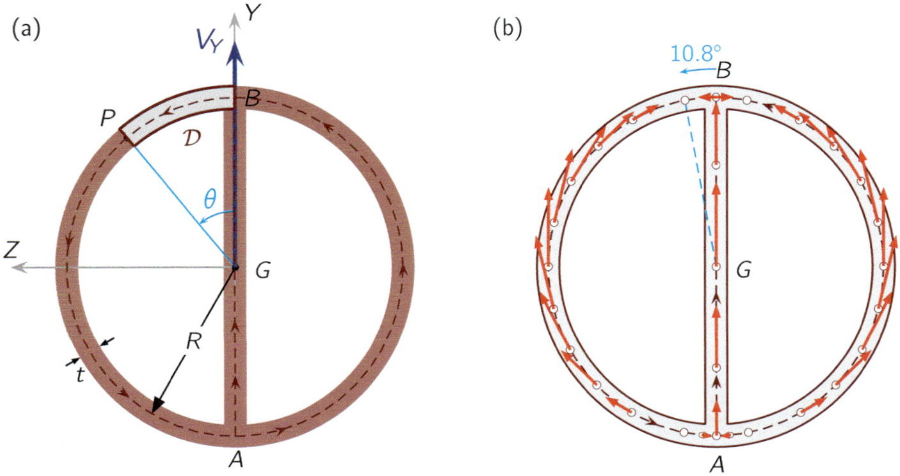

Figure 9.29 Solution for the circular section with a central web. (a) Application of the Bredt equation. (b) Shear flow.

On the straight part, the shear is maximum at G:

$$\tau^G \simeq 1.900\frac{V_Y}{S} = \tau^{\max}$$

It decreases steadily in magnitude toward cuts A and B while keeping the same orientation as that of the applied shear force V_Y.

The overall maximum shear stress is therefore at the centroid, i.e, at mid-height of the straight portion. ♡

Example 9.11: Empennage under Arbitrary Shear

Consider a part of an empennage, say a rudder under general shear, as shown in Figure 9.30. Using superposition, the problem may be decomposed in two parts. In the first part, we recover the problem treated in Example 9.9. In the second part, the shear is vertical in the figure along the Z-axis. This problem is far more difficult because the shear breaks symmetry.

The methodology is in many ways similar to that adopted in approaching Example 9.10 since there are two cells separated by a web. Both the node rule at D (or C) as well as the cell rule for both cells must be used. Details are to be worked out under Problem 9.18. ♡

Remark 9.13 *If the thin-walled section has a mix of open and closed portions then the calculation is amenable to the methods exposed above. The shear flow is known and equal to zero on any open cut. Such a cut is therefore used as the origin and the shear flow is determined from there using the three basic rules pertaining to multi-cell closed sections.* $*$

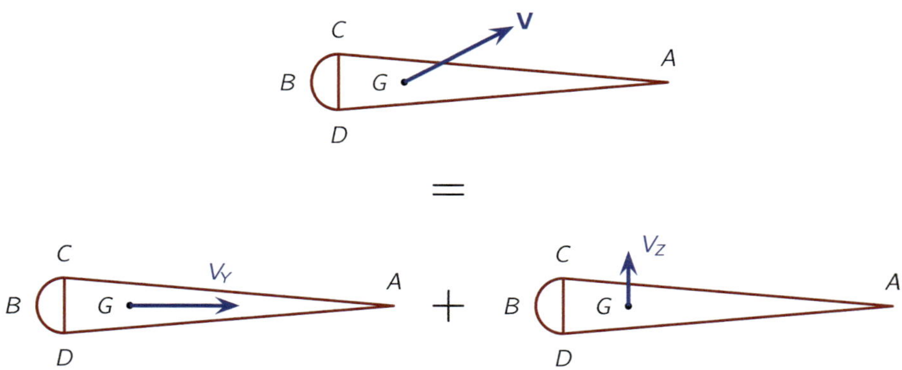

Figure 9.30 Empennage under shear: nonsymmetric case.

9.7 SHEAR CENTER

DEFINITION 9.2 The center of torsion, also called center of twist, of a given cross-section S is the point C about which the rotation (twist) induced by torsion occurs. ♣

The above definition came out naturally from examining the kinematics of torsion in Chapter 8; see Section 8.2.6.

In general, the force field $\vec{\tau}\,\mathrm{d}S$ prevailing in S admits $\mathbf{V}$ as general resultant and $\mathbf{M}^t$ as a resultant moment in C. Recall that $\mathbf{M}_X$ is the resultant moment in the centroid G and the two may be obtained from each other by the moment transport equation, see Section 8.7.

An important consequence of what precedes is that if one considers the force field $\vec{\tau}\,\mathrm{d}S$ that is *solely* induced by the shear force $\mathbf{V}$, then this field has a zero resultant moment in C. In other words, this force field must satisfy the following equation:

$$\iint_S \overrightarrow{CP} \times \vec{\tau}\,\mathrm{d}S = \mathbf{0} \tag{9.62}$$

with P denoting the current point on S.

In practice, Equation (9.62) allows to determine the position of point C, using the theory of transverse shear presented in this chapter.

Remark 9.14 *On the terminology of center of shear versus center of torsion* *

Example 9.12: Shear Center of D-Shaped Thin Profile

Consider the beam analyzed in Example 9.8. With reference to Figure 9.22, the center of torsion lies necessarily on the symmetry axis GZ. It is defined by the following equation:

$$\mathbf{M}^t = \iint_S \overrightarrow{CP} \times \vec{\tau}\,\mathrm{d}S = \mathbf{M}^O + \overrightarrow{CO} \times \mathbf{V} = \mathbf{0}$$

That is, using the algebraic measure on the X-axis:

$$M^O + \overline{OC}\,V_Y = 0$$

so that the sign of $\overline{OC}$ will determine, after calculation, what side is C on with respect to the origin O. Also, M^O denotes the moment of the shear traction field about O and may be expressed using polar coordinates:

$$\mathbf{M}^O = \iint_S \overrightarrow{OP} \times \vec{\tau}\,\mathrm{d}S = \oint_{\mathscr{C}} \overrightarrow{OP} \times \vec{\tau}\,t\,\mathrm{d}l = \int_0^{\pi} R\mathbf{e}_r \times \tau\mathbf{e}_\theta\,tR\,\mathrm{d}\theta$$

since the moment vanishes on the straight part where $\vec{\tau} \parallel \overrightarrow{OP}$. Hence, using the results obtained in Example 9.8 for $\phi(\theta) = t\tau(\theta)$:

$$M^O = R^2 \int_0^{\pi} \phi(\theta)\,\mathrm{d}\theta = -\frac{2}{3}\frac{\pi+6}{\pi+2} tR^4 \frac{V_Y}{I_Z}$$

With I_Z as given under Example 9.8, one finds

$$\overline{OC} = 4\frac{\pi+6}{(\pi+2)(3\pi+4)}R \simeq 0.530R$$

Thus, the center of torsion lies farther away from the origin than the centroid G, compare with $\overline{OG} \simeq 0.389R$. ♡

9.8 NON-PRISMATIC BEAMS

So far in the chapter, only prismatic beams have been considered. For a beam of arbitrary shape, recall the assumptions of the theory of long beams:

- On the one hand, the diameter of each cross-section S is small in comparison with the total length of the median line, its radius of curvature and its radius of torsion in G.
- On the other hand, that the geometry and the washout angle of S vary continuously and slowly with arclength s along the median line.

Under such circumstances, every thin slice of the beam limited by sections S and S', neighboring each other and located at arclengths s and $s+\mathrm{d}s$, can be considered as prismatic, and one can apply the previous results to that slice, in particular,

- Saint-Venant's method for determining the stresses, strains, rotations, warpings, reduced sections.
- the formulae giving shear flux, circulation and energy density.

Furthermore, all formulae obtained for $\mathbf{V} = \mathbf{V}_Y$ are easily transposed to the case where $\mathbf{V} = \mathbf{V}_Z$ by simple permutation of indices Y and Z.

Let us summarize these formulae in a table:

Table 9.2
Formulae for Shearing of Non-Prismatic Beams

Formula	$\mathbf{V} = \mathbf{V}_Y$	$\mathbf{V} = \mathbf{V}_Z$
Bredt	$\phi_{\mathscr{C}} = -\frac{V_Y}{I_Z}Q_Z(\mathcal{D})$	$\phi_{\mathscr{C}} = -\frac{V_Z}{I_Y}Q_Y(\mathcal{D})$
Circulation	$\oint_{\mathscr{C}} \vec{\tau} \cdot \mathrm{d}l = -\frac{\nu}{1+\nu}\frac{V_Y}{I_Z}Q_Y(\mathcal{D})$	$\oint_{\mathscr{C}} \vec{\tau} \cdot \mathrm{d}l = -\frac{\nu}{1+\nu}\frac{V_Z}{I_Y}Q_Z(\mathcal{D})$
Energy	$\frac{\mathrm{d}W^V}{\mathrm{d}s} = \frac{1}{2}k_Y\frac{V_Y^2}{GS}$	$\frac{\mathrm{d}W^V}{\mathrm{d}s} = \frac{1}{2}k_Z\frac{V_Z^2}{GS}$
Deflection (straight beam)	$\frac{\mathrm{d}v^V}{\mathrm{d}x} = k_Y\frac{V_Y}{GS}$	$\frac{\mathrm{d}v^V}{\mathrm{d}x} = k_Z\frac{V_Z}{GS}$

9.9 THE SAINT-VENANT PROBLEM REVISITED

9.9.1 PRELIMINARY REMARKS

The method followed in Section 9.1 is different, in both style and details, from the original treatment by Saint-Venant. It is also different from the account of Love or that, more succinct, of Timoshenko, who follow closely the original exposé by Saint-Venant. In all of these, different choices of the stress function pair (φ, f) were made. In particular, their choice of function $f(y,z)$ is problem-specific.

The method presented in this book has two didactic advantages. First, the function $f(y,z)$ is independent of the specific shape of the cross-section; it is given by Equation (9.9). The instructor may eventually elect to introduce its expression as early as Equation (9.4), as done by Laroze. A more deductive analytical approach was followed above. The second advantage of the choices made is that the main stress function $\varphi(y,z)$ is harmonic. By way of contrast, the governing partial differential equation in the treatments by Saint-Venant, Love, and Timoshenko is not homogeneous, as it has a problem-specific right-hand side. The only simplification that follows from the classical treatments is that the boundary condition on $\varphi(y,z)$ on the contour of the cross-section is simple with $\varphi = 0$ therein, as in the torsion problem. It must be noted, however, that in the proposed method none of the governing equations is problem-specific, which includes the boundary condition on the contour, Equation (9.16).

As observed under Remark 9.1, other choices for the function $g(z)$, hence for $f(y,z)$, are possible. In fact, the classical treatments also keep that flexibility in choosing $g(z)$ subject to $\varphi = 0$ on the boundary. However, not all boundary conditions seem to be given due consideration in classical texts.

9.9.2 GENERAL APPROACH

Equilibrium considerations suggest solutions taking the form of Equation (9.4). Strains are given by Equation (9.6) and their compatibility requires that Equation (9.8) be satisfied. In Section 9.1 a didactic choice was made for $g(z)$ that led to the format of Equation (9.9). That choice is not necessary. Here, a more general approach is outlined.

The first compatibility equation still implies that the Laplacian of φ must be independent of y (in the direction of shear). On account of that, Equation $(9.8)_2$ implies the existence of a class $\mathscr{C}^1$ function $h(z)$ such that

$$\Delta\varphi = h(z) + C \tag{9.63}$$

where $C = 0$ if $M^t = 0$, which will be assumed. The relation between h and g is specified by integrating Equation $(9.8)_2$, subject to $h(0) = 0$:

$$h(z) = \frac{V_Y}{2I_Z}\left\{\frac{2\nu}{1+\nu}z - g'(z)\right\} \tag{9.64}$$

where the constant of integration is dismissed on the same premise that $M^t = 0$.

Interestingly, integration of the rotation–displacement relations leaves Equation (9.11) unchanged, i.e. independent of function $g(z)$. The same applies to Equation (9.12).

The traction-free boundary condition on the lateral surface S_2, Equation (9.16), generalizes into

$$\mathrm{d}\varphi = -f(y,z)\,\mathrm{d}z = \frac{V_Y}{2I_Z}\left(y^2 - g(z)\right)\,\mathrm{d}z \qquad \text{on } \Gamma, \text{boundary of } S \tag{9.65}$$

Next, consider the boundary conditions on end sections S_0 and S_1. Equation (9.17) and Equation (9.18) are shown to hold, irrespective of the choice made for $g(z)$.

Proof *First, consider Equation* (9.17).

$$\begin{aligned}
\iint_S \tau_{xy}\,\mathrm{d}S &= \iint_S \left[f + \frac{\partial \varphi}{\partial z}\right] \mathrm{d}y\,\mathrm{d}z && \textit{using (9.4)} \\
&= \iint_S f\,\mathrm{d}S + \oint_\Gamma -\varphi\,\mathrm{d}y && \textit{(Riemann's formula)} \\
&= \iint_S f\,\mathrm{d}S + \oint_\Gamma y\,\mathrm{d}\varphi && \textit{(integration by parts)} \\
&= \iint_S f\,\mathrm{d}S - \oint_\Gamma y f(y,z)\,\mathrm{d}z && \textit{using (9.65)} \\
&= \iint_S f\,\mathrm{d}S - \iint_S \frac{\partial}{\partial y}(y f(y,z))\,\mathrm{d}S && \textit{(Riemann's formula)} \\
&= \iint_S f\,\mathrm{d}S - \iint_S y\frac{\partial f}{\partial y}\,\mathrm{d}S - \iint_S f\,\mathrm{d}S \\
&= -\frac{V_Y}{2I_Z}\iint_S y(-2y)\,\mathrm{d}S = \frac{V_Y}{I_Z}\iint_S y^2\,\mathrm{d}S = V_Y
\end{aligned}$$

Similarly, the proof of Equation (9.18) *proceeds as follows:*

$$\begin{aligned}
\iint_S \tau_{xz}\,\mathrm{d}S &= -\iint_S \frac{\partial \varphi}{\partial y}\,\mathrm{d}S && \textit{using (9.4)} \\
&= \oint_\Gamma \varphi\,\mathrm{d}z && \textit{(Riemann's formula)} \\
&= \oint_\Gamma z\,\mathrm{d}\varphi && \textit{(integration by parts)} \\
&= -\oint_\Gamma z f(y,z)\,\mathrm{d}z && \textit{using (9.65)} \\
&= -\iint_S \frac{\partial}{\partial y}(z f(y,z))\,\mathrm{d}S && \textit{(Riemann's formula)} \\
&= -\iint_S z\frac{\partial f}{\partial y}\,\mathrm{d}S = \frac{V_Y}{I_Z}\iint_S yz\,\mathrm{d}S = 0
\end{aligned}$$

The last integral vanishes because positions, along with other quantities (resultants, second moments of area) are defined with respect to the central principal *axes of inertia.* ∎

The third boundary condition on the extreme sections, Equation (9.19), which holds on current cross-section S, is expected to lead to a constraint on the choice of $g(z)$. Projecting that equation onto the x-axis of the prism, obtain

$$\frac{V_Y}{2I_Z}\iint_S z\left(g(z)-y^2\right)\mathrm{d}S-2\iint_S \varphi\,\mathrm{d}S=V_Y z^C \qquad (9.66)$$

where z^C denotes the z-coordinate of the center of twist C. The location of C being tacitly given by the very way in which the Saint-Venant problem is posed, Equation (9.66) does provide indeed a relationship between φ and g.

Proof *With G the centroid of the section, Equation* (9.19) *may be split as follows:*

$$\overrightarrow{CG}\times\mathbf{V}+\iint_S \overrightarrow{GP}\times\vec{\tau}\,\mathrm{d}S=\mathbf{0}$$

The x-projection then gives:

$$\iint_S\left[y\left(-\frac{\partial\varphi}{\partial y}\right)-z\left(\frac{\partial\varphi}{\partial z}+f\right)\right]\mathrm{d}S=-z^C V_Y$$

$$\iint_S \varphi\,\mathrm{d}S+\iint_S \varphi\,\mathrm{d}S-\iint_S z f\,\mathrm{d}S=-z^C V_Y \qquad \textit{(integration by parts)}$$

which then yields Equation (9.66) *by substituting for f its expression in Equation* (9.5). ■

Remark 9.15 *Equations* (9.65) *and* (9.66) *provide two relations between $\varphi(y,z)$ and $g(z)$. Equation* (9.65) *applies pointwise on the boundary Γ of every cross-section S whereas Equation* (9.66) *expresses a constraint on average values over the entire cross-section. The two relations are independent of each other since Equation* (9.65) *expresses that the lateral surface is traction-free whereas Equation* (9.66) *expresses that the shear tractions have a zero net torque (about C).* *

Remark 9.16 *If the cross-section is symmetric about the axis of shearing (the y-axis) then center of twist and centroid are identified with each other so that $z^C=0$. Equation* (9.66) *is then identically verified as both integrals vanish separately.* *

Remark 9.17 *The issue of completeness of stress functions is worth emphasizing in the present context. While the solution to the elastostatics problem is unique, multiple choices for the (φ,g) pair are possible. One may choose to make φ harmonic by setting $h(z)=0$ in Equation* (9.63). *This would determine $g(z)$ through Equation* (9.64), *hence the function $f(y,z)$ from Equation* (9.5). *This is the didactic choice made in Section 9.1 and discussed under preliminary remarks. Another choice consists of choosing $g(z)$ such that the function $f(y,z)$ would vanish on the boundary Γ thus simplifying the traction-free boundary condition, Equation* (9.65) *into $\varphi=0$ on Γ. This is the classical choice made by Saint-Venant, followed by Love and Timoshenko.* *

Remark 9.18 *In practice, when a choice is made for auxiliary function g, Equation* (9.66) *serves to determine the position of the center of twist, i.e.* z^C. *

Remark 9.19 *The term* $\iint_S \varphi \, \mathrm{d}S$ *in Equation* (9.66) *is reminiscent of the case of pure torsion, Equation* (8.17). *The counterpart of Equation* (9.66) *in the case of pure torsion is precisely Equation* (8.17).

In general, a shear traction field would admit V_Y *as a resultant force and* M^t *as a resultant moment in C. Superposition applies and the stress function* φ *may be thought of as consisting of two additive parts, one that obeys Equation* (8.17) *(torsion) and one that obeys Equation* (9.66) *(shear).* *

9.9.3 SHEAR FLUX AND CIRCULATION

The flux and circulation of the shear traction vector $\vec{\tau}$ vanish on the contour Γ of cross-section S.

1. That the flux of $\vec{\tau}$ across Γ vanishes follows from the fact that the integrand itself $\vec{\tau} \cdot \mathbf{n}$ vanishes pointwise on Γ. Indeed, on the boundary, the traction free condition imposes that $\vec{\tau}$ be parallel (tangential) to the boundary.
2. The flux of $\vec{\tau}$ across Γ is balanced by the net axial force $\iint_S \sigma_x \, \mathrm{d}S$, which must therefore be nil. Over the entire cross-section, the normal stresses result in a bending moment only. The situation is obviously different over an arbitrarily chosen subpart $\mathcal{D}$ of S.
3. From the above two observations, it follows that the vanishing of the net flux expresses equilibrium of any slice of the beam along with associated traction-free conditions on the lateral surface.
4. The vanishing of the circulation of $\vec{\tau}$ along Γ is not as trivial. Indeed, the proof of Theorem 9.2 involves the z-derivative of the Saint-Venant function $f(y,z)$ and may suggest that the theorem may not be valid for more general choices of the pair φ, g. It can be shown that the result stands for any such choice.
5. Examination of the above proof shows that the vanishing of the net circulation of $\vec{\tau}$ along Γ appears to be an expression of geometric compatibility.

9.10 DOMAINS OF VALIDITY

1. Recall that Saint-Venant's theory, which served as a basis, assumes that all cross-sections can warp freely; if, on certain sections such warping is prevented or altered, then the results reported in Table 9.2 are not exact in the neighborhood of the cross-sections in question. On the other hand, the Saint-Venant principle states that sufficiently far from such sections, the results in Table 9.2 are indeed exact.
2. The main limitation of thin-wall theory is that it does not apply to single walls when subjected to a shear force normal to the walls. This is so because the simplified Bredt equation, Equation (9.39), does not apply to wide beams. The general Bredt equation, Equation (9.34) or Equation (9.46), applies but delivers a trivial result, which is useless in this case.

PROBLEMS

Problem 9.1 (Saint-Venant Principle) *The theory presented in Section 9.1 assumes specific traction boundary conditions. In Saint-Venant's theory of general bending, more general tractions are considered. State the conditions that must be satisfied by the traction distributions such that the solution given in Section 9.1 remains valid sufficiently far away from end sections. Sketch some traction distributions that meet the stated conditions.* ⋆

Problem 9.2 (Pure Bending versus General Bending) *(i) Sufficiently far away from end sections, compare the displacements and rotations for pure bending versus the bending of a cantilever. (ii) Quantify your answers for a beam with a circular section of radius R.* ⋆

Problem 9.3 (Shearing of a Triangular Section) *The cross-section of a solid beam is a slender isosceles triangle (basis 2b; height 3h with $h > b$), Figure 9.31. It is subjected to a transverse shearing load V_Y parallel to the (vertical) axis of symmetry of the section, GY. The following (non-rigorous) assumption is made: the component τ_{xy} of the shear traction $\vec{\tau}$ is independent of Z. (i) Calculate the function $\tau_{xy}(Y)$. (ii) Find the maximum shear stress, $\tau^{\max}$, and its location. (iii) Calculate the reduced section coefficient k_Y while neglecting the horizontal component τ_{xz}.* ⋆

Problem 9.4 (Shearing of an Angle Beam) *The angle beam of Figure 9.32 has a thin profile of uniform thickness $t \ll \ell$. (i) Under the action of shear force $V = V_Y$, determine the shear stress evolution in the walls. (ii) Determine the center of twist of the section.* ⋆

Problem 9.5 (Shear Flow in U Channel) *The thin profile shown in Figure 9.33 is subjected to a shear force $V = V_Y$. (i) Find the location of the centroid G. (ii) Calculate the shear flow in the walls and draw the diagram $\tau(s)$ giving the shear stress τ as a function of the arclength s. (iii) Calculate the reduced section coefficient k_Y.* ⋆

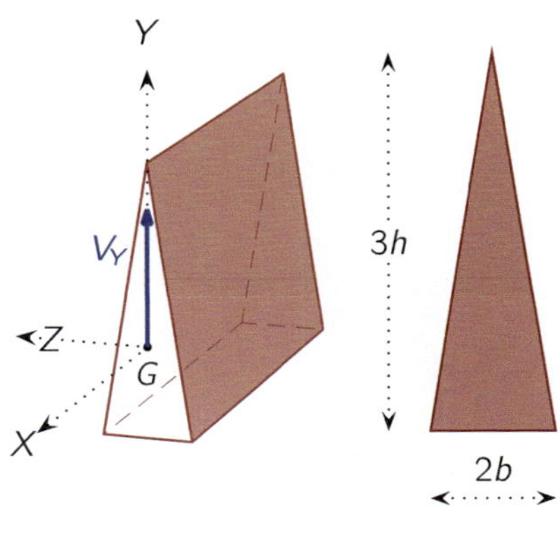

Figure 9.31 Triangular section under shear.

Problem 9.6 (Shearing of a C Channel) *A prismatic beam has the thin profile shown in Figure 9.34. It consists of an open section with uniform thickness t, small before a. (i) Determine the shear traction vector $\vec{\tau}$ at each point of the profile under the action of transverse shear V_Y. Draw the shear stress diagram $\tau(s)$. (ii) Calculate the position of the shear center.* ⋆

Problem 9.7 (Shear Flow in a Wide-Flange I Beam) *Consider a prismatic beam having a thin-walled section in the shape of a symmetric I-beam, Figure 9.35. On the current cross-section acts a transverse shear force V_Y. The wall thickness t is*

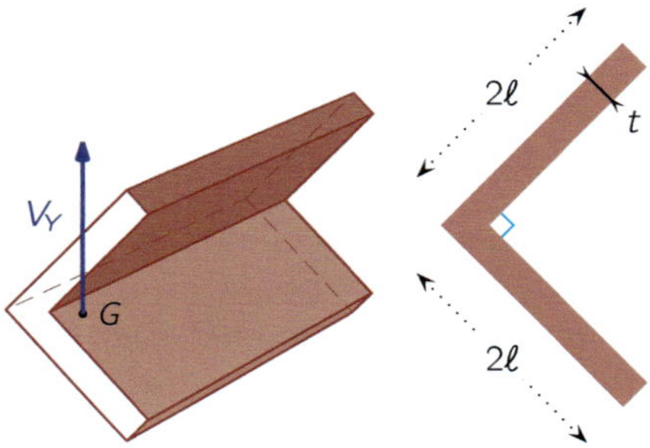

Figure 9.32 Angle beam under shear.

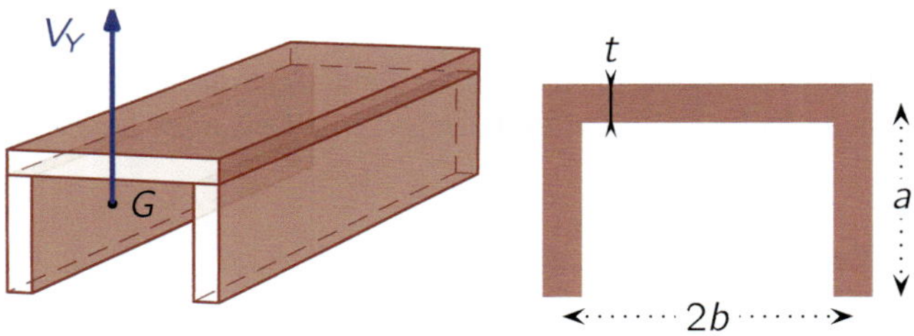

Figure 9.33 U channel under shear.

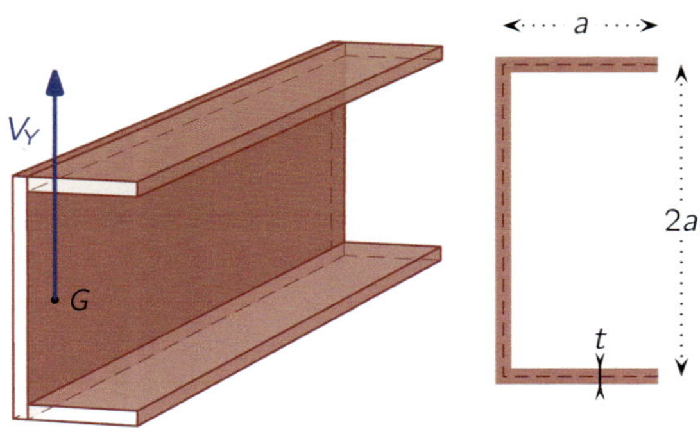

Figure 9.34 C channel under shear.

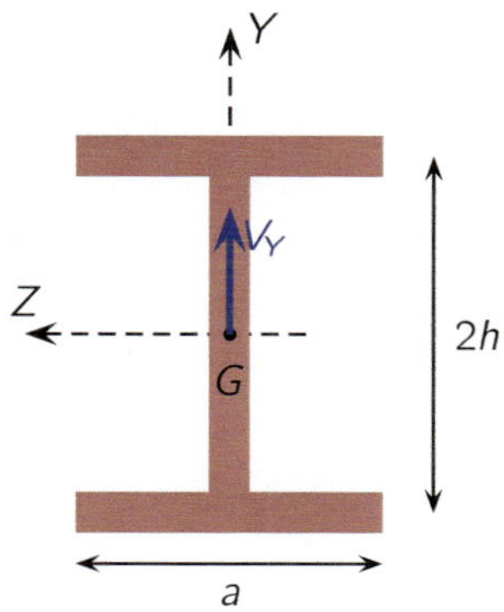

Figure 9.35 I section under shear.

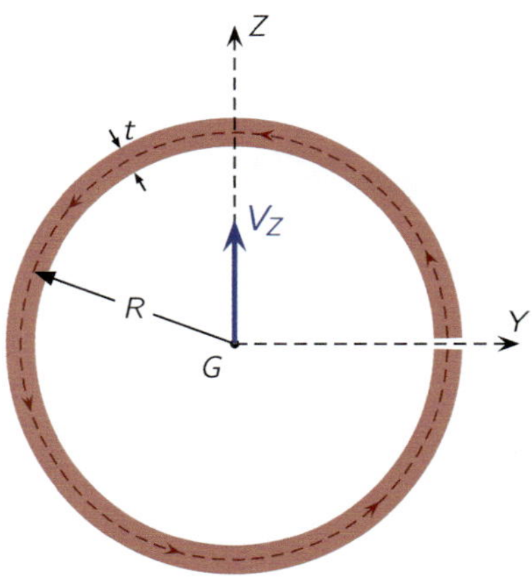

Figure 9.36 Open circular section under shear.

assumed to be uniform throughout the flanges and central web. (i) Determine the shear flow in the walls. (ii) Draw the diagram giving the shear stress τ as a function of the arclength. (iii) Find the location and magnitude of maximum shear. ⋆

Problem 9.8 (Shearing of a Circular Tube with a Slit) *Consider the thin open circular profile depicted in Figure 9.36 under the action of a transverse shear force V_Z. (i) Determine the shear flow in the walls. (ii) Calculate the reduced section coefficient k_Z. (iii) Make quantitative comparisons with the case of a circular closed section.* ⋆

Problem 9.9 (Shearing of a Semi-Circular Tube) *A prismatic beam is made by cutting a circular thin-walled tube of thickness t and radius R following a meridian plane, Figure 9.37. The shear force V_Y induces a field of shear tractions $\vec{\tau}$. (i) Determine the shear flow in the profile. (ii) Find the position of the shear center. (iii) Cal-*

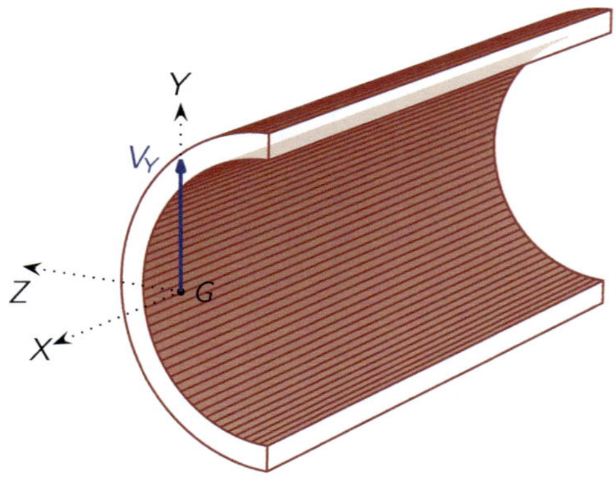

Figure 9.37 Semi-circular profile under shear.

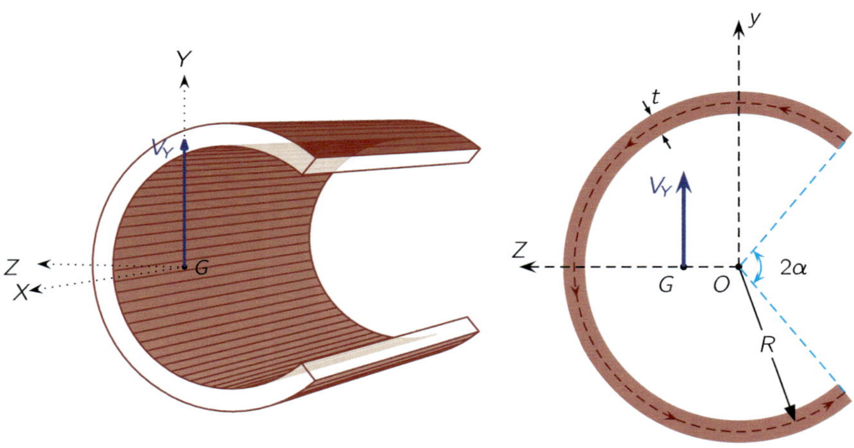

Figure 9.38 Arbitrary open circular tube under shear.

culate the reduced section coefficient. (iv) Make quantitative comparisons with the semi-circular profile treated in Example 9.6. ⋆

Problem 9.10 (Shearing of an Open Circular Tube) *Consider the truncated thin-walled crown-like beam shown in Figure 9.38. The thin profile is under the action of shear force V_Y. (i) Calculate the area and second moment of area I_Z of the section. (ii) Determine the position of the centroid G. (iii) Determine the shear flow in the profile. (iv) Find the position of the shear center. (v) Study the special cases $\alpha = 0$ and $\alpha = \frac{\pi}{2}$.* ⋆

Problem 9.11 (Shearing of a Rectangular Tube with a Slit) *A thin-walled rectangular beam is slit open as shown in Figure 9.39. Under the action of transverse shear V_Z determine the shear flow and the shear center.* ⋆

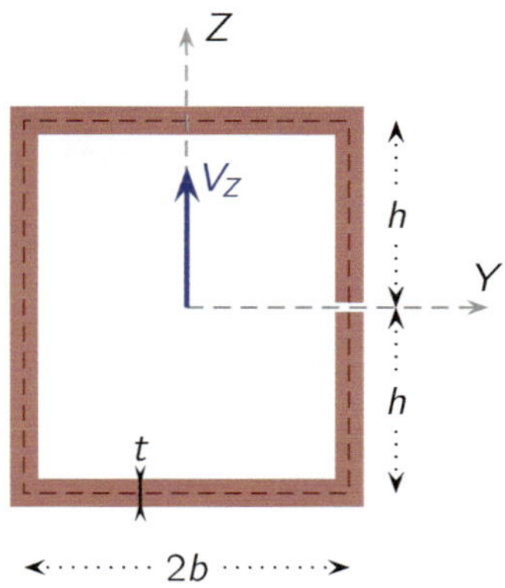

Figure 9.39 Rectangular section with a slit.

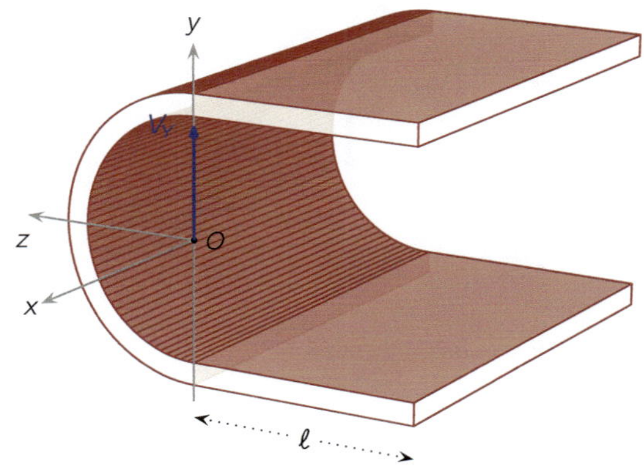

Figure 9.40 Lyre profile under shear.

Problem 9.12 (Shearing of a Lyre-Shaped Thin Profile) *A prismatic beam with thin walls of uniform thickness t is under the action of a shear force V_Y, Figure 9.40. The semi-circular portion has radius R. The straight parts have each length ℓ. Determine the shear flow in the walls and the position of the shear center.* ⋆

Problem 9.13 (Shearing of a Square Tube) *Consider the thin square profile shown in Figure 9.41. Under the action of a shear force V_Y, determine the shear flow in the walls. Calculate the reduced section coefficient.* ⋆

Problem 9.14 (Shearing of a Triangular Tube) *Consider the thin triangular profile shown in Figure 9.42. The median curve is an equilateral triangle of side length 2a. Under the action of a shear force V_Y, determine the shear flow in the walls and the reduced section coefficient.* ⋆

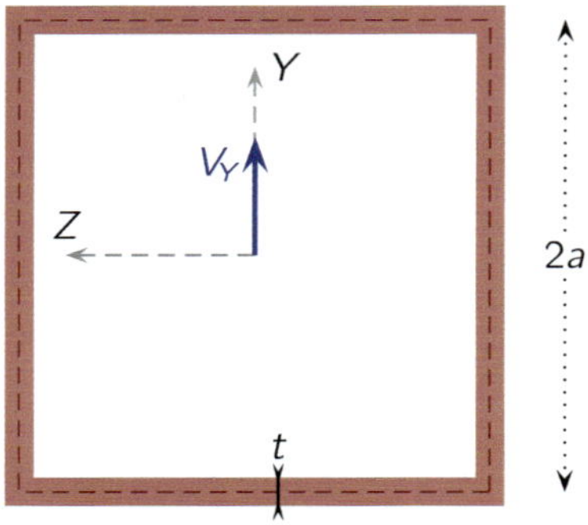

Figure 9.41 Thin rectangular section under shear.

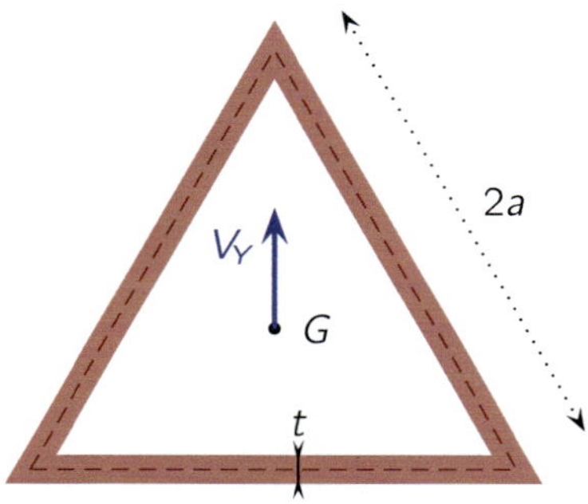

Figure 9.42 Thin triangular section under shear.

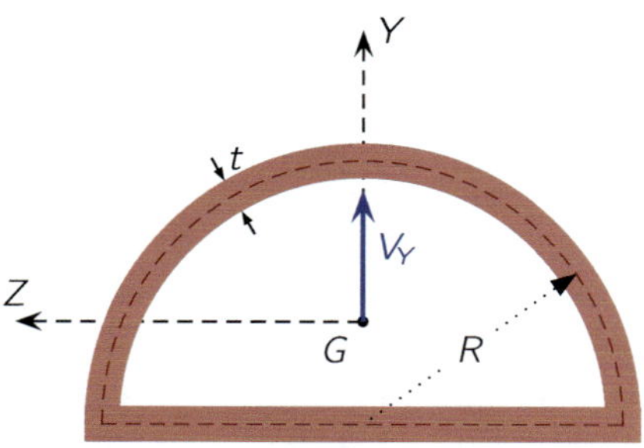

Figure 9.43 Thin rotated D channel.

Problem 9.15 (Symmetric Shearing of a D-Shaped Tube) *A thin-walled close section with a shape of rotated D is under the action of a shear force, as shown in Figure 9.43. (i) Find the position of the centroid G and the area S. (ii) Calculate the second moment of area I_Z. (iii) Determine the shear flow in the walls. (iv) Calculate the reduced section coefficient.* ⋆

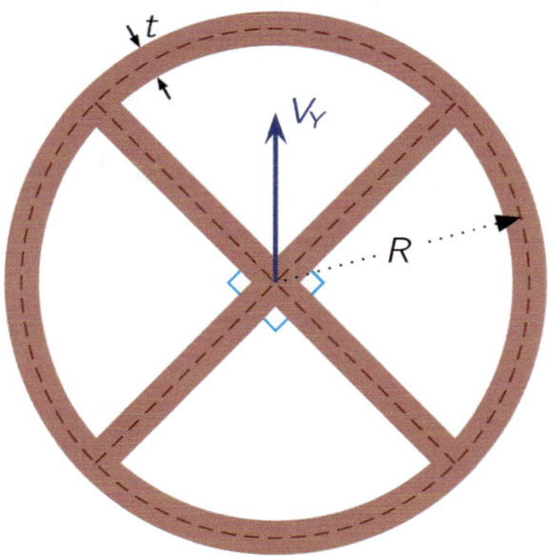

Figure 9.44 Multi-cell tube under shear.

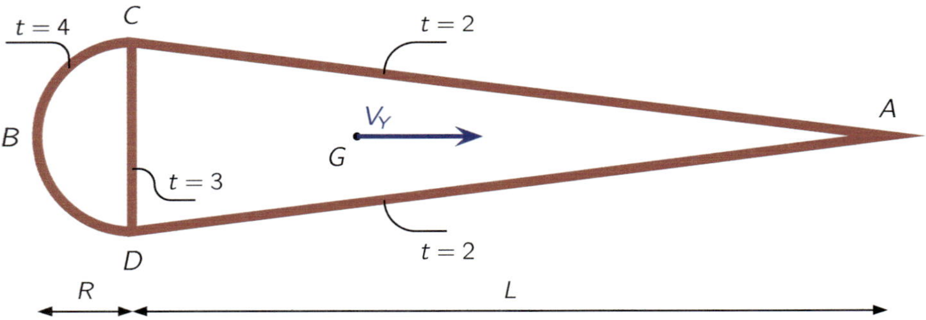

Figure 9.45 Elevator under shear.

Problem 9.16 (Symmetric Shearing of a Multi-Cell Closed Tube) *Consider the multi-cell section of Figure 9.44. The walls have uniform thickness t, small before the tube radius R. Under the action of transverse shear V_Y, determine the shear flow in the walls.* ⋆

Problem 9.17 (Analysis of an Empennage) *The elevator depicted in Figure 9.45 in a planar section, is subjected to a shearing load V_Y, as shown. The shear flows in the walls DAC, $\widehat{CBD}$ and DC are denoted ϕ_1, ϕ_2, ϕ_3, respectively. All dimensions are given in mm. Also, $L = 600$ and $R = 75$.*

(i) Find the position of the centroid G and the area S. (ii) Calculate the second moments of area I_Y and I_Z. (iii) Calculate the shear flows ϕ_1, ϕ_2, ϕ_3. ⋆

Problem 9.18 (Empennage under Nonsymmetric Shear) *Consider an elevator idealized as depicted in Figure 9.45 of Problem 9.17 but now having uniform thickness t and subjected to vertical shear V_Z. Calculate the shear flow in the walls.* ⋆

Problem 9.19 (Shear Center of an Open Circular Tube) *Consider the thin open circular profile depicted in Figure 9.36 of Problem 9.8 under the action of a transverse shearing load V_Z. Find the position of the shear center C.* ⋆

Problem 9.20 (Compatibility in General Bending) *(i) Show that Equation* (9.7) *is obtained by setting $i = x,\ j = y,\ k = z,\ l = y$ then $i = x,\ j = z,\ k = z,\ l = y$ in Equation* (1.18). *(ii) Work out the details for obtaining Equation* (9.8) *using Hooke's law and equilibrium, both specialized to the general Saint-Venant problem of bending.* ⋆

Problem 9.21 (Shearing of a Solid Circular Section) *Consider a prismatic beam having a full circular cross-section with radius R, as shown in Figure 9.4 of Example 9.1. Within the framework of Saint-Venant's theory, determining the solution amounts to finding the stress function $\varphi(y,z)$, which obeys a Dirichlet boundary-value problem, Equation* (9.20)*, with V_Y the shear force, I_Z the second moment of area, and ν Poisson's ratio; see Section 9.1.6.*

1. *Write the equation defining the contour Γ of the cross-section.*
2. *Calculate the Laplacian of the stress function φ given by Equation* (9.21) *of Example 9.1. Conclude.*
3. *Show that the above stress function verifies boundary condition $(9.20)_2$:*
4. *Determine the shear stresses τ_{xy} and τ_{xz} at a given point of the cross-section having coordinates (y,z).*
5. *Find the location and magnitude of the maximum shear stress.*
6. *Determine the warping function $\chi(y,z)$.*
7. *Complete the resolution to find the strains, rotations and displacements.* ⋆

Problem 9.22 (Shearing of an Auxetic Beam) *(i) Consider a beam with a circular cross-section made up of a linear elastic material with Poisson's ratio $\nu = -0.5$. Evaluate the shear distribution along the horizontal diameter. Compare with the case of a standard material having $\nu = 0.3$. Conclude. (ii) More generally, study the variation of the shear stress along the horizontal diameter for the class of auxetic materials (materials with negative Poisson's ratio).* ⋆

Problem 9.23 (Shearing of a Rectangular Section) *Using the stress function given in Example 9.2 determine the expressions of the shear stresses τ_{xy} and τ_{xz}. Give the expression of maximum shear and simplify it in the case of $b \ll a$.*

Answer

$$\tau_{XY} = \frac{V_Y}{2I_Z}\left(a^2 - Y^2\right) + \frac{8}{\pi^3}\frac{\nu}{1+\nu}\frac{b^2 V_Y}{I_Z}\sum_{m=0}^{\infty}\sum_{n=1}^{\infty}\frac{(-1)^{m+n}\cos\frac{(2m+1)\pi Y}{2a}\cos\frac{n\pi Z}{b}}{(2m+1)\left[(2m+1)^2\frac{b^2}{4a^2}+n^2\right]}$$

$$\tau_{XZ} = \frac{4}{\pi^3}\frac{\nu}{1+\nu}\frac{b^3 V_Y}{aI_Z}\sum_{m=0}^{\infty}\sum_{n=1}^{\infty}\frac{(-1)^{m+n}\sin\frac{(2m+1)\pi Y}{2a}\sin\frac{n\pi Z}{b}}{n\left[(2m+1)^2\frac{b^2}{4a^2}+n^2\right]}$$

⋆

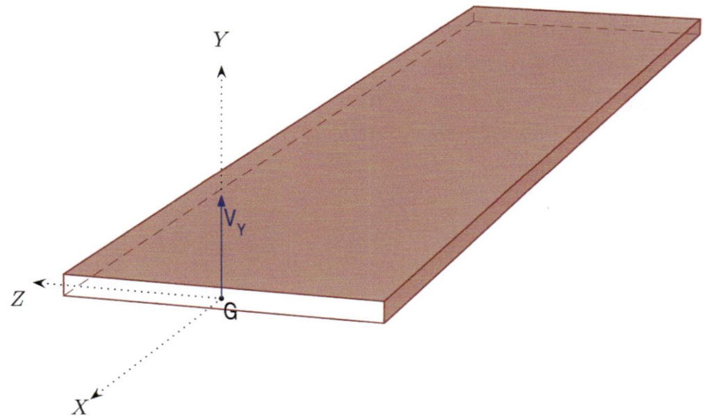

Figure 9.46 Plate under transverse shear.

Problem 9.24 (Limitation of Thin-Wall Approximation) *Consider the thin rectangular plate of Figure 9.46 under the action of transverse shear V_Y, as depicted in the figure. Using the theory of thin-walled tubes in Section 9.6, determine the shear flow in the plate. Conclude.* ⋆

Problem 9.25 (Rectangular Plate) *Consider a rectangular plate, as shown in Figure 9.46. Study the location and orientation of shear stress maxima. Hint: See Figure 9.8.* ⋆

Problem 9.26 (Maximum Deflection in Bending with Shear) *Consider a simply supported beam under its own weight, as in Example 9.4, and let k denote the reduced section coefficient (the shape of the cross-section need not be specified). (i) Show that the bending and shear deflection components v^M and v^V, taken separately, are each maximum at mid-span $X = L/2$, L being the length of the beam. (ii) Together, the two components make up the total deflection v. Show that v is maximum at mid-span, regardless of the shape of the cross-section.* ⋆

Problem 9.27 (Reduced Section Coefficient for Semi-Circular Beam) *Consider the semi-circular thin-walled section of Figure 9.16 treated in Example 9.6. Calculate the reduced section coefficient of the profile. Comment on the result.* ⋆

Problem 9.28 (Topology of Multi-Cell Sections) *If C, B and N denote the number of cells, branches and nodes, respectively, show that (i) $B > N$; (ii) $B > C$. (iii) What relationship is there, if any, between C and N? (iv) Show that $C > B - N$. (v) Is there an explicit formula relating C, B and N?* ⋆

Problem 9.29 (Timoshenko Beam Theory) *As an alternative to the theory of elasticity approach to beam bending with shear, Timoshenko's theory proceeds from relaxing the first and last of the following three assumptions regarding the current* *cross-section: ($\mathcal{A}_1$) it is infinitely rigid in its own plane; ($\mathcal{A}_2$) it remains planar after deformation; ($\mathcal{A}_3$) it remains normal to the centroidal axis of the beam. For reference, the above assumptions are those of the classical Euler–Bernoulli theory; see Problem 7.3. Under the action of a bending moment M about the z-axis and transverse shear V, show the following propositions in succession.*

1. *The axial displacement is given by*

$$u(x,y,z) = -y\theta_z \neq -y\frac{\mathrm{d}v_G}{\mathrm{d}x} \tag{9.67}$$

2. *If the slope is written as*

$$\begin{cases} \dfrac{\mathrm{d}v_G}{\mathrm{d}x} = \theta_z - \dfrac{1}{\kappa GS}\dfrac{\mathrm{d}M}{\mathrm{d}x} \\ \dfrac{\mathrm{d}\theta_z}{\mathrm{d}x} = \dfrac{M}{EI} \end{cases} \tag{9.68}$$

then the shear coefficient κ is the inverse of the reduced section coefficient k.

Above, $\theta_z(x)$ denotes the overall rotation of the section, taken on average and assumed to be small, $v_G(x)$ is the y displacement of the centroid, E is Young's modulus and $I \equiv I_z$ is the second moment of area. Show that the theory is consistent provided an effective shear strain defined through

$$\varepsilon_{xy} = \frac{1}{2}\kappa\left(\frac{\partial u}{\partial y} + \frac{\partial v}{\partial x}\right) \tag{9.69}$$

Discuss and comment. ⋆

10 Combined Loading

The previous four chapters considered separate analyses of stresses, strains, displacements, and rotations related to each of the four basic types of loading that can operate on a beam's cross-section: axial force, bending moment, torque, and transverse shear. The general case where all types of loading are eventually applied at once is now examined.

10.1 STRESSES AND STRAINS

Consider an arbitrary beam, a portion of which is depicted in Figure 10.1. The current cross-section S, centered about G, admits GX, GY, GZ as principal axes. Let dS denote the section on S of a longitudinal fiber and $P(Y,Z)$ the center of dS.

On dS applies a surface force $\vec{t}\,\mathrm{d}S$, the traction vector $\vec{t}$ having a normal component $\vec{\sigma}$ (due to the axial force and bending moment) and a tangential component $\vec{\tau}$ (due to the torque and transverse shear), as illustrated in Figure 10.1. For $\sigma = \sigma_X$ one has the very simple linear expression:

$$\sigma = \frac{N}{S} - \frac{M_Z}{I_Z}Y + \frac{M_Y}{I_Y}Z \tag{10.1}$$

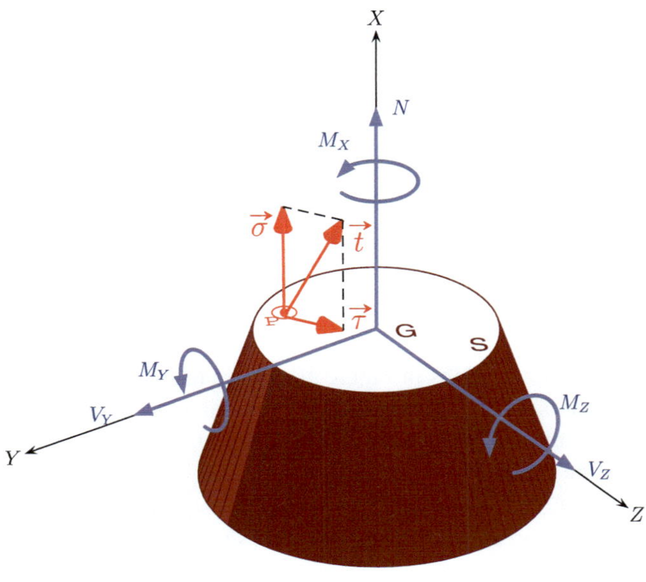

Figure 10.1 Screwer on current cross-section.

DOI: 10.1201/9781003537229-12

where I_Y and I_Z stand for the usual second moments of area of S about the Y- and Z-axes, respectively.

For the shear traction $\vec{\tau}(\tau_{XY}, \tau_{XZ})$, on the other hand, no simple general expression is available in terms of the coordinates of P. In the important case of thin-walled structures, finding the field $\vec{\tau}$ is relatively easy, by means of the shear flux and circulation. In the other cases, τ_{XY} and τ_{XZ} are determined by means of a Saint-Venant function $\varphi(Y,Z)$ or a warping function $\chi(Y,Z)$.

The extensions ε_X, ε_Y, ε_Z of the fiber in P are proportional to σ_X with $\varepsilon_X = \sigma_X/E$ and $\varepsilon_Y = \varepsilon_Z = -\nu\varepsilon_X$.

The shear strains γ_{XY} and γ_{XZ} are proportional to τ_{XY} and τ_{XZ} with $2\gamma_{XY} = \tau_{XY}/G$, $2\gamma_{XZ} = \tau_{XZ}/G$ where $G = \mu = \frac{E}{2(1+\nu)}$.

10.2 DISPLACEMENTS AND STIFFNESS MEASURES

Consider a thin slice of the beam of thickness ds, comprised between cross-sections S and S'. Let us imagine that the slice is decomposed into infinitely thin sheets, as illustrated in Figure 10.2, and examine the displacements (translations and rotations) of S' relative to S:

- Under the action of normal force N, Figure 10.2a, all the sheets have their thickness uniformly multiplied by $(1+\varepsilon_X)$ with $\varepsilon_X = \frac{N}{ES}$ where E is Young's modulus. Section (S') undergoes a translation, parallel to the longitudinal axis, of vector-director $\mathrm{d}\mathbf{\Lambda} = \frac{1}{ES}\mathbf{N}\,\mathrm{d}s$.
- Under the action of bending moment M_Z, Figure 10.2b, the sheets rotate relative to each other, about axes parallel to GZ, such that certain parts contract while others extend. Section (S') undergoes a rotation of angle: $\mathrm{d}\boldsymbol{\omega} = \frac{1}{EI_Z}\mathbf{M}_Z\,\mathrm{d}s$.
- Under the action of the torque $\mathbf{M}^t$, Figure 10.2c, the sheets rotate relative to each other about the axis of twist. Section (S') undergoes a rotation of angle: $\mathrm{d}\boldsymbol{\omega} = \frac{1}{GJ}\mathbf{M}^t\,\mathrm{d}s$, where G is the shear modulus
  Under the action of transverse shear V_Y, Figure 10.2d, the sheets glide past each other, parallel to axis GY. Section (S') undergoes a translation, parallel to GY, of vector-director $\mathrm{d}\mathbf{\Lambda} = k_Y\frac{1}{GS}\mathbf{V}_Y\,\mathrm{d}s$ where k_Y denotes the reduced section coefficient and G as above.

To simplify the drawings, the cross-sections in Figure 10.2 were taken to be rectangular such that the axis of twist is identified with the longitudinal axis GX. Furthermore, only the overall displacement of the sheets was considered, leaving out their potential warpings and in-plane deformations.

Recall that the coefficients ES, EI_Z, GJ, GS/k_Y denote the stiffnesses per unit length of the beam under normal force, bending, torsion, and shear, respectively.

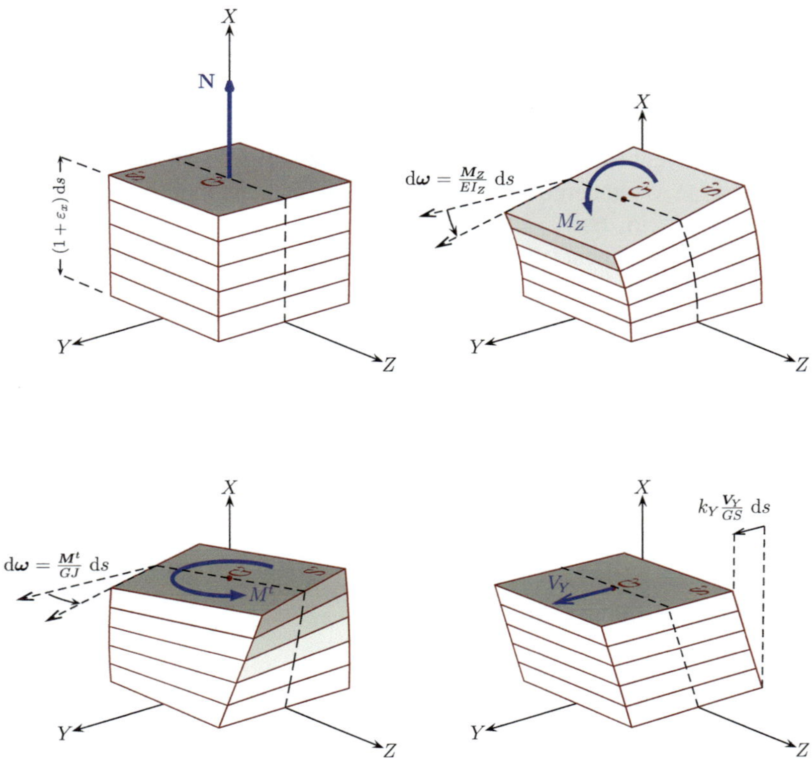

Figure 10.2 Translations and rotations induced by the four types of solicitation in beam theory (a–d).

10.3 BRESSE'S RELATIONS

The basic deflection second-order or fourth-order differential equations are valid for straight beams. They do not apply in general to more complex structures, which are subjected to combined loading. This is in particular the case for isolated curved beams and more generally for an assemblage of beams, whether straight or not.

Bresse's relations permit to express the displacements (translations and rotations) of a cross-section, say S_2, in terms of:

- on the one hand, the displacements of another cross-section, say S_1;
- on the other hand, the screwer (i.e., internal resultants) prevailing on all cross-sections S located between S_1 and S_2.

Consider, for instance, a beam of arbitrary shape, Figure 10.3. Its median line (centroidal axis) is oriented in some way, as shown. Let S_1 be a cross-section located at arclength s_1 and S_2 another cross-section located at s_2 with $s_2 > s_1$. Also, S denotes the current cross-section between S_1 and S_2, s its curvilinear abscissa, $\{G, XYZ\}$ its central principal frame, and C its center of twist. Let S' denote a neighboring cross-section at $s + \mathrm{d}s$. The laboratory (global) frame of reference is denoted $\{O,xyz\}$. It is emphasized that S_1 and S_2 are not necessarily the end sections of the beam. The beam shown is connected to the ground (reference solid) either directly or by means of other structural elements. Under the action of loading, there arises on S the screwer $\{\mathcal{S}\} = \{N, V_Y, V_Z, M_X, M_Y, M_Z\}$. Here, M_X is the longitudinal moment, which is related to, but different from, the torque M^t; see Equation (8.60) in Section 8.7.

Sections S_1, S, S_2 undergo the rotations $\boldsymbol{\omega}_1$, $\boldsymbol{\omega}$, $\boldsymbol{\omega}_2$, and their centers G_1, G, G_2 the translations $\boldsymbol{\Lambda}_1$, $\boldsymbol{\Lambda}$, $\boldsymbol{\Lambda}_2$.

An overall displacement (rigid; with no deformation) of portion S_1–S_2 gives:

$$\boldsymbol{\omega}_2 = \boldsymbol{\omega}_1 \qquad \text{and} \qquad \boldsymbol{\Lambda}_2 = \boldsymbol{\Lambda}_1 + \boldsymbol{\omega}_1 \times \overrightarrow{G_1G_2}$$

When portion S_1–S is taken to be fixed, pure deformation of the sole slice S–S' (refer to Figure 10.3) induces on S_2

- a rotation $\mathrm{d}\boldsymbol{\omega}$ such that

$$\frac{\mathrm{d}\boldsymbol{\omega}}{\mathrm{d}s} = \frac{\mathbf{M}_Y}{EI_Y} + \frac{\mathbf{M}_Z}{EI_Z} + \frac{\mathbf{M}^t}{GJ}$$

- a translation $\mathrm{d}\boldsymbol{\Lambda}_2$ of G_2 such that

$$\frac{\mathrm{d}\boldsymbol{\Lambda}_2}{\mathrm{d}s} = \frac{\mathbf{N}}{ES} + k_Y\frac{\mathbf{V}_Y}{GS} + k_Z\frac{\mathbf{V}_Z}{GS} + \left(\frac{\mathbf{M}_Y}{EI_Y} + \frac{\mathbf{M}_Z}{EI_Z}\right) \times \overrightarrow{GG_2} + \frac{\mathbf{M}^t}{GJ} \times \overrightarrow{CG_2}$$

Under the action of the overall displacement and the deformation of the entire S_1–S_2 portion, it follows then:

$$\boldsymbol{\omega}_2 = \boldsymbol{\omega}_1 + \int_{s_1}^{s_2} \left(\frac{\mathbf{M}_Y}{EI_Y} + \frac{\mathbf{M}_Z}{EI_Z} + \frac{\mathbf{M}^t}{GJ}\right) \mathrm{d}s \tag{10.2}$$

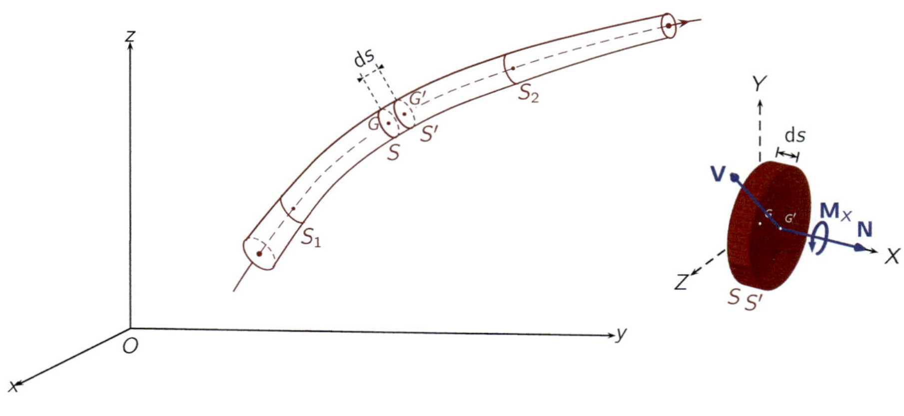

Figure 10.3 Cross-sections of an arbitrary beam entering Bresse's relations.

$$\boldsymbol{\Lambda}_2 = \boldsymbol{\Lambda}_1 + \boldsymbol{\omega}_1 \times \overrightarrow{G_1G_2} + \int_{s_1}^{s_2} \left(\frac{\mathbf{N}}{ES} + k_Y \frac{\mathbf{V}_Y}{GS} + k_Z \frac{\mathbf{V}_Z}{GS} \right) \mathrm{d}s \tag{10.3}$$
$$+ \int_{s_1}^{s_2} \left[\left(\frac{\mathbf{M}_Y}{EI_Y} + \frac{\mathbf{M}_Z}{EI_Z} \right) \times \overrightarrow{GG_2} + \frac{\mathbf{M}^t}{GJ} \times \overrightarrow{CG_2} \right] \mathrm{d}s$$

The vectorial equations (10.2) and (10.3) are known as Bresse's relations; the first concerns the rotations, whereas the second concerns the translations. They are applicable only if the assumption of small displacements is verified on the part S_1–S_2.

Bresse's relations allow to determine $\boldsymbol{\omega}_2$ and $\boldsymbol{\Lambda}_2$ when $\boldsymbol{\omega}_1$ and $\boldsymbol{\Lambda}_1$ are known along with the screwer.

Bresse's relations also allow to lift the indeterminacy in hyperstatic problems when $\boldsymbol{\omega}_1$, $\boldsymbol{\omega}_2$ or (and) $\boldsymbol{\Lambda}_1$, $\boldsymbol{\Lambda}_2$ are known.

10.4 OSSATURES

An assemblage of beams defines an ossature. A general method of calculation of ossatures is now presented, which applies to the simplest of cases (one beam) as well as the most complex ones. The method is presented in five generic steps.

10.4.1 GEOMETRY

The first step consists of analyzing the geometry of the ossature. This involves the global frame of reference, the median lines of all beams and their cross-sections.

1. **Global Frame:** Define an orthonormal, global frame of reference $\{O, xyz\}$, related to the structure and in which the reactions and the displacements will later be expressed.
2. **Median Lines:** For each beam of the ossature, the geometry of the median line is determined by providing:
   - the coordinates of current point G in terms of a parameter;
   - the radii of curvature and torsion in G;
   - Frenet's frame $\{G, \mathbf{t}, \mathbf{n}, \mathbf{b}\}$ in G.
3. **Cross-Sections of the Beams:** For each cross-section, one needs to determine:
   - the principal axes GY and GZ, and the washout angle φ;
   - the position of the center of twist C;
   - the local stiffnesses: ES, $\frac{GS}{k_Y}$, $\frac{GS}{k_Z}$, EI_Y, EI_Z, GJ.

10.4.2 STATICS

The second step involves examining links and applied loads and writing equilibrium statements.

1. **Analysis of Links:** For each link, joint and connection (internal or external), one notes:
   - the degrees of freedom allowed;

  - the degrees of freedom blocked;
  - the non-zero components (forces and moments) at all links;
  - in the case of elastic links, the stiffnesses (i.e., the ratios load/displacement, in general force/translation or couple/rotation);
  - the number of independent scalar reactions to be determined.
2. **Analysis of Loads:** One enumerates the given external forces, which constitute the loading applied to the structure.
3. **Equilibrium:** One writes the equations in forces and moments:
  - of global equilibrium of the ossature;
  - of equilibrium of each beam.

In the isostatic case (statically determinate ossature), one infers at this stage the expressions of all reactions, internal and external, in terms of the sole external forces (loading).

In the hyperstatic case (internally of order p, externally of order q), one expresses all reactions in terms of the given, applied loads and of $(p+q)$ reactions chosen among them as hyperstatic unknowns.

10.4.3 SCREWERS

On the current cross-section S of each beam, one expresses the six components of the screwer, i.e., the internal resultants, namely N, V_Y, V_Z, M_X, M_Y, and M_Z in terms of the loading and, eventually, the hyperstatic unknowns; one then infers the expression of the moment of torsion M^t using the moment transport equation.

Recall the two methods that permit to obtain the screwer:

- the direct method states that the screwer on S equals the torsor of all forces applied to the beam downstream from S (or equivalently, equals minus the torsor of all forces applied to the beam upstream from S);
- the indirect method consists of integrating the system of local equilibrium equations of the beam, Equation (5.43) of Chapter 5.

10.4.4 HYPERSTATIC UNKNOWNS

Various methods are available which allow to lift the indeterminacy (Menabrea's theorem, Bresse's relations, ...) All they do is express the conditions imposed on certain displacements (translations or rotations) by the redundant links. The examples given below will illustrate the use of Bresse's relations to this end. In many cases, however, energy methods are more efficient, as will be illustrated for the same examples elsewhere.

10.4.5 STRESSES, STRAINS, DISPLACEMENTS

After determining the unknown hyperstatic reactions, the components of the screwer can be expressed in terms of the applied loading. Next, the stresses introduced on the cross-sections by each component of the screwer can be inferred using methods from the previous four chapters. Hooke's law then permits the strains to be determined anywhere in the structure.

The translations and rotations undergone by each cross-section can finally be calculated using Bresse's relations; see Section 10.3. Alternatively, the translations and rotations may be obtained using Castigliano's theorem, when appropriate, or the differential equations relating displacements and screwer components.

10.5 PLANAR OSSATURES

Assume that the median lines of the beams lie in the same plane (xOy) that is also a plane of symmetry for both the ossature and the links (or connections). Furthermore, assume that the loading is confined to that plane. In this case, the problem is said to be planar.

The reactions at each link have their resultant force in (xOy) and their resultant moment perpendicular to it.

For the entire ossature, or any subpart thereof, there are at most three independent scalar equilibrium equations:

- force equilibrium along x and y;
- moment equilibrium along z.

Taking the principal axes GX and GY of the current cross-section S to lie in the plane of symmetry (xOy), Figure 10.4, one can assert that screwer components V_Z, M_X, M_Y are nil, hence that M^t is also nil. The three relevant equilibrium equations are then obtained from Equation (5.46) as follows:

$$p_X + \frac{\mathrm{d}N}{\mathrm{d}s} - \frac{V}{R} = 0\,; \qquad p_Y + \frac{N}{R} + \frac{\mathrm{d}V}{\mathrm{d}s} = 0\,; \qquad m + \frac{\mathrm{d}M}{\mathrm{d}s} + V = 0 \tag{10.4}$$

Integration of the above equations permits the determination of the axial force N, transverse shear $V = V_Y$ and the bending moment $M = M_Z$.

The current cross-section S of a beam undergoes a translation $\boldsymbol{\Lambda}$ parallel to (xOy) and a rotation $\boldsymbol{\omega}$ about an axis parallel to Oz.

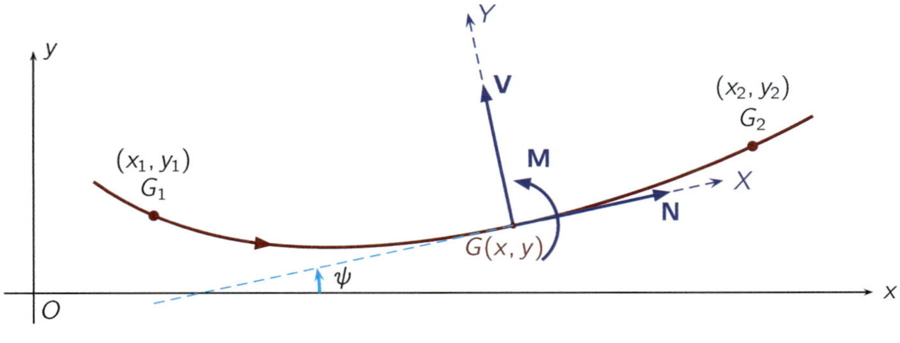

Figure 10.4 Definition of terms entering Bresse's relations for plane ossatures.

Bresse's relations (10.2) and (10.3) are now written by projecting on the axes for any beam in the ossature:

$$\omega_2 = \omega_1 + \int_{s_1}^{s_2} \frac{M}{EI}\,\mathrm{d}s$$

$$u_2 = u_1 - \omega_1(y_2 - y_1) + \int_{s_1}^{s_2} \left[\frac{N\cos\psi}{ES} - k\frac{V\sin\psi}{GS} + (y_2 - y)\frac{M}{EI}\right]\mathrm{d}s \qquad (10.5)$$

$$v_2 = v_1 + \omega_1(x_2 - x_1) + \int_{s_1}^{s_2} \left[\frac{N\sin\psi}{ES} + k\frac{V\cos\psi}{GS} + (x_2 - x)\frac{M}{EI}\right]\mathrm{d}s$$

where u and v denote the components along x and y of the displacement $\mathbf{\Lambda}$ of G, s_1 and s_2 the arclengths of extreme sections S_1 and S_2 of the beam portion considered, with $s_1 < s_2$. Also, as shown in Figure 10.4, ψ denotes the angle $(\overrightarrow{Ox}, \overrightarrow{GX})$. Finally, $k = k_Y$ is the reduced section coefficient.

10.6 EXAMPLES

Some planar problems of beams and ossatures that are quite often encountered in practice are analyzed in this section.

Example 10.1: Ring Dynamometer

A ring of constant cross-section S has a circular median line centered about O, of radius R and lying in the plane of symmetry (xOy). It is subjected to two diametrically opposed forces each of intensity F, as shown in Figure 10.5, which provides the notations and axes.

Owing to the symmetry of the problem with respect to the x-axis, the analysis is restricted to the top half of the ring, ABC. The ring being a closed beam, the problem is internally hyperstatic. This means that one should first determine the screwer on one cross-section in order to determine it then on all others. For this plane problem, the non-zero internal resultants are: N, $V = V_Y$, $M = M_Z$.

Let us first determine the screwer on the section at C, denoted S_C. Cut the ring following S_A and S_C then isolate the semi-ring ABC, as illustrated in Figure 10.5.

Symmetry about the x-axis implies no transverse shear on either S_A or S_C.

Equilibrium along the y-axis of the semi-ring imposes on both S_A and S_C a normal force equal to $F/2$.

Finally, symmetry with respect to the y-axis implies equality of the bending moments acting on S_A and S_C (of opposite sign). Let Γ denote this bending moment, which thus constitutes the hyperstatic unknown of the problem.

Orienting the median line as shown in the figure, the internal resultants are obtained as follows:

$$N = \frac{F}{2}\cos\theta\,; \qquad V = -\frac{F}{2}\sin\theta\,; \qquad M = \frac{FR}{2}(1 - \cos\theta) + \Gamma \qquad (10.6)$$

on part AB $(0 < \theta < \frac{\pi}{2})$, and

$$N = -\frac{F}{2}\cos\theta\,; \qquad V = \frac{F}{2}\sin\theta\,; \qquad M = \frac{FR}{2}(1 + \cos\theta) + \Gamma \qquad (10.7)$$

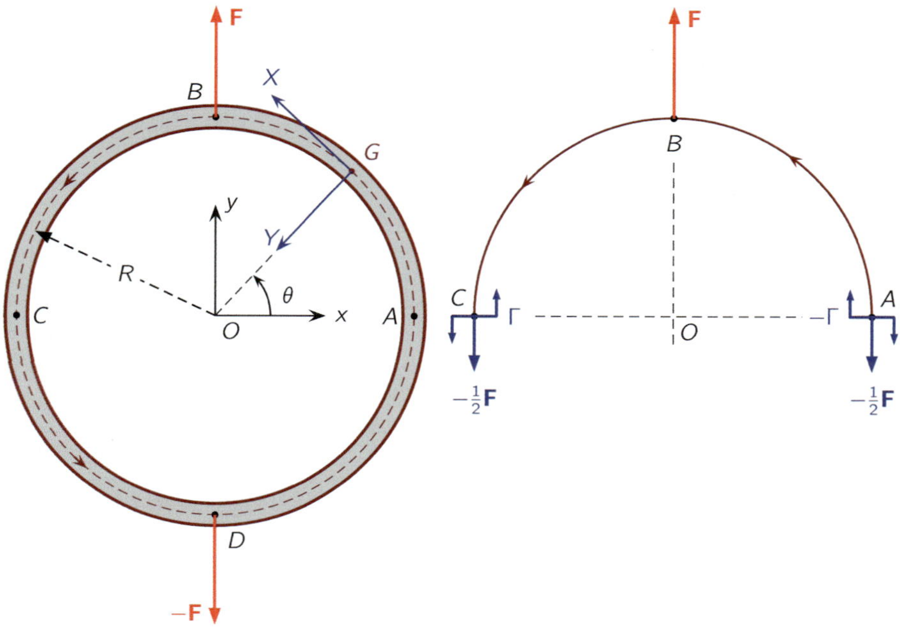

Figure 10.5 (a) Dynamometer under load. (b) Ring cut open along diameter.

on part BC ($\frac{\pi}{2} < \theta < \pi$).

Bresse's relations are useful in lifting the indeterminacy. From symmetry, sections S_A and S_B do not rotate. Applying the first of Bresse's relations (10.5) for plane problems to portion AB, with $\omega_1 = \omega_A = 0$ and $\omega_2 = \omega_B = 0$, it follows:

$$\int_0^{\frac{\pi}{2}} M\,\mathrm{d}\theta = 0 = \int_0^{\frac{\pi}{2}} \left[\frac{FR}{2}(1-\cos\theta)+\Gamma\right]\mathrm{d}\theta$$

from which one gets

$$\Gamma = -M^A = M^C = -\frac{\pi-2}{2\pi}FR = -0.1817\,FR$$

Having determined Γ, Equations (10.6) and (10.7) fully determine the screwer on all sections of the ring. To find the deformed shape of the ring, one would need to integrate the coupled differential equations (7.23) established in Chapter 7. An exercise of this sort is proposed under Problem 10.17. Here, it suffices to illustrate how the second and third of Bresse's planar relations (10.5) applied to the same portion AB of the ring lead to

$$\begin{cases} u^A = -\dfrac{4-\pi}{4\pi}\dfrac{R^3F}{EI} + \left(\dfrac{1}{E}-\dfrac{k}{G}\right)\dfrac{RF}{4S} = -0.068\,\dfrac{R^3F}{EI} + 0.25\left(\dfrac{1}{E}-\dfrac{k}{G}\right)\dfrac{RF}{S} \\ v^B = \dfrac{\pi^2-8}{8\pi}\dfrac{R^3F}{EI} + \dfrac{\pi}{8}\left(\dfrac{1}{E}+\dfrac{k}{G}\right)\dfrac{RF}{S} = 0.074\,\dfrac{R^3F}{EI} + 0.393\left(\dfrac{1}{E}+\dfrac{k}{G}\right)\dfrac{RF}{S} \end{cases} \tag{10.8}$$

having used $\psi = \theta + \frac{\pi}{2}$, $x = R\cos\theta$ and $y = R\sin\theta$. Also, $u^B = 0$ and $v^A = 0$ from symmetry considerations.

The thinner the ring, the more preponderant is the bending term (in $\frac{R^3F}{EI}$). ♡

Example 10.2: Analysis of a Frame

A square frame, of constant cross-section S, admits (xOy) as plane of symmetry. It is made up by welding four identical prismatic beams, each having length $2a$.

The frame is subjected to uniformly distributed opposite loads on sides $A'B'$ and $C'D'$, as shown in Figure 10.6, which gives additional notations and the axes used.

The problem is symmetric with respect to the x-axis and the y-axis. It suffices to analyze the top half of the frame in order to determine the non-zero components of the screwer, namely N, V, M.

Symmetry about the x-axis entails that the shears V are nil on sections S_A and S_C. In addition, force equilibrium along y gives the normal forces on the same sections: $N^A = N^C = -pa$. The bending moments $-\Gamma$ on S_A and Γ on S_C are unknown: the structure is one time hyperstatic.

First, express the internal resultants:

$$\text{on } AA' \begin{cases} N = -pa \\ V = 0 \\ M = \Gamma \end{cases} \quad \text{on } A'B' \begin{cases} N = 0 \\ V = px \\ M = \Gamma - \dfrac{1}{2}p(a^2 - x^2) \end{cases} \quad \text{on } B'C \begin{cases} N = -pa \\ V = 0 \\ M = \Gamma \end{cases} \tag{10.9}$$

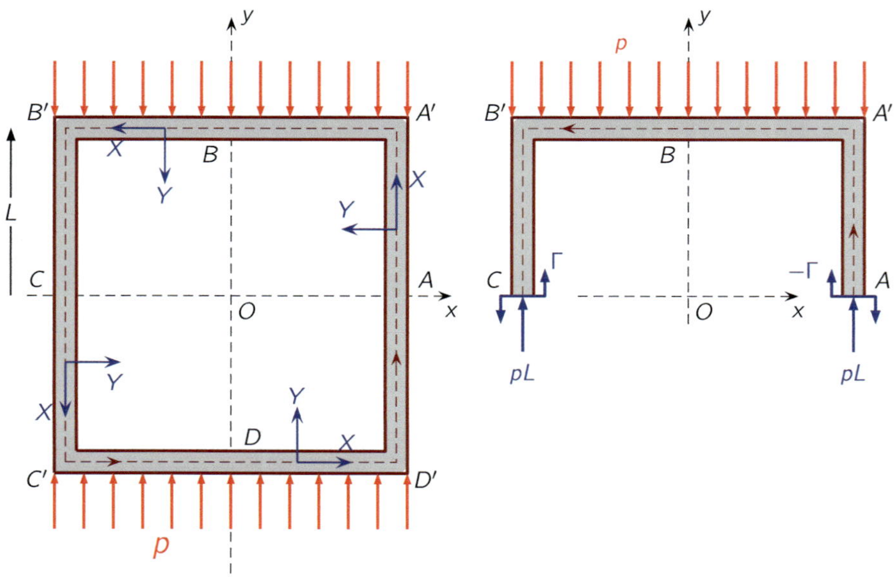

Figure 10.6 Square frame under uniform compression.

Then apply the first of Bresse's relations (10.5) to portion AB of the frame, with $\omega_1 = \omega_A = 0$ and $\omega_2 = \omega_B = 0$ by symmetry:

$$\int_{y=0}^{a} M\,\mathrm{d}y + \int_{x=a}^{0} M(-\mathrm{d}x) = 0 = \int_0^a \Gamma\,\mathrm{d}y + \int_0^a \left[\Gamma - \frac{1}{2}p(a^2 - x^2)\right]\mathrm{d}x$$

from which one gets

$$\Gamma = \frac{1}{6}pa^2$$

Hence, the screwer on the current cross-section is finally

$$D'A' \begin{cases} N = -pa \\ V = 0 \\ M = \frac{1}{6}pa^2 \end{cases} \qquad A'B' \begin{cases} N = 0 \\ V = px \\ M = -\frac{1}{3}pa^2 + \frac{1}{2}px^2 \end{cases} \qquad C'D' \begin{cases} N = 0 \\ V = -px \\ M = -\frac{1}{3}pa^2 + \frac{1}{2}px^2 \end{cases} \tag{10.10}$$

Having so determined the internal resultants, one can calculate the stresses, then the strains by Hooke's law, and finally by Bresse's relations the displacements (translations and rotations) on each section S. ♡

Example 10.3: Analysis of a Gantry

Consider the gantry depicted in Figure 10.7, made up of three identical prismatic beams of length L, welded together. The figure also gives the axes, the links (one clamping, one support), the loading (one force $\mathbf{F}$); (xOy) is a plane of symmetry.

Global equilibrium of the gantry writes

$$X^O = -F\,, \quad Y^O + Y^C = 0\,, \quad M^O - LF + LY^C = 0 \tag{10.11}$$

where X^O and Y^O denote the components of the reaction $\mathbf{R}^O$ at O and M^O is the clamping moment at O.

The problem is therefore externally hyperstatic of degree 1. One may choose to work with Y^C as the hyperstatic unknown.

Using the direct method, it is straightforward to obtain the screwer as follows:

$$\text{on } OA \begin{cases} N = Y^C \\ V = -F \\ M = LY^C - (L-y)F \end{cases} \qquad AB \begin{cases} N = F \\ V = Y^C \\ M = (L-x)Y^C \end{cases} \qquad BC \begin{cases} N = -Y^C \\ V = 0 \\ M = 0 \end{cases} \tag{10.12}$$

To determine Y^C, consider applying the second of Bresse's relations (10.5) to the entire length of the gantry since the vertical displacements of both ends are restricted ($v_O = v_C = 0$) in addition to $\omega_O = 0$. One obtains:

$$\begin{aligned} v_C =& v_O + \omega_O(x_C - x_O) + \int_0^L \left[\frac{N}{ES} + (x_C - x)\frac{M}{EI}\right]\mathrm{d}y \\ &+ \int_0^L \left[k\frac{V}{GS} + (x_C - x)\frac{M}{EI}\right]\mathrm{d}x + \int_L^0 \left[-\frac{N}{ES} + (x_C - x)\frac{M}{EI}\right](-\mathrm{d}y) \end{aligned}$$

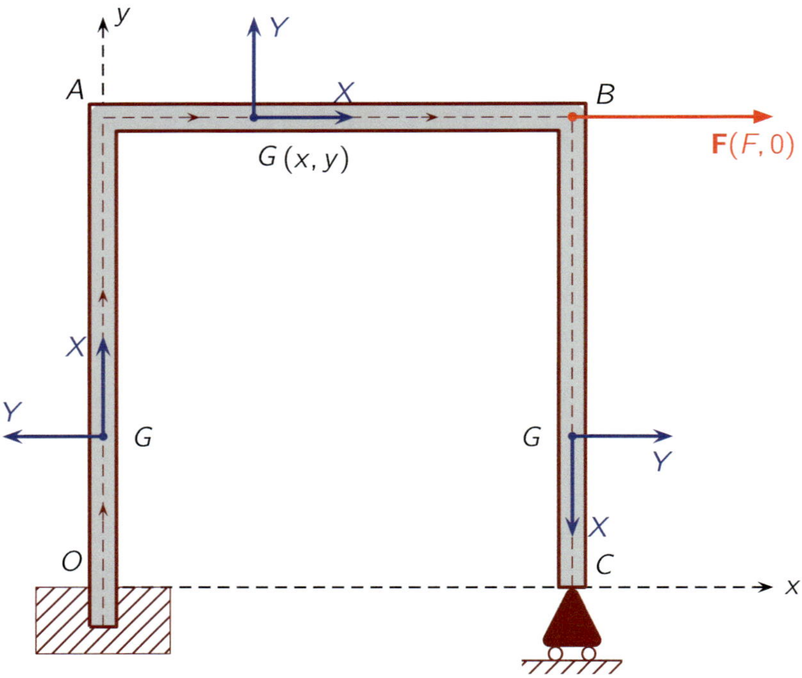

Figure 10.7 Gantry under a concentrated horizontal force.

The first integral corresponds to beam OA ($\psi = \frac{\pi}{2}$, $\mathrm{d}s = \mathrm{d}y$), the second to beam AB ($\psi = 0$, $\mathrm{d}s = \mathrm{d}x$), and the third to beam BC ($\psi = -\frac{\pi}{2}$, $\mathrm{d}s = -\mathrm{d}y$)

Injecting the screwer expressions in Equation (10.12) in the above integrals and noting that $x = 0$ on (OA), $x_C - x = 0$ on (BC), and $x_C = L$, it follows:

$$\frac{Y^C L}{ES} + \frac{L}{EI}\left(L^2 Y^C - \frac{L^2}{2}F\right) + k\frac{Y^C L}{GS} + \frac{1}{3}\frac{Y^C L^3}{EI} + \frac{Y^C L}{ES} = 0$$

This equation permits to determine Y^C. After simplification,

$$Y^C = \frac{3}{8}F \frac{1}{1 + \frac{3}{4}\frac{EI}{L^2 S}\left(\frac{2}{E} + \frac{k}{G}\right)} \tag{10.13}$$

Expressed this way, it becomes clear that the longer the beams ($SL^2 \gg I$), the more negligible before unity becomes the terms due to the axial force ($\frac{3}{2}\frac{I}{L^2 S}$) and to the transverse shear ($\frac{3}{4}k\frac{EI}{L^2 GS}$).

The hyperstatic character being lifted, one obtains the final expression of the screwer, the stresses, strains and displacements, using the methods of the previous four chapters. ♡

Example 10.4: Analysis of a Gallows

Consider the gallows depicted in Figure 10.8, made up of three prismatic beams, and admitting (xOy) as a plane of symmetry. The figure also gives the axes, the links and the loading (force $\mathbf{F}$). The mid-span connections at A and C, as well as the connection at B, are simple articulations (no moment transmission). The three beams have the same section S and the same bending stiffness (per unit length) EI. Beams OB and BD have each length $2L$, whereas beam AC has length $\sqrt{2}L$.

Global equilibrium gives $Y^O = F$ and $M^O = 2LF$, while $X^O = 0$, with the usual notations for the reactions and clamping moment at O; see Figure 10.9.

Since beam AC is simply articulated at its two ends, it must work in tension or compression, like a bar. With N^{AC} denoting the normal force in it, writing separately the equilibrium of beam OB gives (see Figure 10.9)

$$\begin{cases} X^B + \dfrac{\sqrt{2}}{2} N^{AC} = 0 \\ Y^O + Y^B + \dfrac{\sqrt{2}}{2} N^{AC} = 0 \\ M^O - L\dfrac{\sqrt{2}}{2} N^{AC} - 2LX^B = 0 \end{cases} \tag{10.14}$$

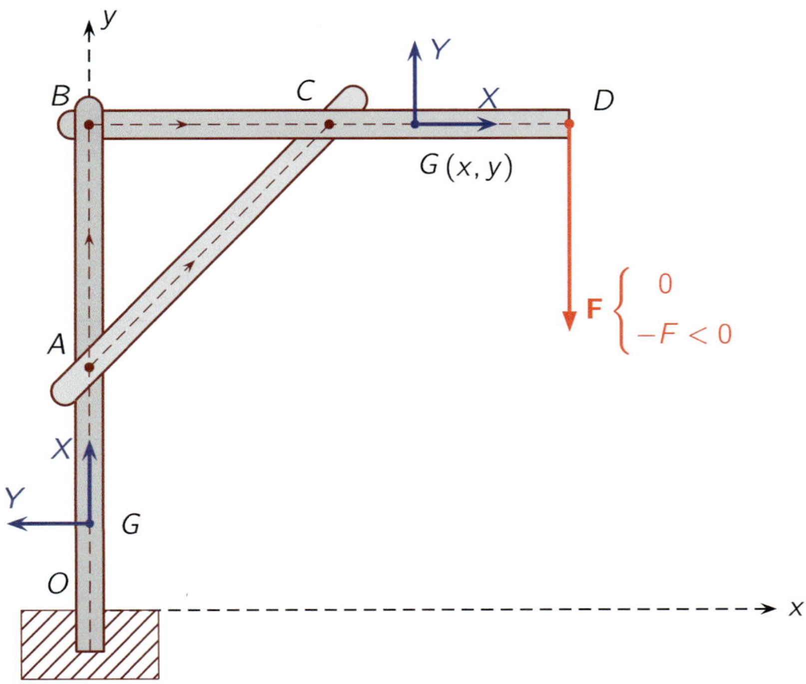

Figure 10.8 Three-beam structure with articulations.

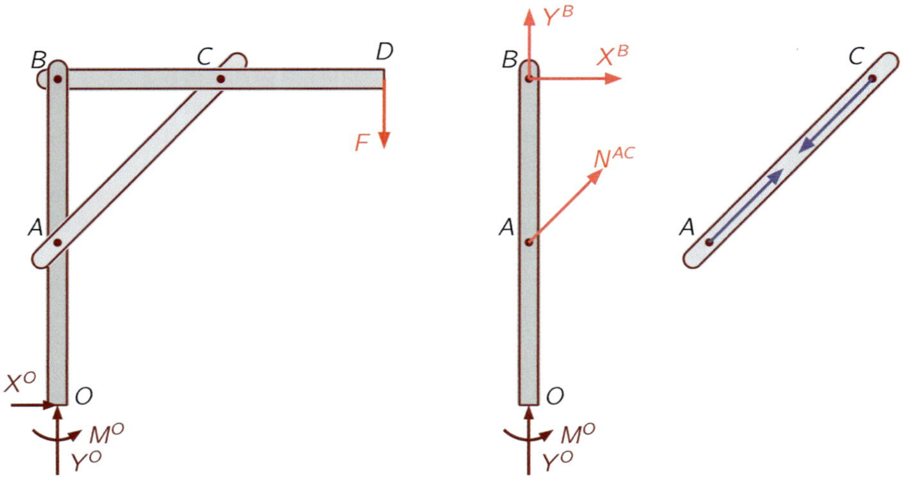

Figure 10.9 Free-body diagrams of entire structure, beam OB and beam AC.

where X^B and Y^B denote the resultant forces describing the action of beam BD on OB. Inserting the reactions at O then leads to

$$X^B = 2F\,, \quad Y^B = F\,, \quad N^{AC} = -2\sqrt{2}F \tag{10.15}$$

One can check that the equilibrium of member BD is then automatically verified.

The screwer on the current cross-section is then obtained by the direct method piecewise:

$$\text{on } OA \begin{cases} N = -F \\ V = 0 \\ M = -2LF \end{cases} \quad AB \begin{cases} N = F \\ V = -2F \\ M = -2F(2L-y) \end{cases} \quad AC \begin{cases} N = -2\sqrt{2}F \\ V = 0 \\ M = 0 \end{cases}$$

$$BC \begin{cases} N = 2F \\ V = F \\ M = -Fx \end{cases} \quad CD \begin{cases} N = 0 \\ V = -F \\ M = -(2L-x)F \end{cases} \tag{10.16}$$

The structure is statically determinate (isostatic), both internally and externally. Stresses and strains may then be evaluated everywhere, as explained in Section 10.1.

In principle, some particular deflections can be evaluated using Bresse's relations. For example, it is straightforward to show that point B does not move vertically, i.e., $v^B = 0$. Calculating the main deflection at D using Bresse's relations would be cumbersome, because the rotations at A, B and C are different in each beam. It will be shown elsewhere that the vertical displacement of point D can be obtained very quickly using an energy method. ♡

10.7 ELASTIC POTENTIAL ENERGY

The energy stored in a slice of a beam, at the level of cross-section S, is given per unit length as follows:

$$\frac{\mathrm{d}W}{\mathrm{d}s} = \frac{1}{2}\left(\frac{N^2}{ES} + k_Y\frac{V_Y^2}{GS} + k_Z\frac{V_Z^2}{GS} + \frac{M_Y^2}{EI_Y} + \frac{M_Z^2}{EI_Z} + \frac{M^{t\,2}}{GJ}\right) \tag{10.17}$$

The above expression is obtained by simply summing up the various energy densities obtained separately for each of the four types of loading, as derived in the previous four chapters.

There is no cross-term; each of the six components does work only against the displacement that it produces. This is due to the choice of principal axes for expressing the components of the screwer $\{\mathcal{S}\}$.

The energy stored by the beam between sections S_1 and S_2 is thus

$$W = \frac{1}{2}\int_{s_1}^{s_2}\left(\frac{N^2}{ES} + k_Y\frac{V_Y^2}{GS} + k_Z\frac{V_Z^2}{GS} + \frac{M_Y^2}{EI_Y} + \frac{M_Z^2}{EI_Z} + \frac{M^{t\,2}}{GJ}\right)\mathrm{d}s \tag{10.18}$$

This expression is useful when applying energy methods based on theorems such Castigliano's, Menabrea's and Maxwell–Betti's. These methods are presented in other specialized texts.

10.8 STRESS CONCENTRATIONS

The basic equations of beam theory, which allow to calculate the stresses at a point P, are only valid if that point is sufficiently far from:

- zones where concentrated loads are applied, external loads or reactions (Saint-Venant principle);
- zones over which abrupt variations of cross-sections occur.

These zones are called stress concentration zones, as they are locations of elevated stresses and large stress gradients.

DEFINITION 10.1 The stress concentration factor is the ratio:

$$K = \frac{\sigma^{\max}}{\sigma^{\text{nom}}}, \text{ for a normal stress} \tag{10.19}$$

$$K = \frac{\tau^{\max}}{\tau^{\text{nom}}}, \text{ for a shear stress} \tag{10.20}$$

$\sigma^{\max}$ and $\tau^{\max}$ denote the maximum normal and tangential stresses, respectively, in the stress concentration zone.

σ^{nom} and τ^{nom} denote the nominal normal and tangential stresses, respectively, at the same 'point', that is the stress calculated using beam theory. ♣

In practice, stress concentration factors are obtained:

- either by calculation (theory of elasticity, finite element method, ...);
- or by experiment (photoelasticity, Moiré, strain gauges, ...)

A thorough and practical treatise on the subject is due to R.E. Peterson. Some illustrative examples are borrowed from that book and modified as needed for the sake of clarity.

Example 10.5: Thin Rectangular Bar with Semi-Circular Notches

Let W be the width of the bar, w the width at the notch root, R the notch radius, Figure 10.10. Also, $t = (W - w)/2$ is the notch depth and $h \ll W$ is the notch thickness (not shown). Peterson's charts cover a wide range of conditions, including for arbitrary notch depths $t \neq R$, as depicted in the three-dimensional drawing of Figure 10.10. For semi-circular notches $t = R$. The end sections are loaded uniformly such that

$$\sigma = \frac{F}{Wh}$$

Beam theory would give on the mid-section (at $x = 0$) a uniform normal stress

$$\sigma_x = \frac{F}{wh} = \sigma^{\text{nom}}$$

In reality, the normal stress on the minimum-width section is not uniform. Its maximum is at the notch root where its value is $\sigma^{\max} = K\sigma^{\text{nom}}$. The factor K is given by the experimental curve of Figure 10.11.

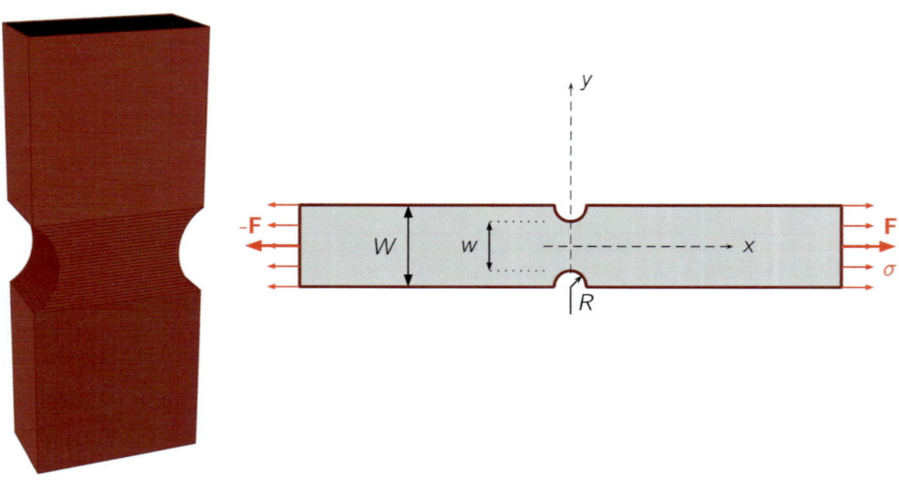

Figure 10.10 Notched rectangular bar under tension.

The behavior in the limit $2R/W \to 0$ corresponds to a high curvature notch. On the other hand, the limit $2R/W \to 1$ corresponds to a bar with a tiny ligament where the nominal stress diverges. What is key is that the maximum stress concentration factor is slightly over 3. ♡

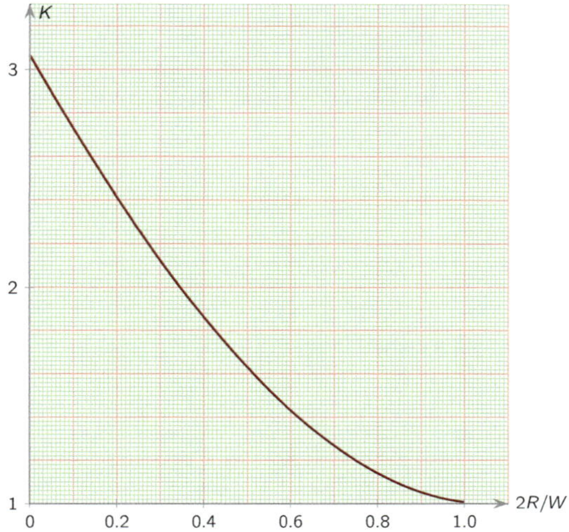

Figure 10.11 Stress concentration factor in tension of a notched rectangular bar versus the ratio of notch diameter to bar width.

Example 10.6: Thin Rectangular Plate with Central Circular Hole

Let W be the width of the plate, R the hole radius, Figure 10.12. Also, $h \ll W$ is the notch thickness (not shown).

The end sections are loaded uniformly such that

$$\sigma = \frac{F}{Wh}$$

Beam theory would give on the mid-section (at $x = 0$) a uniform normal stress

$$\sigma_x = \frac{F}{(W - 2R)h} = \sigma^{\text{nom}}$$

In reality, the normal stress on the minimum-width section is not uniform. Its maximum is at the two points highlighted in red where $\sigma^{\max} = K\sigma^{\text{nom}}$, Figure 10.12. The factor K is given by the experimental curve of Figure 10.13. Experimental data from Peterson's charts is available for $2R/W < 0.5$. The extrapolation (dashed) is based on calculations.

Here too, the maximum value of K is about 3, as in the bar with side notches of Example 10.5.

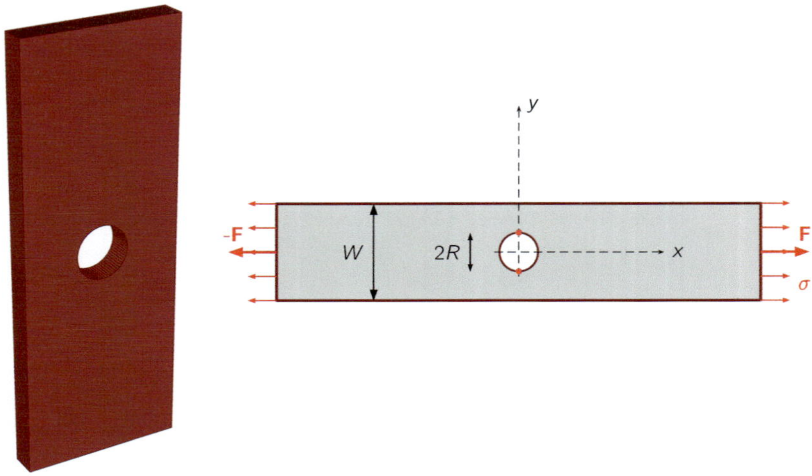

Figure 10.12 Thin plate with a central hole.

The following equation

$$K = 2 + \left(1 - \frac{2R}{W}\right)^3$$

provides a good approximation of the whole curve. ♡

Example 10.7: Solid Shaft with Circumferential Notch

Let D be the shaft's diameter, d its minimum-section diameter and R the notch radius, Figure 10.14. Also, $t = D - d/2$ is the notch depth, which is equal to R for a semi-circular intercept by the meridian. Peterson's charts cover a wide range of conditions, including for arbitrary notch depths $t \neq R$. Here, $t = R$.

The end sections are subjected to torsional loading.

Beam theory would give at the notch root a shear stress

$$\tau^{\text{nom}} = \frac{16}{\pi}\frac{M^t}{d^3}$$

In reality, the maximum shear stress is $\tau^{\max} = K\tau^{\text{nom}}$. The factor K is given by the curve of Figure 10.15.

Interestingly, the maximum value of the stress concentration factor is about 2. This is in contrast with the a maximum value near 3 for stress concentrations under remote tension; see Examples 10.5 and 10.6. ♡

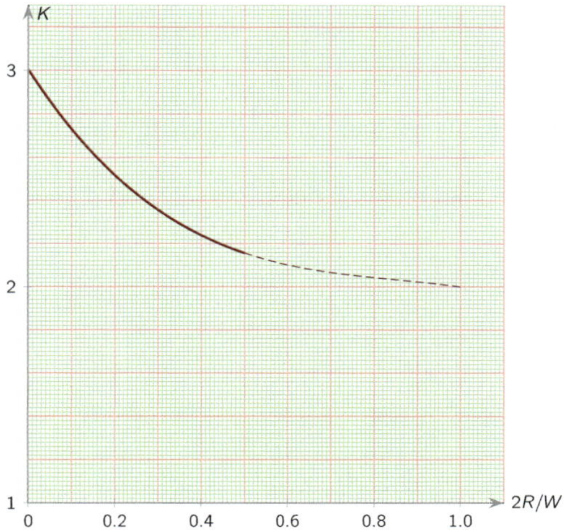

Figure 10.13 Stress concentration factor in tension of a rectangular plate with a central hole versus the ratio of hole diameter to bar width.

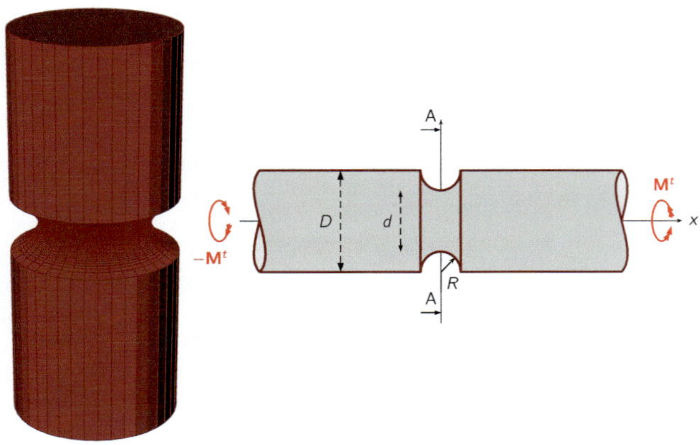

Figure 10.14 Round notched bar under torsion.

Example 10.8: Bending of a Cylindrical Beam with a Fillet

The beam is composed of two solid and coaxial cylindrical parts. The change in cross-section is done by a step fillet (torus of radius R). Let D be the maximum diameter, d the minimum diameter, Figure 10.16. Also, $t = (D - d)/2$ is the fillet depth. Peterson's charts cover a wide range of conditions, including for arbitrary fillet depths $t \neq R$, as depicted in the three-dimensional drawing of Figure 10.16.

The beam is subjected to pure bending. The normal stresses σ_x are maximum at points A and B, Figure 10.16. Beam theory gives

$$\sigma^{\text{nom}} = \pm \frac{32}{\pi} \frac{M}{d^3}$$

with a $-$ sign at A and $+$ sign at B.

In reality, the magnitude of the maximum stress is $\sigma^{\max} = K\sigma^{\text{nom}}$. The factor K is given by the experimental curves of Figure 10.17 as a function of the ratio d/D for various values of the t/R ratio. Some special and limit cases are depicted in the auxiliary drawings to the left and right of the chart, which correspond to select points on the curves.

Note that when the fillet radius R decreases, at fixed d and D, the stress con-

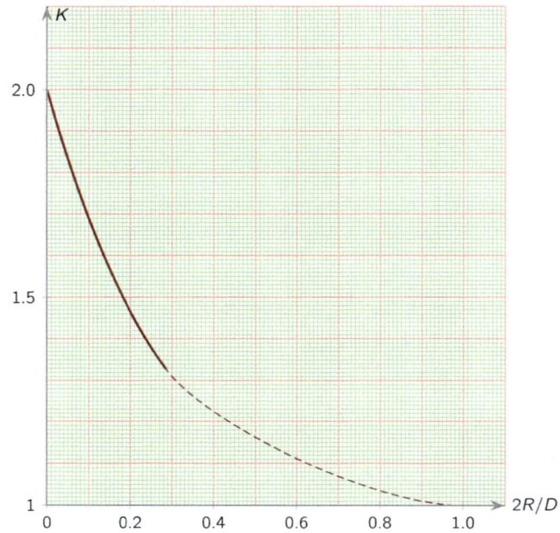

Figure 10.15 Stress concentration factor in torsion of a round notched bar versus the ratio of notch root diameter to the bar's diameter.

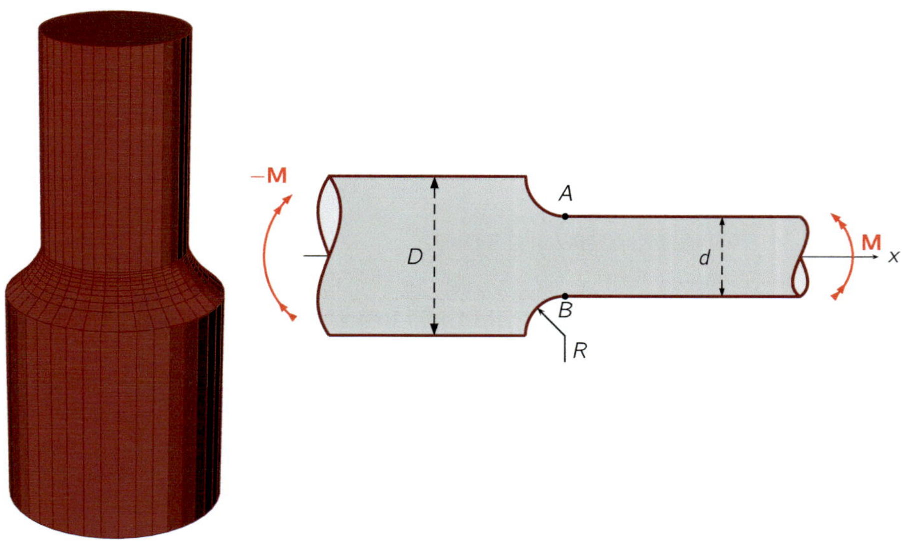

Figure 10.16 Round bar with a fillet under bending.

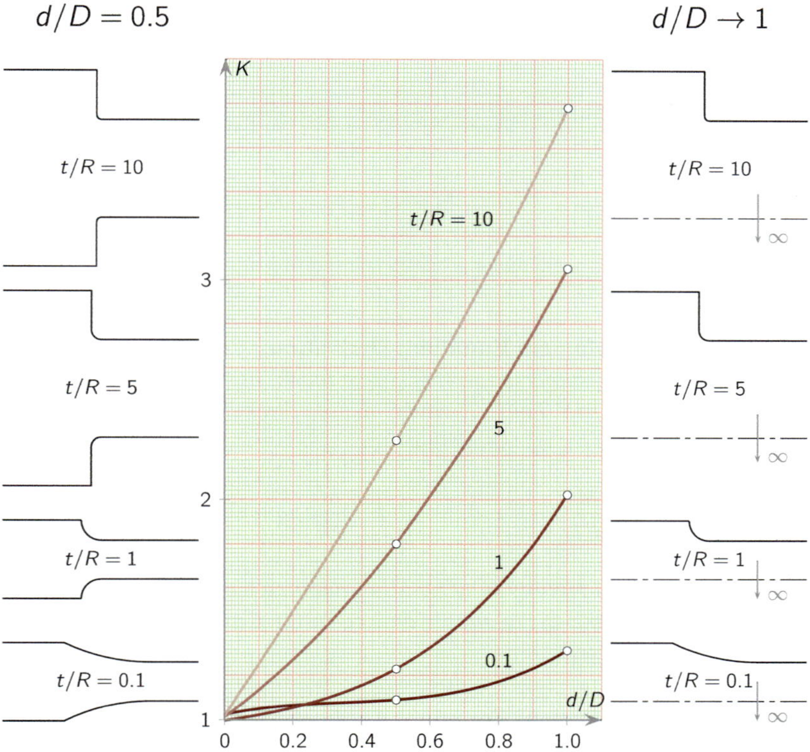

Figure 10.17 Stress concentration factor in bending of a round bar with a fillet versus the ratio of minimum to maximum diameter of the bar for various values of the fillet depth to radius ratio.

centration factor considerably increases.

The case of fillets under bending illustrates situations where the stress concentration factor may reach values in excess of 3, depending on geometric factors. ♡

PROBLEMS

Problem 10.1 (Dimensioning of a Shaft under Torsion and Bending) *Consider a cylindrical, solid transmission shaft working in torsion coupled with bending. The most loaded cross-section is subjected to a couple* $\mathbf{M}$ *having components* $M^t, 0, M^b$ *along principal axes with* M^t *denoting the torque and* M^b *the bending moment; the magnitude of the couple is* $M = 10\,\mathrm{kN \cdot m}$*. The allowable stress is* $\sigma_a = 50$ *MPa and an equivalent (von Mises) stress is defined as* $\sigma_{eq} = \sqrt{\sigma^2 + 4\tau^2}$ *where* σ *and* τ*, respectively, refer to the normal stress and shear stress acting on the cross-section. Dimension the radius R of the shaft.* ⋆

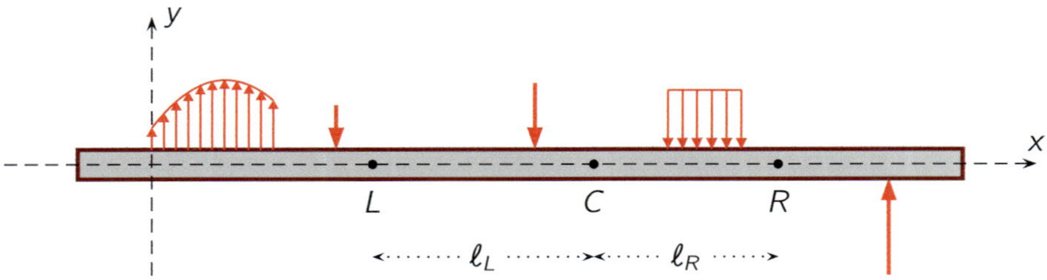

Figure 10.18 Continuous beam under arbitrary transverse loads.

Problem 10.2 (Three-Moment Equation: Derivation) *Consider a straight beam with median plane xOy, subjected to transverse loads, distributed or concentrated, as depicted in Figure 10.18. Let L, R, and C denote three arbitrary sections located at x_L, x_C, x_R with $x_L < x_C < x_R$. Show that*

$$-\int_{x_L}^{x_C} \frac{M}{EI}\,\mathrm{d}x + \frac{1}{\ell_L}\int_{x_L}^{x_C} \frac{M}{EI}(x_C - x)\,\mathrm{d}x - \frac{1}{\ell_R}\int_{x_C}^{x_R} \frac{M}{EI}(x_R - x)\,\mathrm{d}x = \frac{v_C - v_L}{\ell_L} + \frac{v_C - v_R}{\ell_R}$$

where EI is the flexural rigidity in xOy, $M(x)$ denotes the bending moment, v_i the deflection of section i, and ℓ_L, ℓ_R are as defined in the figure. The deflection due to the shear force is neglected. ⋆

Problem 10.3 (Three-Moment Equation: Application) *Determine the reactions at the supports for the beam shown in Figure 10.19 under its own weight.* ⋆

Problem 10.4 (Planar Gantry: Screwer and Diagrams) *Consider the gantry in Figure 10.20 consisting of three beams AB, BC, CD, welded together. End section A is a pin joint while the other end D is simply supported on a horizontal plane. This structure is under the action of a uniformly distributed load of intensity q on the ascending part AB. To describe the structure, point A is taken as the origin of arclengths. Focus is placed on the plane problem. (i) Determine the reactions. 1. Determine on each cross-section the components N, V_Y, M_Z of the screwer. (ii) Draw the corresponding diagrams.* ⋆

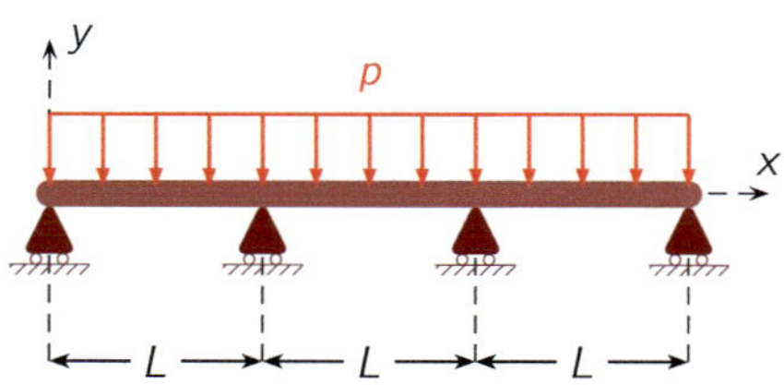

Figure 10.19 Multiply supported beam.

Problem 10.5 (Combined Bending and Torsion) *A single arm sign structure consists of two identical box beams, each having length L and a thin-walled square cross-section of side length 2a, as shown in the inset of Figure 10.21. The sign panel is rigidly attached to the end section at B. Under the action of wind, the panel is subjected to a force F acting in the z direction at the center C of the panel, located at a distance b from the point of attachment B, as depicted in Figure 10.21. The weight of the structure as well as the action of wind on it are negligible in comparison with F. (i) Determine the reactions at O and the screwer components on both portions OA and AB. (ii) Let EI and GJ respectively denote the (lineal) bending stiffness and torsion stiffness of the box beam. Determine the expressions of I and J in terms of a and t. (iii) Calculate all three components of the rotation,* $\boldsymbol{\omega}_B$*, of end section B and the out-of-plane displacement of the panel,* w_C*. The influence of the shear force may be neglected. (iv) Calculate the maximum values taken by the bending moment, internal torque, normal stress* σ_x*, and shear stress* τ*. (v) Carry out numerical applications using:* $t = 3$ *mm,* $a = 20$ *cm,* $L = 7$ *m,* $E = 70$ *GPa,* $\nu = \frac{1}{3}$*,* $F = 2000$ *N,* $b = 50$ *cm.* ⋆

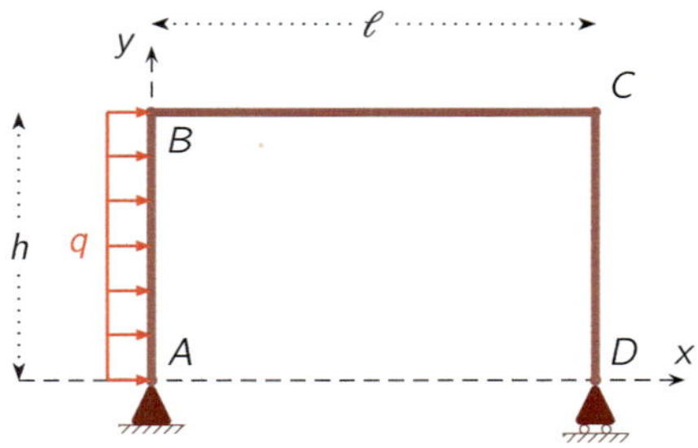

Figure 10.20 Gantry under uniform load.

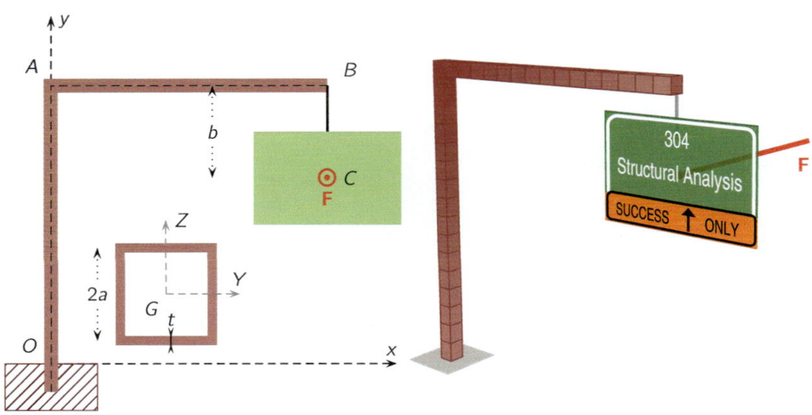

Figure 10.21 Two-beam structure under combined bending and torsion.

Problem 10.6 (Optimal Design) *Consider the single arm sign structure in Figure 10.21. Using the results of Problem 10.5: (i) Analyze the effect of depth b on the panel's deflection w_C. (ii) In particular, determine the ratio b/L that realizes a minimum in w_C in terms of the bending and torsion stiffnesses. Specialize the result for the thin square section by expressing b/L in terms of Poisson's ratio. (iii) Analyze the effect of depth b on the twist ω_x^B of end section B. What value of b minimizes the rotation?* ⋆

Problem 10.7 (Dimensioning a Bollard) *A boat exerts a horizontal force F on a cylindrical bollard, as depicted in Figure 10.22. (i) On the most loaded section at O, calculate the components of the screwer. (ii) Calculate the maximum normal stress and the maximum shear stress; indicate the locations where these values are attained. (iii) If σ_a denotes the allowable stress, determine the values of the radius R and the maximum value of h. Here, the allowable stress is used such that everywhere: $\sigma \leq \sigma_a$ and $\tau \leq \sigma_a/2$. (iv) Carry out a numerical application for: $F = 1000$ kN, $\nu = 0.25$, $\sigma_a = 500$ MPa.* ⋆

Problem 10.8 (Analysis of a Square Frame) *A square frame of side length 2L is loaded by two equal and opposite forces of magnitude F, Figure 10.23. The cross-section is a full square of side length 2a, as shown in the inset. The vertical plane is a plane of symmetry of the frame. (i) Determine the non-zero internal resultants. (ii) Draw the bending moment diagram. (iii) Calculate the displacements of the points of application of the forces.* ⋆

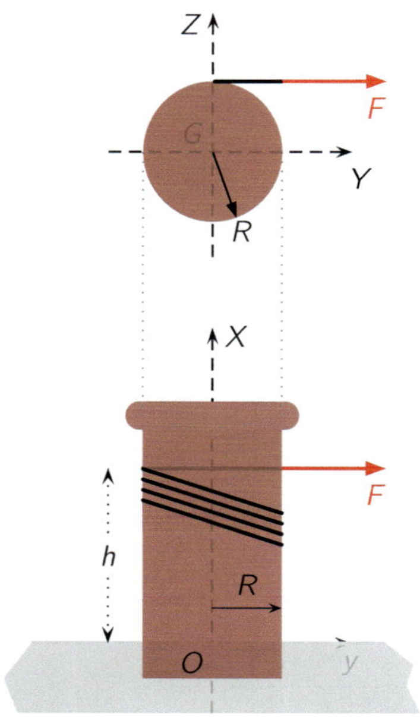

Figure 10.22 Top and front views of bollard under load.

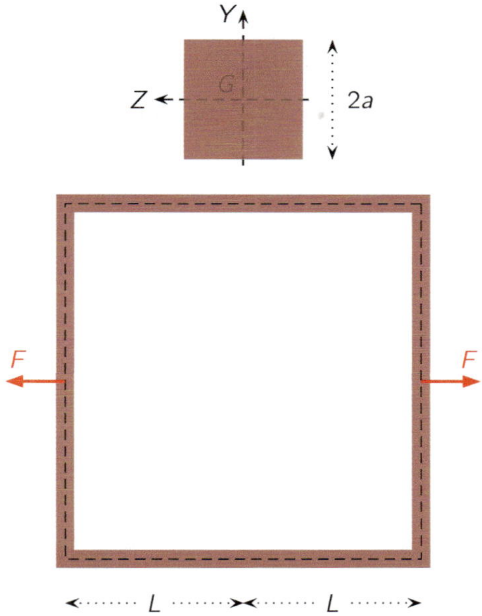

Figure 10.23 Square frame with square cross-section.

Problem 10.9 (Mechanics of a Two-Beam Structure) *Two identical prismatic beams sharing the same median plane are welded together at a right angle in B, Figure 10.24. The characteristics of each beam are: length L, cross-section S, second moment of area I, and Young's modulus E. The two beams are attached to supports by ball joints at A and A′. The structure is loaded by a vertical force F at B. Displacements induced by the shear force are neglected in comparison with those due to bending or normal force. (i) Determine the reaction/connection loads: external at A and A′, internal at B. (ii) Calculate the non-zero internal resultants. (iii) Calculate the displacement of B.* ⋆

Problem 10.10 (Bending by Torsion of a Ring) *A torus (circular ring with a solid circular cross-section) is subjected to a uniformly distributed torque with intensity m (torsion couple per unit length of the ring); see Figure 10.25. The radius of the ring is R and the radius of the circular section is a, as shown in the inset. The flexural and torsional rigidities are EI and GJ. (i) Determine the components of the screwer on the current cross-section. (ii) Show that there is no twisting and calculate the rotation of the current cross-section.* ⋆

Problem 10.11 (Combined Loading of a Fuselage Frame) *A fuselage frame has a shape of a ring of radius R. Two equal and opposite bending moments* $\mathbf{\Gamma}$ *and* $-\mathbf{\Gamma}$ *are applied as shown in Figure 10.26. The bending stiffness is EI. (i) Determine the degree to which the frame is internally hyperstatic. (ii) Determine the components*

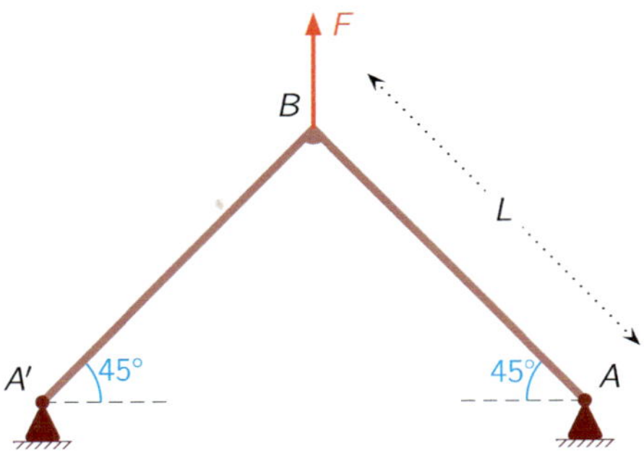

Figure 10.24 Two welded beams.

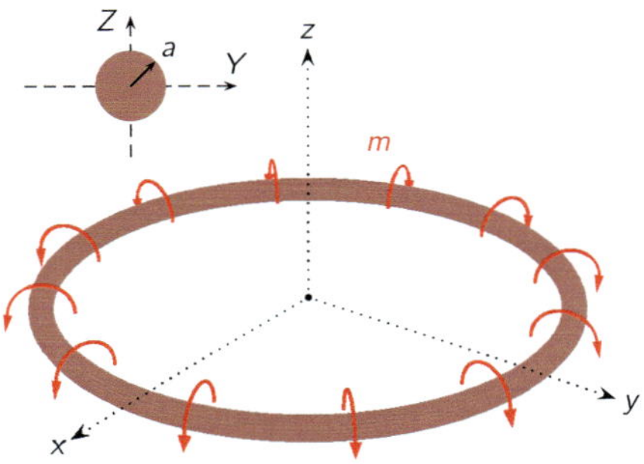

Figure 10.25 Torsion of a torus.

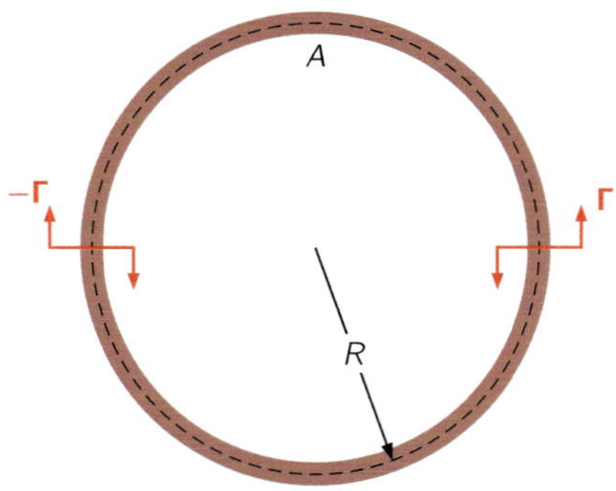

Figure 10.26 Fuselage frame.

of the screwer on the current cross-section. (iii) Calculate the rotations of sections where the bending couples are applied, assuming that section A is fixed. ⋆

Problem 10.12 (Ring under Two Torsion Couples) *A circular ring is subjected to two opposite torsion couples* $\mathbf{\Gamma}$ *and* $-\mathbf{\Gamma}$ *on diametrically opposed sections A and A′, Figure 10.27. (i) Determine the components of the screwer on the current cross–*

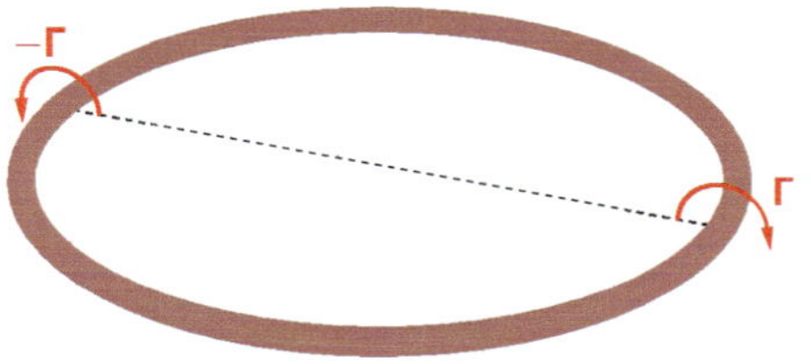

Figure 10.27 Torsion of a ring.

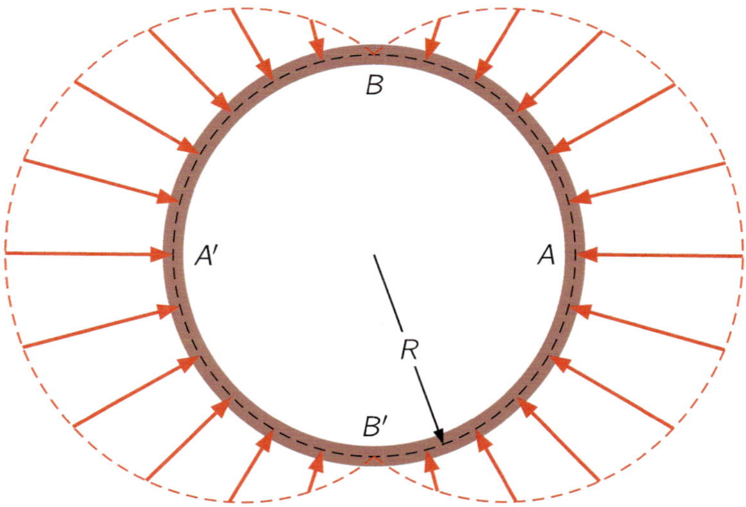

Figure 10.28 Ring under external pressure.

section. (ii) Draw the diagrams of M_X and M_Y as function of an appropriate angle. (iii) Calculate the rotation of section A where $\mathbf{\Gamma}$ *is applied.* ⋆

Problem 10.13 (Ring under Nonuniform Pressure) *A ring of radius R is acted upon by a radial, centripetal, distributed force of density* $p\cos\theta$ *on its right half and* $-p\cos\theta$ *on its left half, Figure 10.28. (i) Determine the shear and normal forces on sections A, A′, B, B′. (ii) Determine the components of the screwer on the current cross-section. (iii) Calculate the deflections of sections A, A′, B, B′. The deformations due to normal and shear forces are neglected in comparison with the bending moment.* ⋆

Problem 10.14 (Ring under Its Own Weight) *A circular ring with a square solid cross-section lies on a horizontal floor under its own weight* $W = 2\pi aw$*. Here, a denotes the ring's radius and b half the side length of the cross-section, Figure 10.29. (i) Determine the shear and normal forces as well as the bending moment on the current cross-section. (ii) Draw their diagrams. (iii) Calculate the deflection along the radial direction while neglecting the contributions of the normal and shear forces.* ⋆

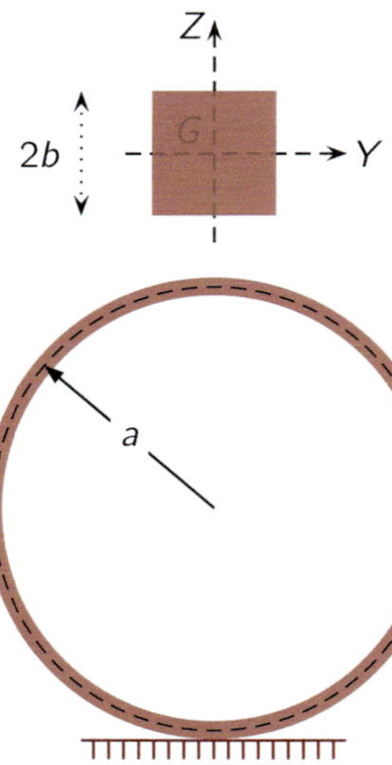

Figure 10.29 Ring supporting its weight.

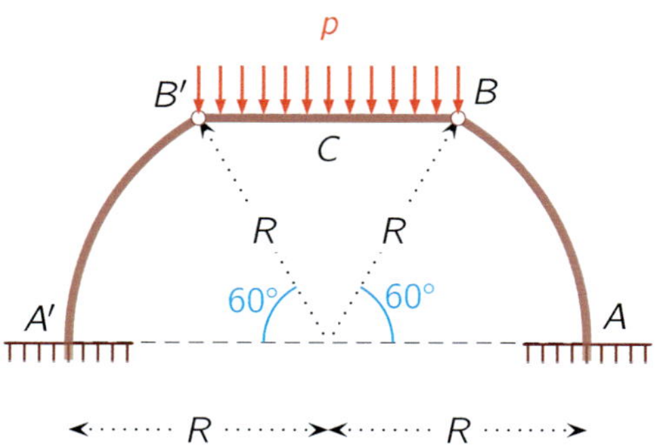

Figure 10.30 Curved gantry under distributed load.

Problem 10.15 (Gantry under Distributed Load) *A gantry with curved legs comprises three beams of a uniform cross-section having bending stiffness EI, Figure 10.30. The rising curved beams are fixed at the ground and linked to the horizontal beam BB′ via ball joints. The gantry is loaded by a uniformly distributed force of intensity p. (i) Determine the components N, V, M of the screwer on the current cross-section in all three beams. (ii) Calculate the deflection of the midsection at C. Deflections due to the normal and shear forces can be neglected.* ⋆

Problem 10.16 (Gantry with Curved Legs) *Consider the gantry of Figure 10.31. The vertical plane of the figure is a plane of symmetry, and the gantry is loaded in that plane by a concentrated downward force F, as shown. The cross-section is constant with properties S and EI.*

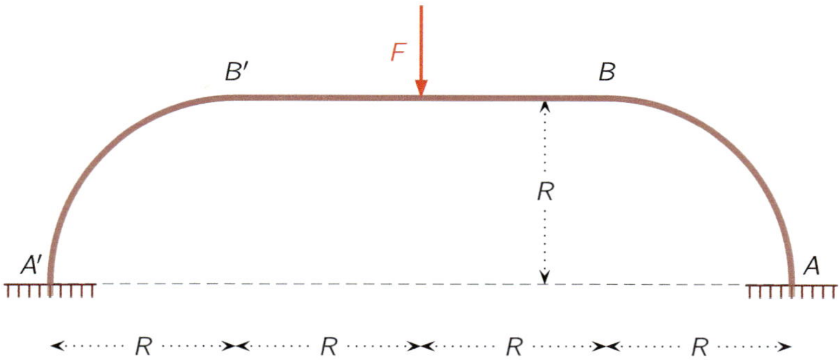

Figure 10.31 Curved gantry under a point load.

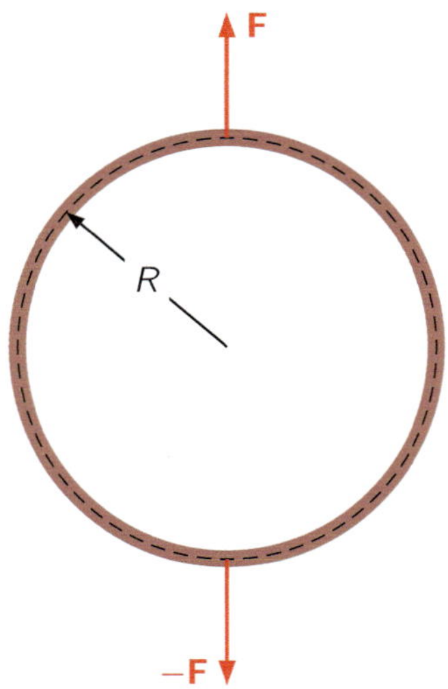

Figure 10.32 Ring dynamometer.

(i) Determine the reactions at A and A′. (ii) Determine the internal resultants acting on the current cross-section. (iii) Calculate the deflection of the mid-point C. Deflections due to the normal and shear forces can be neglected. ⋆

Problem 10.17 (Deflections of a Ring) *Consider the ring dynamometer of Figure 10.32. The internal resultants were determined in Example 10.1. Calculate the deflections and plot the deformed shape. Repeat the exercise by reverting the sign of the applied force.* ⋆

Problem 10.18 (Design of a Frame) *A square frame is subjected to uniformly distributed loads on the top and bottom parts, as shown in Figure 10.33. The components of the screwer were determined in Example 10.2. (i) Determine the maximum*

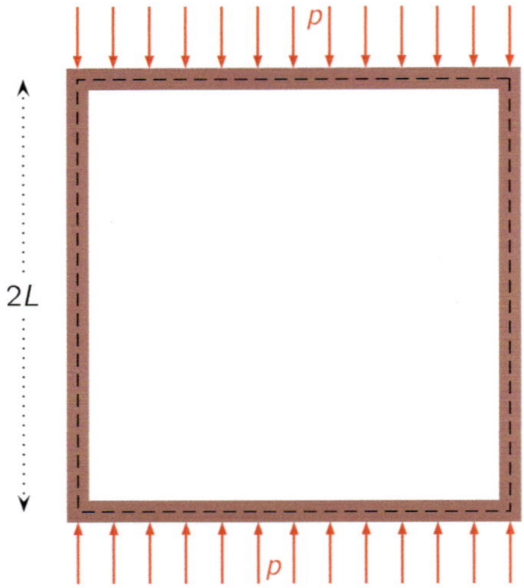

Figure 10.33 Frame under distributed load.

deflections. (ii) Determine the maximum normal stress and maximum shear stress in the frame and their locations. (iii) If the allowable stress is $\sigma_a = 100$ *MPa and the maximum displacement is 1 cm what is the critical value of p for* $L = 50$ *cm for a safe operation of the frame? An equivalent stress is defined as* $\sigma_{eq} = \sqrt{\sigma^2 + 4\tau^2}$ *where* σ *and* τ *respectively refer to the normal and shear stresses acting on the cross-section.* ⋆

11 Buckling

11.1 DISCRETE STRUCTURES

11.1.1 EQUILIBRIUM – STABILITY

A discrete structure is any assemblage of *non-deformable* solids, which is connected to a built structure that constitutes the reference solid, the latter being itself reported to some orthonormal frame $\{O,xyz\}$.

It is assumed that all links, internal and external, are bilateral, conservative and that some links are endowed with some linear stiffness.

It is also assumed that the given external forces (loading) derive from a time-independent scalar potential.

Any configuration of such a structure is determined by n independent parameters, say $q_1, q_2, \ldots, q_n$, named configuration parameters (or generalized coordinates).

Under the considered loading, a particular configuration is said to be an *equilibrium configuration* if when the structure is abandoned therein, with zero velocity, it remains in that configuration. Such a configuration is characterized by the values q_i^0 of the generalized coordinates, which will be taken equal to zero any time that is possible.

DEFINITION 11.1 An equilibrium configuration is said to be *stable* if, when the structure is taken away sufficiently little from that configuration and the initial velocities given to it are sufficiently small, one is assured that the motion of the structure will occur in the neighborhood of that equilibrium. ♣

Mathematically, the above definition may be specified by introducing the notion of deviation D of the structure relative to the equilibrium under consideration:

$$D = \sum m\left[(x-x_0)^2 + (y-y_0)^2 + (z-z_0)^2\right]$$

where the summation is extended over all the particles P that make up the structure, each having mass m, current coordinates x,y,z, and coordinates x_0,y_0,z_0 at equilibrium.

DEFINITION 11.2 The equilibrium configuration is said to be *stable* if, for all ε positive, given and arbitrarily small, one can find η_1 and η_2 such that, taking $D_0 < \eta_1$ (initial deviation) and $K_0 < \eta_2$ (initial kinetic energy), one is assured that D stays smaller than ε. ♣

DOI: 10.1201/9781003537229-13

11.1.2 LEJEUNE-DIRICHLET THEOREM

The total potential energy of the structure has two contributions: external and internal:

$$E_{\mathrm{T}} = V(q_1, q_2, \dots, q_n) + W(q_1, q_2, \dots, q_n)$$

The external potential energy V denotes the potential of external forces and the internal potential energy W is the elastic potential energy of the structure. In the present case of a discrete structure, W is entirely stored in the elastic links.

In Mechanics, the following theorem is demonstrated.

Theorem 11.1: Lejeune-Dirichlet

A configuration corresponding to a strict minimum of the total potential energy E_{T} is a stable equilibrium configuration. ♢

This theorem is also widely known as the principle of minimum total potential energy.

Remark 11.1 *Starting from an equilibrium configuration, denoted* (0), *consider virtual variations of the configuration parameters,* $(\delta q_1, \delta q_2, \dots, \delta q_n)$. *The corresponding variation of the total energy is:*

$$\delta E_{\mathrm{T}} = \left(\frac{\partial E_{\mathrm{T}}}{\partial q_1}\right)_0 \delta q_1 + \cdots + \left(\frac{\partial E_{\mathrm{T}}}{\partial q_n}\right)_0 \delta q_n$$

From the theorem of virtual work, one has $\delta E_{\mathrm{T}} = 0$, $\forall \delta q_1, \dots, \delta q_n$. *It then follows that:*

$$\left(\frac{\partial E_{\mathrm{T}}}{\partial q_1}\right)_0 = 0, \left(\frac{\partial E_{\mathrm{T}}}{\partial q_2}\right)_0 = 0, \dots, \left(\frac{\partial E_{\mathrm{T}}}{\partial q_n}\right)_0 = 0$$

*

Remark 11.2 *In the equilibrium configuration* (0), *take* $q_i^0 = 0$ *and* $E_{\mathrm{T}}(\mathbf{0}) = 0$, *which is always possible.*

A second-order Taylor development of E_{T} *in the neighborhood of* (0) *then writes*

$$E_{\mathrm{T}} = \frac{1}{2} \sum_{i=1}^{n} \sum_{j=1}^{n} \left(\frac{\partial^2 E_{\mathrm{T}}}{\partial q_i \partial q_j}\right)_0 q_i q_j \tag{11.1}$$

If the above quadratic form is positive definite, that is, if its eigenvalues are all strictly positive, then E_{T} *attains a strict minimum in* (0), *which is then a stable equilibrium configuration.* *

Remark 11.3 *The converse of Theorem 11.1 (Lejeune-Dirichlet) is not exact in its entire generality; one finds indeed examples where stable equilibrium does not correspond to a strict minimum of* E_T*. Such cases are scarce in practice and can only be encountered when the potential V presents some singularities near* (0).

On the other hand, if E_T *is expressible under the quadratic form of Equation* (11.1)*, the stability condition of Lejeune-Dirichlet is then necessary and sufficient (strict minimum of* E_T*).* ∗

Remark 11.4 *If the structure is subjected, in addition to the conservative loading, to dissipative forces (dry or viscous friction) and to forces perpendicular to velocities (Coriolis forces for example), then the Lejeune-Dirichlet theorem remains valid.* ∗

11.1.3 INSTABILITY – BUCKLING

Consider a discrete structure in some equilibrium configuration denoted (0). Increase the loading applied to it. For a *critical* value of the load, it is possible that the equilibrium becomes *unstable*: a *buckling* of the structure occurs subsequently in that the structure deviates abruptly from configuration (0).

A critical buckling load constitutes a limit case between stability and instability; it corresponds then to an *indifferent* equilibrium.

For a structure that has n degrees of freedom, subjected to some type of loading, there exist n (eigen) *modes of buckling*, each corresponding to an indifferent equilibrium configuration, near configuration (0), and to a value of the load.

In general, one is mostly interested in the (eigen) mode associated with the smallest value of the critical load. Indeed, it is that limit load that must not be exceeded. This mode is called the first eigen-mode of buckling or simply the *fundamental mode.*

In the following, the principal methods that permit the determination of critical buckling loads and associated modes are presented along with illustrative examples.

Direct Method: It consists of placing the structure in a (kinematically possible) configuration near (0) and determining the loading that is capable of maintaining it therein, once abandoned with zero initial velocity.

Energy Method: Based on Theorem 11.1 (Lejeune-Dirichlet), this method permits the stability conditions to be found, hence in the limit, the indifferent equilibria and corresponding critical loads.

Example 11.1: Stability of Rigid Bar under Its Own Weight

Consider the planar structure depicted in Figure 11.1, comprising a prismatic rigid beam (length L, weight $P = pL$) and a spring (stiffness k, negligible weight). The beam is connected to the ground through a pivot (or hinge) at O. Figure 11.1a represents the equilibrium configuration (0), the beam being vertical and the spring at rest, and Figure 11.1b a neighboring, kinematically admissible configuration.

This structure possesses one degree of freedom and one may choose the angle θ as the configuration parameter. Let us study the stability of the equilibrium.

The Direct Method consists of seeking the value P_c of the weight P such that the structure, abandoned in the configuration of Figure 11.1b with no initial velocity, would stay therein (neutral equilibrium). For that, it is necessary and sufficient that the moment about O of the weight and the restoring force in the spring vanish, that is

$$\frac{1}{2}P_c L\theta - kL^2\theta = 0 \tag{11.2}$$

Hence,

$$P_c = p_c L = 2kL \Longrightarrow p_c = 2k \tag{11.3}$$

If the weight per unit length of the bar is greater than p_c, the configuration of Figure 11.1a is unstable, the restoring couple being smaller than the moment due to the weight, which tends to move the bar further away from its vertical position.

The Energy Method consists of calculating the total potential energy $E_T = W(\theta) + V(\theta)$ in the neighboring configuration (θ). Here,

$$W = \frac{1}{2}kL^2 \sin^2\theta \simeq \frac{1}{2}kL^2\theta^2 \tag{11.4}$$

and (taking $V(0) = 0$)

$$V = \frac{1}{2}PL(1-\cos\theta) \simeq -\frac{1}{2}pL^2\frac{\theta^2}{2} \tag{11.5}$$

so that

$$E_T = \frac{1}{2}L^2\left(k - \frac{p}{2}\right)\theta^2 \tag{11.6}$$

The configuration $(\theta = 0)$ is stable if it corresponds to a strict minimum of E_T, that is if p is smaller than $p_c = 2k$; it is unstable if p is greater than $p_c = 2k$. The limit case, for $p = p_c$ corresponds to neutral (or indifferent) equilibrium. ♡

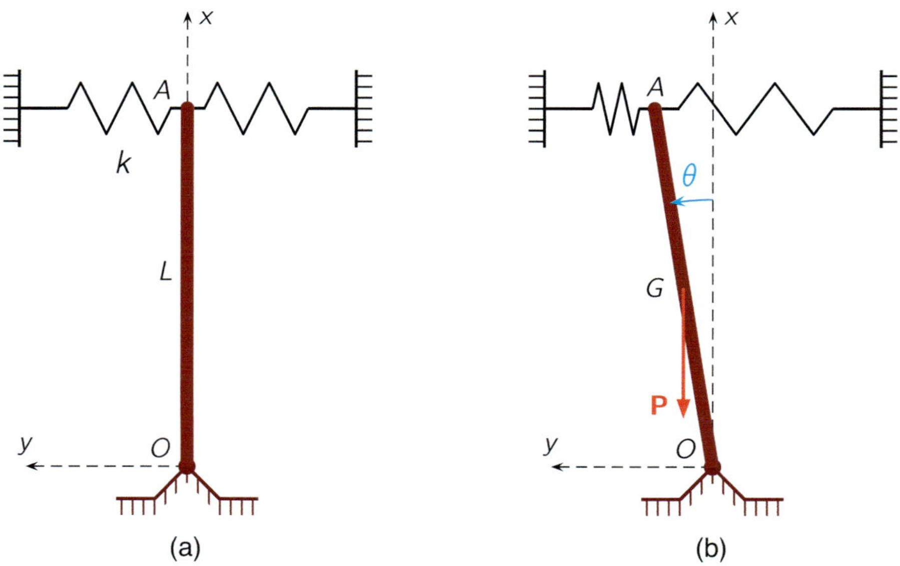

Figure 11.1 Buckling of a system comprising a rigid bar (a) and a spring (b).

Example 11.2: Two-Degree-of-Freedom Structure

Consider the planar structure shown in Example 11.2. It consists of three identical rigid bars of length L. The links at the two extremities O and C are hinges and the three bars are connected at A and B by two spiral springs of identical stiffness k (units of moment per radian). All weights are neglected.

The aim is to study the stability of this structure under the action of a compressive force of intensity F.

Figure 11.2a represents the equilibrium configuration (0), the springs being at rest, and Figure 11.2b a neighboring, kinematically possible configuration.

The Direct Method consists of expressing equilibrium in the perturbed configuration of Figure 11.2b:

$$\begin{cases} LF\theta_1 + k(\theta_2 - \theta_1) = 0 & \text{equilibrium of bar } OA \\ LF\theta_3 + k(\theta_2 - \theta_3) = 0 & \text{equilibrium of bar } BC \\ L\theta_1 + L\theta_2 + L\theta_3 = 0 & \text{relationship between } \theta_1, \theta_2, \theta_3 \end{cases} \quad (11.7)$$

The latter of these equations expresses kinematic admissibility in that points O and C must remain aligned (so that $\overrightarrow{OC} \perp \mathbf{j}$, thus $\overrightarrow{OA} \cdot \mathbf{j} + \overrightarrow{AB} \cdot \mathbf{j} + \overrightarrow{BC} \cdot \mathbf{j} = 0$ using $\sin\theta_i \approx \theta_i$ for θ_i small and algebraic). The first two equations are moment equilibrium equations about points O and C, respectively; see Problem 11.1.

One has in Equation (11.7) a linear system that is homogeneous in θ_1, θ_2, θ_3. Its trivial solution is $\theta_1 = \theta_2 = \theta_3 = 0$; it hints at the fact that a configuration such as that shown in Figure 11.2b cannot be an equilibrium configuration for an arbitrary value of F. On the other hand, to each value F_c of F that renders zero for the determinant of the system, corresponds a non-trivial solution $\{\theta_1, \theta_2, \theta_3\}$, defined up to a multiplicative constant, that is, a neutral equilibrium configuration of the type shown in Figure 11.2b.

Making the determinant vanish writes

$$(LF - k)(LF - 3k) = 0 \quad (11.8)$$

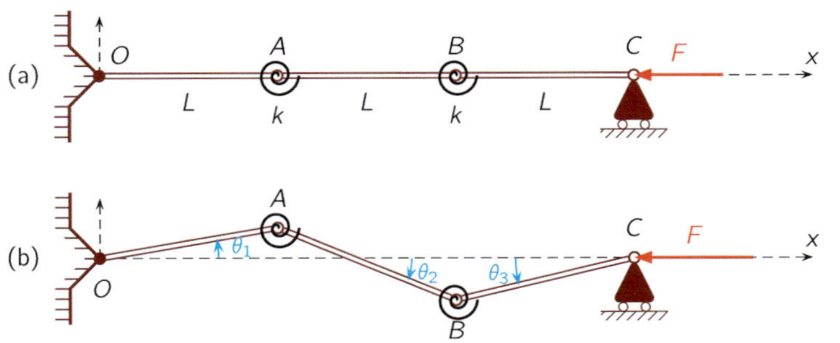

Figure 11.2 Buckling of a system comprising three rigid bars and two spiral springs. (a) Initial or stable equilibrium configuration. (b) Perturbed configuration.

The critical loads are therefore: $F_{c1} = k/L$ and $F_{c2} = 3k/L$. To the fundamental critical load F_{c1} corresponds the fundamental mode ($\theta_2 = 0$, $\theta_1 + \theta_3 = 0$) of Figure 11.3a; to F_{c2} corresponds the mode ($\theta_1 = \theta_3 = -\frac{1}{2}\theta_2$), which is represented in Figure 11.3b.

The Energy Method proceeds by calculating first the total potential energy $E_T = W(\theta_1, \theta_3) + V(\theta_1, \theta_3)$ taking θ_1 and θ_3 as the configuration parameters. One finds (see Problem 11.1)

$$\begin{cases} W = \frac{1}{2}k\left[(2\theta_1 + \theta_3)^2 + (\theta_1 + 2\theta_3)^2\right] = \frac{1}{2}k\left[5\theta_1^2 + 8\theta_1\theta_3 + 5\theta_3^2\right] \\ V = -\frac{1}{2}FL(2\theta_1^2 + 2\theta_1\theta_3 + 2\theta_3^2) \end{cases} \tag{11.9}$$

Hence,

$$E_T = \frac{1}{2}\left[(5k - 2FL)\theta_1^2 + 2(4k - FL)\theta_1\theta_3 + (5k - 2FL)\theta_3^2\right] \tag{11.10}$$

Diagonalizing this quadratic form leads to

$$E_T = \frac{1}{2}\left[(k - FL)\left(\frac{\theta_1 - \theta_3}{\sqrt{2}}\right)^2 + 3(3k - FL)\left(\frac{\theta_1 + \theta_3}{\sqrt{2}}\right)^2\right] \tag{11.11}$$

its eigenvalues being $k - FL$ and $3(3k - FL)$.
If both eigenvalues are positive, i.e., if one has $FL < k$ and $FL < 3k$, then the quadratic form is positive definite and, by way of consequence, configuration (0) is stable since it corresponds then to a strict minimum of E_T.
One finds the two neutral equilibria that make E_T identically nil:

$k = FL$ for $\theta_1 + \theta_3 = 0$ (fundamental mode)

$3k = FL$ for $\theta_1 = \theta_3$ (harmonic mode)

♡

The topic of discrete structures is now left aside to focus in the remainder of the chapter on *continuous structures*, comprising of *deformable beams*. The definitions, theorems, and previous methods remain valid, but the number of degrees of freedom becomes infinite. One would expect, in particular, an infinite suite of eigenmodes of buckling, the most important of which is the fundamental one, which is associated with the smallest critical load.

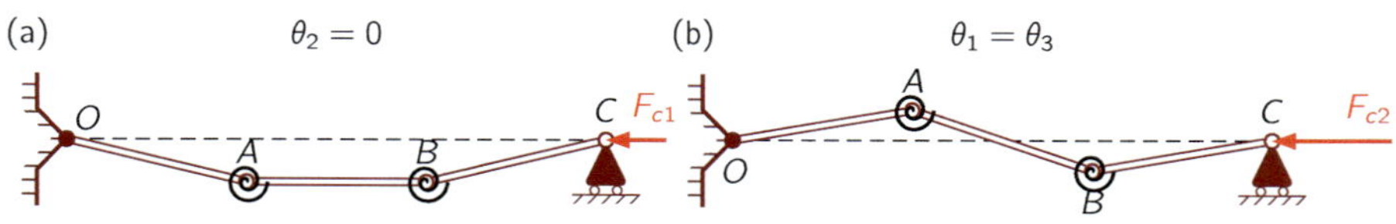

Figure 11.3 Two modes of buckling for the structure of Figure 11.2a (a) fundamental mode; (b) harmonic mode.

11.2 EULER BUCKLING

DEFINITION 11.3 The buckling by plane flexure of a slender prismatic beam, loaded in pure compression with simply-supported ends, is known as *Euler buckling.* ♣

The phenomenon may occur in columns under an axial load, in the compressed bars of truss structures, etc. It represents a type of structural failure.

11.2.1 IDEAL CASE

Let L denote the length of the beam, I_Y and I_Z its principal second moments of area, with $I_Y > I_Z$. The material has Young's modulus E. Figure 11.4a shows the beam in stable equilibrium under the action of a compressive force **F**.

The problem is posed as follows: having imposed to the beam a small bending deformation, in the (xOy) plane, with n half-waves (three in Figure 11.4b, to fix ideas), what is the magnitude F_c of the compressive load capable of maintaining such deformation?

One must therefore express the configuration of Figure 11.4b, which is infinitely close to Figure 11.4a, as an equilibrium configuration, under load **F**.

On the current cross-section of abscissa x, which undergoes a deflection $v(x)$, acts a bending moment $\mathbf{M} = \overrightarrow{GA} \times \mathbf{F}$, the reaction $\mathbf{R}^A$ being zero. Given that $\overrightarrow{GA} \cdot \mathbf{j} = -v(x)$ and $\mathbf{F} = -F\mathbf{i}$, one obtains

$$M_Z = -vF \tag{11.12}$$

On the other hand, the deflection must satisfy the fundamental equation for plane

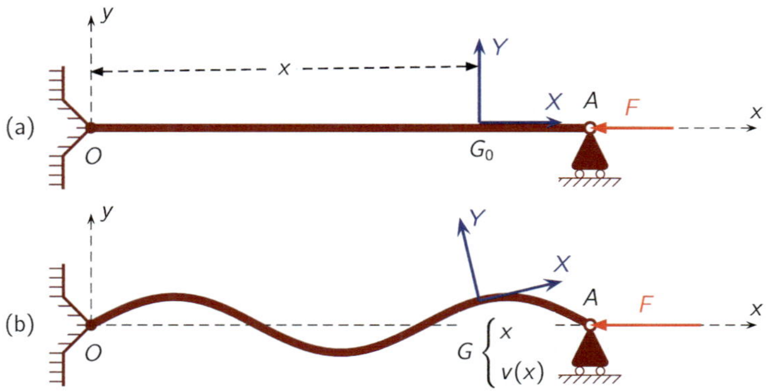

Figure 11.4 Pinned-supported beam under compression (a) in stable equilibrium; (b) in a kinematically admissible perturbed configuration.

bending (see Chapter 7):

$$\frac{d^2v}{dx^2} = \frac{M}{EI} \tag{11.13}$$

with $M = M_Z$ and $I = I_Z$, and can thus be determined by integrating the equation:

$$\frac{d^2v}{dx^2} + \frac{F}{EI}v = 0 \tag{11.14}$$

Equation (11.14) is a homogeneous, second-order linear differential equation in v with constant coefficients. Its general solution takes the form:

$$v = A\cos\sqrt{\frac{F}{EI}}x + B\sin\sqrt{\frac{F}{EI}}x \tag{11.15}$$

The boundary conditions impose:

- For $x = 0$, $v = 0$, hence $A = 0$
- For $x = L$, $v = 0$, hence $B\sin\sqrt{\frac{F}{EI}}L = 0$

Two possible cases emerge: either $B = 0$, in which case the deflection is identically equal to zero (trivial solution corresponding to stable equilibrium) or

$$\sin\sqrt{\frac{F}{EI}}L = 0 \tag{11.16}$$

The coefficient $\sqrt{\frac{F}{EI}}L$ must therefore be an integer multiple of π, namely $n\pi$ since the deformed median line of the beam was taken to have n half-waves.

The median line thus admits the following equation:

$$v = B\sin n\pi\frac{x}{L} \qquad \text{since} \qquad \sqrt{\frac{F}{EI}} = n\frac{\pi}{L} \tag{11.17}$$

The corresponding critical load is

$$F_c^n = n^2\pi^2\frac{EI}{L^2} \tag{11.18}$$

11.2.1.1 Conclusions

1. The fundamental buckling mode corresponds to $n = 1$ (arc) and a critical load:

$$F_c = F_c^1 = \pi^2\frac{EI}{L^2} \tag{11.19}$$

This is Euler's critical force.

2. To each buckling mode corresponds a critical load, but the deformation of the median line is defined up to a multiplicative constant B, which remains undetermined: the equilibrium is indeed neutral (or indifferent).

3. In practice, if the compressive force F is increasingly applied to the straight beam of Figure 11.4a then:
   - For $F < \pi^2 \frac{EI}{L^2}$, equilibrium is stable.
   - For $F = \pi^2 \frac{EI}{L^2}$, equilibrium is neutral: if the deflection of any cross-section were to be imposed, that deflection would remain and a small transverse load would suffice to modify it.
   - For $F > \pi^2 \frac{EI}{L^2}$, the bending deformation increases until eventual rupture of the beam: that is instability.

In what precedes, Euler buckling was analyzed in the ideal case, that is assuming that the beam was perfectly straight before loading and that the compressive forces were rigorously aligned with the median line of the beam; in addition, in calculating the deflection, the influence of the transverse shear was neglected.

Let us examine in what follows the deviations from the ideal case, caused by not satisfying any of the above assumptions.

11.2.2 INFLUENCE OF INITIAL DEFORMATION

Suppose that the median line of the beam prior to loading possesses some deflection, in the (xOy) plane, denoted $v_0(x)$, Figure 11.5a. To fix ideas, take, for example,

$$v_0(x) = B_0 \sin \pi \frac{x}{L} \tag{11.20}$$

and find the equilibrium configuration taken by the beam under the action of a compressive force F smaller in magnitude than $F_c = \pi^2 \frac{EI}{L^2}$.

In that configuration, Figure 11.5b, one has on the section at x, a bending moment $M = -vF$.

On the other hand, the change in curvature, at abscissa x, is

$$\frac{d^2(v - v_0)}{dx^2} = \frac{M}{EI} \tag{11.21}$$

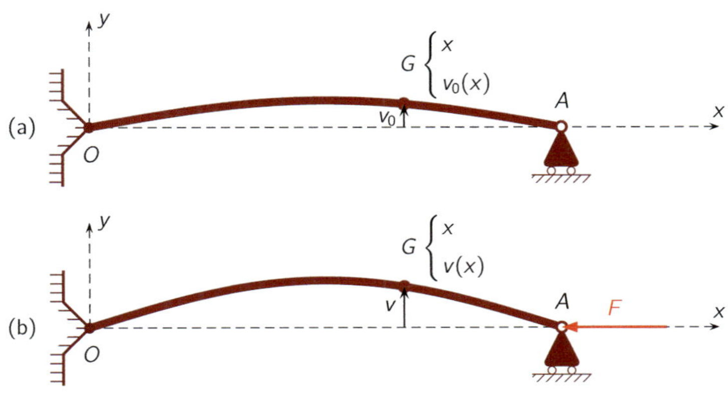

Figure 11.5 Pinned-supported beam with initial geometric imperfection under compression (a) at rest; (b) in a kinematically admissible perturbed configuration.

The deflection satisfies therefore the equation:

$$\frac{d^2v}{dx^2}+\frac{F}{EI}v=-\frac{\pi^2}{L^2}B_0\sin\pi\frac{x}{L} \tag{11.22}$$

which admits the general solution

$$v=A\cos\sqrt{\frac{F}{EI}}x+B\sin\sqrt{\frac{F}{EI}}x+\frac{B_0}{1-\dfrac{F}{F_c}}\sin\pi\frac{x}{L} \tag{11.23}$$

The imposed boundary conditions write:

- For $x=0$, $v=0$, hence $A=0$
- For $x=L$, $v=0$, hence $B\sin\sqrt{\frac{F}{EI}}L=0$

The latter equation is exactly the same as in the ideal case except that here $\sqrt{\frac{F}{EI}}L<\pi$ since the compressive force is taken to be smaller than F_c; therefore, the only meaningful solution now corresponds to $B=0$. Thus, the deflection in the configuration of Figure 11.5b is given by

$$v=\frac{1}{1-\dfrac{F}{F_c}}B_0\sin\pi\frac{x}{L} \tag{11.24}$$

Let us consider the maximum deflection $\delta=v(\frac{L}{2})$ to graph it in terms of the load F, which varies between 0 and F_c, Figure 11.6.

One finds that the critical buckling load is the same as in the case of a perfectly prismatic beam, but that each value of the load F induces a deflection, the latter increasing very rapidly when F approaches F_c.

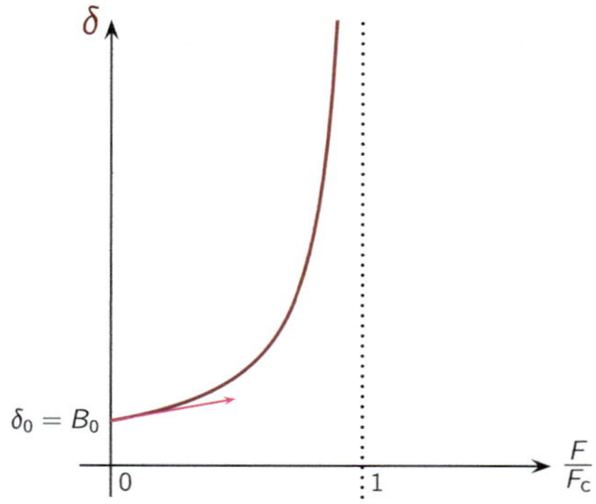

Figure 11.6 Maximum deflection of the beam versus the applied force normalized by Euler's critical force.

11.2.3 INFLUENCE OF LOAD ECCENTRICITY

Another source of imperfection lies in load misalignment. Suppose now that the median line of the beam is straight prior to loading, but that the compressive forces **F** and −**F** are applied at some offset distance h from that line, as sketched in Figure 11.7.

Equilibrium, under the action of the compressive force F taken to be smaller than $F_c = \pi^2 \frac{EI}{L^2}$, Figure 11.7b, leads to the following differential equation for the deflection:

$$\frac{d^2v}{dx^2} = \frac{M}{EI} = -v\frac{F}{EI} \tag{11.25}$$

The general solution is

$$v = A\cos\sqrt{\frac{F}{EI}}x + B\sin\sqrt{\frac{F}{EI}}x \tag{11.26}$$

The boundary conditions are now:

- For $x = 0$, $v = h$, hence $A = h$
- For $x = L$, $v = h$, hence $B = h\tan\left(\sqrt{\frac{F}{EI}}\frac{L}{2}\right)$

Thus, the maximum deflection is

$$\delta = v\left(\frac{L}{2}\right) = \frac{h}{\cos\sqrt{\frac{F}{EI}}\frac{L}{2}} = \frac{h}{\cos\frac{\pi}{2}\sqrt{\frac{F}{F_c}}} \tag{11.27}$$

Figure 11.8 gives the variations of this deflection as a function of F varying between 0 and Euler's critical load F_c.

The critical load is still $\pi^2\frac{EI}{L^2}$; Any load F would produce a deflection; the latter would rapidly increase if F were to approach F_c. Note that the slope of the curve at the origin is $\frac{\pi^2}{8}h$.

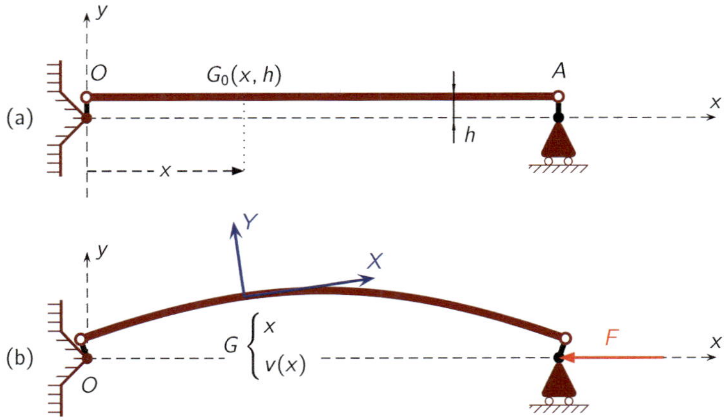

Figure 11.7 Pinned-supported beam with initial offset under compression (a) at rest; (b) in a kinematically admissible perturbed configuration.

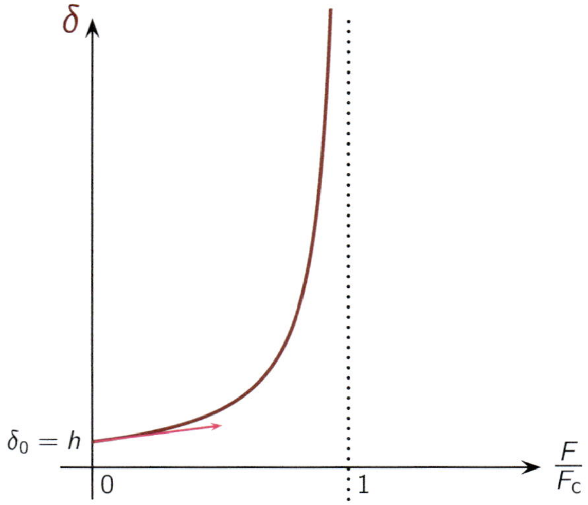

Figure 11.8 Maximum deflection of the beam versus the applied force normalized by Euler's critical force.

11.2.4 INFLUENCE OF TRANSVERSE SHEAR

Let us revisit the ideal case of Section 11.2.1 to express the fact that the deflection $v(x)$ in the neighboring configuration of Figure 11.4b is induced by a bending moment $M_Z = M$ as well as a transverse shear $V_Y = V$ (see Equation (9.33) of Theorem 9.1):

$$\frac{d^2v}{dx^2} = \frac{M}{EI} + \frac{k}{GS}\frac{dV}{dx} \tag{11.28}$$

with

$$M = -vF \qquad \text{and} \qquad V = -\frac{dM}{dx} = F\frac{dv}{dx} \tag{11.29}$$

The governing equation is therefore

$$\frac{d^2v}{dx^2} + \frac{F}{EI}\frac{1}{1 - \dfrac{kF}{GS}}v = 0 \tag{11.30}$$

Account being taken of the boundary conditions ($v = 0$ for both $x = 0$ and $x = L$), one obtains:

$$v = B\sin n\frac{\pi x}{L}$$

with

$$F_c^n = n^2\frac{\pi^2 EI}{L^2}\frac{1}{1 + n^2\dfrac{\pi^2 EI}{L^2}\dfrac{k}{GS}} \tag{11.31}$$

Therefore, taking into account the transverse shear, one finds a critical buckling load that is slightly smaller than Euler's.

To fix ideas, let us compute the correcting coefficient in the case of a cylindrical bar (length $L = 2$ m), of circular solid cross-section (radius $R = 1$ cm), made of steel ($E = 210$ GPa, $\nu = 0.29$). The reduced section coefficient is then $k = 1.175$ (see Equation (9.32) of Example 9.3). For the fundamental mode ($n = 1$) one finds:

$$\left(1 + \frac{\pi^2 EI}{L^2}\frac{k}{GS}\right)^{-1} = 0.9998$$

so that

$$F_c \approx \frac{\pi^2 EI}{L^2} = 4070\text{N}$$

Conclusion: Having examined the above-mentioned deviations from the ideal case, including geometric imperfections, one can only confirm the validity and importance of the critical buckling load found by Euler:

$$\frac{\pi^2 EI}{L^2}$$

11.3 GENERALIZATION OF EULER'S BUCKLING

This section concerns the buckling by plane bending of a prismatic beam, loaded in pure compression, but considering various types of links. Euler's problem consists of the special case of a simply supported beam.

11.3.1 CASES AMENABLE TO EULER'S PROBLEM

In some cases, one can obtain the critical load without any calculation. Let us illustrate the idea on a few examples limiting the focus on the first mode.

Example 11.3: Beam on Three Supports

The beam of length L now sits on a pivot at mid-span and is simply supported at both ends, Figure 11.9.

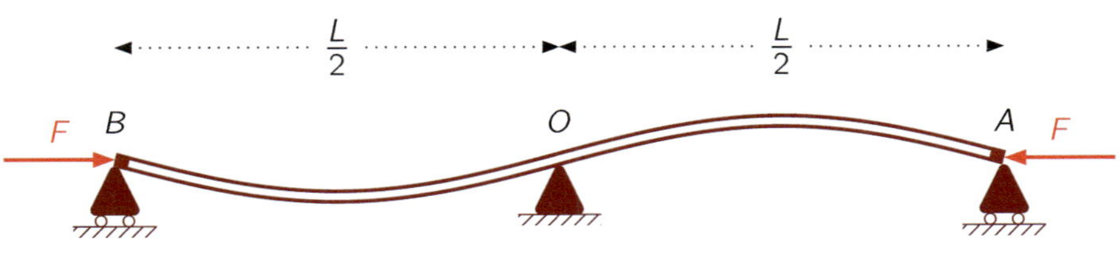

Figure 11.9 Compressed beam on three hinges.

The figure represents the fundamental mode, which is identical to Euler's second mode for a beam of length L. Here, one thus has

$$F_c = 4\pi^2 \frac{EI}{L^2} \tag{11.32}$$

Adding a middle support multiplies the buckling load by a factor 4 with respect to Euler's fundamental mode. ♡

Example 11.4: Buckling of Clamped–Free Beam

Consider a prismatic beam of length L, clamped at one end and loaded at the free end. Figure 11.10, which represents the beam and its symmetrical counterpart (dashed) with respect to the clamping plane (yOz), shows that its fundamental critical load is identical to Euler's for a beam of length $2L$. In other words,

$$F_c = \frac{\pi^2}{4} \frac{EI}{L^2} \tag{11.33}$$

Buckling of a clamped-free beam occurs at a quarter of the force needed to buckle a simply supported beam of the same length. ♡

Example 11.5: Doubly-Clamped Beam

The beam of length L is clamped at one end (the section at the origin O); the end section S^A has only one degree of freedom in translation along the axial direction, Figure 11.11.

As shown in the figure, which represents the fundamental mode, there are two inflexion points B and C, where the curvature $\frac{d^2v}{dx^2}$, hence the bending moment M, both vanish.

Letting $OB = CA = L'$ and $BC = L''$, one can assert the following:

- the portion BC is equivalent to a simply supported beam of length L'' (Euler's fundamental case); hence,

$$F_c = \pi^2 \frac{EI}{L''^2} \tag{11.34}$$

- the portion OB constitutes a clamped–free beam of length L', treated in Example 11.4; hence,

$$F_c = \frac{\pi^2}{4} \frac{EI}{L'^2} \tag{11.35}$$

The same applies to CA.

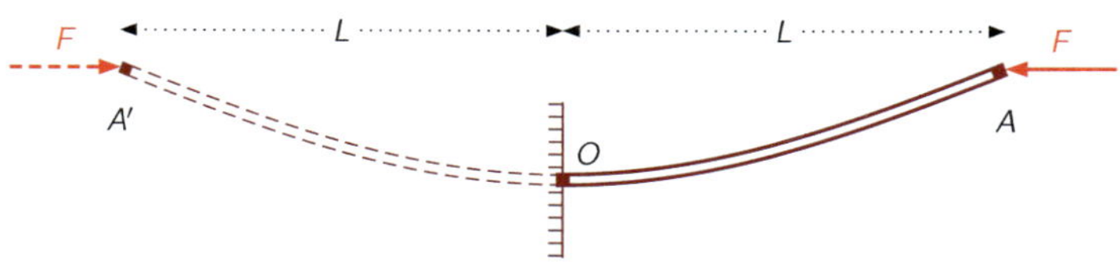

Figure 11.10 Clamped-free beam under compression.

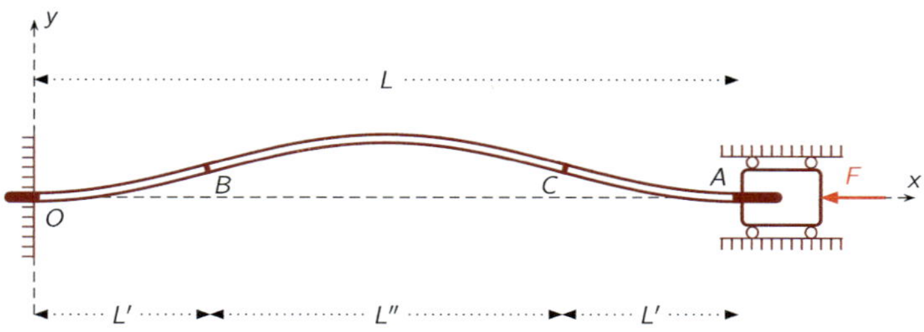

Figure 11.11 Compressed beam with no hinges.

All portions are subjected to the same uniform compression of magnitude F. Therefore, one may write, when buckling occurs,

$$L'' = \pi \sqrt{\frac{EI}{F_c}} \tag{11.36}$$

directly from Equation (11.34), and

$$L' = \frac{\pi}{2} \sqrt{\frac{EI}{F_c}} \tag{11.37}$$

from Equation (11.35). Along with the geometric relation $L = 2L' + L''$, the latter two equations give

$$L = 2\pi \sqrt{\frac{EI}{F_c}} \tag{11.38}$$

Finally, one gets

$$F_c = 4\pi^2 \frac{EI}{L^2}, \qquad L' = \frac{1}{4}L, \qquad L'' = \frac{1}{2}L \tag{11.39}$$

The buckling load for a doubly–clamped beam is the same as that of a beam on three supports; see Example 11.3. The fundamental mode shapes are, of course, different. ♡

11.3.2 GENERAL CASE

The number of cases that are amenable to Euler's problem, as treated in the previous section, is limited. In general, the methods introduced in Section 11.1.3 must be employed. In the examples below as well as in the remainder of this chapter, the direct method is used exclusively.

Example 11.6: Clamped–Pinned Beam

A beam of length L is clamped at one end and is simply supported at the other end, as illustrated in Figure 11.12. The beam is reported to a set of axes, as shown, with the xOy plane being a median plane. The origin of coordinates is placed at section O. The end section S^A has two degrees of freedom (axial translation and rotation).

One seeks to determine the magnitude of the compressive force capable of maintaining a configuration of the type shown in Figure 11.12. The structure is externally hyperstatic of degree one since there are three reactions (two forces Y^O, Y^A and one moment $\mathcal{M}^O$) for only two independent equilibrium equations:

$$Y^O + Y^A = 0, \qquad \mathcal{M}^O + LY^A = 0 \tag{11.40}$$

Take Y^A as a hyperstatic unknown and express the deflection $v(x)$.

At a distance x from the origin, the bending moment acting on the current cross-section centered about point G, Figure 11.12, is

$$M = -vF + (L-x)Y^A \tag{11.41}$$

where the first term emerges as in the classical Euler case (see Section 11.2.1) and the second term represents the moment of the reaction Y^A about G (the free-body diagram of the part GA downstream from G is not shown for that purpose.) The curvature–moment relation, Equation (11.13) or Equation (7.24), then provides the differential equation for the deflection:

$$\frac{\mathrm{d}^2v}{\mathrm{d}x^2} + \frac{F}{EI}v = \frac{Y^A}{EI}(L-x) \tag{11.42}$$

Its general solution may be written as follows:

$$v = A\cos\sqrt{\frac{F}{EI}}x + B\sin\sqrt{\frac{F}{EI}}x + \frac{Y^A}{F}(L-x) \tag{11.43}$$

where the last term represents a particular solution. The rotation of the current cross-section, identified with the slope, is then given by

$$v'(x) = -A\omega\sin\omega x + B\omega\cos\omega x - \frac{Y^A}{F} \qquad \text{with} \quad \omega \equiv \sqrt{\frac{F}{EI}} \tag{11.44}$$

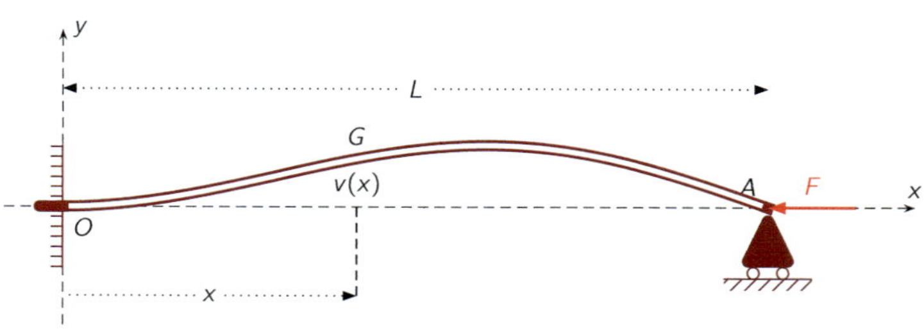

Figure 11.12 Compressed beam with one hinge.

The boundary conditions are as follows:

$$v(L) = 0 \quad \Rightarrow \quad A\cos\omega L + B\sin\omega L = 0 \tag{11.45}$$

$$v(0) = 0 \quad \Rightarrow \quad A + \frac{Y^A}{F}L = 0 \tag{11.46}$$

$$v'(0) = 0 \quad \Rightarrow \quad B\omega - \frac{Y^A}{F} = 0 \tag{11.47}$$

Noting that $F = EI\omega^2$, the above three conditions may be expressed as a homogeneous linear system:

$$\begin{bmatrix} \cos\omega L & \sin\omega L & 0 \\ EI\omega^2 & 0 & L \\ 0 & EI\omega^3 & -1 \end{bmatrix} \begin{Bmatrix} A \\ B \\ Y^A \end{Bmatrix} = \begin{Bmatrix} 0 \\ 0 \\ 0 \end{Bmatrix} \tag{11.48}$$

A trivial solution of the above system is $A = B = Y^A = 0$, which corresponds to the stable equilibrium configuration. The latter naturally emerges as a possible solution. For the above system to admit solutions other than the trivial one, it is necessary and sufficient that its determinant be nil, a condition that may be expressed after development as follows:

$$EI\omega^2(\sin\omega L - \omega L\cos\omega L) = 0 \tag{11.49}$$

The force F must therefore satisfy the transcendental equation:

$$\tan\omega L = \omega L \tag{11.50}$$

The smallest positive root of the above equation is $\omega L = 4.4934$ or

$$F_c = 20.1906\frac{EI}{L^2} \simeq \pi^2\frac{EI}{(0.7L)^2} \tag{11.51}$$

It corresponds to the fundamental mode of Figure 11.12 with

$$B = -0.2226A; \qquad Y^A = -20.20\frac{EI}{L^3}A \tag{11.52}$$

It can be checked that the inflexion point lies at a distance equal to $0.05L$ from the clamped end (i.e., et $0.95L$ from the support).

The subsequent critical loads are

$$F_c^2 = 59.72\frac{EI}{L^2} \simeq \pi^2\frac{EI}{(0.407L)^2}, \qquad F_c^3 = 118.9\frac{EI}{L^2} \simeq \pi^2\frac{EI}{(0.288L)^2} \tag{11.53}$$

and subsequently,

$$F_c^n \simeq \left(\frac{2n+1}{2}\right)^2\pi^2\frac{EI}{L^2} \tag{11.54}$$

♡

In the above example, there was no point load applied along the beam and the complexity arose because of the hyperstatic character. In a second example below, the beam is isostatic (statically determinate) but with a point load at mid-span arising from a middle support. Let us examine how the buckling analysis proceeds in such a case.

Example 11.7: Beam Pinned at the Origin with a Middle Support

Consider a beam with supports at the origin and at mid-span, Figure 11.13. Express that the figure represents a neutral equilibrium configuration under the tip compressive load F.

Global equilibrium, in addition to $X^O = F$, gives

$$Y^O + Y^A = 0 \qquad \text{and} \qquad LY^A + 2Fv^B = 0 \quad \Rightarrow \quad Y^A = -2\frac{v^B}{L}F \tag{11.55}$$

The bending moment, in the new configuration, writes

- For $0 < x < \frac{L}{2}$:

$$M = Y^A\left(\frac{L}{2} - x\right) - (v - v^B)F = F\left[2\frac{x}{L}v^B - v(x)\right] \tag{11.56}$$

where the expression of Y^A in terms of v^B in Equation (11.55) was used. Obtain then,

$$\frac{d^2v}{dx^2} + \frac{F}{EI}v = \frac{2F}{EI}v^B\frac{x}{L} \tag{11.57}$$

The general solution is

$$v = A_1\cos\omega x + B_1\sin\omega x + 2v^B\frac{x}{L} \qquad \text{with } \omega^2 = \frac{F}{EI} \tag{11.58}$$

The boundary conditions are

$$v(0) = 0 \quad \Rightarrow \quad A_1 = 0 \tag{11.59}$$

$$v\left(\frac{L}{2}\right) = 0 \quad \Rightarrow \quad B_1\sin\frac{\omega L}{2} + v^B = 0 \tag{11.60}$$

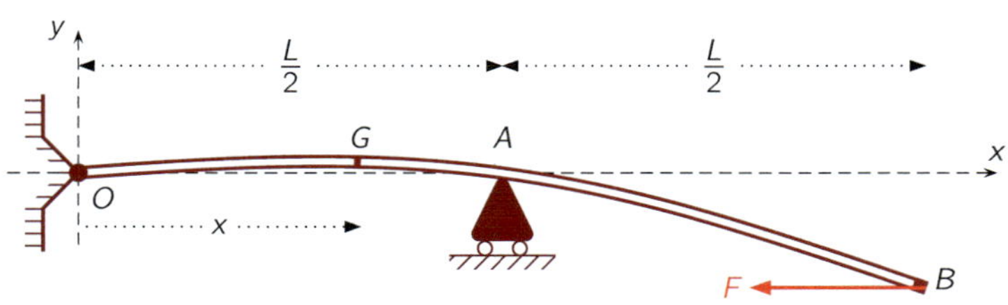

Figure 11.13 Beam with two hinges compressed at the free end.

- For $\frac{L}{2} < x < L$:

$$M = F\left[v^B - v(x)\right] \tag{11.61}$$

so that

$$\frac{\mathrm{d}^2 v}{\mathrm{d}x^2} + \frac{F}{EI}v = \frac{F}{EI}v^B \tag{11.62}$$

The general solution is

$$v = A_2 \cos\omega x + B_2 \sin\omega x + v^B \tag{11.63}$$

The boundary conditions are

$$v\left(\frac{L}{2}\right) = 0 \quad \Rightarrow \quad A_2 \cos\frac{\omega L}{2} + B_2 \sin\frac{\omega L}{2} + v^B = 0 \tag{11.64}$$

$$v'\left(\frac{L^-}{2}\right) = v'\left(\frac{L^+}{2}\right) \quad \Rightarrow$$

$$B_1 \cos\frac{\omega L}{2} + A_2 \sin\frac{\omega L}{2} - B_2 \cos\frac{\omega L}{2} + \frac{2}{\omega L}v^B = 0 \tag{11.65}$$

$$v(L) = v^B \quad \Rightarrow \quad A_2 \cos\omega L + B_2 \sin\omega L = 0 \tag{11.66}$$

The integration constants satisfy therefore the following linear system:

$$\begin{bmatrix} \sin\frac{\omega L}{2} & 0 & 0 & 1 \\ 0 & \cos\frac{\omega L}{2} & \sin\frac{\omega L}{2} & 1 \\ 0 & \cos\omega L & \sin\omega L & 0 \\ \cos\frac{\omega L}{2} & \sin\frac{\omega L}{2} & -\cos\frac{\omega L}{2} & \frac{2}{\omega L} \end{bmatrix} \begin{Bmatrix} B_1 \\ A_2 \\ B_2 \\ v^B \end{Bmatrix} = \begin{Bmatrix} 0 \\ 0 \\ 0 \\ 0 \end{Bmatrix} \tag{11.67}$$

This homogeneous system admits non-trivial solutions if its determinant Δ vanishes. For that, one must have

$$\Delta = \frac{2\sin\frac{\omega L}{2}}{\omega L}\left(\sin\frac{\omega L}{2} - \omega L\cos\frac{\omega L}{2}\right) = 0 \tag{11.68}$$

This equation admits two sets of positive solutions:

- those of $\sin\frac{\omega L}{2} = 0$
- those of $\tan\frac{\omega L}{2} = \omega L$

The smallest of all is the first of the second set:

$$\frac{\omega L}{2} = 1.1655 \quad \Rightarrow \quad F_c^1 = 5.4336\frac{EI}{L^2} = \pi^2\frac{EI}{(1.3477L)^2} \tag{11.69}$$

It corresponds to the fundamental mode represented in Figure 11.13.

The next solution is the first of the first set:

$$\omega L = 2\pi \quad \Rightarrow \quad F_c^2 = 4\pi^2\frac{EI}{L^2} \tag{11.70}$$

It corresponds to buckling following Euler's second mode, with two half-waves and A as center of symmetry. ♡

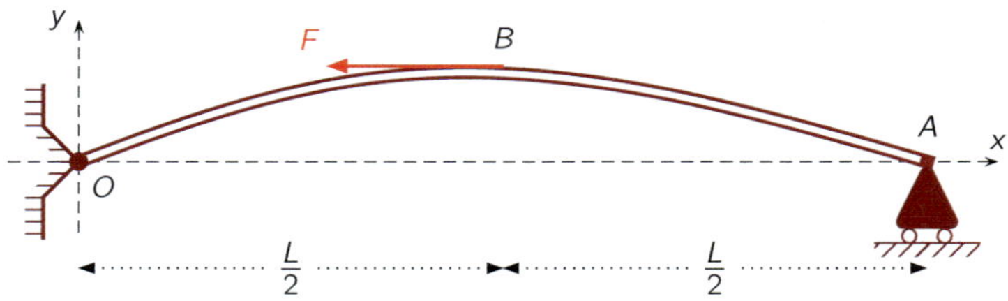

Figure 11.14 Beam with two hinges compressed at mid-span.

Remark 11.5 *The above problem is statically determinate (isostatic). However, the beam's deflection in neutral equilibrium can take an arbitrary magnitude. Correspondingly, the reactions at the supports would be commensurate with tip deflection v^B as per Equation* (11.55). $*$

Example 11.8: Doubly Supported Beam Loaded in Compression at Mid-Span

Figure 11.14 shows the links, axes and loading. Write the conditions under which it represents a neutral equilibrium configuration under the action of force F. In stable equilibrium (straight beam), portion OB is under uniform compression of magnitude F whereas portion BA is not loaded.

In the configuration of Figure 11.14, global equilibrium writes

$$X^O = F, \quad Y^O + Y^A = 0, \quad LY^A + Fv^B = 0 \quad \Rightarrow \quad Y^A = -\frac{v^B}{L}F \tag{11.71}$$

The bending moment, in the new configuration, writes

- on portion OB:

$$M = Y^A(L-x) - (v - v^B)F = -Fv + Fv^B\frac{x}{L} \tag{11.72}$$

so that

$$\frac{\mathrm{d}^2v}{\mathrm{d}x^2} + \frac{F}{EI}v = \frac{F}{EI}v^B\frac{x}{L} \tag{11.73}$$

Using the usual notation $\omega^2 = \frac{F}{EI}$ and boundary conditions $v(0) = 0$ and $v\left(\frac{L}{2}\right) = v^B$, the general solution of Equation (11.73) is

$$v = B\sin\omega x + v^B\frac{x}{L} \qquad \text{with} \quad B\sin\frac{\omega L}{2} = \frac{v^B}{2} \tag{11.74}$$

- on portion BA:

$$M = Y^A(L-x) = -Fv^B\left[1 - \frac{x}{L}\right] \tag{11.75}$$

so that

$$\frac{\mathrm{d}^2v}{\mathrm{d}x^2} = -\frac{F}{EI}v^B\left[1 - \frac{x}{L}\right] \tag{11.76}$$

which admits the general solution

$$v = v^B (\omega L)^2 \left[\frac{1}{6}\frac{x^3}{L^3} - \frac{1}{2}\frac{x^2}{L^2} + \left(\frac{11}{24} - \frac{2}{(\omega L)^2} \right) \frac{x}{L} + \frac{2}{(\omega L)^2} - \frac{1}{8} \right] \tag{11.77}$$

having used boundary conditions

$$v\left(\frac{L}{2}\right) = v^B \quad \text{and} \quad v(L) = 0 \tag{11.78}$$

It also follows that

$$\frac{\mathrm{d}v}{\mathrm{d}x} = \frac{v^B}{L}(\omega L)^2 \left[\frac{1}{2}\frac{x^2}{L^2} - \frac{x}{L} + \frac{11}{24} - \frac{2}{(\omega L)^2} \right] \tag{11.79}$$

Slope continuity in B finally gives the characteristic equation:

$$\tan \frac{\omega L}{2} = \frac{3\frac{\omega L}{2}}{\left(\frac{\omega L}{2}\right)^2 - 9} \tag{11.80}$$

The smallest (non-trivial) positive solution is

$$\frac{\omega L}{2} = 2.16 \qquad \Rightarrow \qquad F_c = 18.66\frac{EI}{L^2} = \pi^2 \frac{EI}{(0.727L)^2} \tag{11.81}$$

This critical load corresponds to the fundamental mode of Figure 11.14. ♡

11.4 METHODS OF CALCULATION OF CRITICAL LOADS

A general direct method is first presented for analyzing buckling in structures.An example is provided which concerns the buckling of a frame. Often, it is impossible to implement the method exactly, for reasons that will be made clear below. Therefore, an approximate method is introduced and illustrated through an example. The approximate method is fully analytical but hints at a key step needed in implementing more systematic approximate methods based on energy principles. The latter go beyond the scope of this text.

11.4.1 GENERAL APPROACH

Before analyzing other structural instability problems, the direct method followed to tackle the Euler problem is extended to the general case of an assemblage of beams. The method consists of five steps.

Step 1: starting from the structure in equilibrium, consider it under load and in a neighboring (infinitely close) configuration, compatible with the links.

Step 2: in this new configuration, write the equilibrium of the structure, then express, on each cross-section, the screwer (i.e., internal resultants), which is function of the loading, any potential hyperstatic unknowns and the new position of the that section.

Step 3: Infer the system of differential equations that must be satisfied by the displacements (translations and/or rotations) of the current cross-section.

Step 4: Integrate the system, then write the conditions imposed by the links.

Step 5: Determine the particular values of the loading giving rise to non-trivial solutions, i.e., solutions making it possible for equilibrium to be neutral (or indifferent): one thus obtains a series of critical loads. To each of these corresponds a buckling mode.

Remark 11.6 *In general, one is only interested in the fundamental mode, associated with the smallest critical load.* ∗

Remark 11.7 *Each buckling mode corresponds to a neutral equilibrium and hence is defined up to a multiplicative constant.* ∗

Remark 11.8 *The critical loads constitute the eigenvalues of the problem; the corresponding buckling modes are the associated eigenvectors.* ∗

Example 11.9: Buckling by Plane Bending of a Gantry under Compressive Loading

Figure 11.15a represents the gantry at rest; Figure 11.15b represents it under load in a deformed, kinematically possible neighboring configuration. The top inset shows a magnified deformation of beam CB.

Global equilibrium gives:

$$X^A = F\left(1-\frac{2h}{L'}\right) \simeq F; \quad X^D = F\left(1+\frac{2h}{L'}\right) \simeq F; \quad Y^D = -Y^A \tag{11.82}$$

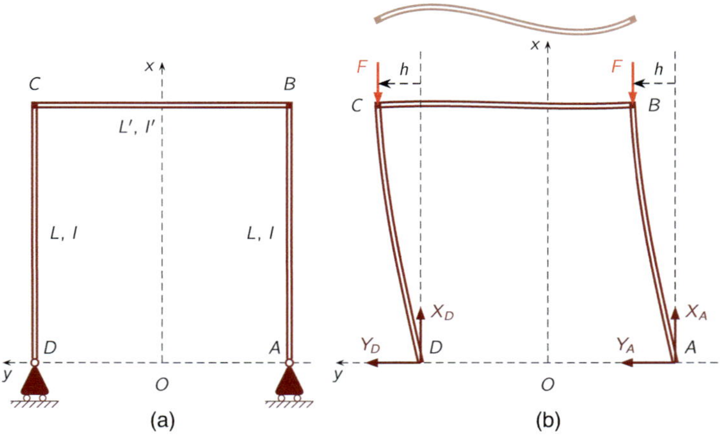

Figure 11.15 Gantry at rest (a) and free-body diagram in kinematically admissible neighboring configuration (b).

Orienting the structure in the sense $ABCD$, the bending moment has the following expressions:

$$\text{on } AB: \quad M = -Fv + Y^A x \quad \Rightarrow \quad M^B = -Fh + Y^A L \tag{11.83}$$

$$\text{on } BC: \quad M = 2\frac{h}{L'}Fy + Y^A L \tag{11.84}$$

$$\text{on } CD: \quad M = Fv + Y^A x \quad \Rightarrow \quad M^C = Fh + Y^A L \tag{11.85}$$

where Y^A is a hyperstatic unknown.

Let us express the deflection $v(x)$ on the two legs AB and CD, taking into account the condition $v(0) = 0$ on each, and denoting $\omega^2 = F/EI$:

On vertical member AB:

$$\frac{d^2v}{dx^2} + \omega^2 v = \frac{Y^A}{EI}x \quad \Rightarrow \quad v(x) = a\sin\omega x + \frac{Y^A}{F}x \tag{11.86}$$

with

$$v(L) = a\sin\omega L + \frac{Y^A}{F}L = h \tag{11.87}$$

$$\omega^B = v'(L) = a\omega\cos\omega L + \frac{Y^A}{F} \tag{11.88}$$

On vertical member CD:

$$\frac{d^2v}{dx^2} + \omega^2 v = -\frac{Y^A}{EI}x \quad \Rightarrow \quad v(x) = b\sin\omega x - \frac{Y^A}{F}x \tag{11.89}$$

with:

$$v(L) = b\sin\omega L - \frac{Y^A}{F}L = h \tag{11.90}$$

$$\omega^C = b\omega\cos\omega L - \frac{Y^A}{F} \tag{11.91}$$

On the transversal BC, Bresse's relations for rotations and deflections, Equation (10.5), give

$$\omega^B = \frac{L'}{6EI'}(Fh - 3Y^A L) \quad \text{and} \quad \omega^C = \frac{L'}{6EI'}(Fh + 3Y^A L) \tag{11.92}$$

By equating these expressions of ω^B and ω^C with those calculated above, Equations (11.88) and (11.91), obtain the two equations:

$$a\,\omega\cos\omega L + Y^A\left(\frac{1}{F} + \frac{1}{2}\frac{LL'}{EI'}\right) - h\frac{L'F}{6EI'} = 0 \tag{11.93}$$

$$b\,\omega\cos\omega L - Y^A\left(\frac{1}{F} + \frac{1}{2}\frac{LL'}{EI'}\right) - h\frac{L'F}{6EI'} = 0 \tag{11.94}$$

Equations (11.87), (11.90), (11.93) and (11.94) involve the four constants a, b, Y^A, and h that characterize the neutral equilibrium configuration of Figure 11.15b.

They are assembled in the following homogeneous linear system:

$$\begin{bmatrix} \sin\omega L & 0 & \frac{L}{F} & -1 \\ 0 & \sin\omega L & -\frac{L}{F} & -1 \\ \omega\cos\omega L & 0 & \frac{1}{F}+\frac{LL'}{2EI'} & -\frac{L'F}{6EI'} \\ 0 & \omega\cos\omega L & -\frac{1}{F}-\frac{LL'}{2EI'} & -\frac{L'F}{6EI'} \end{bmatrix} \begin{Bmatrix} a \\ b \\ Y^A \\ h \end{Bmatrix} = \begin{Bmatrix} 0 \\ 0 \\ 0 \\ 0 \end{Bmatrix} \tag{11.95}$$

This system admits non-trivial solutions if its determinant is nil, which writes after some algebraic manipulations:

$$\omega L\frac{IL'}{I'L}\left(2+\omega^2L^2\frac{IL'}{I'L}\right)\tan^2\omega L-4\left(3+2\omega^2L^2\frac{IL'}{I'L}\right)\tan\omega L+12\omega L=0 \tag{11.96}$$

Put in this form, the equation appears as one of second-order in $\tan\omega L$, which facilitates resolution. Two sets of roots thus emerge

- those of: $\omega L\tan\omega L = 6\dfrac{I'L}{IL'}$
- those of: $\tan\omega L = \dfrac{2\omega L}{2+\omega^2L^2\frac{IL'}{I'L}}$

The smallest positive root is the first of the first set.

In the special case where all three bars are identical ($I = I'$ and $L = L'$), one gets

$$\omega L = 1.3496 \qquad \Rightarrow \qquad F_c^1 = 1.8216\frac{EI}{L^2} = 0.1846\pi^2\frac{EI}{L^2} \tag{11.97}$$

which corresponds to the buckling mode depicted in Figure 11.15b. The transversal beam BC has some mild waviness, as magnified in the top inset. Note that the angles at junctions B and C are right angles. The shorter the legs the larger the waviness of the transversal. ♡

11.4.2 APPROXIMATE METHODS

Integrating the differential equations of the problem, object of step 4 of the methodology of Section 11.4.1, is often impossible analytically. To carry out the integration, indeed, one needs to express the screwer on each beam cross-section in the deformed configuration, which is *a priori* unknown.

To alleviate such difficulty, one can determine the critical loading, with very good accuracy, by assuming *a priori* an approximate deformed configuration that respects the conditions imposed by the links, the loading, the limits. This approximation is typically implemented in energy-based methods but it can also be used in a direct method, as illustrated in the below example.

Example 11.10: Buckling of a Mast Subject to Its Own Weight

Consider the stability of a vertical prismatic mast, clamped at its base, under the sole action of its weight, Figure 11.16a.

Let L be the height of the mast, p its (uniform) weight per unit length, and $I = I_Z$ its cross-sectional second moment of area, with $I_Z < I_Y$.

Figure 11.16a represents the mast at rest (undeformed, before loading); Figure 11.16b represents it in a kinematically admissible neighboring configuration (deformed, under load).

The problem is first solved using the direct method of Section 11.4.1 with no approximation. Mast equilibrium, in the neutral configuration of Figure 11.16b, writes

$$X^O = pL, \quad Y^O = 0, \quad M^O = -p\int_0^L v(\xi)\,\mathrm{d}\xi \tag{11.98}$$

The bending moment acting on the cross-section at height x, admits the expression:

$$M(x) = p\int_x^L [v(\xi) - v(x)]\,\mathrm{d}\xi \tag{11.99}$$

The deflection then satisfies the integro-differential equation:

$$EI\frac{\mathrm{d}^2 v}{\mathrm{d}x^2} = p\int_x^L [v(\xi) - v(x)]\,\mathrm{d}\xi = -p\left[\int_L^x v(\xi)\,\mathrm{d}\xi + (L-x)v(x)\right] \tag{11.100}$$

Differentiating both sides, obtain the differential equation:

$$\frac{\mathrm{d}^3 v}{\mathrm{d}x^3} + \frac{p}{EI}(L-x)\frac{\mathrm{d}v}{\mathrm{d}x} = 0 \tag{11.101}$$

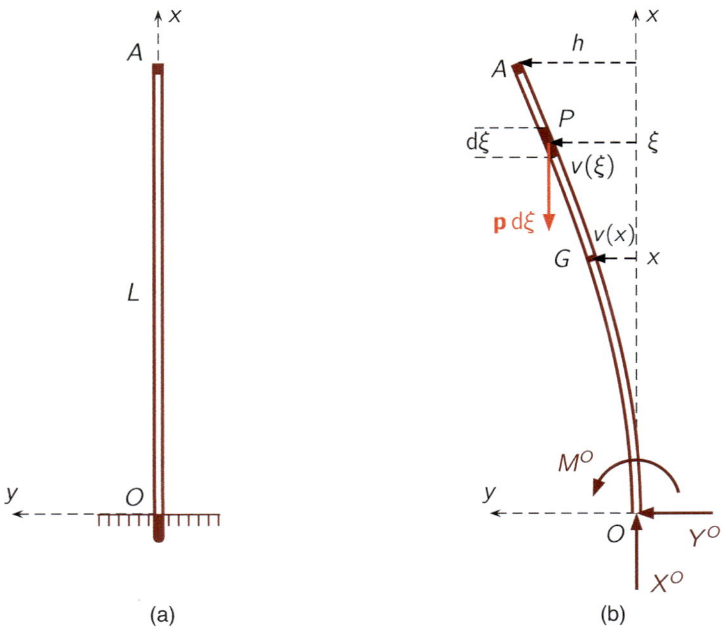

Figure 11.16 Mast at rest (a) and free-body diagram in kinematically admissible neighboring configuration (b).

The function $u(x) = \frac{\mathrm{d}v}{\mathrm{d}x}$ satisfies the linear, homogeneous second-order differential equation:

$$\frac{\mathrm{d}^2u}{\mathrm{d}x^2} + \frac{p}{EI}(L-x)u = 0 \tag{11.102}$$

The latter can be solved using Bessel functions (see Problem 11.22). The sought critical loads are found to be

$$p_c = 7.837\frac{EI}{L^3} \qquad \Rightarrow \qquad L_c = 1.986\sqrt[3]{\frac{EI}{p}} \tag{11.103}$$

The latter equation gives the critical height of the mast having a certain weight density p, beyond which equilibrium loses stability.

Next, the critical loads are recalculated using a direct method, but now an approximate expression, say $v_1(x)$, is assumed for the deflection in the neighboring configuration. The approximation must satisfy the four imposed conditions:

$$v_1(0) = 0 \tag{11.104}$$

$$v_1'(0) = 0 \tag{11.105}$$

$$v_1(L) = h \tag{11.106}$$

$$v_1''(L) = 0 \tag{11.107}$$

Equations (11.104) and (11.105) are kinematic boundary conditions; Equation (11.106) expresses a limit; Equation (11.107) is a loading condition (no moment imposed at A).

The lowest-degree polynomial that satisfies the above conditions is

$$v_1(x) = \frac{h}{2L^3}\left[3Lx^2 - x^3\right] \tag{11.108}$$

To this approximate deflection v_1 corresponds the bending moment:

$$M_1(x) = p\int_x^L [v(\xi) - v(x)]\,\mathrm{d}\xi = \frac{3}{8}\frac{ph}{L^3}\left[L^4 - 4L^2x^2 + 4Lx^3 - x^4\right] \tag{11.109}$$

If one then calculates a second approximate value $v_2(x)$ for the deflection by integrating the equation:

$$\frac{\mathrm{d}^2v_2}{\mathrm{d}x^2} = \frac{M_1}{EI} = \frac{3}{8}\frac{ph}{EIL^3}\left[L^4 - 4L^2x^2 + 4Lx^3 - x^4\right] \tag{11.110}$$

one obtains, account being taken of the conditions $v_2(0) = 0$ and $v_2'(0) = 0$:

$$v_2(x) = \frac{1}{80}\frac{ph}{EIL^3}\left[15L^4x^2 - 10L^2x^4 + 6Lx^5 - x^6\right] \tag{11.111}$$

The condition $v_2(L) = h$ then provides a first approximation of critical values:

$$p_c = 8\frac{EI}{L^3} \qquad \Rightarrow \qquad L_c = 2\sqrt[3]{\frac{EI}{p}} \tag{11.112}$$

The error made, in excess, relative to the exact solution of Equation (11.103) is 2% on the critical weight p_c and 0.7% on the critical length L_c.

A second iteration, starting from $v_2(x)$ and following the same methodology, would lead to a second, even better, approximation for p_c and L_c. ♡

11.5 LATERAL BUCKLING

11.5.1 PRESENTATION OF THE PHENOMENON

Consider the prismatic beam of Figure 11.17a having a rectangular cross-section, such that the bending stiffness EI_Z is large relative to the bending stiffness EI_Y and relative to the torsional stiffness GJ.

To the end section centered about A is applied a downward force F parallel to Oy; the beam's own weight is comparatively negligible.

The beam is thus loaded in plane bending with a moment $M_Z = -(L-x)F$ and a transverse shear $V_Y = -F$ acting on the cross-section at x, L being the beam's length. Upon increasing the magnitude of the applied load F, the latter may reach a critical value F_c for which the bending ceases to be planar to 'compose' with a bending deformation in the (xOz) plane and a torsion, as depicted in Figure 11.17b: that is lateral buckling.

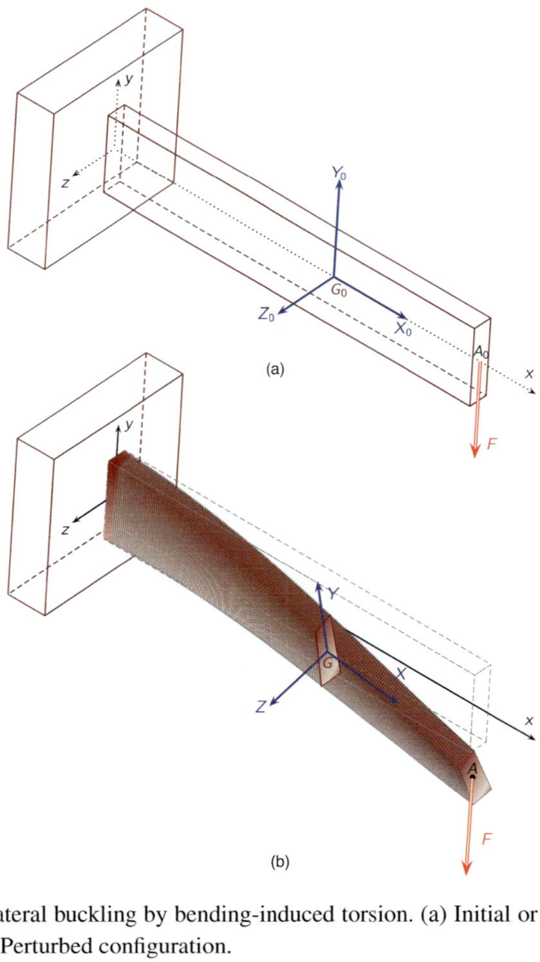

Figure 11.17 Lateral buckling by bending-induced torsion. (a) Initial or stable equilibrium configuration. (b) Perturbed configuration.

DEFINITION 11.4 Lateral buckling is the instability that occurs by slouchy bending with torsion of a beam admitting a median plane, say (xOy), loaded in that plane and having stiffnesses EI_Y and GJ small relative to EI_Z ♣

This type of instability can be analyzed by the general method of Section 11.4.1. Let $\{G_0, X_0Y_0Z_0\}$ and $\{G, XYZ\}$ denote the principal frame of cross-section S at abscissa x, in the configuration at rest (Figure 11.17a) and in a neighboring deformed configuration (Figure 11.17b), respectively. Also, let $[P]$ denote the transformation matrix allowing to pass from the first frame to the second.

Reducing each term to its principal component, one has

$$[P] = \begin{bmatrix} 1 & -\frac{dv}{dx} & -\frac{dw}{dx} \\ \frac{dv}{dx} & 1 & -\omega_x \\ \frac{dw}{dx} & \omega_x & 1 \end{bmatrix} \quad \text{and} \quad [P]^{-1} = \begin{bmatrix} 1 & \frac{dv}{dx} & \frac{dw}{dx} \\ -\frac{dv}{dx} & 1 & \omega_x \\ -\frac{dw}{dx} & -\omega_x & 1 \end{bmatrix} \tag{11.113}$$

where $v(x)$ and $w(x)$ are the components of the deflection $\overrightarrow{G_0G}$ along Oy and Oz, and $\omega_x(x)$ the angle of twist; see Problem 11.23. These matrices permit the differential equations of the problem to be written in $\{G, XYZ\}$, that is in the deformed configuration.

11.5.2 EXAMPLES

In what follows, some important cases are treated limiting the attention to prismatic beams and neglecting any influence of transverse shear.

Example 11.11: Lateral Buckling under "Pure" Bending

A cantilever rectangular beam of length L, height $2a$ and width $2b$ $(a > b)$ is bent along the cross-section's greatest dimension. The height over width ratio is taken as $a/b = 4$. With the axes as shown in Figure 11.18, the second moments of area are

$$I_Z = \frac{4}{3}a^3b, \quad I_Y = \frac{4}{3}ab^3 \quad \Rightarrow \quad \frac{I_Y}{I_Z} = \frac{b^2}{a^2} = \frac{1}{16} = 0.0625 \tag{11.114}$$

Thus, the flexural rigidity in the vertical xOy plane, where the beam is bent, is much greater than out of plane. Likewise, from the tabulated results in Example 8.3, the torsion constant J for $a/b = 4$ is estimated as follows:

$$J = 0.28a^2b^2 \quad \Rightarrow \quad \frac{J}{I_Z} = 0.21\frac{b}{a} = 0.0525 \tag{11.115}$$

The torsion rigidity is, therefore, small compared with the flexural rigidity in the vertical plane. We have a situation prone to lateral buckling.

With respect to the $\{G_0, X_0Y_0Z_0\}$ frame associated with the undeformed beam, the internal moment acting on the current cross-section is simply the applied moment $\mathbf{M} = M\mathbf{e}_z$. The undeformed beam is shown dashed in Figure 11.18. Using the transformation matrix $[P]^{-1}$ in Equation (11.113), the components of $\mathbf{M}$ are

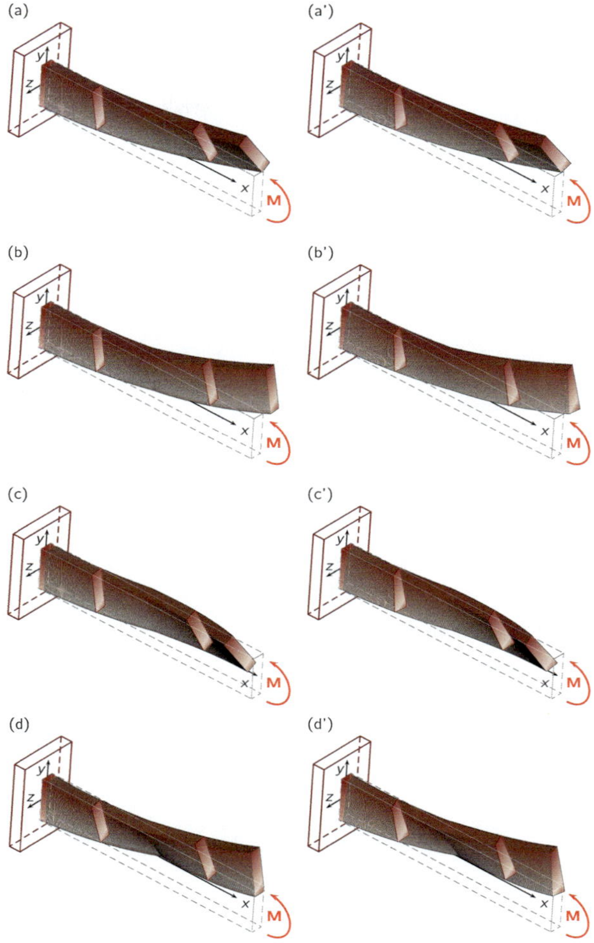

Figure 11.18 Buckled configurations under "pure" bending for various end conditions at $x = L$: (a) six DOFs allowed, (b) $\omega_X = 0$, (c) $\omega_Y = 0$, (d) $\omega_X = 0$ and $\omega_Y = 0$. (a')–(d') With warping included.

obtained in the $\{G, XYZ\}$ frame, associated with the perturbed configuration, as follows:

$$\begin{cases} M_X = M^t = M\dfrac{\mathrm{d}w}{\mathrm{d}x} \\ M_Y = M\omega_x \\ M_Z = M \end{cases} \tag{11.116}$$

The torsion-bending coupling is noteworthy. The secondary bending rotation, $\omega_Y = w'$, implies a torque while the twist ω_x implies an out-of-plane bending moment.

Recall the following relations from the analysis of bending (Chapter 7) and torsion (Chapter 8):

$$\begin{cases} M^t = GJ\dfrac{\mathrm{d}\omega_x}{\mathrm{d}x} \\ M_Y = -EI_Y\dfrac{\mathrm{d}^2 w}{\mathrm{d}x^2} \\ M_Z = EI_Z\dfrac{\mathrm{d}^2 v}{\mathrm{d}x^2} \end{cases} \tag{11.117}$$

The third of these equations is immediately integrated using $v(0) = v'(0) = 0$ and gives

$$\frac{\mathrm{d}v}{\mathrm{d}x} = \omega_Z = \frac{M}{EI_Z}x \quad \text{and} \quad v(x) = \frac{M}{2EI_Z}x^2 \tag{11.118}$$

The first two relations in Equation (11.117) give, on account of the expressions of M^t and M_Y in Equation (11.116):

$$\begin{cases} GJ\dfrac{\mathrm{d}\omega_x}{\mathrm{d}x} = M\dfrac{\mathrm{d}w}{\mathrm{d}x} \\ EI_Y\dfrac{\mathrm{d}^2 w}{\mathrm{d}x^2} = -M\omega_x \end{cases} \tag{11.119}$$

Differentiating the first gives

$$GJ\frac{\mathrm{d}^2\omega_x}{\mathrm{d}x^2} = M\frac{\mathrm{d}^2 w}{\mathrm{d}x^2} \tag{11.120}$$

Eliminating w from the above equation using Equation $(11.119)_2$ leads to

$$\frac{\mathrm{d}^2\omega_x}{\mathrm{d}x^2} + \lambda^2\omega_x = 0 \qquad \text{with} \quad \lambda^2 = \frac{M^2}{EI_Y\,GJ} \tag{11.121}$$

which is the governing differential equation for the twist ω_X.

The general solution of Equation (11.121) is

$$\omega_x = A\cos\lambda x + B\sin\lambda x \tag{11.122}$$

where A and B are integration constants, to be determined based on boundary conditions.

The solution depends on details of boundary conditions to a great extent. With no claim of precision in the drawn deformed configurations, Figure 11.18 illustrates some situations for a cantilever beam. For each set of boundary conditions, two special sections are depicted at a quarter and at three-quarters of the span in order to have a better sense of the overall deformation. The latter is more easily visualized when omitting warping. For completeness, warping is included in the primed version of each subfigure. The warping is represented using results from Example 8.3 keeping terms up to seventh order.

In Figure 11.18a, it is assumed that all six degrees of freedom are allowed at the tip ($x = L$). In Figure 11.18b, the tip section is not allowed to rotate about its axis. In Figure 11.18c, the tip section is not allowed to bend out of plane. In Figure 11.18d, the tip section has both $\omega_x = 0$ and $\omega_Y = 0$.

In practice, simple solutions are obtained when all cross-sections are free to warp. For example, if the tip rotation about the beam's axis is not allowed, then the boundary conditions associated with Equation (11.121) are

$$\omega_x(0) = 0 \quad \Rightarrow \quad A = 0 \tag{11.123}$$

$$\omega_x(L) = 0 \quad \Rightarrow \quad B \sin \lambda L = 0 \tag{11.124}$$

A nontrivial solution is thus obtained for $\lambda L = n\pi$ with n an integer, such that the critical moment is obtained as follows:

$$\lambda L = \pi \quad \Rightarrow \quad M_c = \pi \frac{\sqrt{EI_Y\, GJ}}{L} \tag{11.125}$$

The buckled shape depicted in Figure 11.17b is an approximation.

The critical moment given by Equation (11.125) scales inversely with the beam's length and is proportional to the geometric average of the torsion and secondary flexural rigidities. The prefactor depends on specific boundary conditions.

It is not common to bend a rectangular beam along its greatest in-plane dimension. However, lateral buckling is a concern for I beams among others. ♡

Note the similarity between the differential equation (11.121) governing the twist in lateral buckling and that governing the deflection v in beam-column buckling, Equation (11.14) of Section 11.2.1. The case of pure bending is the only one that leads to such a simple differential equation. Lateral buckling problems usually lead to second-order ODEs with a non-constant coefficient for the ω_x term. Their resolution generally involves implementing Bessel's functions. This is illustrated in the following example.

Example 11.12: Lateral Buckling of a Cantilever Beam

Consider the rectangular beam of Figure 11.17a. Determine the critical load F_c.

The moment of the screwer on S is $\mathbf{M} = \overrightarrow{GA} \times \mathbf{F}$. In component form, one has

$$\begin{Bmatrix} M_{X_0} \\ M_{Y_0} \\ M_{Z_0} \end{Bmatrix} = \begin{Bmatrix} L - x \\ v(L) - v(x) \\ w(L) - w(x) \end{Bmatrix} \times \begin{Bmatrix} 0 \\ -F \\ 0 \end{Bmatrix} = \begin{Bmatrix} F\,[w(L) - w(x)] \\ 0 \\ -F(L-x) \end{Bmatrix} \tag{11.126}$$

with respect to $\{G_0, X_0Y_0Z_0\}$. Using $[P]^{-1}$, the components of $\mathbf{M}$ are obtained in $\{G, XYZ\}$, only retaining the principal component:

$$\begin{cases} M_X = M^t = F\,[w(L) - w(x)] - F(L-x)\dfrac{\mathrm{d}w}{\mathrm{d}x} \\ M_Y = -F(L-x)\omega_x \\ M_Z = -F(L-x) \end{cases} \tag{11.127}$$

As in Example 11.11, the following differential equations hold:

$$M^t = GJ\frac{\mathrm{d}\omega_x}{\mathrm{d}x}, \qquad M_Y = -EI_Y\frac{\mathrm{d}^2 w}{\mathrm{d}x^2}, \qquad M_Z = EI_Z\frac{\mathrm{d}^2 v}{\mathrm{d}x^2} \tag{11.128}$$

Integration of the third equation is straightforward and gives

$$\frac{\mathrm{d}v}{\mathrm{d}x} = \omega_Z = -\frac{F}{2EI_Z}x(2L-x) \quad \text{and} \quad v(x) = -\frac{F}{6EI_Z}x^2(3L-x) \tag{11.129}$$

using boundary conditions $v(0) = v'(0) = 0$. The other two relations write

$$\begin{cases} GJ\dfrac{\mathrm{d}\omega_x}{\mathrm{d}x} = F\left[w(L) - w(x)\right] - F(L-x)\dfrac{\mathrm{d}w}{\mathrm{d}x} \\ EI_Y\dfrac{\mathrm{d}^2 w}{\mathrm{d}x^2} = F(L-x)\omega_x \end{cases} \tag{11.130}$$

They show the coupling between the secondary bending and torsion. Differentiating the first yields

$$GJ\frac{\mathrm{d}^2\omega_x}{\mathrm{d}x^2} = -F(L-x)\frac{\mathrm{d}^2 w}{\mathrm{d}x^2} \tag{11.131}$$

Eliminating w from the above equation using the second part of Equation (11.130) yields then

$$\frac{\mathrm{d}^2\omega_x}{\mathrm{d}x^2} + k^2(L-x)^2\omega_x = 0 \qquad \text{with} \quad k^2 = \frac{F^2}{EI_Y\, GJ} \tag{11.132}$$

The general solution of the above differential equation is

$$\omega_x = \sqrt{L-x}\left[A J_{0.25}\left(\frac{k}{2}(L-x)^2\right) B J_{-0.25}\left(\frac{k}{2}(L-x)^2\right)\right] \tag{11.133}$$

where $J_{0.25}$ and $J_{-0.25}$ denote Bessel functions of the first kind, of order $\frac{1}{4}$ and $-\frac{1}{4}$, respectively.

The boundary conditions associated with Equation (11.132) are

- For $x = L$, one has $M^t = 0$ hence $\frac{\mathrm{d}\omega_x}{\mathrm{d}x} = 0$, which imposes $A = 0$.
- For $x = 0$, one has $\omega_x = 0$ so that

$$BJ_{-0.25}\left(\frac{kL^2}{2}\right) = 0 \tag{11.134}$$

Excluding the trivial solution ($B = 0$), the smallest root of the above equation is

$$\frac{kL^2}{2} = 2.0063 \qquad \Rightarrow \qquad F_c = 4.0126\frac{\sqrt{EI_Y\, GJ}}{L^2} \tag{11.135}$$

The buckling mode shown in Figure 11.17 is an approximate depiction. ♡

Remark 11.9 *In this section, it has been tacitly assumed that the torsional stiffness GJ is constant along the axis of the beam. In reality, this is true only if all cross-sections are free to warp. If certain sections cannot warp freely, then the stiffness GJ becomes, in all rigor, a function of x and, on average, it increases with respect to the free-warping scenario. As a result, the resistance to lateral buckling is expected to be higher than calculated above. Therefore, the cross-sectional warping shown in Figure 11.17, which is associated with torsion, does not affect the above analysis.* ∗

Remark 11.10 *An approximate treatment of warping-affected lateral buckling consists of introducing a constant "warping rigidity" that would augment the net torque by a term proportional to the second derivative of the twist angle* ω_x. *This eventually results in a third order differential equation for* ω_x. *Other secondary effects, namely transverse shear effects, may also be included. Covering these aspects goes beyond the scope of this elementary text.* *

Example 11.13: Buckling of a Doubly Clamped Beam under Its Own Weight

A horizontal beam, subjected to its own weight, is connected to the ground by two sliding links at both ends, as shown in Figure 11.19a.

The asymmetric cross-section has its center of twist C lying on the principal axis GZ, as depicted in Figure 11.19b, with $\overline{GC} = \delta$.

In a neighboring, kinematically possible configuration, the beam may twist and deflect into the horizontal (xOz) plane, Figure 11.19b. In this new configuration, the reactions have the following components, in $\{O,xyz\}$:

$$\mathbf{R}^A \begin{cases} 0 \\ p\dfrac{L}{2} \\ 0 \end{cases} \qquad \mathbf{M}^A \begin{cases} M_x^A \\ M_y^A \\ M_z^A \end{cases} \qquad \mathbf{R}^B \begin{cases} 0 \\ p\dfrac{L}{2} \\ 0 \end{cases} \qquad \mathbf{M}^B \begin{cases} M_x^A \\ -M_y^A \\ -M_z^A \end{cases} \tag{11.136}$$

Three is therefore the degree to which the problem is externally hyperstatic.

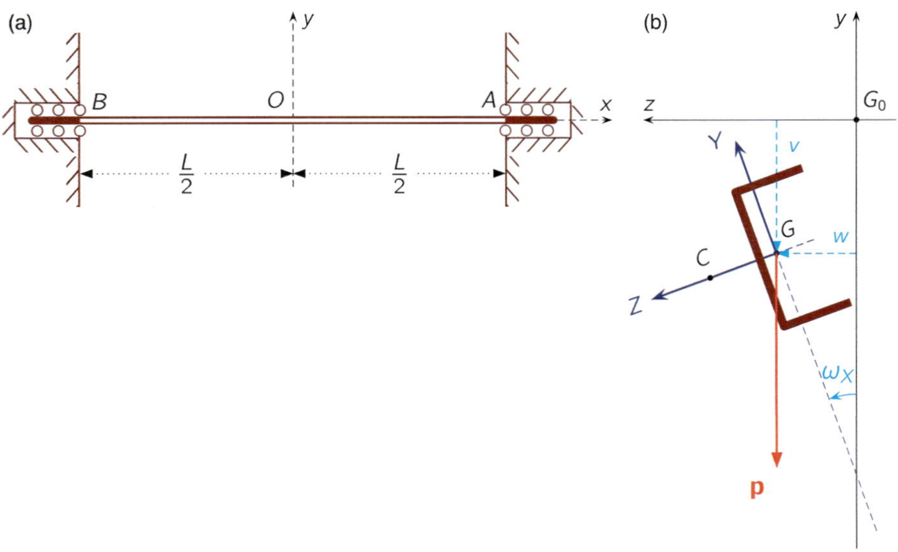

Figure 11.19 Beam with two hinges under its own weight (a) in stable equilibrium (b) cross-section in buckled state after bending and twisting.

The screwer at x admits as a resultant:

$$\mathbf{R} = \mathbf{R}^A + \left(\frac{L}{2} - x\right)\mathbf{p} \tag{11.137}$$

and as a resultant moment:

$$\mathbf{M} = \mathbf{M}^A + \overrightarrow{GA} \times \mathbf{R}^A + \int_x^{\frac{L}{2}} \overrightarrow{GG'} \times \mathbf{p}\,\mathrm{d}\xi \tag{11.138}$$

where G' denotes the center of the section located at abscissa ξ.

Using the matrix $[P]^{-1}$ of Equation (11.113), the components of $\mathbf{M}$ are obtained in the basis of $\{G, XYZ\}$, again only retaining the principal components:

$$\left\{\begin{array}{l} M_X = M_X^A + pxw(x) + p\displaystyle\int_x^{\frac{L}{2}} w(\xi)\,\mathrm{d}\xi + \frac{\mathrm{d}w}{\mathrm{d}x}\left[M_Z^A + \frac{p}{2}\left(\frac{L^2}{4} - x^2\right)\right] \\ M_Y = M_Y^A + \omega_X\left[M_Z^A + \dfrac{p}{2}\left(\dfrac{L^2}{4} - x^2\right)\right] \\ M_Z = M_Z^A + \dfrac{p}{2}\left(\dfrac{L^2}{4} - x^2\right) \end{array}\right. \tag{11.139}$$

The torque is thus inferred as

$$M^t = M_X + px\delta \tag{11.140}$$

The bending portion of the problem within the vertical (xOy) plane is decoupled; determining M_Z^A, $v(x)$ and $\frac{\mathrm{d}v}{\mathrm{d}x} = \omega_Z$ is then immediate and gives

$$M_Z^A = -\frac{pL^2}{12}, \quad v(x) = -\frac{pL^4}{24EI_Z}\left(\frac{1}{4} - \frac{x^2}{L^2}\right)^2, \quad \frac{\mathrm{d}v}{\mathrm{d}x} = \frac{pL^3}{6EI_Z}\frac{x}{L}\left(\frac{1}{4} - \frac{x^2}{L^2}\right) \tag{11.141}$$

For the other coupled bending–torsion problem, the relations between internal resultants and displacements write

$$\left\{\begin{array}{l} M^t = GJ\dfrac{\mathrm{d}\omega_x}{\mathrm{d}x} = M_X^A + px[\delta + w(x)] + p\displaystyle\int_x^{\frac{L}{2}} w(\xi)\,\mathrm{d}\xi + \frac{\mathrm{d}w}{\mathrm{d}x}\frac{p}{2}\left[\frac{L^2}{12} - x^2\right] \\ M_Y = -EI_Y\dfrac{\mathrm{d}^2 w}{\mathrm{d}x^2} = M_Y^A + \omega_x\dfrac{p}{2}\left[\dfrac{L^2}{12} - x^2\right] \end{array}\right. \tag{11.142}$$

Differentiating the first, one obtains

$$\frac{GJ}{p}\frac{\mathrm{d}^2\omega_x}{\mathrm{d}x^2} = \delta + \frac{1}{2}\left[\frac{L^2}{12} - x^2\right]\frac{\mathrm{d}^2 w}{\mathrm{d}x^2} \tag{11.143}$$

or, account being taken of the second relation in Equation (11.142):

$$\frac{\mathrm{d}^2\omega_x}{\mathrm{d}x^2} + \frac{p^2}{4EI_Y\,GJ}\left[\frac{L^2}{12} - x^2\right]^2 \omega_X = \frac{p}{GJ}\left[\delta - \frac{M_Y^A}{2EI_Y}\left(\frac{L^2}{12} - x^2\right)\right] \tag{11.144}$$

The general solution of the above equation is not known.

Implementing the approximate method used above (see Example 11.10), start from

$$\omega_X^1 = \omega_0 \left(1 - 4\frac{x^2}{L^2}\right) \tag{11.145}$$

as a first approximation of the function $\omega_X(x)$ to obtain

$$p_c = 110.33\,\frac{\sqrt{EI_Y\,GJ}}{L^3} \qquad \Rightarrow \qquad L_c = 4.80\left(\frac{EI_Y\,GJ}{p^2}\right)^{\frac{1}{6}} \tag{11.146}$$

♡

11.6 BUCKLING OF CURVED BEAMS

An exhaustive study of buckling of curved beams cannot possibly be carried out nor can it be done for prismatic beams. The modalities of buckling, the types of loadings, the shapes of the beams are too numerous. On the other hand, the general method expounded in Section 11.4.1 is always valid. It will be illustrated below in two analyses of buckling by plane bending of curved beams admitting a median plane and loaded in pure compression.

Example 11.14: Ring under Uniform External Pressure

Consider a circular ring loaded as shown in Figure 11.20a. Let p be the compressive force per unit length of the median curve and assume $I_Y > I_Z = I$.

In order to determine the fundamental critical load p_c, impose to the ring a very small ovalization, of principal axes Ox and Oy (duly amplified in Figure 11.20b), and calculate the value p_c of p capable of maintaining it. Figure 11.20a and Figure 11.20b respectively represent the ring in the initial and final configurations. Between the

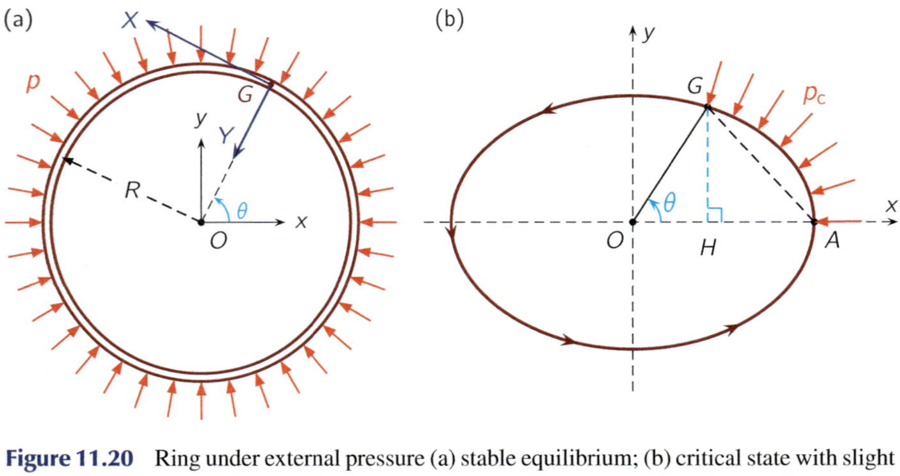

Figure 11.20 Ring under external pressure (a) stable equilibrium; (b) critical state with slight ovalization and partially depicted pressure.

two, the normal force N is left unchanged but a bending moment $M(\theta)$ has appeared, causing for the current cross-section a translation of components $U(\theta)$ and $V(\theta)$, along the axes GX and GY, respectively.

Specializing Equation (7.23) from Chapter 7 to the deflections due to bending *alone*, one gets

$$\begin{cases} \dfrac{\mathrm{d}^2V}{\mathrm{d}\theta^2} + V = \dfrac{R^2M}{EI} \\ \dfrac{\mathrm{d}U}{\mathrm{d}\theta} = V \end{cases} \tag{11.147}$$

To calculate $M(\theta)$, operate a cut (as a thought experiment) following section S^A at the origin over which one has

$$N = -p\overline{OA}, \qquad \mathcal{V} = 0, \qquad M = M^O \tag{11.148}$$

with $\mathcal{V}$ the transverse shear, not to be confused with the deflection V. The moment M^O is a hyperstatic unknown.

The force per unit length p on arc $\widehat{AG}$ is statically equivalent to the same force distributed over the chord AG, Figure 11.20b. Thus, one has

$$M = M^O + \frac{1}{2}p\left(\overline{AG}^2 - 2\,\overline{OA}\,\overline{HA}\right) \tag{11.149}$$

or also

$$M = M^O - \frac{1}{2}p\left(\overline{OA}^2 - \overline{OG}^2\right) \tag{11.150}$$

where account has been taken of the metric relation

$$\overline{AG}^2 = \overline{OG}^2 + \overline{OA}^2 - 2\,\overline{OH}\,\overline{OA} \tag{11.151}$$

Finally, with $\overline{OA} = R - V(0)$ and $\overline{OG} = R - V(\theta)$, and to the neglect of second-order terms:

$$M = M^O + pR\left[V(0) - V(\theta)\right] \tag{11.152}$$

where M^O is still undetermined.

To lift the indeterminacy, use the first Bresse formula, Equation (10.5):

$$0 = \int_0^{\frac{\pi}{2}} M\,\mathrm{d}\theta = \frac{\pi}{2}\left[M^O + pRV(0)\right] - pR\int_0^{\frac{\pi}{2}} V(\theta)\,\mathrm{d}\theta \tag{11.153}$$

There is no need to compute the last integral explicitly. Indeed, the second relation in Equation (11.147) gives upon integration between A and B:

$$\int_0^{\frac{\pi}{2}} V(\theta)\,\mathrm{d}\theta = U\left(\frac{\pi}{2}\right) - U(0) = 0 \tag{11.154}$$

Therefore,

$$M^O = -pRV(0) \qquad \text{and} \qquad M = -pRV(\theta) \tag{11.155}$$

The differential equation governing the deflection V in Equation (11.147) is thus written:

$$\frac{\mathrm{d}^2V}{\mathrm{d}\theta^2} + k^2V = 0 \qquad \text{with} \qquad k^2 = 1 + \frac{pR^3}{EI} \tag{11.156}$$

Its general solution is

$$V = A\cos k\theta + B\sin k\theta \tag{11.157}$$

The conditions $V'(0) = 0 = V'\left(\frac{\pi}{2}\right)$ impose $B = 0$ and

$$\sin k\frac{\pi}{2} = 0 \tag{11.158}$$

so that $k = 2$ (smallest positive solution). One finally gets

$$p_c = 3\frac{EI}{R^3}, \qquad V(\theta) = A\cos 2\theta, \qquad U(\theta) = \frac{A}{2}\sin 2\theta \tag{11.159}$$

♡

Example 11.15: Arch under Distributed Loading

Consider the parabolic arch $\widehat{AB}$ of Figure 11.21, having a median line in the xOy plane and a constant cross-section with $I = I_Z < I_Y$. The arch is connected to the ground through hinges about the z-axis.

The ascending vertical is Oy; the median line is oriented from B to A. The loading consists of a downward vertical force of constant intensity p per unit length of the x-axis. The weight is negligible.

Such an arch works in pure compression: on each cross-section the only non-zero component of the screwer is the normal force $\mathbf{N}$. Indeed, as seen in Section 6.6 of Chapter 6, the equilibrium shape of a cable fixed at both ends and subjected to a similar vertical loading is a parabolic arc; see Example 6.7. In other words, the parabolic arc is the funicular curve for vertical pressure loading; see Remark 6.9.

Here, the equation of the parabola is:

$$y = h\left(1 - \frac{x^2}{a^2}\right) \tag{11.160}$$

where x and y denote the coordinates of current point G on the median line, $2a$ the length of chord BA and h the rise of the arch.

Using the results obtained for the cable in Example 6.7, one can assert that:

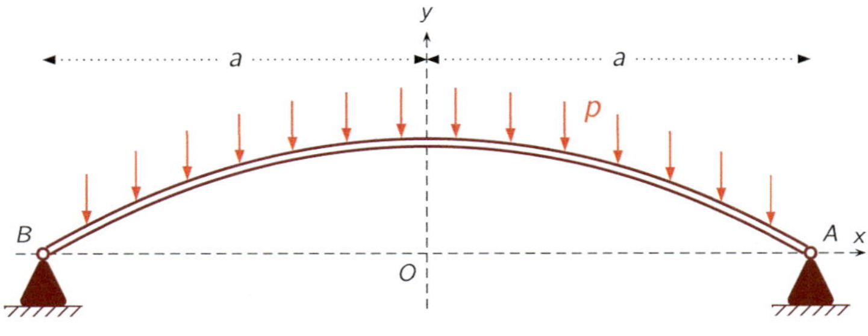

Figure 11.21 Parabolic arch under uniform vertical loading.

- the reaction $\mathbf{R}^A$ has components:

$$X^A = -p\frac{a^2}{2h} \quad \text{and} \quad Y^A = pa \tag{11.161}$$

- the reaction $\mathbf{R}^B$ has components:

$$X^B = p\frac{a^2}{2h} \quad \text{and} \quad Y^B = pa \tag{11.162}$$

- the normal force $\mathbf{N}$ has components:

$$N_x = -p\frac{a^2}{2h} \quad \text{and} \quad N_y = px \tag{11.163}$$

The aim is to determine the fundamental critical load p_c beyond which equilibrium of the arch becomes unstable.

The arch being funicular, the bending moment in G is zero in the stable equilibrium configuration. This leads to a relationship between three moments: that of X^A, that of X^B and that of the distributed load. There is no need to express that.

In an infinitely close configuration, the current point G has coordinates x and $[y(x)+v(x)]$ so that a bending moment emerges due to the change in both the armlength and magnitude of the thrust X^B. If

$$X^B + \xi X^B = \frac{pa^2}{2h}(1+\xi)$$

denotes the new horizontal reaction in B, the bending moment can be expressed as follows:

$$M(x) = -X^B v(x) - \xi X^B y(x) = -\frac{1}{2}p\frac{a^2}{h}\left[v(x) + \xi h\left(1 - \frac{x^2}{a^2}\right)\right] \tag{11.164}$$

Attention is limited to a very low arch, i.e., such that $h \ll a$. In this case, the change in curvature of the median line is

$$\frac{\mathrm{d}^2 v}{\mathrm{d}x^2} = \frac{M}{EI} \tag{11.165}$$

which can also be written as follows:

$$\frac{\mathrm{d}^2 v}{\mathrm{d}x^2} + \omega^2 v = -\omega^2 h\xi\left(1 - \frac{x^2}{a^2}\right) \quad \text{with} \quad \omega^2 = \frac{pa^2}{2hEI} \tag{11.166}$$

The general solution of the above differential equation is

$$v(x) = A\cos\omega x + B\sin\omega x - h\xi\left(1 - \frac{x^2}{a^2} + \frac{2}{a^2\omega^2}\right) \tag{11.167}$$

The coefficients A, B, ξ are determined using the conditions:

$$v(-a) = 0, \quad v(a) = 0, \quad u^B - u^A = 0 \text{ (Bresse)} \tag{11.168}$$

which, all calculations done, may be written as follows:

$$\begin{bmatrix} \cos\omega a & -\sin\omega a & -\frac{2h}{a^2\omega^2} \\ \cos\omega a & \sin\omega a & -\frac{2h}{a^2\omega^2} \\ \sin\omega a & 0 & -2h\left(\frac{a\omega}{3}+\frac{1}{a\omega}\right) \end{bmatrix} \begin{Bmatrix} A \\ B \\ \xi \end{Bmatrix} = \begin{Bmatrix} 0 \\ 0 \\ 0 \end{Bmatrix} \tag{11.169}$$

Note that the last boundary condition above is expressed while neglecting the deformations due to the normal force and transverse shear when using Equation $(10.5)_2$ such that the latter reduces to:

$$\int_{-a}^{a} yv''\,\mathrm{d}x = \int_{-a}^{a} y'v'\,\mathrm{d}x = 0 \tag{11.170}$$

where integration by parts has been used.

For non-trivial solutions to exist, the determinant of the above system must vanish. This condition writes

$$\sin\omega a\left[\sin\omega a - \omega a\left(1+\frac{\omega^2 a^2}{3}\right)\cos\omega a\right] = 0 \tag{11.171}$$

Its solutions are those:

- on the one hand of:

$$\sin\omega a = 0 \tag{11.172}$$

- on the other hand, those of:

$$\tan\omega a = \omega a + \frac{(\omega a)^3}{3} \tag{11.173}$$

The smallest positive solution is:

$$\omega a = \pi \quad \Rightarrow \quad p_c = 2\pi^2\frac{hEI}{a^4} \quad \text{and} \quad v(x) = B\sin\frac{\pi x}{a} \tag{11.174}$$

This is therefore an antisymmetric mode, which does not introduce any additional reaction at the supports ($\xi = 0$). The buckled shape is depicted in Figure 11.22 with the initial arch shown dashed. ♡

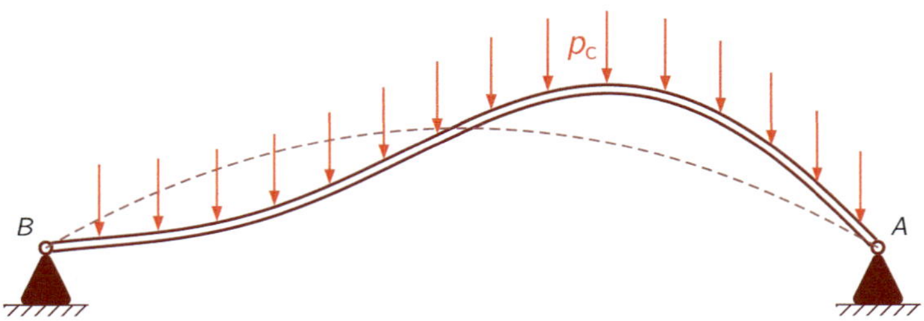

Figure 11.22 Buckled mode of parabolic arch under uniform vertical loading.

$F_c = \frac{\pi^2 EI}{L^2}$	$F_c = 4\frac{\pi^2 EI}{L^2}$	$F_c = \frac{1}{4}\frac{\pi^2 EI}{L^2}$	$F_c = 4\frac{\pi^2 EI}{L^2}$

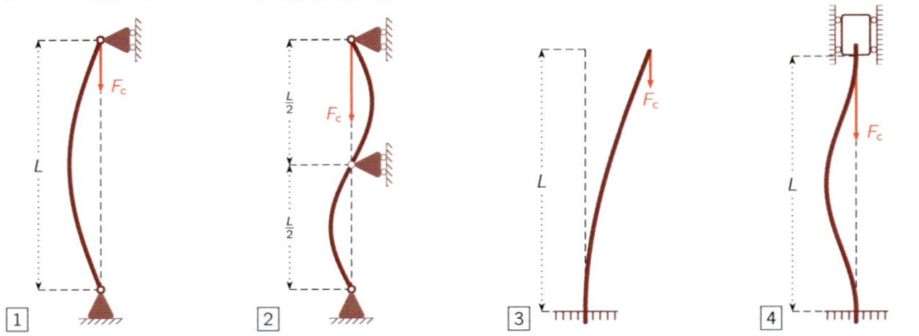

$F_c = 2.046\frac{\pi^2 EI}{L^2}$	$F_c = 0.551\frac{\pi^2 EI}{L^2}$	$F_c = 1.891\frac{\pi^2 EI}{L^2}$	$\tan \omega L = -\frac{\lambda}{L}\omega L$

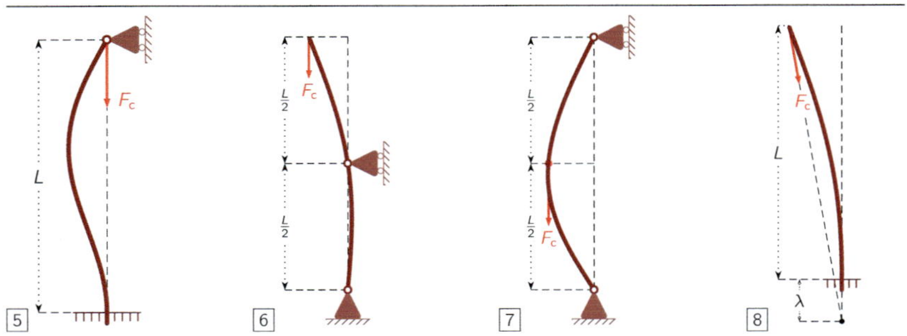

11.7 FORMULAE FOR BEAM BUCKLING

Attention is limited to the fundamental mode of buckling for beams admitting a median plane and having (at least piecewise) constant cross-section.

11.7.1 BUCKLING BY PLANE BENDING OF COMPRESSED BEAMS

Case [1] is Euler's ideal case covered in Section 11.2.1. Cases [2], [3], and [4] are amenable to Euler's ideal case, as treated in Examples 11.3 to 11.5 of Section 11.3.1. Cases [5], [6], and [7] correspond to Examples 11.6–11.8 of Section 11.3.2.

Case [8] is an example of a "follower force" type problem where the force's line of action is made to pass through a fixed point. A realization of this scenario using a rope is shown in Problem 11.15.

In case [9], the cantilevered beam-column is tangentially compressed at the free end. This case belongs to so-called "follower forces", which are non-conservative. The buckling force cannot be obtained by a static method, as described in this book, nor can it be obtained using an energy method. This case requires a dynamic stability analysis that goes beyond the scope of this book. The result is due to Beck.

9	10	11	12
$F_c = 2.008\frac{\pi^2 EI}{L^2}$	$(pL)_c = 0.794\frac{\pi^2 EI}{L^2}$	$(pL)_c = 1.996\frac{\pi^2 EI}{L^2}$	$F_c = 3.390\frac{\pi^2 EI}{L^2}$

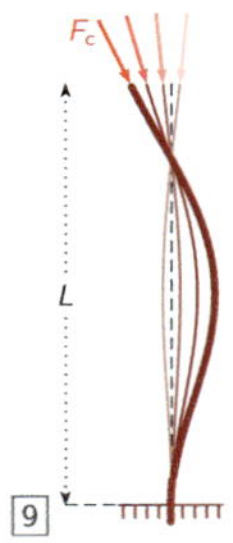

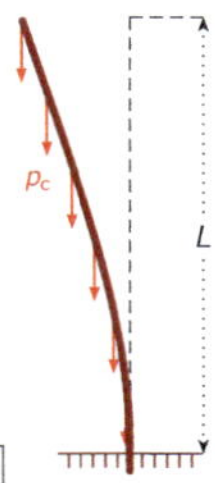

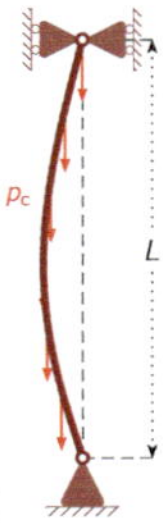

$\frac{ab\omega}{a+b}\sin(a+b)\omega = \sin\omega a \sin\omega b$

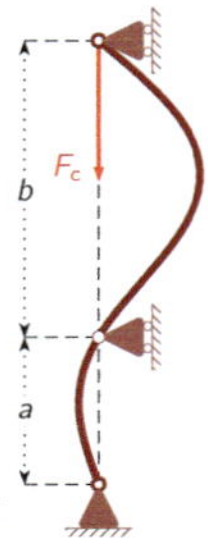

12

13	14	15	16
$F_c = 1.670\frac{\pi^2 EI}{L^2}$	$F_c = 1.020\frac{\pi^2 EI}{L^2}$	$F_c = 0.247\frac{\pi^2 EI}{L^2}$	$F_{1c} = 0.638\frac{\pi^2 EI}{L^2}$

$\tan\omega a + \tan\omega b = \omega(a+b)$

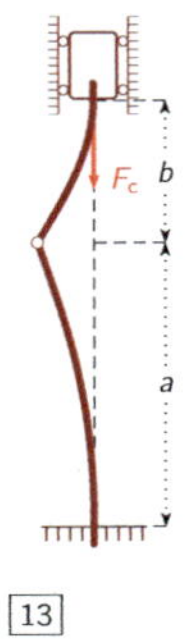

$\sqrt{I_1}\tan\omega_1\ell_1 + \sqrt{I_2}\tan\omega_2\ell_2 = 0$

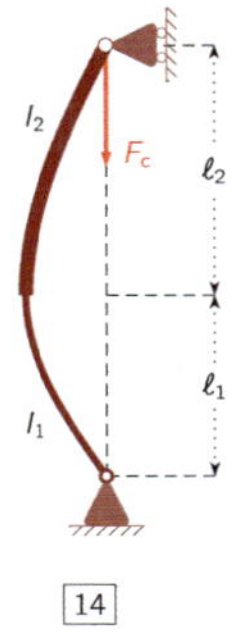

$\sqrt{I_2}\tan\omega_1\ell_1\tan\omega_2\ell_2 = \sqrt{I_1}$

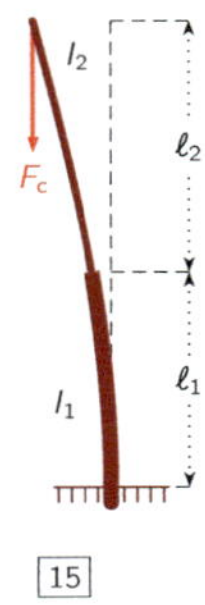

$\omega_2 I_2\tan\omega_1\ell_1\tan\omega_2\ell_2 + \omega_1 I_1 = 0$

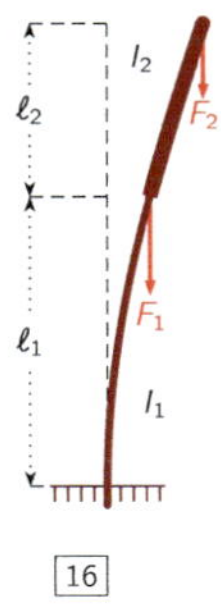

Long after it was obtained, no definite conclusion was made regarding its practical value, since no method had been devised for applying a tangential force to a column during bending. The Beck column problem, and similar problems, have generated immense theoretical interest and are still subject of research in structural dynamics. Experimental realization of this type of loading was finally achieved by aerospace engineers in Japan. Nowadays, follower force loaded beams and plates are typically used as simple models of missiles, rockets and space structures, such as solar panels.

Case [10] was treated in Example 11.10 using two methods. The critical value reported in the table is that obtained by the direct method using Bessel's functions. The mast of case [11] is the subject of Problem 11.20 where the critical load is obtained by approximate iterations.

For cases [12] to [19] where the geometry is parameterized, the characteristic equation for F_c is transcendental and is given rotated next to the mode shape. The

17	18	19	20
$F_c = 2.854\,\frac{\pi^2 EI}{L^2}$	$F_c = 0.795\,\frac{\pi^2 EI}{L^2}$	$F_c = 0.643\,\frac{\pi^2 EI}{L^2}$	$F_c = 3.661\,\frac{\pi^2 EI}{L^2}$
$\left(\frac{1}{\omega L} - \frac{1}{\tan\omega L} + \frac{EI\omega}{k}\right)\left(\frac{1}{\omega L} - \frac{1}{\tan\omega L} + \frac{EI\omega}{k'}\right) = \left(\frac{1}{\omega L} - \frac{1}{\sin\omega L}\right)^2$	$\sin\omega a \sin\omega b = \left(\omega a - EI\omega^3/k\right)\sin\omega(a+b)$	$2(1-\cos\omega a)\tan\omega b = \sin\omega a + (\omega^3 EI/k - \omega a)(\cos\omega a - \sin\omega a \tan\omega b)$	$F_n = \frac{\pi^2 EI}{L^2}\left[n^2 + \frac{kL^4}{n^2\pi^4 EI}\right]$
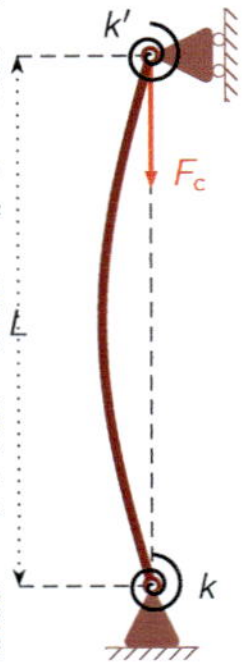	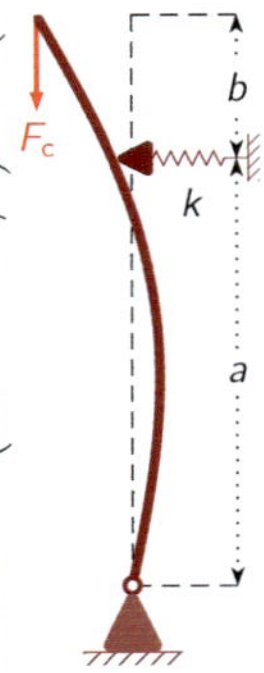	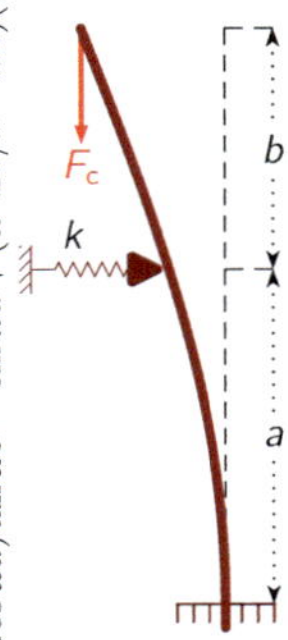	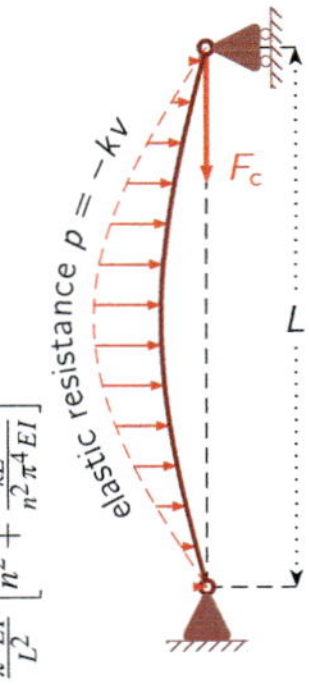

transcendental equation is written in terms of ω defined by:

$$\omega^2 = \frac{F_c}{EI}$$

or, where applicable, in terms of ω_1, ω_2 defined by:

$$\omega_1^2 = \frac{F_c}{EI_1} \qquad \omega_2^2 = \frac{F_c}{EI_2}$$

except for case 16 where

$$\omega_1^2 = \frac{F_{1c}}{EI_1} \qquad \omega_2^2 = \frac{F_{2c}}{EI_2}$$

Each value of F_c given in the table corresponds to the realization shown and is given in units of Euler's critical load with $L = a + b$ or $L = \ell_1 + \ell_2$. When two beams are welded together, cases 14 to 16, the parallel assembly formula is used:

$$\frac{EI}{L} = \left[\frac{\ell_1}{EI_1} + \frac{\ell_2}{EI_2}\right]^{-1}$$

For case 12 the value of 3.390 corresponds to $a/L = 1/3$. For case 13 the value of 1.670 is for $a/L = 2/3$. For case 14 the value of 1.020 is for $a/L = 5/12$ and

$I_2/I_1 = 2$. For case $\boxed{15}$ the value of 0.247 is for $a/L = 11/24$ and $I_2/I_1 = 1/2$. For case $\boxed{16}$ the value of 0.638 is for $a/L = 15/24$ and $I_2/I_1 = 2$. It is the smallest value of F_1 required for buckling when the tip force F_2 is half the critical value when a tip force alone is applied, which is

$$F_c = 0.231\pi^2 \frac{EI}{L^2}$$

For case $\boxed{17}$ the value of 2.854 corresponds to $\bar{k} = kL/EI = k'L/EI = 10$. The prefactor must always lie between 1 and 4 corresponding to $k = k' = 0$ (Euler's ideal case $\boxed{1}$) and $k = k' \to \infty$ (case $\boxed{4}$). For example, for $\bar{k} = 1000$, the prefactor is 3.985.

For the two cases with linear springs, $\boxed{18}$ and $\boxed{19}$, the spring is at rest when the beam is not loaded, its median line being then identified with the dashed line.

For case $\boxed{18}$ the value of 0.795 corresponds to $\bar{k} = kL^3/EI = 50$. Reducing the stiffness to $\bar{k} = 10$ would give a prefactor of 0.470, but increasing it to $\bar{k} = 1000$ would only increase the prefactor to 0.840.

For case $\boxed{19}$ the value of 0.643 is for $\bar{k} = kL^3/EI = 100$. Dividing the stiffness by ten, $\bar{k} = 10$, would give a prefactor of 0.356.

Finally, for case $\boxed{20}$, the integer n corresponds to the number of half-waves of the buckled beam. The realization shown corresponds to $\bar{k} = kL^4/EI = 4^4$, which leads to $n = 1$, as shown in the table. Other values of $\bar{k}$ may lead to different mode shapes, as treated in Problem 11.18.

11.7.2 LATERAL BUCKLING

Results are tabulated below for cases where the center of twist coincides with the centroid. All cross-sections are free to warp; see Remark 11.9. The vertical plane is the X-Y plane and it is assumed that $I_Y < I_Z$. In other words, the flexural rigidity in the vertical plane is much greater than out of plane. This favors buckling by torsion and slouchy bending. It is emphasized that the degrees of freedom (DOFs), allowed or prevented, affect the critical loads.

In cases $\boxed{1}$, $\boxed{2}$, $\boxed{3}$ and $\boxed{10}$, the applied moment $\mathbf{M}$ remains parallel to the fixed out-of-plane z-axis. In cases $\boxed{4}$, $\boxed{8}$ and $\boxed{9}$, the force $\mathbf{F}$ remains parallel to the vertical y-axis and is applied to the center of the section. In cases $\boxed{5}$, $\boxed{6}$ and $\boxed{7}$, the distributed force p remains parallel to the vertical y-axis and is applied to the median line.

$M_{\mathrm{c}} = \frac{\pi}{2}\frac{\sqrt{EI_Y GJ}}{L}$	$M_{\mathrm{c}} = \pi\frac{\sqrt{EI_Y GJ}}{L}$
six DOFs allowed at tip	$\omega_X = 0$ at tip
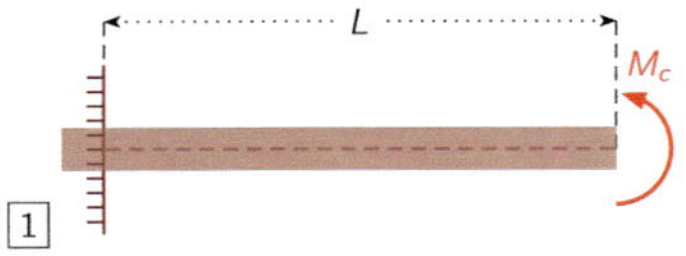	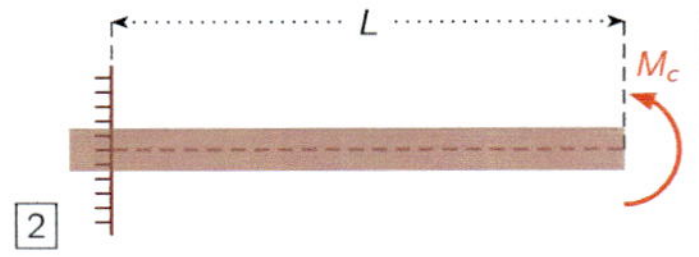
$M_{\mathrm{c}} = 2\pi\frac{\sqrt{EI_Y GJ}}{L}$	$F_{\mathrm{c}} = 4.013\frac{\sqrt{EI_Y GJ}}{L^2}$
$\omega_X = \omega_Y = 0$ at tip	six DOFs allowed at tip
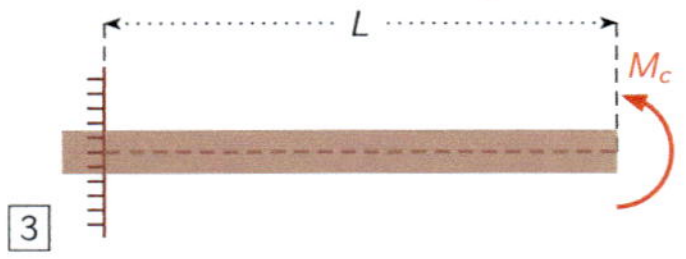	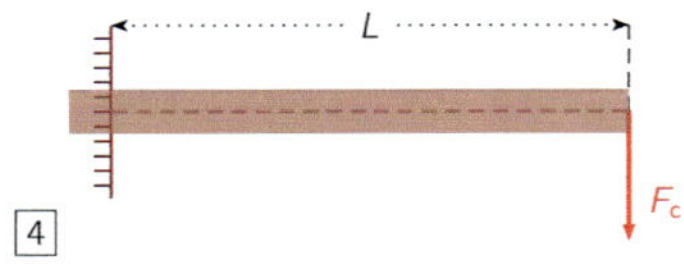
$p_{\mathrm{c}} = 12.85\frac{\sqrt{EI_Y GJ}}{L^3}$	$p_{\mathrm{c}} = 28.30\frac{\sqrt{EI_Y GJ}}{L^3}$
six DOFs allowed at tip	three DOFs at both hinges: u, ω_Y, ω_Z
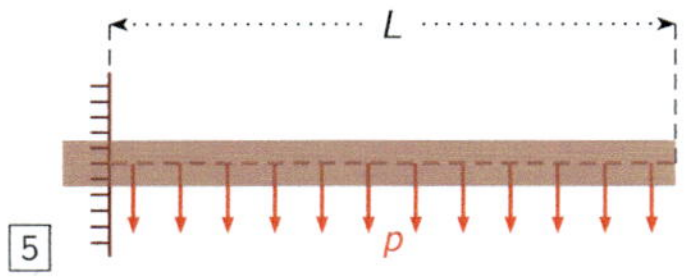	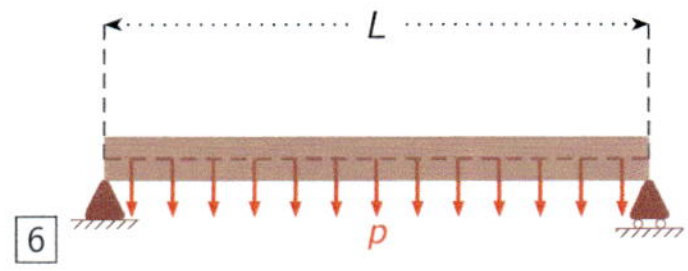

$p_{\mathrm{c}} = 110.33\frac{\sqrt{EI_Y GJ}}{L^3}$	$F_{\mathrm{c}} = 45.77\frac{\sqrt{EI_Y GJ}}{L^2}$
only DOF allowed at tip: u	only DOF allowed at tip: u
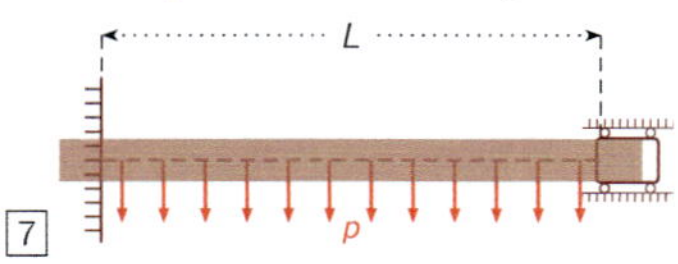	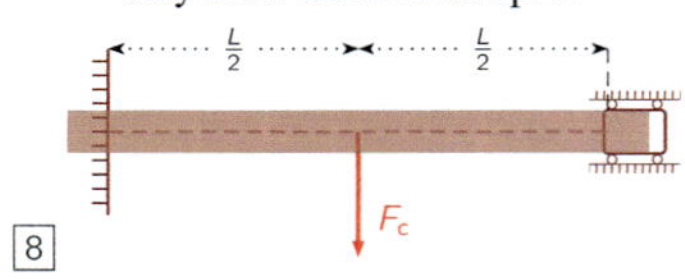
$F_{\mathrm{c}} = 16.93\frac{\sqrt{EI_Y GJ}}{L^2}$	$M_{\mathrm{c}} = \sqrt{\frac{(EI_Y - GJ)^2}{2R^2} + \pi^2\frac{EI_Y GJ}{L^2}}$
three DOFs at both hinges: u, ω_Y, ω_Z	$\omega_X = 0$ at tip
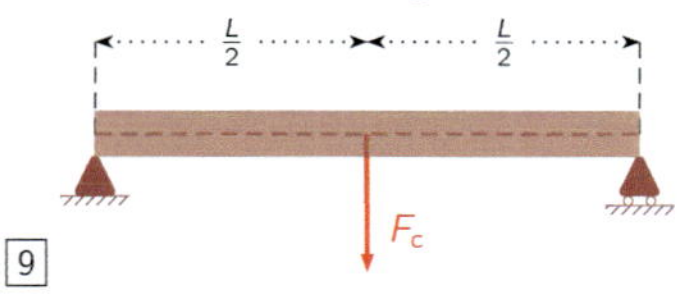	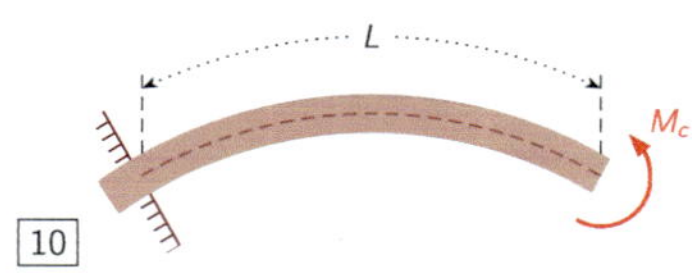

$p_c = 3\frac{EI}{R^3}$	$p_c = 2.94\pi^4 \frac{h}{L}\frac{EI}{L^3}$
1 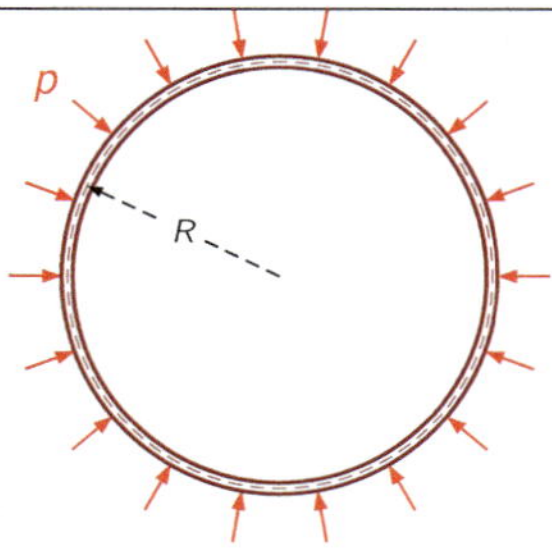	Thrust: $\chi_c = 4\pi^2\frac{EI}{L^2}$ 2 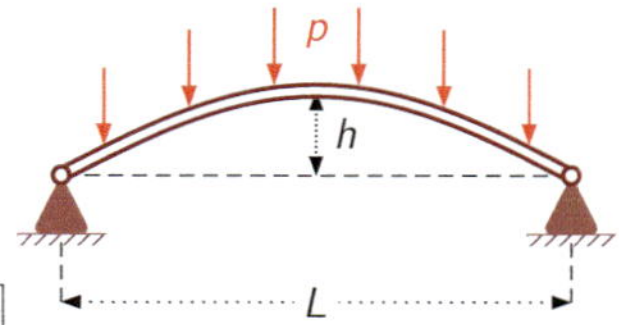
$p_c = \left(\frac{\pi^2}{\alpha^2} - 1\right)\frac{EI}{R^3}$	$p_c = (k^2 - 1)\frac{EI}{R^3}$
Thrust: $Q_c = p_c R$ 3 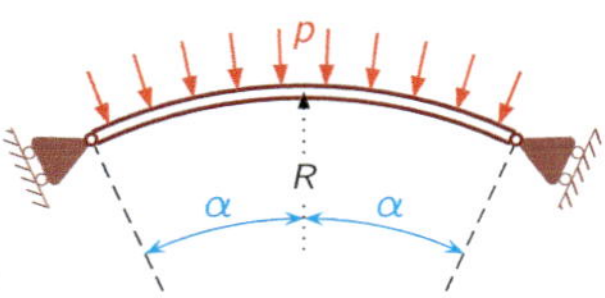	$k \tan\alpha = \tan k\alpha$ Thrust: $Q_c = p_c R$ 4 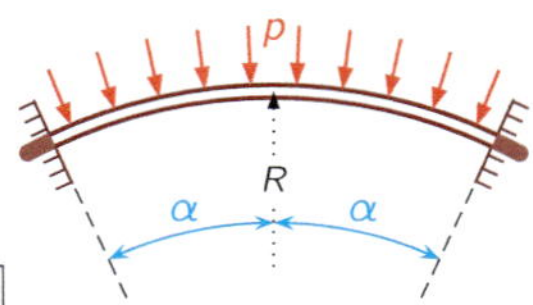

11.7.3 BUCKLING OF RINGS AND ARCHES

Critical loads are expressed in terms of the critical pressure per unit length, p_c, and, where applicable, in terms of the critical thrust. The latter is denoted X_c when the horizontal component matters and Q_c when the direction is along the median line. It is assumed that $I \equiv I_Z < I_Y$ so that the flexural rigidity is smaller in the vertical plane.

Case 1 is the classical ring problem covered in Example 11.14. The arch of case 2 has a sinusoidal shape, which is non-funicular for the uniform vertical pressure loading. The given expressions are valid for a shallow arch where the rise h is small compared with the span L. The arches of cases 3 and 4 are circular, thus funicular for radial pressure; see Example 6.4 and Remark 6.9. The critical loads of case 4 depend on a transcendental constant k, solution of the given characteristic equation. The expressions given are valid for both low and high arches.

Cases 5 to 8 correspond to parabolic arches, which are funicular for vertical pressure loading, as shown for case 5 under Example 11.15. Cases 5 and 6 are for low arches ($h \ll 2a$). The expressions given for high arches, cases 7 and 8, are more general but semi-analytical. They depend on the rise factor $\rho = h/(2a)$ and the *reduced* second moment of area I^*; see Problem 5.16. The expressions are thus valid for a spanwise constant I^*, which implies a variable section. The expressions given for shallow arches are retried by setting $I^* = I$ and letting $\rho \to 0$ in those given for high arches. The low arch expressions are given because, at leat for case 5, an exact analytical solution is obtained; see Example 11.15.

$p_c = 2\pi^2 \frac{h}{a}\frac{EI}{a^3} = \frac{2h}{a^2}X_c$	$p_c = 4.092\pi^2 \frac{h}{a}\frac{EI}{a^3}$
Thrust: $X_c = \pi^2 \frac{EI}{a^2}$	Thrust: $X_c = 2.046\pi^2 \frac{EI}{a^2}$
5 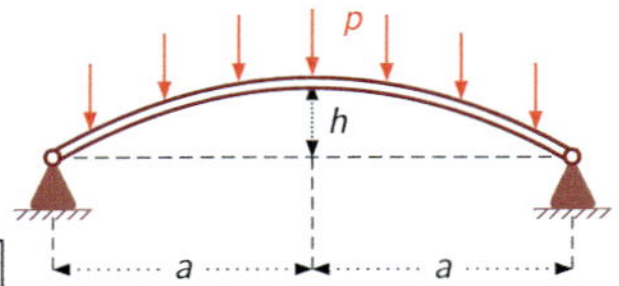	6 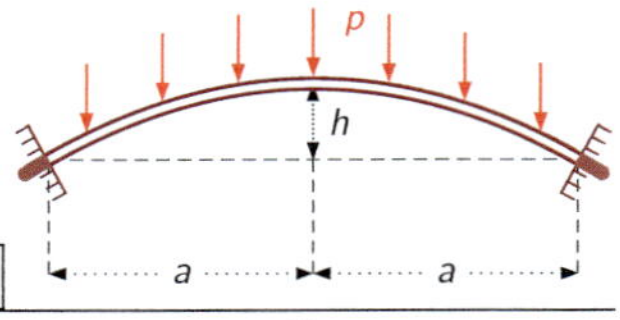
$p_c = \frac{2h}{a^2}X_c$	$p_c = \frac{2h}{a^2}X_c$
Thrust: $X_c = \pi^2 \frac{EI^*}{a^2}\frac{1}{1+6.144\rho^2}$	Thrust: $X_c = 2.046\pi^2 \frac{EI^*}{a^2}\frac{1}{1+3.352\rho^2}$
7 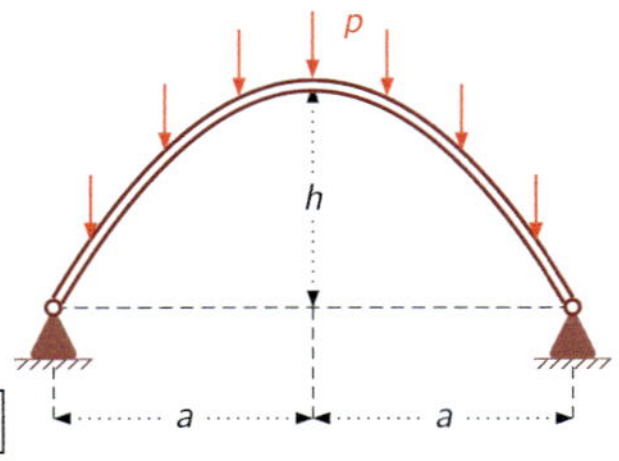	8 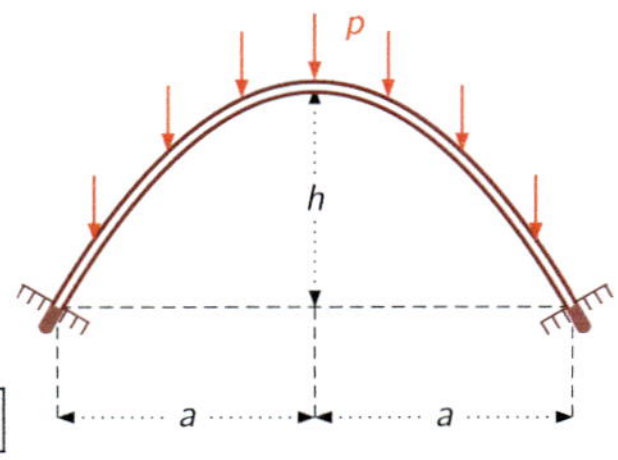

Interestingly, the critical thrust for the hinged arch of case 5 is the same as that of the second mode of a beam-column having the same span. This is consistent with the buckling mode being antisymmetric. All parabolic arches above buckle in an antisymmetric mode.

More generally, it can be shown that the critical horizontal thrust for a parabolic arch is given by an expression that takes the form:

$$X_c = m\pi^2 \frac{EI^*}{a^2}\frac{1}{1+K\rho^2} \tag{11.175}$$

where the numerical factors m and K depend on the number of articulations (or hinges). Interestingly, one finds that the fundamental mode is antisymmetric for 0 or 2 hinges and symmetric for 1 or 3 hinges. Table 11.1 gives the values of m and K for several cases of practical importance and the corresponding buckling modes.

PROBLEMS

Problem 11.1 (Discrete Structure with Spiral Springs) *Consider the two-degree-of-freedom structure of Example 11.2. (i) Establish the equilibrium equations given as Equation* (11.7) *in the example by drawing the free-body diagrams of bars OA and BC (ii) Prove that the buckling modes associated with critical loads* F_{c1} *and* F_{c2}*, are as illustrated in Figure 11.3. (iii) Show that the internal and external potential energies of the structure are given by Equation* (11.9) *of that example. (iv) Find the principal values and principal directions of the quadratic form* E_T *in Equation* (11.10) *of the example.* ⋆

Table 11.1

Numerical Factors Entering the Expression of the Critical Thrust in Parabolic Arches

Type of Arch	m	K	Buckling Mode
Fixed arch	2.046	3.352	Antisymmetric
Two hinges	1	6.144	Antisymmetric
One hinge	1.113	1.145	Symmetric
Three hinges	0.753	1.990	Symmetric

Problem 11.2 (Plane Assembly of Three Rigid Bars) *Three rigid bars are connected by two linear springs to form a plane structure supported and compressed as shown in Figure 11.23. The figure depicts the structure "at rest". The links at O, A, B, C are pin joints with axes perpendicular to the plane of the figure. Determine the buckling modes of the structure and the corresponding critical loads.* ⋆

Problem 11.3 (Structure–Spring System) *A planar structure consists of two rigid bars AB and BD, pin-jointed at B, and a spring connected to bar BD at midspan, as shown in Figure 11.24, which also gives the supports. The structure is loaded in its plane of symmetry by an axial force F applied at A. (i) Show that the structure possesses a single buckling mode. (ii) Determine the corresponding critical force.* ⋆

Problem 11.4 (Shear Correction to Euler's Critical Load) *Consider a solid circular cylindrical bar of length L and radius R, made up of a material with elastic constants E, ν. Let F_c^V denote the fundamental buckling load accounting for the*

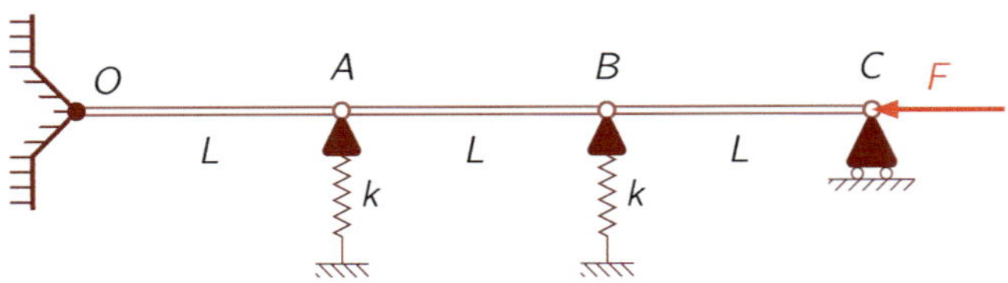

Figure 11.23 Bars connected by springs.

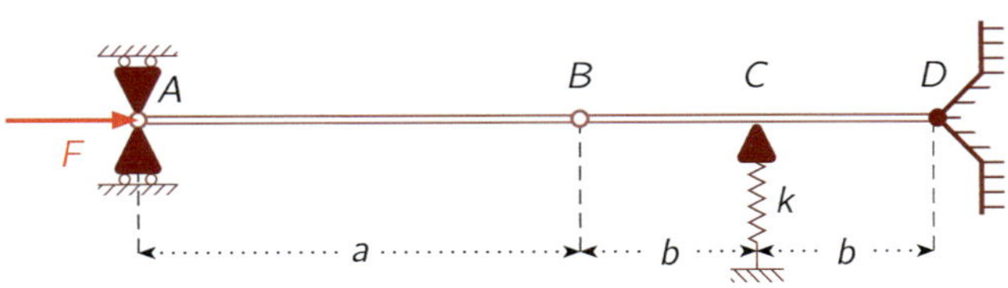

Figure 11.24 Two pin-jointed rigid bars and a spring.

shear correction, as inferred from Equation (11.31)*, and* F_c *Euler's critical load. (i) Expand* F_c^V/F_c *in terms of the ratio* R/L *treated as a small parameter (beam assumption). (ii) Evaluate the shear correction factor for* $R/L = 10$ *then* $R/L = 5$ *using* $\nu = 0.29$*, as in Section 11.2.4. (iii) Discuss the values in the case of an auxetic material.* ⋆

Problem 11.5 (Influence of Own Weight on Euler Critical Load) *A prismatic horizontal beam, of median plane xOy and characteristics L (length), E and I* ($I = I_Z \ll I_Y$) *is subjected to its own weight* $P = pL$*, Figure 11.25. Calculate the critical compression force leading to buckling by plane bending (fundamental mode only).* ⋆

Problem 11.6 (Compressed Beam with Varying Inertia) *A straight beam admitting a median plane consists of two parts welded together at B, Figure 11.26. The two parts OB and BA have flexural rigidities* EI_1, EI_2 *and lengths* ℓ_1, ℓ_2*, respectively. Derive the transcendental equation that defines the critical force for buckling in terms of* E, I_1, I_2, ℓ_1, ℓ_2 ⋆

Problem 11.7 (Beam with Varying Inertia: Applications) *Consider the same beam in Figure 11.26. Starting from the equation that defines the critical loads obtained in Problem 11.6, we wish to analyze some features of the solution. (i) For the special family of cases where beams OB and BA have the same lengths* ($\ell_1 = \ell_2$)*, study the variation of the buckling force with the ratio of moments of inertia, i.e., with* I_2/I_1*. (ii) In particular, state which scenario would lead to the lowest critical force:* $I_1 < I_2$ *or* $I_1 > I_2$*. (iii) Consider now an even smaller family of cases satisfying:* $\ell_2^2/\ell_1^2 = I_2/I_1$*. Determine explicitly the critical loads and their corresponding bucking modes. (iv) Discuss how variable inertia may be taken advantage of in mitigating buckling.* ⋆

Problem 11.8 (Flexible Beam with Spiral Spring) *Consider a prismatic beam, of median plane xOy, as represented in Figure 11.27 and having characteristics L*

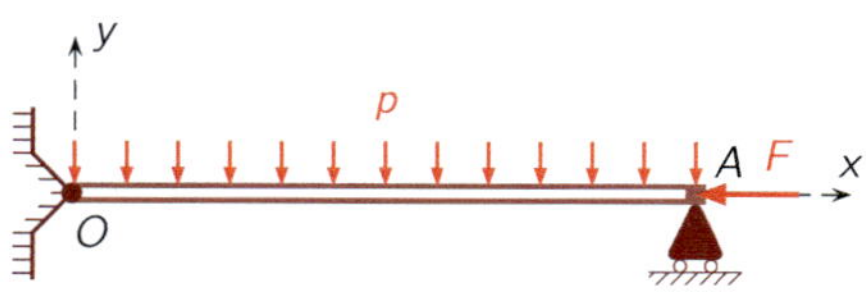

Figure 11.25 Compressed beam under its own weight.

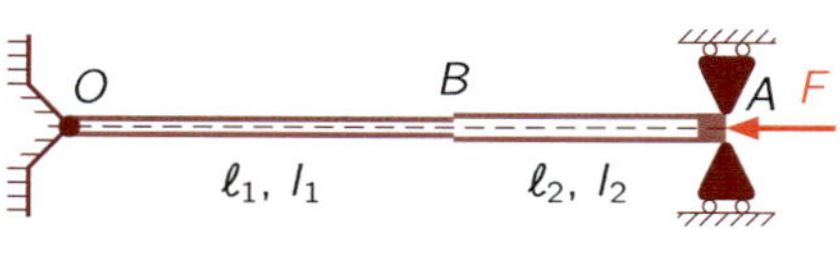

Figure 11.26 Beam of variable inertia.

(length), E and I ($I = I_Z \ll I_Y$). The stiffness of the spiral spring is denoted k. Calculate the critical loads for buckling by plane bending of this compressed beam. ⋆

Problem 11.9 (Buckling of a Compressed Structure) *A planar structure consists of two prismatic beams OA and AB connected together by a pin joint and to the exterior by clamping the left end and a simple support at B, Figure 11.28. Beam AB is modeled as rigid, whereas OA is made up of an elastic material having modulus E and second moments of area I_Y, I_Z with $I_Y \gg I_Z = I$. The structure is compressed by a force F, as shown. Determine the critical force for buckling.* ⋆

Problem 11.10 (Flexible Beam on Elastic Support) *A prismatic beam of length 2L sits on three supports, one of them being elastic having stiffness k, Figure 11.29. The bending stiffness is denoted EI with $I = I_Z \ll I_Y$. Calculate the critical force (fundamental mode) for buckling by plane bending.* ⋆

Problem 11.11 (Two Degree-of-Freedom Link) *A prismatic beam is compressed as depicted in Figure 11.30. The cross-section at O is fixed, but the section at A has two translational degrees of freedom (horizontal and vertical) while its rotation is prevented. Calculate the critical buckling loads F_n and their modes.* ⋆

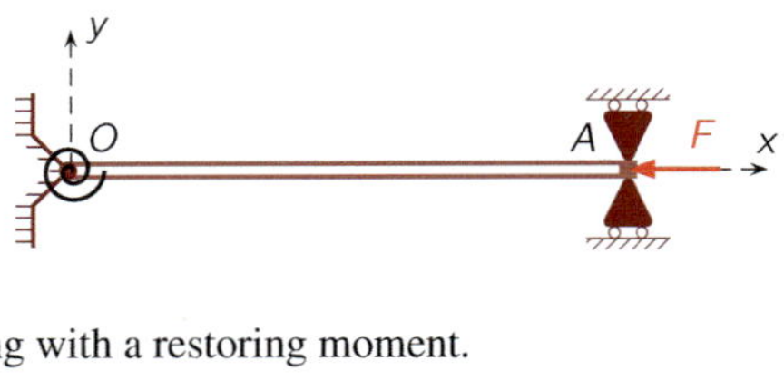

Figure 11.27 Buckling with a restoring moment.

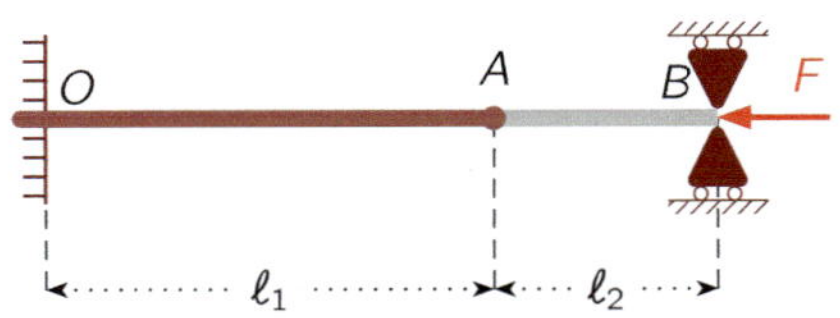

Figure 11.28 Compressed structure.

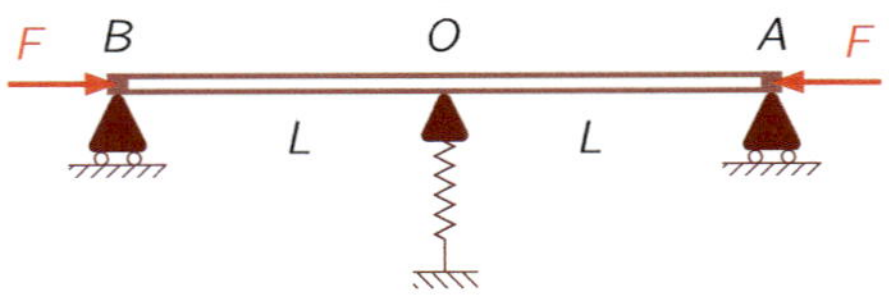

Figure 11.29 Beam on three supports, one being elastic.

Problem 11.12 (Beam on Elastic Support) *Consider a prismatic beam, of median plane xOy, as represented in Figure 11.31 and having characteristics L (length), E and I ($I = I_Z \ll I_Y$). The beam is fixed at one end and is supported by an elastic support at the other end. The stiffness of the spring is denoted k. Calculate the critical loads for buckling by plane bending of this compressed beam.* ⋆

Problem 11.13 (Buckling by Offset Compression) *A beam similar to that of Problem 11.12 is now compressed using a mechanism that involves a rigid bar AB, as shown in Figure 11.32. The bar has length $L' \ll L$ and remains perpendicular to the beam after deflection. Find the characteristic equation that defines the critical forces for buckling.* ⋆

Problem 11.14 (Constrained Buckling of a Beam) *A beam OAB of total length 3L is fixed at O, Figure 11.33. In addition, the beam is attached at A and B to a rigid bar CAB that can slide parallel to the left wall. Initially, point C is identified with O. However, upon compression induced buckling, the section at O remains fixed whereas C can move up or down. The motion of bar CAB constrains the beam deflections at A*

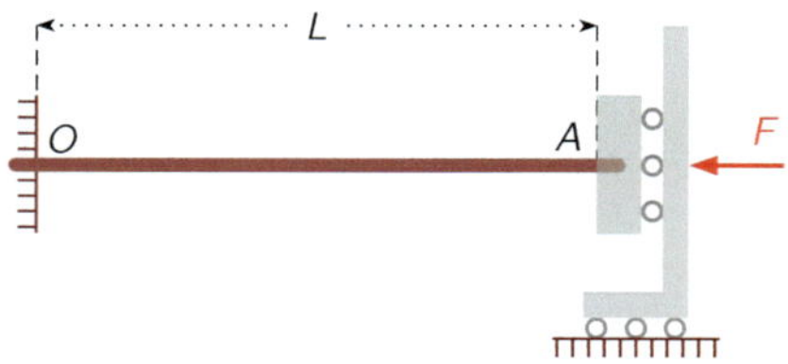

Figure 11.30 Compressed beam.

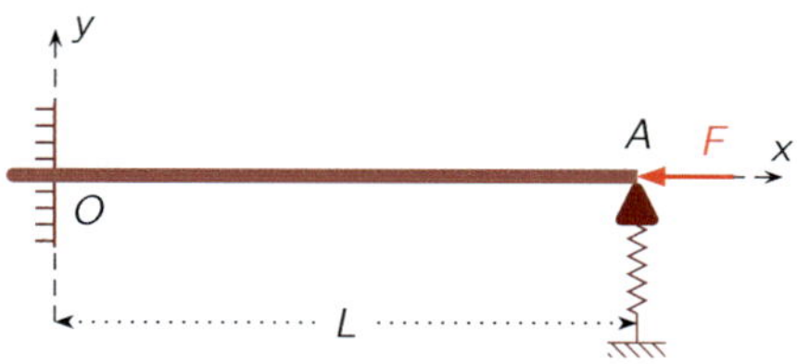

Figure 11.31 Buckling of hyperstatic beam.

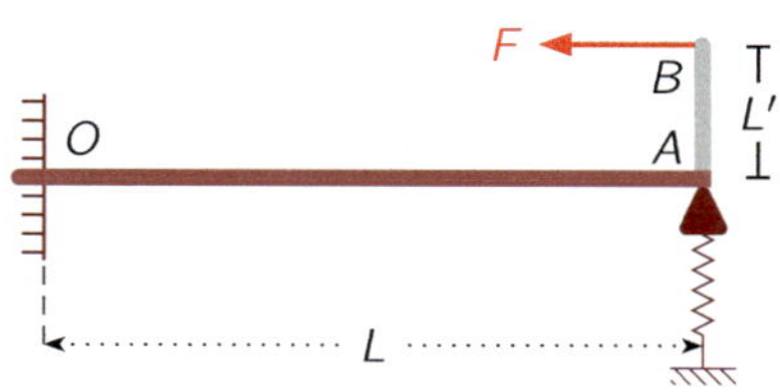

Figure 11.32 Bar-mediated compression.

and B to be equal. The bending stiffness is EI and the locations of attachment points A and B are given in Figure 11.33. Calculate the critical forces for buckling by plane bending. ⋆

Problem 11.15 (Beam–Rope System) *A prismatic beam has characteristics L and EI (in its median plane) with $I = I_Z \ll I_Y$. At one end, it is clamped; at the other end, it is pulled on by a rope that is constrained to pass through point B, Figure 11.34. The rope holds a weight such that the pulling force has magnitude F as shown. Initially, the beam–rope system lies horizontally on the dashed line. The deformed configuration in Figure 11.34 corresponds to a putative buckled state. (i) Neglecting the deflection due to shear, derive the transcendental equation that defines the suite of values of the critical forces $F_{c1} \cdots F_{c2}$. (ii) Study the limits $\lambda = 0$ and $\lambda / L \to \infty$ (that is $\lambda \gg L$). (iii) Calculate F_{c1} for the fundamental mode using: $L = 10$ m, $\lambda = 20$ m, $E = 210$ GPa, $I = 6 \cdot 10^5$ mm^4.* ⋆

Problem 11.16 (Beam on Three Supports with Hinged Ends) *The beam represented in Figure 11.35 has bending stiffness EI ($I = I_Z \ll I_Y$). The middle support is such that $\ell_2 = 2\ell_1$.*

1. *Calculate the critical load for buckling, F_c.*
2. *Show that:*

$$\pi^2 \frac{EI}{\ell_2^2} < F_c < \pi^2 \frac{EI}{\ell_1^2}$$

⋆

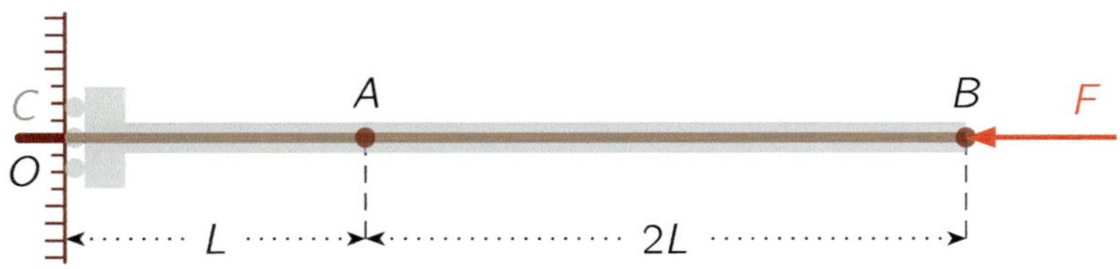

Figure 11.33 Beam connected to a sliding bar.

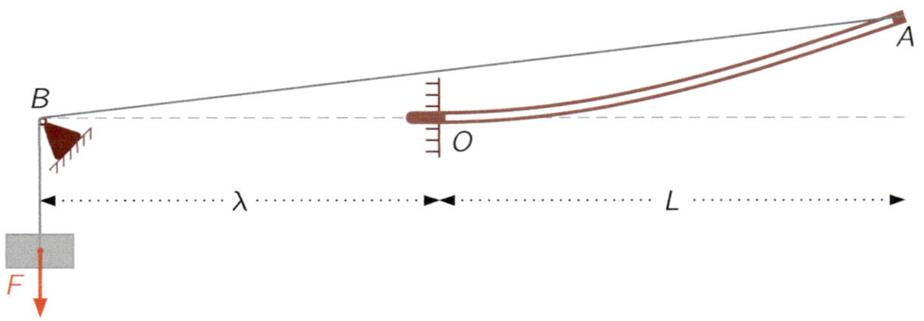

Figure 11.34 Rope pulling on a beam.

Problem 11.17 (Two-Beam Assembly) *Two prismatic beams OA and AB are hinged together, as shown in Figure 11.36, which also gives fixity conditions at both ends. Both beams have flexural rigidity EI with $I_Y \gg I_Z = I$. The assembly is compressed by a force F, as shown. Determine the critical force for buckling when $\ell_1 = 2\ell_2$.* ⋆

Problem 11.18 (Beam Immersed in Elastic Medium) *A prismatic beam of length L and flexural rigidity EI with $I = I_Z \ll I_Y$ is immersed in an elastic medium characterized by stiffness k; in other words, if an element* $\mathrm{d}x$ *of the beam undergoes the deflection $v(x)$, the medium exerts on it a restoring (vertical) force equal to $-kv\,\mathrm{d}x$, as sketched in Figure 11.37. The beam is shown in some deformed configuration to help visualize the restoring force. (i) Calculate the critical force F_n corresponding to a mode with n half-waves. (ii) Determine the smallest value of F_n and the corresponding mode. (iii) Work out a numerical example using: $L = 10$ m, $E = 200$ GPa, $k = 10^5$ N/m^2 for a circular section of radius $r = 2$ cm.* ⋆

Problem 11.19 (Buckling of a Gantry) *For the gantry in Figure 11.15, draw the mode shape for various values of the ratio L'/L.* ⋆

Problem 11.20 (Buckling of a Mast) *Consider a vertical prismatic beam connected to its surrounding by a pin-joint and simply supported at the top, the support plane being vertical, Figure 11.38. The beam is only subjected to its weight, be it w per unit length. The aim is to determine the critical buckling length ℓ_{cr} using iterative approximations. The proposed procedure is as follows:*

1. *Determine the reactions at the supports by expressing equilibrium in some deformed configuration.*
2. *Find the fourth degree polynomial in x/ℓ that satisfies all boundary conditions.*

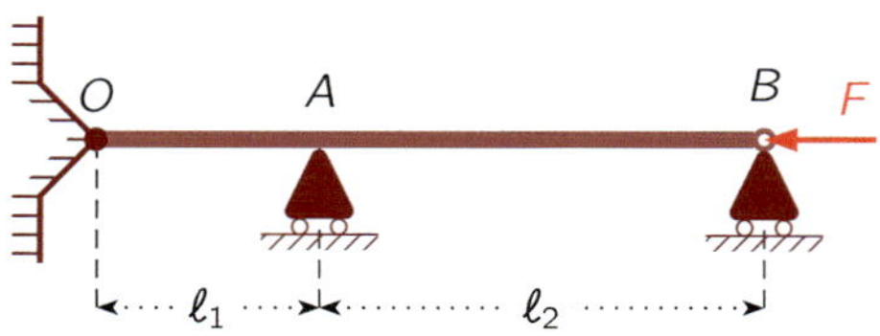

Figure 11.35 Beam on three supports.

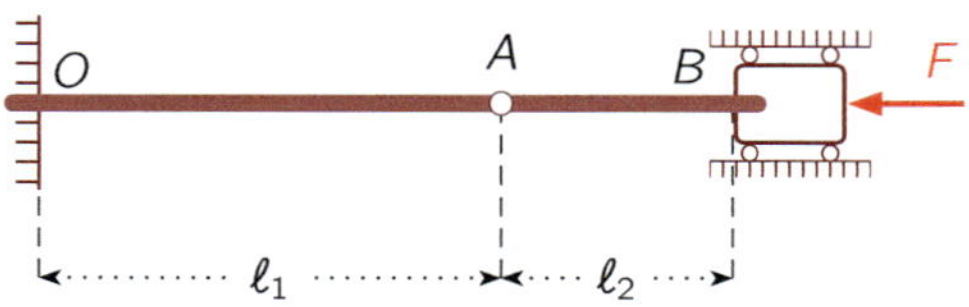

Figure 11.36 Two-beam structure.

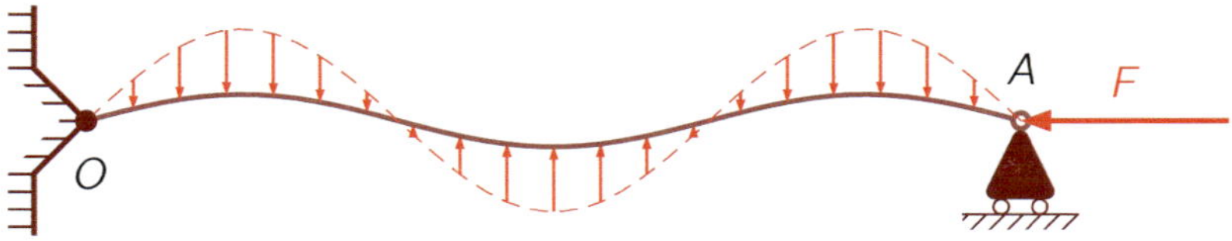

Figure 11.37 Beam swimming in an elastic medium.

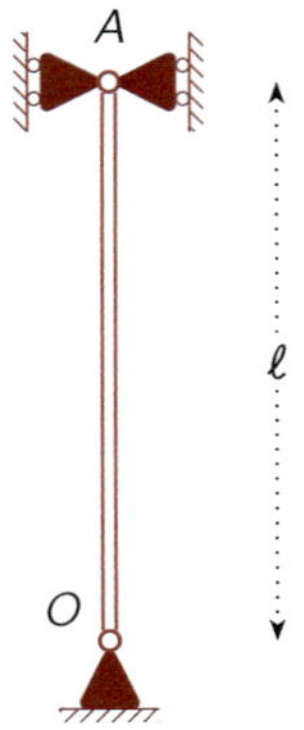

Figure 11.38 Mast under its own weight.

3. *Deduce an approximate value of the bending moment* $M_0(x)$ *acting on the current cross-section.*
4. *Integrate the differential equation of the problem*

$$\frac{d^2 v_1}{dx^2} = \frac{M_0}{EI}$$

5. *Determine an approximate value of* ℓ_{cr}. *Iterate.* ⋆

Problem 11.21 (Lateral Buckling under Pure Bending) *A beam, fixed at one end, is subjected to a bending moment M, about the z-axis, as depicted in Figure 11.39. The tip A is free to twist around the beam's axis, as sketched in the inset showing a detail at A. The second moment of the area* I_Z *is much greater than* I_Y. *Determine the critical value* M_c *of the bending moment that would cause lateral buckling of this beam.* ⋆

Problem 11.22 (Buckling of a Mast) *Consider the mast under its own weight in Example 11.10. Obtain the exact solution by solving Equation* (11.102) *using Bessel functions.* ⋆

Problem 11.23 (Transformation Matrix between Two Orthonormal Frames) *Prove Equation* (11.113). ⋆

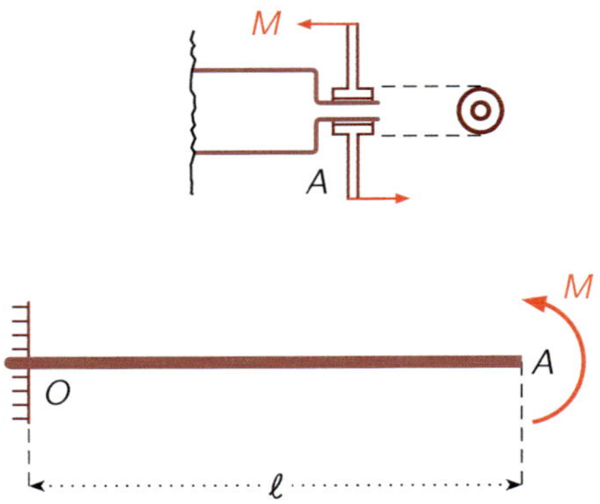

Figure 11.39 Bent beam and detail.

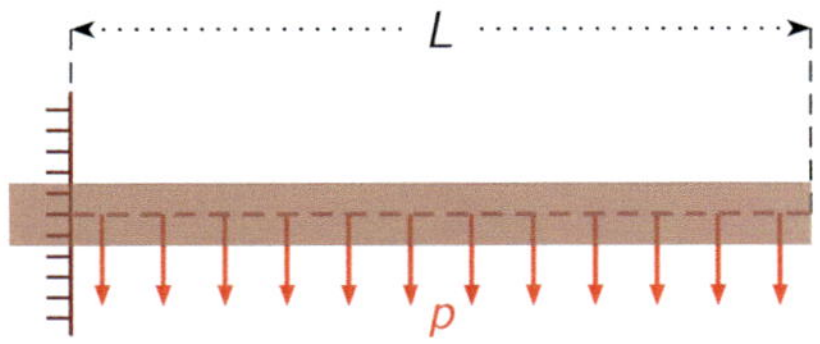

Figure 11.40 Cantilever under its own weight.

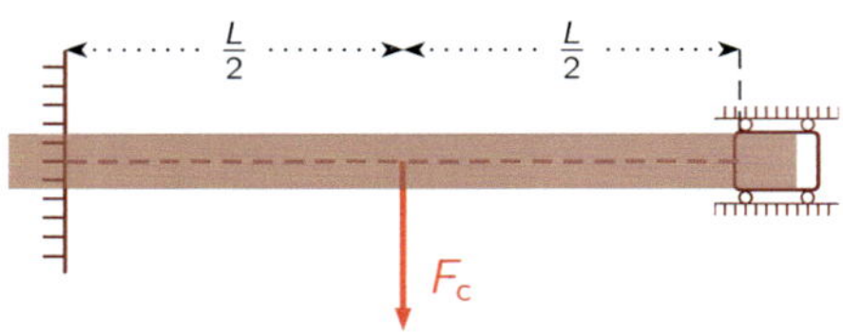

Figure 11.41 Beam and concentrated weight.

Problem 11.24 (Lateral Buckling of a Prismatic Beam under Its Own Weight) *Consider a prismatic beam with a rectangular cross-section, Figure 11.40. The notation is as in Example 11.12, but now the beam is only subjected to its own weight $P = pL$ in vertical plane (xOy). Calculate the critical length of the beam at the onset of instability. Give its expression in terms of the data of the problem (appropriate stiffness measures and p.)* ⋆

Problem 11.25 (Lateral Buckling of a Doubly-Clamped Beam) *A horizontal beam with end conditions as shown in Figure 11.41 is loaded by a weight F attached in O, center of the mid-section, such that the weight of the beam is relatively negligible. The asymmetric cross-section has its center of twist C lying on the principal axis GZ, as depicted in Figure 11.19b of Example 11.13, with $\overline{GC} = \delta$. Determine the critical weight F_c for buckling.* ⋆

12 Experimental Measurements

This edition is closed with a brief and elementary discussion of basic experimental methods in structural mechanics. Emphasis is laid on measuring internal stress resultants, i.e., the components of the screwer acting on the current cross-section in a beam. When only one component is present (e.g., normal force alone, torque alone, etc.), the methods are applicable to arbitrary cross-sections. Under combined loading, however, sufficient conditions must be sought to guarantee the independence of various measurements. These conditions pertain to symmetries imposed on the cross-section.

This chapter may serve as the theory basis for a laboratory course in structures. Field techniques such as photoelasticity and Moiré interferometry (for stresses) or digital image correlation (for strains) are not covered.

12.1 PRINCIPLE AND METHODS

Forces and moments cannot be measured directly. Therefore, internal resultants are determined indirectly using surface strain measurements and theory. Any measurement in experimental science is based on some theory. To obtain reliable measurements, the theory must be *unquestionable*. All measurements discussed below are based on the theory of elasticity. In many cases, approximate theories, such as those prevailing under the name of "Strength of Materials", cannot be used as basis for experimental measurements.

12.1.1 STRAIN GAUGES

To measure strain, some differential displacement measurement is needed. In general, two types of methods are used to measure strain. In contact-based methods, the measuring device is physically attached to the body of the structure at the point of interest. In this case, mechanical motion is converted into electrical current. The science of such measurements falls under electromechanics. Examples of contact displacement measurement devices include linear and rotary variable differential transformers (LVDT and RVDT), extensometers, and strain gauges.

Non-contact methods, on the other hand, rely on optics and eventually lasers. Laser-based extensometry has seen interesting developments in recent decades. It is only nominally non-contact since reflective tapes must be attached to the specimen. Video-based extensometry has also seen tremendous developments. A pervasive method of surface strain measurement is nowadays based on digital image correlation. Here too, some surface preparation is assumed in terms of speckles painted on the surface.

DOI: 10.1201/9781003537229-14

What is of particular importance to structural design is that, in the vast majority of cases, deformations are in the elastic regime. For a wide range of materials, this implies strain levels of the order of 10^{-6} or below. Neither laser-based extensometry nor optical methods have enough resolution to yield good estimates of surface strains. Furthermore, contact extensometers and LVDTs provide too crude measures. For this reason, the only truly reliable measure of surface strains in a structure undergoing elastic and reversible deformation is the good old strain gauge. Modern strain gauge technology allows for strains of the order of 10^{-2} to be reached, thereby covering the widest (decade-based) range of strains among all measurement methods.

A strain gauge consists of a metallic foil pattern supported by an insulating flexible carrier or backing, shown yellow in Figure 12.1. The gauge is attached to the surface of a structural member by an adhesive. Assuming perfect bonding, any deformation of the member is transmitted to the foil. When the latter is deformed, its electrical resistance changes. If ε denotes the normal strain at the point of attachment and R is the electrical resistance of the gauge, then the relative change in resistance is related to the strain by the relation:

$$\frac{\delta R}{R} = K\varepsilon \tag{12.1}$$

where the proportionality constant K is a characteristic of the gauge, called the gauge factor. The main strain direction in Figure 12.1 is indicated by horizontal arrows. Resistance is measured between the two large maroon pads. The purpose of markings outside the active area is to align the gauge during installation.

Nowadays, various materials are used for the foil, backing, and adhesive depending on the manufacturer. Materials for the metallic foil include constantan, a binary nearly equiatomic copper–nickel alloy and isoelastic (iron-based, nickel-rich) alloys. These materials have in common stable electrical resistance across a wide range of temperatures, high ductility, and good fatigue life. Values around 2 are typical for the gauge factor K in Equation (12.1); see Problems 12.1 and 12.2. The backing material of choice is polyimide. A typical material used as adhesive is cyanoacrylate.

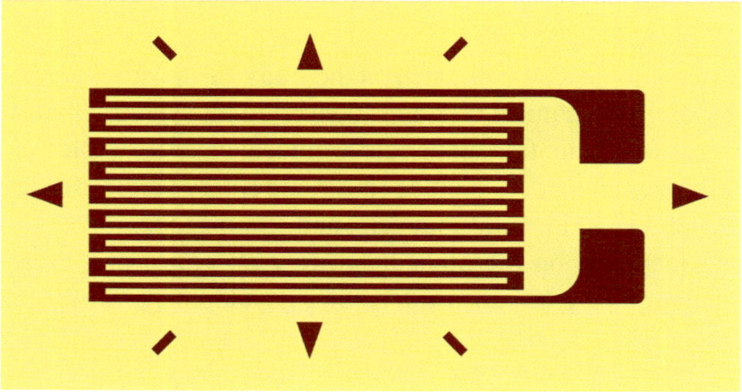

Figure 12.1 Typical strain gauge.

The most common model is the linear strain gauge shown in Figure 12.1. The metallic foil consists of a long, thin conductive strip arranged in a zig-zag pattern of parallel lines. The number of parallel lines defines the width of the grid. A single line would undergo the same strain, hence the same relative change in resistance. However, it would need to be operated at a much lower voltage, making it difficult to measure resistance changes accurately. On the other hand, a wide grid assumes negligible strain gradients in the direction of the width. Thus, in practice, the number of line traces of the foil pattern would depend on the application.

Temperature effects need consideration from various perspectives. What is of particular concern is to isolate thermal dilatation of the structure from the strain induced by mechanical loading. Thermal dilatation of the structure at the location of strain measurement generally affects the strain gauge's deformation, hence its resistance variation. On the other hand, mere temperature variation affects the resistance of the grid. In practice, the concurrence of both phenomena (the dilatation of structure and gauge as well as the temperature dependence of electrical resistance) is exploited to minimize, if not nullify spurious thermal effects over a temperature window. A strain gauge that compensates for the apparent temperature-induced strain at the point of measurement is known as a self-compensated strain gauge. It is important to note that a self-compensated strain gauge is adapted to the material of the structure; see Problem 12.3.

Models other than the linear strain gauge of Figure 12.1 have been developed over the decades, which are all based on variations around the linear gauge. That includes strain gauge rosettes, which are particular arrangements of strain gauges. Rosettes are particularly useful under biaxial load states. In what follows, we shall exclusively refer to linear strain gauges leaving operational details to technological implementations. Where appropriate, we shall also make reference to self-compensated strain gauges.

12.1.2 WHEATSTONE BRIDGE CIRCUIT

Electrical circuits pervade various engineering disciplines. They are used throughout aerospace engineering, from avionics (navigation, flight and engine control systems, cockpit instrumentation, ...) and health monitoring to wind tunnel instrumentation and operation. In this section, we discuss a particular circuit that is essential to measuring resistance variations, the Wheatstone bridge. Nowadays, digital multimeters are commonly used to measure a resistance. A key feature of the Wheatstone bridge circuit is that it allows very low values of resistance, down to the milli-Ohms (mΩ) range, to be accurately measured. This range is particularly relevant to strain measurements in structures operating in the elastic regime.

The Wheatstone bridge circuit consists of four resistors, $R_1, \ldots, R_4$, connected in a square, and a power source of electromotive force V. The power source is connected to the square of resistors through the input terminals, nodes A and C in Figure 12.2. A galvanometer G (or any potentiometer) measures the voltage difference between points B and D, which are the output terminals. No current nominally flows in the wire connecting B and D, which are not nodes per se. Therefore, R_1 and R_2 are in series; R_3 and R_4 likewise. Thus, the Wheatstone bridge consists of two parallel pairs of resistors in series.

12.1.2.1 Circuit Analysis

Let I be the electric current produced by the power source, as shown in Figure 12.2. At A, the node rule of electric circuits gives:

$$I = I_B + I_D \tag{12.2}$$

where I_B and I_D, respectively, denote the currents that flow from A to B and from A to D, Figure 12.2.

The potentiometer G acts as an ideal voltmeter that does not draw any current from the circuit. Applying Ohm's law to resistors R_1 and R_2 in series gives

$$I_B = \frac{V_A - V_C}{R_1 + R_2} = \frac{V}{R_1 + R_2} \tag{12.3}$$

and to resistors R_3 and R_4 in series:

$$I_D = \frac{V_A - V_C}{R_3 + R_4} = \frac{V}{R_3 + R_4} \tag{12.4}$$

The voltage difference between B and D is then

$$G = V_B - V_D = V_B - V_A + V_A - V_D \tag{12.5}$$

Now, apply Ohm's law to resistor R_1 alone to get

$$V_A - V_B = R_1 I_B \tag{12.6}$$

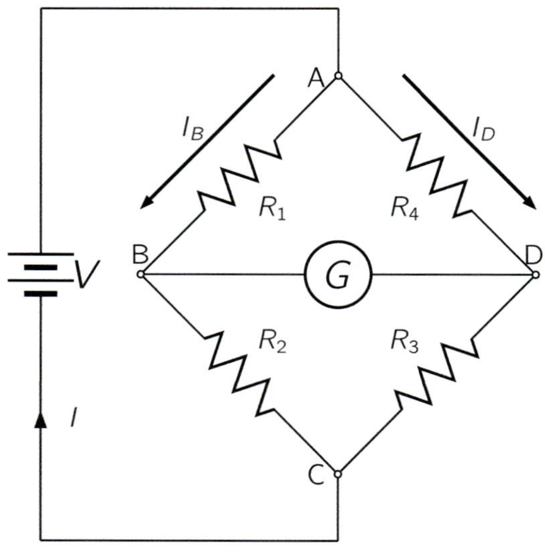

Figure 12.2 Wheatstone bridge circuit.

then to resistor R_4 alone

$$V_A - V_D = R_4 I_D \tag{12.7}$$

Combining Equations (12.5) to (12.7) gives

$$G = -R_1 I_B + R_4 I_D \tag{12.8}$$

Finally, substituting for the currents I_B, I_D their expressions given by Equations (12.3) and (12.4) into Equation (12.8) provides the voltage read by the galvanometer as follows:

$$\frac{G}{V} = \frac{R_4}{R_4 + R_3} - \frac{R_1}{R_1 + R_2} \tag{12.9}$$

12.1.2.2 Balanced Condition

The Wheatstone bridge is said to be balanced when no voltage is detected between the output terminals B and D. Rearranging Equation (12.9) as:

$$\frac{G}{V} = \frac{R_4 R_2 - R_3 R_1}{(R_1 + R_2)(R_4 + R_3)} \tag{12.10}$$

then setting $G = 0$ in the above equation leads to

$$R_1 R_3 = R_2 R_4 \tag{12.11}$$

which is the balanced condition of the bridge.

12.1.2.3 Out-of-Balance Analysis

A typical situation that arises in many applications of the Wheatstone bridge circuit is one where the bridge is initially balanced. This would be the case, for example, if all resistances were the same. If any resistance, say a strain gauge, were to undergo a change, say because of strain, then a voltage would be recorded between the output terminals. A basic question is how the change in voltage measured by the potentiometer, δG, is related to the change in resistance. More generally, all four resistances may vary. Let δR_i denote the variation of resistance R_i from a balanced condition. A more general question is that of which relationship there is between δG and the four δR_i.

The analysis of the out-of-balance Wheatstone bridge proceeds as follows. Viewing the voltage G given by Equation (12.9) as a function of the R_i's, its variation is given by

$$\delta G = \sum_{i=1}^{4} \frac{\partial G}{\partial R_i} \delta R_i \tag{12.12}$$

Introduce the two auxiliary functions

$$f(x;\alpha) = \frac{x}{x + \alpha} \qquad \text{and} \qquad g(x;\beta) = \frac{\beta}{\beta + x} = f(\beta;x) \tag{12.13}$$

where x is the variable while α and β are parameters. Their derivatives are given by

$$f'(x;\alpha) = \frac{\alpha}{(x+\alpha)^2} \qquad \text{and} \qquad g'(x;\beta) = -f'(x;\beta) \tag{12.14}$$

where the minus sign in the latter comes from the fact that the Jacobian of the transformation $(x,\beta) \mapsto (\beta,x)$ has a determinant equal to -1.

It then follows from Equation (12.9) that

$$\frac{G}{V} = f(R_4;R_3) - f(R_1;R_2) = g(R_3;R_4) - g(R_2;R_1) \tag{12.15}$$

such that

$$\frac{\delta G}{V} = -f'(R_1;R_2)\delta R_1 - g'(R_2;R_1)\delta R_2 + g'(R_3;R_4)\delta R_3 + f'(R_4;R_3)\delta R_4 \tag{12.16}$$

The derivatives in the above equation are evaluated using Equation (12.14). Then rearranging each term so as to exhibit relative variations of resistance leads to

$$\begin{aligned} \frac{\delta G}{V} = &- \frac{R_2}{R_1\left(1+\frac{R_2}{R_1}\right)^2}\frac{\delta R_1}{R_1} + \frac{R_1}{R_2\left(\frac{R_1}{R_2}+1\right)^2}\frac{\delta R_2}{R_2} \\ &- \frac{R_4}{R_3\left(1+\frac{R_4}{R_3}\right)^2}\frac{\delta R_3}{R_3} + \frac{R_3}{R_4\left(\frac{R_3}{R_4}+1\right)^2}\frac{\delta R_4}{R_4} \end{aligned} \tag{12.17}$$

Since the bridge is initially balanced, Equation (12.11) gives

$$\frac{R_1}{R_2} = \frac{R_4}{R_3} \equiv u \tag{12.18}$$

Noting that

$$\frac{1/u}{(1+1/u)^2} = \frac{u}{(u+1)^2}$$

Equation (12.17) can be written as

$$\boxed{\frac{\delta G}{V} = \frac{u}{(1+u)^2}\left[-\frac{\delta R_1}{R_1} + \frac{\delta R_2}{R_2} - \frac{\delta R_3}{R_3} + \frac{\delta R_4}{R_4}\right]} \tag{12.19}$$

A more general equation may be derived, which accounts for electric current flow through the potentiometer. Equation (12.19) is sufficient for our purposes and is essential to what follows.

12.1.3 PRINCIPLE

The theory of elasticity provides expressions relating internal resultants to surface strains. Internal resultants in beams are, therefore, indirectly measured using strain gauges. Being a contact method of strain measurement, a strain gauge allows mechanical motion to be converted into electrical current. Strain becomes a change in resistance, which is measured using an electrical circuit known as the Wheatstone bridge. The latter consists of four resistors one, two, or four of which are associated with strain gauges depending on what is measured and how it can be measured. The art of measuring internal resultants amounts to determining a suitable number of strain gauges and suitable locations on the structural member. Self-compensated strain gauges may be used in some cases to correct for temperature effects. In some cases, the test rig itself is set up to compensate for spurious effects that may emerge under combined loading.

This overarching principle is illustrated in the next section for each component of the screwer acting on the current cross-section of the beam. Then the case of combined loading is treated in some detail.

12.2 MEASUREMENT OF SCREWER COMPONENTS

12.2.1 NORMAL FORCE

Consider the beam depicted in Figure 12.3 and subjected to some axial loading. The aim is to measure the magnitude of the normal force N acting on a cross-section S.

In the most general case, the cross-section does not possess any axis of symmetry. Let GY, GZ denote the principal axes of S, as shown in Figure 12.3. It is further assumed that one of the axes (GY in the figure) intersects the beam's outer surface at two points P and P'. Two identical strain gauges g and g' are glued to the beam at

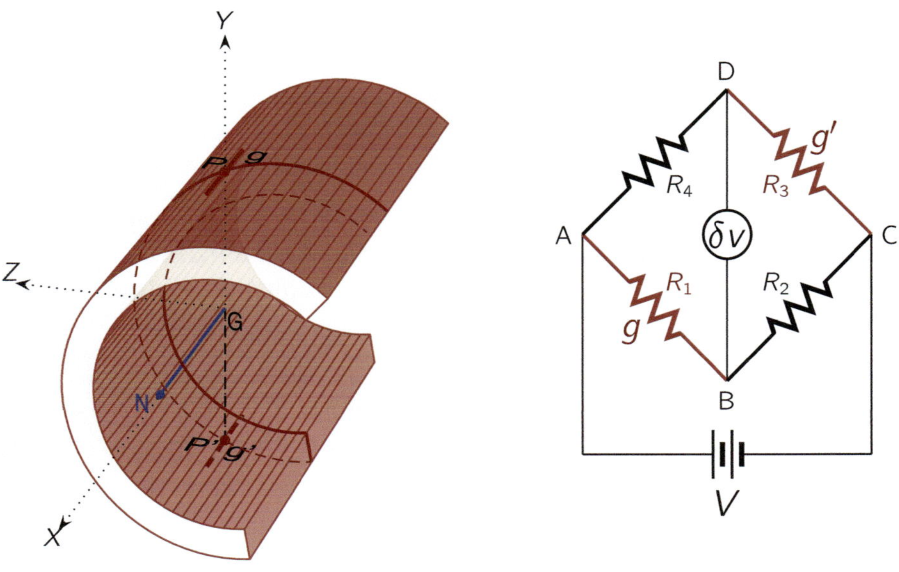

Figure 12.3 Principle of measurement of normal force in a beam.

these two points. Each gauge is oriented along the axial X direction. The two strain gauges are made of the two resistors R_1 and R_3 of the Wheatstone bridge depicted in Figure 12.3. Equal "ballast" resistors are placed in R_2 and R_4, which are identical to R_1 and R_3. The bridge is thus balanced when the beam is not under load.

Let V denote the voltage of the source power and $\delta v = V_B - V_D$ the out-of-balance voltage that is detected when the strain gauges are stretched under the action of loading. The resistances R_2 and R_4 being constant, Equation (12.19) becomes

$$\frac{\delta v}{V} = \frac{1}{4}\left[-\frac{\delta R_1}{R_1} - \frac{\delta R_3}{R_3}\right] \tag{12.20}$$

where $u = 1$ since the four resistances are initially the same. Using Equation (12.1), the above equation leads to

$$\frac{\delta v}{V} = -\frac{K}{4}(\varepsilon_1 + \varepsilon_3) \tag{12.21}$$

where ε_1, ε_3 are the axial strains at P and P', respectively, and K is the gauge factor. From Chapter 6, the strains are related to the normal force N through

$$\varepsilon_1 = \varepsilon_3 = \frac{N}{ES} \tag{12.22}$$

where E is Young's modulus of the material of the beam. It follows then that

$$\frac{\delta v}{V} = -\frac{K}{2}\frac{N}{ES} \tag{12.23}$$

which allows the magnitude of the normal force to be determined from the out-of-balance voltage reading on the potentiometer as

$$N = -\frac{2ES}{K}\frac{\delta v}{V} \tag{12.24}$$

Remark 12.1 *Temperature effects can be effectively eliminated by using for R_2 and R_4 self-compensated strain gauges for the material of the beam, identical to R_1 and R_3. Equation* (12.19) *shows that the dilatation parts would then cancel out.* $*$

Remark 12.2 *The configuration illustrated in Figure 12.3 consists of two active strain gauges R_1, R_3 and two completion resistors R_2, R_4. It is referred to as a half-bridge configuration. A variant of the half-bridge configuration is based on rebalancing the bridge, as treated in Problem 12.4.* $*$

Remark 12.3 *If the asymmetric cross-section is such that the GY axis intersects the surface at one point only (see Problem 12.5) then a quarter-bridge configuration may be used. This configuration involves a single active strain gauge.* $*$

Example 12.1: Normal Force in a Blade

Consider the helicopter blade in Figure 12.4 having constant cross-sectional area and rotating at constant angular velocity ω. Two identical wireless turning contacts

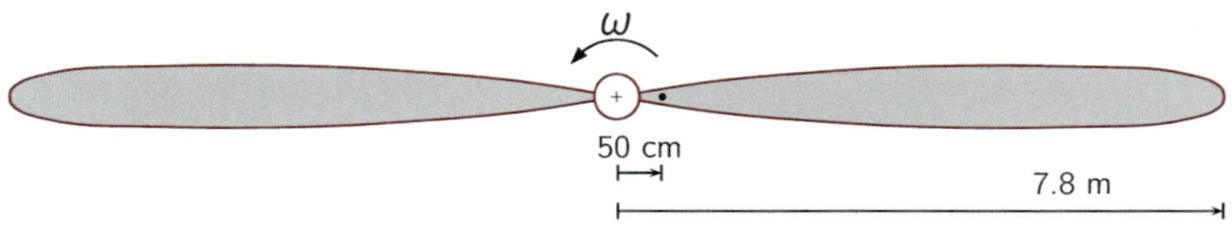

Figure 12.4 Helicopter blade with two sensors near the hub.

are attached to the blade's root, near its attachment to the hub. The two contacts act as remote strain gauges with gauge factor $K = 2$, mounted to a Wheatstone bridge in a half-bridge configuration. The power source voltage is 120 V and the voltage of the potentiometer reads 4 V. If the length of the blade is 7.8 m, its mass is 110 kg, cross-sectional area is 0.0006 m^2 and the distance between the contacts and the axis of the hub is 50 cm, what is the angular velocity ω? Assume that the material of the blade is some aluminum alloy with Young's modulus $E = 70$ GPa.

The rotation of the blade induces a nonuniform centrifugal force in the blade. As a result, the normal force acting on a cross-section located at a distance x from the hub is given by the expression (see Equation (6.39) of Example 6.2):

$$N(x) = \frac{1}{2} m\omega^2 (L^2 - x^2) \tag{12.25}$$

where m denotes the blade's mass per unit length and L its length. Here, $m =$ 14.1 kg/m and $L = 7.8$ m. Also, the wireless contacts are located at $x = 0.5$ m. The reading δv of the potentiometer is related to N by Equation (12.24). Combining the two equations gives

$$\omega^2 = \frac{4}{m(L^2 - x^2)} \frac{ES}{K} \frac{|\delta v|}{V} \tag{12.26}$$

This leads to $\omega = 57.2$ rad/s or 547 rpm. ♡

12.2.2 BENDING MOMENT

Let GY, GZ denote the principal axes of current cross-section S of an arbitrary beam, as depicted in Figure 12.5. The aim is to measure the M_Z component of the bending moment acting on S.

It is assumed that one of the axes (GY in the figure) intersects the beam's outer surface at two points P and P' having ordinates a and $-b$, respectively. Two identical strain gauges g and g', oriented along the axial X direction, are glued to the beam at rest at these two points. These two strain gauges are made the two resistors R_1 and R_4 of the Wheatstone bridge depicted in Figure 12.5. Completion resistors R_2 and R_3 consist of calibrated decade boxes and are given equal resistances to balance the bridge when the beam is at rest.

Under the action of loading, a bending moment $M_Z = M$ acts on S and the strain gauges undergo the following axial strains:

$$\varepsilon = -\frac{Ma}{EI} \quad \text{and} \quad \varepsilon' = \frac{Mb}{EI} \tag{12.27}$$

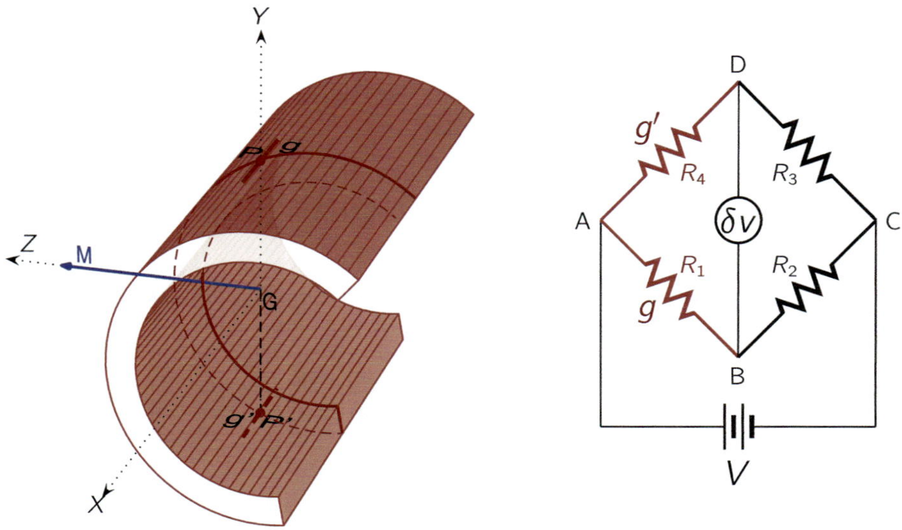

Figure 12.5 Principle of measurement of bending moment in a beam.

with $I \equiv I_Z$. This induces the relative variations of resistance:

$$\frac{\delta R_1}{R_1} = K\varepsilon \quad \text{and} \quad \frac{\delta R_4}{R_4} = K\varepsilon' \tag{12.28}$$

where K is the gauge factor common to both strain gauges. The bridge is then out of balance and the voltage at the potentiometer reads, in accordance with Equation (12.19):

$$\frac{\delta v}{V} = \frac{1}{4}\left[-\frac{\delta R_1}{R_1} + \frac{\delta R_2}{R_2} - \frac{\delta R_3}{R_3} + \frac{\delta R_4}{R_4}\right] \tag{12.29}$$

if R_2 is taken equal to R_1, which is assumed for simplicity. Combining the above relations gives

$$\frac{\delta v}{V} = \frac{1}{4}K(a+b)\frac{M}{EI} + \frac{1}{4}\left[\frac{\delta R_2}{R_2} - \frac{\delta R_3}{R_3}\right] \tag{12.30}$$

Here, V is the power source voltage and δR_2, δR_3 are eventually imposed resistances to the calibration resistors R_2 and R_3.

Two methods may be envisaged to determine M. The first consists of reading the out-of-balance voltage without touching R_2 or R_3, such that $\delta R_2 = 0$ and $\delta R_3 = 0$. Equation (12.30) then gives

$$M = 4\frac{EI}{K(a+b)}\frac{\delta v}{V} \tag{12.31}$$

The second method consists of rebalancing the bridge by seeking variations of calibration resistances δR_2 and δR_3 such that a null voltage is obtained. Setting $\delta v = 0$

in Equation (12.30) then gives

$$M = \frac{EI}{K(a+b)}\left[\frac{\delta R_3}{R_3} - \frac{\delta R_2}{R_2}\right] \tag{12.32}$$

Remark 12.4 *The strain gauges g and g′ need not be self-compensated for the material of the beam to factor out thermal effects; see Problem 12.8. This is unlike the measurement of the normal force N in Section 12.2.1.* ∗

Remark 12.5 *A full-bridge setup may also be used for measuring the bending moment; see Problem 12.9.* ∗

12.2.3 TORQUE

As a preliminary to the measurement of a torque in a shaft, it is useful to analyze the state of strain prevailing at a surface point of the shaft. The analysis is generally valid for a prismatic beam with arbitrary cross-section, loaded at both ends with the lateral surface being free of any tractions. To fix ideas, however, consider a circular cylindrical shaft, as depicted in Figure 12.6.

The strain tensor $\boldsymbol{\varepsilon}$ at a surface point P is most suitably represented by the matrix given by Equation (8.3) in Chapter 8 with $\gamma_{x\theta}$ denoting the shear strain. By symmetry of the strain tensor, the hoop strain $\gamma_{x\theta}$ is the same as $\gamma_{\theta x}$, which is the shear strain acting in a meridian plane of normal $\mathbf{e}_\theta$ along the x direction, as shown in Figure 12.6.

The radial direction (i.e., normal to the surface at P) is a principal strain direction associated with a zero eigenvalue since $\boldsymbol{\varepsilon}\,\mathbf{e}_r = \mathbf{0}$. In the plane tangent to the surface at P, which is a principal plane, the directions at $\pm 45°$ to the generatrix are principal

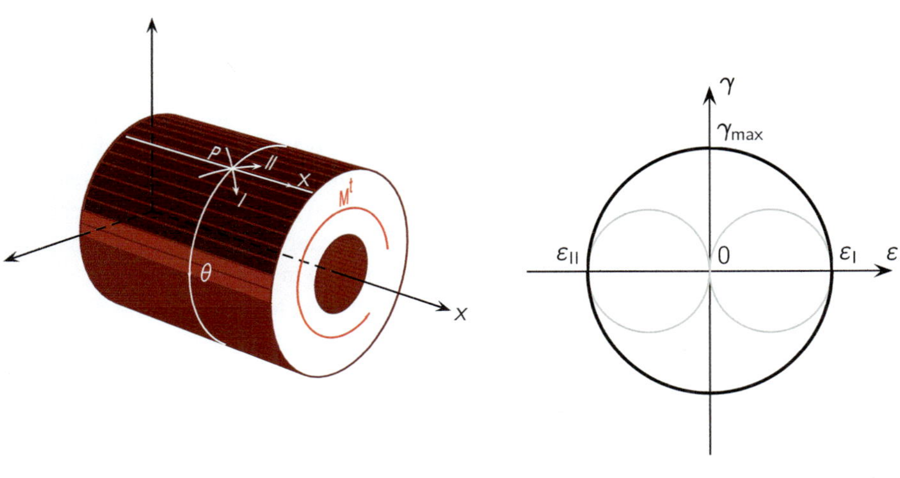

Figure 12.6 Principal strains at a surface point in a shaft under torsion.

strain directions associated with the principal strains:

$$\begin{cases} \varepsilon_{\mathrm{I}} = \gamma_{x\theta} = \dfrac{1}{2G}\dfrac{M^t R}{I_x} \\ \varepsilon_{\mathrm{II}} = -\gamma_{x\theta} = -\dfrac{1}{2G}\dfrac{M^t R}{I_x} \end{cases} \tag{12.33}$$

where M^t is the applied torque, R the external radius, I_x the polar moment of area, and G the shear modulus. Here, the notation ε_{I}, $\varepsilon_{\mathrm{II}}$ is used, as opposed to ε_1, ε_2 from Chapter 1, to avoid confusion with strain gauge dilatations, $\varepsilon_1, \varepsilon_2, \ldots$.

Mohr's circle corresponding to the principal tangent plane at P is shown with a thick solid line in Figure 12.6 by considering the case $M^t > 0$ for definiteness. The maximum shear strain is $\gamma_{\max} = \varepsilon_{\mathrm{I}} = \gamma_{x\theta}$. The gray circles correspond to the other two principal planes.

Neither the axial fibers nor the circumferential ones undergo any dilatation. In other words, the normal strains along those directions, which lie at $\pm 45^\circ$ from the principal directions, are nil. The largest linear dilatations (normal strains) are the principal dilatations ε_{I} and $\varepsilon_{\mathrm{II}}$ of the helicoidal fibers that are inclined at $\pm 45^\circ$ from the generatrices.

Assume now that a strain gauge is attached to the surface at point P along principal direction I, Figure 12.6. From Equation $(12.33)_1$, obtain the torque as

$$M^t = 2G\frac{I_x}{R}\varepsilon_{\mathrm{I}} \tag{12.34}$$

For a solid shaft, $I_x = \frac{\pi}{2}R^4$ such that

$$M^t = \pi G R^3 \varepsilon_{\mathrm{I}} \tag{12.35}$$

In practice, to obtain an accurate measurement of the torque, a full bridge configuration is used whereby four strain gauges are glued to the beam so as to eliminate thermal dilatation effects and, as will be seen in Section 12.3, any dilatations that may come from other internal resultants.

The full bridge consists of four identical strain gauges, denoted $g_1, \ldots, g_4$, having the same nominal resistance and gauge factor K. Gauges g_1 and g_2 are glued at P along principal directions I and II, Figure 12.7. On the other hand, gauges g_3 and g_4 are glued at a point P', which is diametrically opposite to P, along principal directions I and II at that point; see Figure 12.7. The four strain gauges are wired so as to make the four resistances of a Wheatstone bridge, as depicted in the figure.

When the shaft is at rest, the voltage is zero between terminals B and D. Under the action of a torque M^t, the shaft deforms and the strain gauges undergo relative elongations $\varepsilon_1, \ldots, \varepsilon_4$ each of which is related to a relative variation of resistance in accordance with Equation (12.1), that is,

$$\frac{\delta R_i}{R_i} = K\varepsilon_i, \quad i = 1 \cdots 4 \tag{12.36}$$

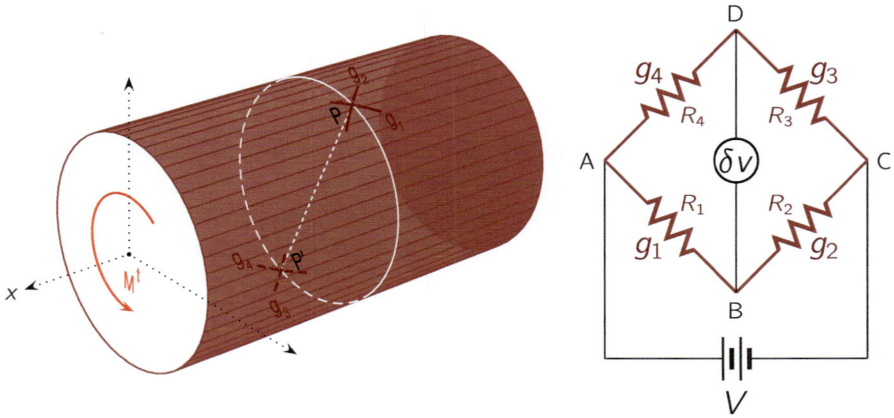

Figure 12.7 Principle of measurement of torque in a beam.

If $V = V_A - V_C$ denotes the power source voltage of the bridge and $\delta v = V_B - V_D$ the out-of-balance voltage detected by the galvanometer as a result of resistance variations, then Equation (12.19) gives

$$\frac{\delta v}{V} = \frac{K}{4}(-\varepsilon_1 + \varepsilon_2 - \varepsilon_3 + \varepsilon_4) \tag{12.37}$$

The strains recorded by gauges g_1 and g_3 correspond to the first principal strain at their respective locations P and P'. From Equation $(12.33)_1$ they are related to the torque by

$$\varepsilon_1 = \varepsilon_3 = \frac{1}{2G}\frac{M^t R}{I_x} \tag{12.38}$$

The strains recorded by gauges g_2 and g_4, on the other hand, correspond to the second principal strain at P and P', respectively. They are equal in magnitude to ε_1 and ε_3 but of the opposite sign. From Equation $(12.33)_2$ they are related to the torque by

$$\varepsilon_2 = \varepsilon_4 = -\frac{1}{2G}\frac{M^t R}{I_x} \tag{12.39}$$

Inserting Equations (12.38) and (12.39) into Equation (12.37) provides the torque in terms of the reading δv at the galvanometer:

$$M^t = -2\frac{GI_x}{KR}\frac{\delta v}{V} \tag{12.40}$$

The measurement given by Equation (12.40) is insensitive to temperature variations because the reading given by Equation (12.37) would not be altered by changes in temperature. To see this, let

$$(\varepsilon_1)_{\text{th}} = (\varepsilon_2)_{\text{th}} = (\varepsilon_3)_{\text{th}} = (\varepsilon_4)_{\text{th}} = \alpha(T - T_0) \tag{12.41}$$

be the relative elongations induced by a temperature change from some reference T_0, the symbol α denoting the coefficient of thermal expansion of the material of the shaft. It is clear that all thermal strains being equal their contributions would cancel out in Equation (12.37).

Remark 12.6 *If the shaft is in rotary motion, then it is impossible to wire the strain gauges to a recording apparatus. In such cases, one may use either turning wireless contacts or a system of remote extensometry such that a signal proportional to δv is emitted by an antenna connected to the rotating shaft, then received and transmitted to the recorders by a fixed antenna.* *

Remark 12.7 *The principle of measurement of a torque in a shaft was illustrated above for a circular cylindrical case, which is common. To measure torque in an arbitrary beam, the principle of the full bridge holds but the measurement hinges on the availability of a relationship between the torque and surface strains. This relationship is provided by a theory of elasticity solution based on the Saint-Venant theory of torsion developed in Chapter 8, e.g., see Problem 12.11, or on thin foil shear flow theory developed in Section 8.5, also in Chapter 8; see Problem 12.12.* *

12.2.4 TRANSVERSE SHEAR

Let GY and GZ denote the principal axes of current cross-section S of an arbitrary prismatic beam. These may be axes of symmetry, as depicted in Figure 12.8 for simplicity. The aim is to measure the magnitude of the transverse shear (or shear force) V_Y acting on S. Unlike all previous cases, the cross-section must also be under the action of some bending moment M_Z; see Equation (5.56) in Section 5.5.4 or Equation (9.1) in Chapter 9. It is therefore important to verify that the measurement protocol does factor out any possible contribution of the bending moment M_Z. It is further assumed that no point load is applied to the section S where measurement is to be carried out.

Under these circumstances, the GZ axis is the neutral axis. It intersects the outer boundary of cross-section S at two points P and P', as shown in Figure 12.8. Two pairs of strain gauges are attached to the beam at the two locations: g_1, g_2 at P and g_3, g_4 at P'. The orientations are at $\pm 45°$ from the X-axis. These four identical gauges are made the four resistances $R_1, \dots, R_4$ of a complete Wheatstone bridge, also depicted in the figure. Since all resistances are nominally the same, the bridge is initially balanced.

With V denoting the voltage of the power source, the out-of-balance voltage δv that appears when the strain gauges deform under the action of the loading satisfies Equation (12.19), which in view of Equation (12.1) and the initially balanced condition ($u = 1$), reads

$$\frac{\delta v}{V} = \frac{K}{4}(-\varepsilon_1 + \varepsilon_2 - \varepsilon_3 + \varepsilon_4) \tag{12.42}$$

where K is the gauge factor and ε_i denotes the linear dilatation of the ith gauge.

Recall that the general format of the strain tensor at a surface point under general bending (M_Z and V_Y together) is given by Equation (9.6). At an arbitrary point, not only are there shear strains, γ_{xy} and γ_{xz}, but also three normal strains. However, since points P and P' lie on the neutral axis, which corresponds to $y = 0$ in Equation (9.6), the normal strains are zero. Therefore, one has the following strain matrix at P and P':

$$[\boldsymbol{\varepsilon}] = \begin{bmatrix} 0 & \gamma_{xy} & \gamma_{xz} \\ \gamma_{xy} & 0 & 0 \\ \gamma_{xz} & 0 & 0 \end{bmatrix} \tag{12.43}$$

for a general section S and with $\gamma_{xz} = 0$ for the special rectangular section of Figure 12.8 because the surface is traction-free; see Figure 9.7. The only non-zero shear strain is given by

$$\gamma_{xy} = \frac{1}{2G}\tau_{xy} \tag{12.44}$$

It is always possible to express the shear stress at P and P' as

$$\tau_{xy} = \beta \frac{V_Y}{S} \tag{12.45}$$

where β is a dimensionless constant to be determined.

Diagonalizing the above matrix, in a way similar to what was done in torsion in the previous Section 12.2.3, the principal strains, which are identified with the linear dilatations of the strain gauges, are deduced as

$$\varepsilon_1 = \varepsilon_3 = \frac{\beta}{2G}\frac{V_Y}{S} \quad \text{and} \quad \varepsilon_2 = \varepsilon_4 = -\frac{\beta}{2G}\frac{V_Y}{S} \tag{12.46}$$

Now, Equation (12.42) becomes

$$\frac{\delta v}{V} = -\frac{\beta K}{2G}\frac{V_Y}{S} \tag{12.47}$$

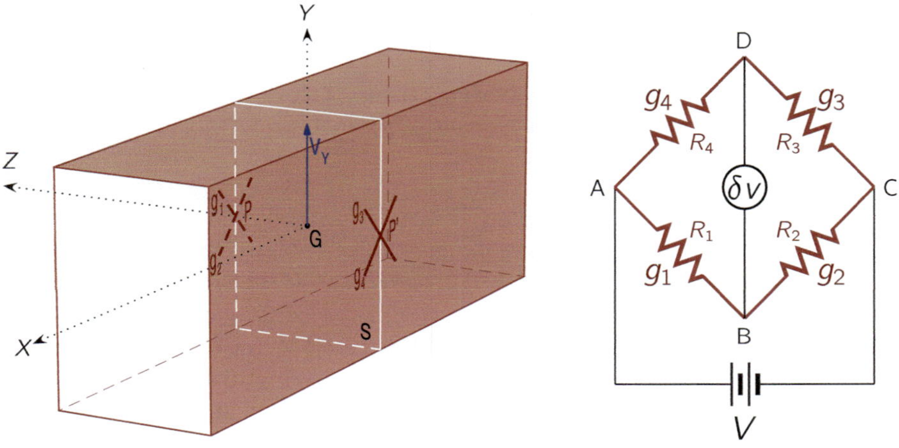

Figure 12.8 Principle of measurement of shear force in a beam.

from which the sought shear force is obtained as

$$V_Y = -\frac{2GS}{\beta K}\frac{\delta v}{V} \tag{12.48}$$

from the reading on the potentiometer if β is known.

For a host of cross-section shapes, the Saint-Venant problem of general bending, which results in a Dirichlet boundary-value elasticity problem (see Equation (9.20)), the coefficient β is known analytically; see Problem 12.14. If not, a numerical solution to the problem may be obtained first.

The measurement given through Equation (12.48) is insensitive to temperature variations for the same reasons discussed in Section 12.2.3 for the measurement of torque. The measurement is also insensitive to the bending moment M_Z. This will be shown in a more general setting in the following section.

Example 12.2: Circular Section under Shear

Let us determine the prefactor β in Equation (12.45) for a circular section of radius R. The points of attachment of the gauges P and P' have in-plane coordinates $(0,R)$ and $(0,-R)$, respectively. Based on the solution found in Example 9.1, the shear stress at either point is

$$\tau_{xy}(0,\pm R) = \frac{1}{8}\frac{3+2\nu}{1+\nu}\frac{V_Y}{I_Z}\left(R^2 - \frac{1-2\nu}{3+2\nu}R^2\right) = \frac{1}{4}\frac{1+2\nu}{1+\nu}\frac{V_Y}{I_Z}R^2 \tag{12.49}$$

Using $I_Z = \frac{\pi}{4}R^4 = \frac{1}{4}R^2 S$ the above equation simplifies into

$$\tau_{xy}(0,\pm R) = \frac{1+2\nu}{1+\nu}\frac{V_Y}{S} \tag{12.50}$$

which shows that the factor β introduced in Equation (12.45) is given by

$$\beta = \frac{1+2\nu}{1+\nu} \tag{12.51}$$

and only depends on Poisson's ratio of the material of the beam. In general, β also depends on relative in-plane dimensions; e.g., see Problem 12.14. ♡

12.3 CASE STUDIES

In devising a method for measuring the normal force N in Section 12.2.1, bending moment M_Z (or M_Y) in Section 12.2.2, and torque M^t in Section 12.2.3, it was assumed that each internal resultant was acting alone on the current cross-section. Also, in devising a method for measuring the shear force V_Y (or V_Z) in Section 12.2.4 it was assumed that only V_Y and M_Z (or only V_Z and M_Y) were acting on S. In general, however, the cross-section of a beam may be subjected to combined loading. This is for example the case of a wing in a wind tunnel. This is more generally the case of any curved beam, even with slight curvature, not admitting a median plane.

In that case, all components of the screwer may be active at once even if the material is homogeneous.

The key issue addressed in this section is the independence of various measurements. Conditions are sought such that the measurement of a given screwer component is insensitive to other components. There is no claim of completeness in addressing this question. In other words, sufficient conditions are sought, more specifically in what pertains to the geometry of the cross-section, the median line itself being eventually a space curve.

The situations that are addressed in subsequent sections may be viewed as accounts of real engineering projects, i.e., as case studies. The situations involve increasingly complex issues where there is often more than one correct answer. The below case studies enable the reader to implement knowledge from the previous chapters by integrating information and key theoretical results for the purpose of finding a simple solution aimed at measuring internal resultants.

Remark 12.8 *The effect of temperature has already been considered in the context above and will not be considered again in this section. In summary, self-compensating strain gauges are needed for measuring N whereas specific bridge configurations are taken advantage of in mitigating temperature variation effects for measuring M_Z, M^t or V_Y.* *

12.3.1 INDEPENDENT MEASUREMENT OF *N*

Consider an arbitrary cross-section subjected to a normal force N, as in Figure 12.3, as well as a bending moment M_Z, as in Figure 12.5. The same half-bridge Wheatstone setup shown in Figure 12.3 is used.

Under load, the potentiometer reading is given by

$$\frac{\delta v}{V} = -\frac{K}{4}(\varepsilon_1 + \varepsilon_3), \tag{12.52}$$

which is Equation (12.21), but now the strain recorded by either gauge has two components, one due to the normal force N and one due to the bending moment M_Z:

$$\begin{cases} \varepsilon_1 = \dfrac{N}{ES} - \dfrac{M_Z a}{EI_Z} \\ \varepsilon_3 = \dfrac{N}{ES} + \dfrac{M_Z b}{EI_Z} \end{cases} \tag{12.53}$$

Recall that a and b respectively denote the distances from the centroid G to P and P', the points of intersection of the GY-axis with the beam's outer surface.

Thus, one sufficient condition for the two bending contributions to cancel out, once inserted into Equation (12.52), is that $a = b$, i.e., that the cross-section be symmetric about the GZ axis. This shows that the half-bridge configuration of Figure 12.3 would be insensitive to a superposed bending moment provided that the section admits GZ as an axis of symmetry, as illustrated in Figure 12.9.

Assume now that in addition to M_Z, there is also bending about the Y-axis. The situation is more complex because, even in the absence of N, the neutral axis does not coincide with either principal axis of inertia. However, referring to the results obtained in Section 7.2 for 'slouchy' bending, it is clear that the bending-induced strains at the locations of gauges g and g' are equal and opposite, precisely because of symmetry about the GZ axis. Interestingly, the result holds even if GY is not an axis of symmetry of the cross-section.

Next, assume that in addition to N and bending moment, the cross-section of Figure 12.9, now assumed to be symmetric about GZ, is also under the action of some torque M^t. The analysis of Section 12.2.3 about Figure 12.6 can be generalized to show that axial fibers do not undergo any torque-induced dilatation. The theory developed in Chapter 8 shows that to be a general result, which is applicable to the cross-section of interest here. In other words, the superposed torque does not contribute to the strains ε_1 and ε_3 of the two gauges g and g' in Figure 12.9.

Likewise, the theory developed in Chapter 9 shows that a superposed transverse shear V_Y would not lead to any axial strains other than those induced by the bending moment M_Z, which have already been shown to cancel out.

In summary, if the cross-section admits GZ as an axis of symmetry then the Wheatstone half-bridge configuration of Figure 12.9 is insensitive to all screwer components other than the sought normal force N.

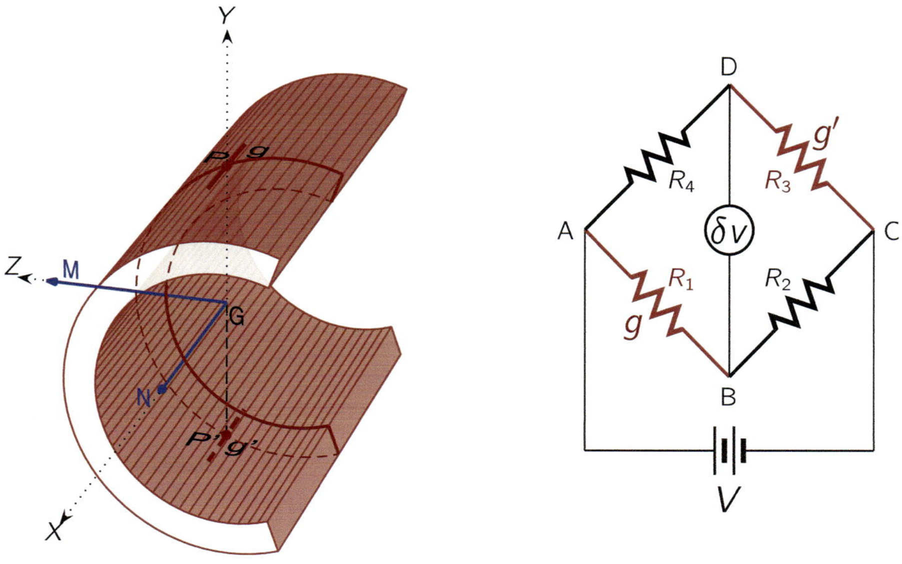

Figure 12.9 Principle of measurement of normal force in a beam loaded in combined extension and bending.

12.3.2 INDEPENDENT MEASUREMENT OF M_Z

Consider an arbitrary cross-section subjected to bending moment M_Z, as well as a normal force N, as shown in Figure 12.10.

The same Wheatstone half-bridge setup shown in Figure 12.5 is used. Under load, the potentiometer reading is given by

$$\frac{\delta v}{V} = \frac{K}{4}(-\varepsilon_1 + \varepsilon_4) \tag{12.54}$$

which is formally the same as Equation (12.30) without touching to resistors R_2 and R_3. However, here the strain recorded by either gauge has an additional component coming from the normal force N:

$$\begin{cases} \varepsilon_1 = \dfrac{N}{ES} - \dfrac{M_Z a}{EI_Z} \\ \varepsilon_4 = \dfrac{N}{ES} + \dfrac{M_Z b}{EI_Z} \end{cases} \tag{12.55}$$

When inserted into Equation (12.54), the strains due to N cancel out and the same Equation (12.31) is recovered, irrespective of any symmetry property of the cross-section. This shows that the half-bridge configuration of Figure 12.10 is insensitive to a superposed normal force.

The only restriction on the shape of the cross-section is implied by the existence of P and P'. If the shape is such that the GY axis intersects the beam at only one point,

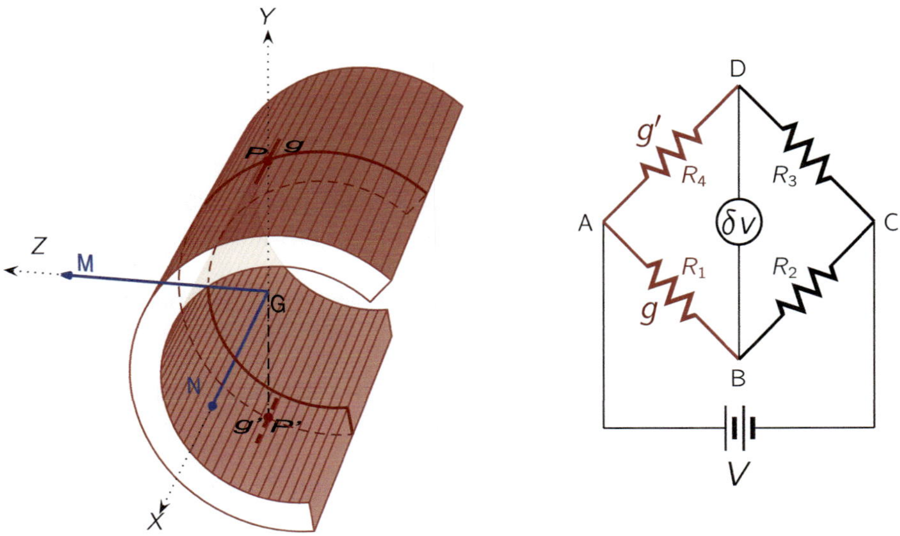

Figure 12.10 Principle of measurement of bending moment in a beam loaded in combined extension and bending.

a quarter bridge setup may be used to measure M_Z but that measurement would not be independent of a possibly superposed normal force N.

Assume now that in addition to N and M_Z, there is also bending about the Y-axis. By superposition, the axial stress anywhere in the beam is

$$\sigma_X = \frac{N}{S} - \frac{M_Z}{I_Z} Y + \frac{M_Y}{I_Y} Z \tag{12.56}$$

If strain gauges g and g' are glued at the points of intersection of the GY-axis with the beam (points P and P' in Figure 12.10) then $Z = 0$ at both locations. By way of consequence, the axial strains due to M_Y are zero at the locations of gauges g and g'. Therefore, the half-bridge configuration of Figure 12.10 is insensitive to both N and M_Y. Unlike in the case of normal force measurement, there is no symmetry requirement on the cross-section.

Next, assume that in addition to bending and extension, the cross-section of Figure 12.10 is also under the action of some torque M^t as well as shear force $\mathbf{V}$ with possible components V_Y and V_Z. The theories of torsion and transverse shearing developed in Chapters 8 and 9 show that axial fibers do not undergo any dilatation in excess of that induced by bending. As a result, neither the torque M^t nor the shears V_Y or V_Z contribute to the strains ε_1 and ε_4 of the two gauges g and g' in Figure 12.10.

In summary, so long as the cross-section is intersected at two points by principal axis GY, the Wheatstone half-bridge configuration of Figure 12.10 is insensitive to all screwer components other than the sought bending moment M_Z.

12.3.3 INDEPENDENT MEASUREMENT OF M^T

Consider first the shaft with a circular cross-section, as in Figure 12.7, but now subjected to a torque M^t as well as a normal force N, a bending moment M and a shear force V, as shown in Figure 12.11. The bending moment and shear force are oriented arbitrarily relative to the radial line PP', P and P' being the surface points where strain gauges are attached. Let θ and ϕ respectively denote the polar angles of vectors $\mathbf{M}$ and $\mathbf{V}$ from $\overrightarrow{GP}$, Figure 12.11. Conditions are examined so that the same Wheatstone full-bridge configuration of Figure 12.7 provides an independent measurement of the torque M^t.

Under load, the out-of-balance reading on the potentiometer is formally still given by Equation (12.37), rewritten here as

$$\frac{\delta v}{\mathcal{E}} = \frac{K}{4}(-\varepsilon_1 + \varepsilon_2 - \varepsilon_3 + \varepsilon_4) = \frac{(\delta v)_N}{\mathcal{E}} + \frac{(\delta v)_M}{\mathcal{E}} + \frac{(\delta v)_V}{\mathcal{E}} + \frac{(\delta v)_T}{\mathcal{E}} \tag{12.57}$$

where the electromotive force of the power source voltage is denoted $\mathcal{E}$ to avoid confusion with the shear force V. Also, the indices N, M, V and T respectively refer to the *a priori* contributions of the normal force, the bending moment, the shear force and the torque to the output of the potentiometer.

Here, however, the strain recorded by each strain gauge has additional components due to other internal resultants:

$$\begin{cases} \varepsilon_1 = \dfrac{1-\nu}{2}\dfrac{N}{ES} - (1-\nu)\dfrac{MR\sin\theta}{EI_x} + \dfrac{\beta}{2G}\dfrac{V\sin\phi}{S} + \dfrac{1}{2G}\dfrac{M^tR}{I_x} \\ \varepsilon_2 = \dfrac{1-\nu}{2}\dfrac{N}{ES} - (1-\nu)\dfrac{MR\sin\theta}{EI_x} - \dfrac{\beta}{2G}\dfrac{V\sin\phi}{S} - \dfrac{1}{2G}\dfrac{M^tR}{I_x} \\ \varepsilon_3 = \dfrac{1-\nu}{2}\dfrac{N}{ES} + (1-\nu)\dfrac{MR\sin\theta}{EI_x} - \dfrac{\beta}{2G}\dfrac{V\sin\phi}{S} + \dfrac{1}{2G}\dfrac{M^tR}{I_x} \\ \varepsilon_4 = \dfrac{1-\nu}{2}\dfrac{N}{ES} + (1-\nu)\dfrac{MR\sin\theta}{EI_x} + \dfrac{\beta}{2G}\dfrac{V\sin\phi}{S} - \dfrac{1}{2G}\dfrac{M^tR}{I_x} \end{cases} \tag{12.58}$$

where the coefficient $\beta = \frac{1+2\nu}{1+\nu}$ is taken from Example 12.2.

The justification of each term in Equation (12.58) is as follows. Consider the strain due to the normal force alone. Anywhere in the beam, the strain matrix is diagonal when expressed in any set of axes sharing the x-axis with

$$\varepsilon_y = \varepsilon_z = -\nu\varepsilon_x \qquad \text{and} \qquad \varepsilon_x = \frac{N}{ES} \tag{12.59}$$

see Equation (6.2). At point P, strain gauges g_1, g_2 measure strains ε_1, ε_2 at $\pm 45°$ from the x-axis. A Mohr circle construction shows quite simply that

$$(\varepsilon_1)_N = (\varepsilon_2)_N = \frac{1}{2}(\varepsilon_x + \varepsilon_y) \tag{12.60}$$

since a physical 45° angle becomes a right angle from the x-axis in the x-y plane in a Mohr circle; see Figure 1.4 in Chapter 1. Combining the above two equations gives

$$(\varepsilon_1)_N = (\varepsilon_2)_N = \frac{1-\nu}{2}\varepsilon_x = \frac{1-\nu}{2}\frac{N}{ES} \tag{12.61}$$

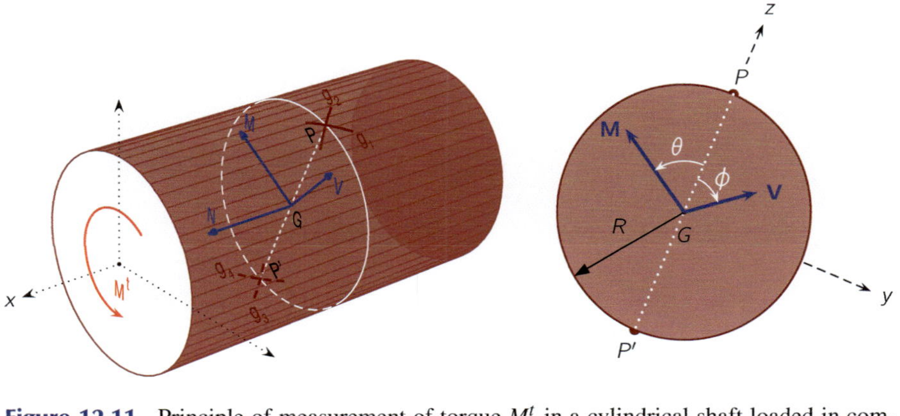

Figure 12.11 Principle of measurement of torque M^t in a cylindrical shaft loaded in combined extension, bending, and torsion.

Likewise, at point P' the strains $(\varepsilon_3)_N$ and $(\varepsilon_4)_N$ are equal to $(\varepsilon_1)_N$ and $(\varepsilon_2)_N$. This explains the first term in each row statement of Equation (12.58).

Next, consider the sole effect of the bending moment $\mathbf{M}$ having magnitude M. For the circular section, $I_y = I_z = I$ and the neutral axis coincides with the orientation of $\mathbf{M}$; see Equation (7.7) in Chapter 7. The strain matrix is again diagonal with

$$\varepsilon_y = \varepsilon_z = -\nu\varepsilon_x \qquad \text{and} \qquad \varepsilon_x = -\frac{M}{EI}y^* \tag{12.62}$$

see Equation (7.3) in Chapter 7. Here, y^* refers to the ordinate perpendicular to the neutral axis, which is the line of action of the bending moment, Figure 12.12. In particular, the location P of strain gauges g_1, g_2 has ordinate $y_P^* = R\sin\theta$, whereas P' has ordinate $-y_P^*$. Alternatively, the bending moment M may be decomposed into a component parallel to PP', which does not lead to any strains at P or P', and a component $M\sin\theta$ perpendicular to PP', which leads to compression at P and tension at P'. Either way, the same result is obtained. Then, the same Mohr circle construction shows that the linear dilatations of g_1, g_2 are

$$(\varepsilon_1)_M = (\varepsilon_2)_M = \frac{1}{2}(\varepsilon_x + \varepsilon_y) = \frac{1-\nu}{2}\varepsilon_x = -\frac{1-\nu}{2}\frac{M}{EI}R\sin\theta \tag{12.63}$$

Likewise, the linear dilatations of strain gauges g_3, g_4 at P' are

$$(\varepsilon_3)_M = (\varepsilon_4)_M = \frac{1-\nu}{2}\frac{M}{EI}R\sin\theta \tag{12.64}$$

The bending contribution to Equation (12.58) is so explained on account of the relation $I_x = 2I$, valid for the circular section.

Now, the contribution of the shear force $\mathbf{V}$ may be decomposed into two components, Figure 12.12. Due to symmetry, the component V_z parallel to PP' does not lead

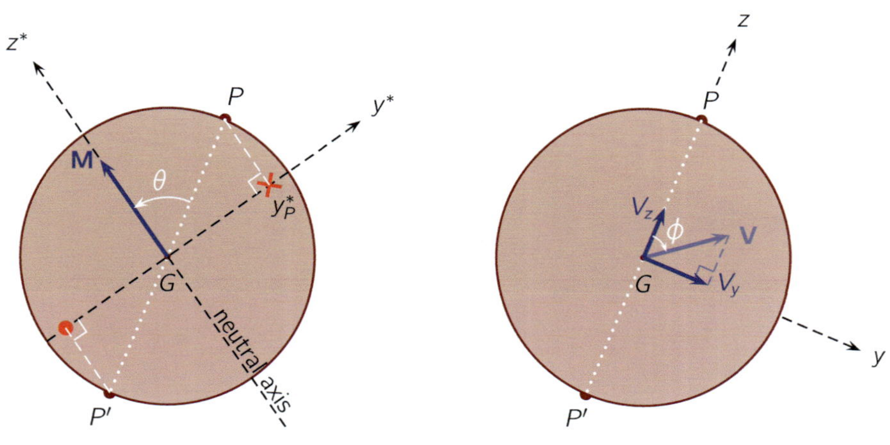

Figure 12.12 (a) Contribution of bending moment to strain gauge dilatations under combined loading. (b) Decomposition of shear force.

to any shear stress or strain at either location P or P'; see Example 9.1 in Chapter 9. On the other hand, the component $V_y = V \sin\phi$ perpendicular to PP' generates equal shear strains at P and P' such that the dilatations of the strain gauges are

$$(\varepsilon_1)_V = (\varepsilon_4)_V = \frac{\beta}{2G}\frac{V_y}{S} \qquad \text{and} \qquad (\varepsilon_2)_V = (\varepsilon_3)_V = -\frac{\beta}{2G}\frac{V_y}{S} \tag{12.65}$$

similar to the relations given in Equation (12.46) but with flipping the relative locations of strain gauges g_3 and g_4; compare Figure 12.11 with Figure 12.8. The coefficient β for the circular section was calculated in Example 12.2. This justifies the shear contribution to Equation (12.58).

The last term in each strain gauge dilatation is due to the torque, as previously explained in Section 12.2.3; see Equations (12.38) and (12.39).

Now that the origin of each term in Equation (12.58) is clear, conclusions on their respective contributions to the potentiometer reading per Equation (12.57) may be drawn. Indeed, when inserted into Equation (12.57), the strains due to N cancel out since:

$$(\varepsilon_1)_N = (\varepsilon_2)_N = (\varepsilon_3)_N = (\varepsilon_4)_N \tag{12.66}$$

and there are two pairs of positive and negative signs in Equation (12.57). Therefore, $(\delta v)_N = 0$. Likewise, the bending-induced strains cancel out because

$$-(\varepsilon_1)_M = -(\varepsilon_2)_M = (\varepsilon_3)_M = (\varepsilon_4)_M \tag{12.67}$$

such that the dilatations of each pair of strain gauges at P and P' cancel out separately. Thus, $(\delta v)_M = 0$ in Equation (12.57). Also, the shear-induced strains cancel out and $(\delta v)_V = 0$ because

$$(\varepsilon_1)_V = -(\varepsilon_2)_V = -(\varepsilon_3)_V = (\varepsilon_4)_V \tag{12.68}$$

so that pairs of strain gauges cancel out which are symmetrical of each other about the centroid G, i.e., g_1, g_4 on the one hand and g_2, g_3 on the other.

Finally, the torque induced dilatations satisfy:

$$(\varepsilon_1)_T = -(\varepsilon_2)_T = (\varepsilon_3)_T = -(\varepsilon_4)_T \tag{12.69}$$

such that the only surviving term in Equation (12.57) is

$$\frac{\delta v}{\mathcal{E}} = \frac{(\delta v)_T}{\mathcal{E}} = -K(\varepsilon_1)_T = -\frac{K}{2G}\frac{M^t R}{I_x} \tag{12.70}$$

which leads to the same Equation (12.40) for the torque versus voltage relationship.

This concludes the proof that the Wheatstone full-bridge configuration of Figure 12.11 is insensitive to any other internal resultant superposed onto the torque M^t.

Consider now a more general shape of the cross-section, for example as shown in Figure 12.13, with a full-bridge setup. The reading on the potentiometer is given by the general relation of Equation (12.57). The strains due to the normal force N are still given formally by Equations (12.61) and (12.66) and would cancel out for any cross-sectional shape. Likewise, the bending-induced strains would cancel out by pair because the relations in Equation (12.67) remain valid pairwise, i.e.,

$$-(\varepsilon_1)_M = -(\varepsilon_2)_M \neq (\varepsilon_3)_M = (\varepsilon_4)_M \tag{12.71}$$

A restriction on the shape of the cross-section comes in consideration of the shear induced strains. If the shear force is applied along a symmetry axis, say GY as in Figure 12.13, then the shear strains are zero at the locations of the strain gauges, which undergo zero dilatation due to V_Y.

For such a symmetric cross-section, the applied torque induces shear strains of different magnitude on the top and bottom surfaces. Thus, at P the shear stress due to M^t must take the form:

$$\tau_{XZ} = \beta \frac{M^t}{J} a \tag{12.72}$$

while at P' it would take the form:

$$\tau_{XZ} = -\beta' \frac{M^t}{J} b \tag{12.73}$$

Here, a and $-b$ denote the coordinates of the points of intersection of the symmetry axis GY with the top and bottom surfaces, respectively. Also, J is the cross-section

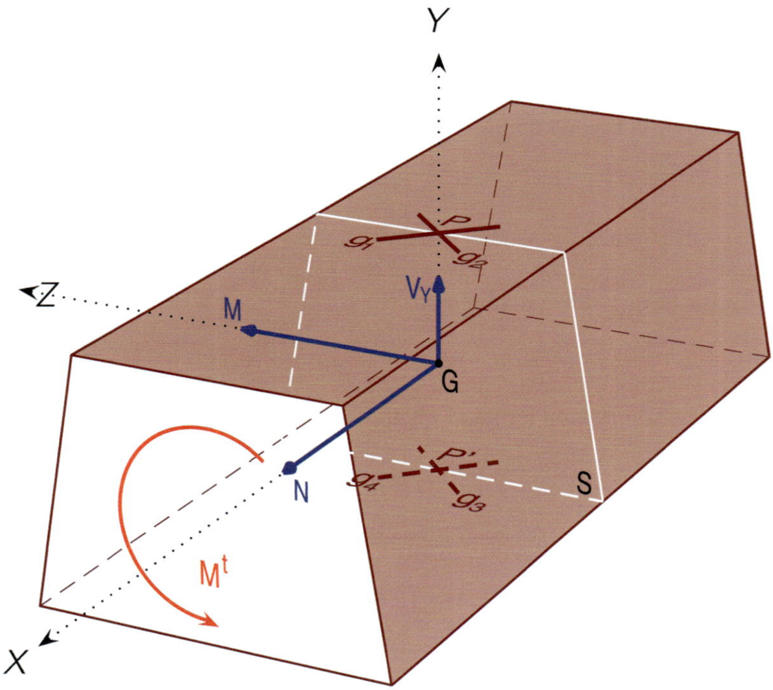

Figure 12.13 Principle of measurement of torque in a beam loaded in combined extension, bending, and torsion.

shape dependent torsion constant whereas β, β' are dimensionless coefficients to be determined, also dependent on the shape of the cross-section. The strain gauge dilatations are then

$$(\varepsilon_1)_T = (\varepsilon_3)_T = \frac{\beta}{2G}\frac{M^t a}{J} \tag{12.74}$$

and

$$(\varepsilon_2)_T = (\varepsilon_4)_T = -\frac{\beta'}{2G}\frac{M^t b}{J} \tag{12.75}$$

By way of consequence, Equation (12.57) gives

$$\frac{\delta v}{\mathcal{E}} = \frac{(\delta v)_T}{\mathcal{E}} = \frac{K}{2}(-(\varepsilon_1)_T + (\varepsilon_2)_T) = -\frac{K}{4G}(\beta a + \beta' b)\frac{M^t}{J} \tag{12.76}$$

The above relation provides a measurement of the torque that is insensitive to all other screwer components provided that the cross-section admits an axis of symmetry and that the transverse shear loading does not break this symmetry.

12.3.4 INDEPENDENT MEASUREMENT OF V_Y

The last case to be considered is that of a cross-section admitting GY as an axis of symmetry and subjected to all screwer components, as illustrated in Figure 12.14. The torque M^t is shown on the end section for clarity but is the same on cross-section S. A full Wheatstone bridge configuration is used. As shown in the figure, the two pairs of strain gauges are placed at P and P', points of intersection of the GZ principal axis with the beam's outer surface.

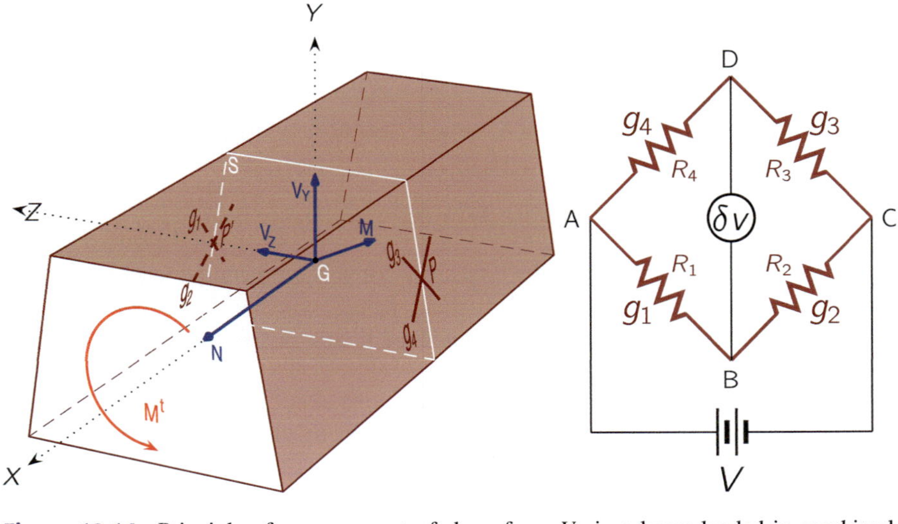

Figure 12.14 Principle of measurement of shear force V_Y in a beam loaded in combined extension, bending, and torsion.

The reading on the potentiometer is given by the general relation of Equation (12.57). The strains due to the normal force N are still given formally by Equations (12.61) and (12.66) and would cancel out for any cross-sectional shape. The bending moment **M** has two components. The component parallel to the Z-axis, M_Z, does not generate any normal stress or strain at locations P or P'. Hence, M_Z does not affect the output voltage. On the other hand, the bending component parallel to the Y-axis, M_Y, would lead to compression at P and tension at P'. Because of the symmetry about the Y-axis, the normal strains at P and P' are equal and opposite to each other. Therefore, the bending-induced strains would cancel out by pair because the relations in Equation (12.67) remain valid, i.e.,

$$-(\varepsilon_1)_M = -(\varepsilon_2)_M = (\varepsilon_3)_M = (\varepsilon_4)_M \tag{12.77}$$

The applied torque induces shear strains of equal magnitude on the left and right surfaces. Thus, at P the shear stress due to M^t must take the form:

$$\tau_{XZ} = \beta \frac{M^t}{J} a \tag{12.78}$$

while at P' it would take the form:

$$\tau_{XZ} = -\beta \frac{M^t}{J} a \tag{12.79}$$

where $\pm a$ denote the coordinates of the points of intersection of principal axis GZ with the side surfaces. Also, J is the torsion constant and β is a dimensionless coefficient, both being dependent on the shape of the cross-section. The strain gauge dilatations are then

$$(\varepsilon_2)_T = (\varepsilon_3)_T = \frac{\beta}{2G} \frac{M^t a}{J} \tag{12.80}$$

and

$$(\varepsilon_1)_T = (\varepsilon_4)_T = -\frac{\beta}{2G} \frac{M^t a}{J} \tag{12.81}$$

Therefore, the torque-induced strains would also cancel out by pair because the relations in Equation (12.69) remain valid, i.e.,

$$-(\varepsilon_1)_T = (\varepsilon_2)_T = (\varepsilon_3)_T = -(\varepsilon_4)_T \tag{12.82}$$

Finally, consider the effect of the total shear force **V** with components V_Y and V_Z. Unless the cross-section is symmetrical about the GZ axis, the V_Z component would generate shear strains at P and P' that would be difficult (not impossible) to filter out.

In what follows, we assume that the cross-section has two axes of symmetry, which must therefore be identified with the principal axes of inertia, GY, GZ. In that case, only V_Y would generate shear strains at the locations of the strain gauges. It is always possible to express the shear stress at P and P' as

$$\tau_{xy} = \beta \frac{V_Y}{S} \tag{12.83}$$

where β is a dimensionless constant to be determined. Here the y direction is parallel to the lateral surface at P or P'. The principal strains are identified with the linear dilatations of the strain gauges, given by:

$$\varepsilon_1 = \varepsilon_3 = \frac{\beta}{2G}\frac{V_Y}{S} \quad \text{and} \quad \varepsilon_2 = \varepsilon_4 = -\frac{\beta}{2G}\frac{V_Y}{S} \tag{12.84}$$

By way of consequence, Equation (12.57) gives

$$\frac{\delta v}{\mathcal{E}} = \frac{(\delta v)_V}{\mathcal{E}} = \frac{K}{2}(-(\varepsilon_1)_V + (\varepsilon_2)_V) = -\beta\frac{K}{2G}\frac{V_Y}{S} \tag{12.85}$$

which by inversion gives the desired shear force measure

$$V_Y = -\frac{2GS}{\beta K}\frac{\delta v}{\mathcal{E}} \tag{12.86}$$

in terms of the material property (shear modulus G), cross-sectional properties (S and β), the gauge factor K and the output voltage $\delta v/\mathcal{E}$.

This measurement is insensitive to internal resultants N, M_Y, M_Z and M^t provided that the cross-section is symmetrical about the GY axis. If, in addition, the cross-section is symmetrical about the GZ axis, the measurement is also insensitive to the other shear component V_Z.

PROBLEMS

Problem 12.1 (Gauge Factor and Poisson's Ratio) *The aim is to estimate the mechanical part of the gauge factor of a strain gauge (piezoresistance or strain-induced resistance change). To this end, consider a wire of length L and cross-sectional area A. If the electrical resistance of the wire is given by the relation:*

$$R = \rho\frac{L}{A}$$

where ρ is the resistivity of the material of the wire, determine the effect of piezoresistance on the gauge factor K in terms of Poisson's ratio ν of the wire's material. ★

Problem 12.2 (Electronics of Gauge Factor) *The aim is to revisit the estimate of Problem 12.1, which cannot explain why some semi-conductors have gauge factors a hundred times more than metals such as constantan. To this end, we will estimate the electronic part of the gauge factor (piezoresistivity, or strain-induced resistivity change). Consider the same wire of length L and cross-sectional area A. From atomic physics, the resistivity is given by:*

$$\rho = \frac{2m\bar{v}}{ne^2\lambda}$$

where m is the mass of the electron, $\bar{v}$ its average velocity, n the number of conduction electrons per unit volume, e the electron charge, and λ the mean free path, i.e., the *average distance traveled by an electron between collisions with ions. Show that the gauge factor (i) is independent of Poisson's ratio; and (ii) can be written as* $K = 2 + f(\varepsilon)$ *where* f *is a function of strain to be determined.* ⋆

Problem 12.3 (Self-Compensating Strain Gauge) *A strain gauge is made up of a material with a coefficient of thermal expansion* α_G *and gauge factor K. The strain gauge is attached to a structure made up of a material with a coefficient of thermal expansion* α. *The structure is subjected to a temperature variation* ΔT. *If the relative change of electrical resistance of the gauge is linear in* ΔT, *i.e.,*

$$\frac{\Delta R}{R} = \beta_G \Delta T$$

where β_G *is a temperature coefficient of resistance, determine the expression of the strain produced in the gauge in terms of* α, α_G, β_G, K *and* ΔT. *Discuss scenarios where this 'apparent strain' can be eliminated.*

⋆

Problem 12.4 (Alternative Half-Bridge Analysis) *Consider the half-bridge setup for normal force measurement shown in Figure 12.3. Assume that* R_2 *and* R_4 *are now calibrated resistor decade boxes and that strain gauges* g *and* g' *are self-compensated for the material. Show that the normal force N can be measured using a zero method based on rebalancing the bridge.* ⋆

Problem 12.5 (Quarter-Bridge Configuration) *The angle section beam shown in Figure 12.15 is under the action of a normal force N. A single strain gauge g is glued to the top surface and forms the resistance of a Wheatstone bridge in a quarter bridge setup. Explain how the setup may be used to measure the magnitude of N.*

⋆

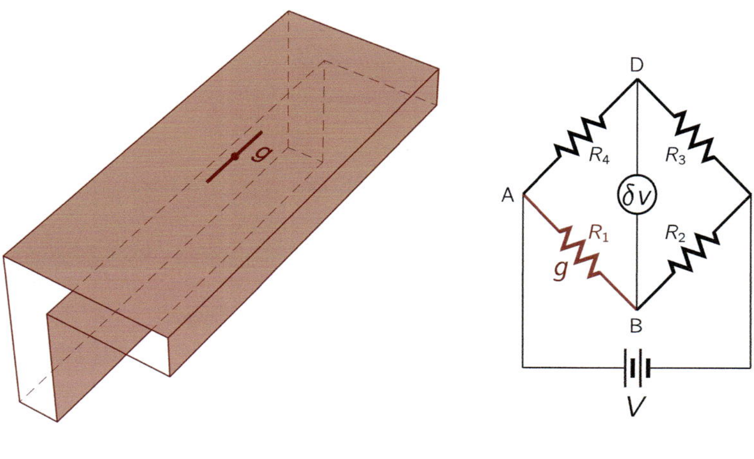

Figure 12.15 Quarter bridge configuration of Wheatstone bridge.

Problem 12.6 (Aircraft Propeller) *A Hamilton Standard blade with dimensions $L \times D \times h$ equal to 7 ft. 9 1/4 in. $\times$ 1 ft. 2 in. $\times$ 5 3/4 in. and weight 117 lb. is made of duralumin (Young's modulus 70 GPa) and turns at 2700 rpm. Two identical wireless turning contacts are attached to the blade's root, at 5 in. from the hub. The two contacts act as remote strain gauges with gauge factor $K = 2$, mounted to a Wheatstone bridge in a half-bridge configuration. The cross-sectional area is approximated as $S = Dh/6$. If the power source voltage is 120 V what is the reading on the galvanometer?* ⋆

Problem 12.7 (Apparent Stiffness of Composite Blade) *A 4.5 m long plane propeller blade with cross-sectional area $S = 0.011\ m^2$ and total weight 80 kg turns at angular velocity 2750 rpm. Two identical wireless turning contacts are attached to the blade's root, at 25 cm from the hub. The two contacts act as remote strain gauges with gauge factor $K = 2$, mounted to a Wheatstone bridge in a half-bridge configuration. The power source voltage is 120 V and the voltage of the potentiometer reads 0.7 V. What is Young's modulus of the material of the blade?* ⋆

Problem 12.8 (Temperature Effect in Bending) *Consider the half-bridge configuration set up to measure the bending moment M_Z in Figure 12.5. Let α denote the coefficient of thermal expansion of the material of the beam. Show that the configuration is insensitive to temperature variations.* ⋆

Problem 12.9 (Full-Bridge Setup in Bending) *Consider the beam in Figure 12.5. At point P attach end-to-end two strain gauges R_1 and R_3. Likewise, at P' attach end-to-end two strain gauges R_2 and R_4. Show that this full-bridge configuration has double the sensitivity to M_Z of the half-bridge configuration in Figure 12.5.* ⋆

Problem 12.10 (Alternative Formulas for Bending) *Consider a beam admitting GY as axis of symmetry. Determine the equations needed to measure the bending moment when the beam is bent in the X-Z plane (bending moment M_Y) for both the half- and full-bridge configurations.* ⋆

Problem 12.11 (Measurement of a Torque in a Square Shaft) *Devise a method for measuring torque in a solid shaft having a square cross-section with side length 2a, Figure 12.16. The measurement must be insensitive to temperature and loads other than the torque M^t.*

⋆

Problem 12.12 (Measurement of a Torque in a Thin-Walled Section) *Consider the thin-walled square profile shown in Figure 12.17. Determine the explicit relationship between the applied torque and the output voltage of a Wheatstone bridge in a full bridge configuration.*

⋆

Problem 12.13 (Error Estimation for Jourawski Formula) *Determine the value of the prefactor β in Equation (12.45) based on Jourawski's formula for the average*

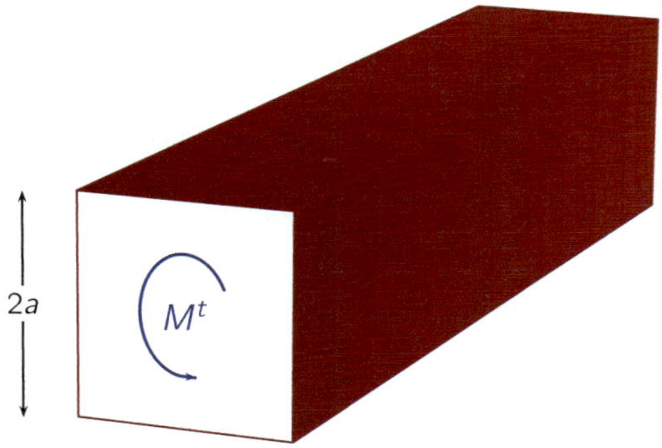

Figure 12.16 Square beam under torsion.

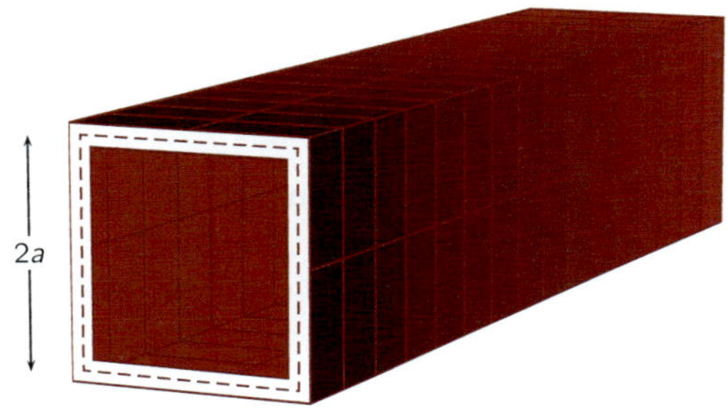

Figure 12.17 Thin-walled square section under torsion.

shear stress. Compare with the value obtained for a circular section in Example 12.2 for values of Poisson's ratio ranging between −1 to 1/2. ⋆

Problem 12.14 (Rectangular Section under Shear) *Consider a rectangular section under the action of a shear force V_Y, as depicted in Figure 12.8. Let a and b respectively denote the half-height and half-width of the section. Plot or tabulate the variation of prefactor β in Equation* (12.45) *as a function of the section's aspect ratio a/b. You may limit the analysis to the range* $0.1 \leq a/b \leq 3$. ⋆

Problem 12.15 (Measurement of Shear Force) *Consider the rectangular beam in Figure 12.18 made of aluminum (Young's modulus $E = 70$ GPa and Poisson's ratio $\nu = 0.33$). Four identical strain gauges are glued to the beam and arranged in a full bridge configuration. Their common gauge factor is $K = 2$. If the power source voltage is 120 V and the voltage of the potentiometer reads 6 V what is the magnitude of the shear force V_Y acting on the cross-section of interest?* ⋆

Problem 12.16 (Measurements under Combined Extension and Bending) *We aim to measure the normal force and the two components M_Y, M_Z of the bending moment on a cross-section of a thin-walled circular beam, Figure 12.19. To this end, three*

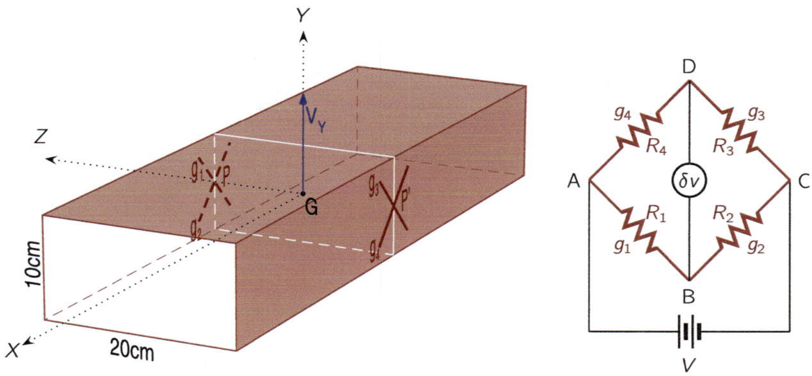

Figure 12.18 Rectangular beam and corresponding full bridge setup.

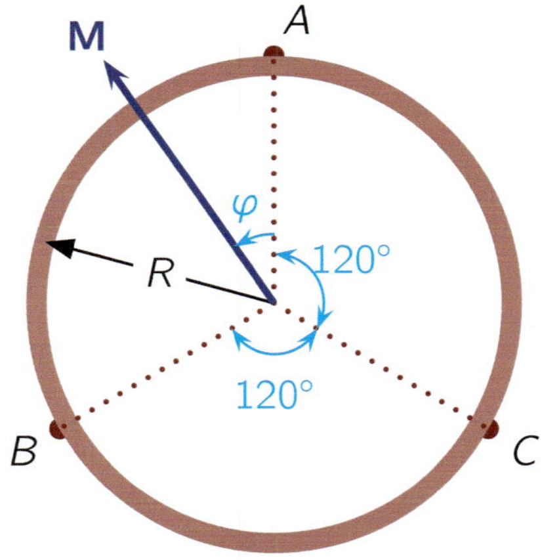

Figure 12.19 Three strain gauges on a thin-walled circular section.

strain gauges are glued in the axial direction at three locations labeled A, B, C in Figure 12.19. The three points are equally spaced as shown. The cross-section has mean radius R and thickness t. Calculate the magnitude of the normal force N, the magnitude of the total bending moment M and the polar angle φ that defines the orientation of the bending moment **M** *in terms of the three dilatations* ε_A, ε_B, ε_C. ⋆

Bibliography

S. Laroze. *Mécanique des Structures: Tome II: Poutres.* Ecole Nationale Supérieure de l'Aéronautique et de l'Espace, 1992.

S. Laroze and J. Barrau. *Mécanique des Structures: Tome II-bis: Exercices.* Ecole Nationale Supérieure de l'Aéronautique et de l'Espace, 1992.

S. Laroze. *Mécanique des Structures: Tome I: Solide élastique, Plaques et Coques.* Ecole Nationale Supérieure de l'Aéronautique et de l'Espace, 1992.

L. Navier. *Résumé des Lecons (données* à *l'Ecole Royale des Ponts et Chaussées)–Première Partie.* Firmin Didot Pères et Fils, Paris, 1826.

L. Navier. *Résumé des Lecons (données* à *l'Ecole des Ponts et Chaussées)–Première Partie*, Second edition. Carilian–Goeury, Paris, 1833.

L. Navier and B. Saint-Venant. *Résumé des Lecons (données* à *l'Ecole des Ponts et Chaussées)–Première Partie*, Third edition. Dunod, Paris, 1864.

S. Timoshenko. *Strength of Materials – Part I: Elementary Theory and Problems*, Second edition. D. Van Nostrand Company, 1940.

B. de Saint-Venant. *De la torsion des prismes avec des considérations sur leur flexion ainsi que sur l'équilibre des solides élastiques en général et des formules pratiques pour le calcul de leur résistance* à *divers efforts s'exercant simultanément.* Extrait du tome XIV des mémoires présentés par divers savants à l'Académie des Sciences. Victor Dalmont, Imprimerie Impériale, Paris, 1855. 332 pages, exclusive of front matter.

B. de Saint-Venant. Mémoire sur la flexion des prismes, sur les glissements transversaux et longitudinaux qui l'accompagnent lorsqu'elle ne s'opère pas uniformément ou en arc de cercle, et sur la forme courbe affectée alors par leurs sections transversales primitivement planes. *Journal de mathématiques pures et appliquées*, 2ème série, Tome I:89–189, 1856.

J. Bresse. *Cours de Mécanique Appliquée. Part I: Résistance des Matériaux et Stabilité des Constructions*, Third edition. Gauthier-Villars, Paris, 1880.

G. Pigeaud. *Résistance des Matériaux et Elasticité*, Second edition. Gauthier-Villars, Paris, 1934.

S. Timoshenko. *History of Strength of Materials.* McGraw-Hill, New York, 1953.

T. H. G. Megson. *Aircraft Structures for Engineering Students*, Fourth edition. Elsevier Aerospace Series. Elsevier, Oxford, UK, 2007.

D. H. Allen and W. E. Haisler. *Introduction to Aerospace Structural Analysis.* Wiley, New York, 1985.

B. K. Donaldson. *Analysis of Aircraft Structures*, Second edition. Cambridge Aerospace Series. Cambridge University Press, New Delhi, India, 2008.

O. A. Bauchau and J. I. Craig. *Structural Analysis.* Springer, Dordrecht, Heidelberg, London, New York, 2009.

R. E. Peterson. *Stress Concentration Factors*, First edition. John Wiley, New York, 1974.

W. D. Pilkey and D. F. Pilkey. *Peterson's Stress Concentration Factors*, Third edition. John Wiley & Sons, Inc., Hoboken, NJ, 2008.

S. Timoshenko and J. M. Gere. *Theory of Elastic Stability*, Second edition. McGraw-Hill, Singapore, 1961.

M. Beck. Die knicklast des einseitig eingespannten, tangential gedrückten stabes. *Zeitschrift für Angewandte Mathematik und Physik*, 3:225–228, 1952.

Y. Sugiyama, J. Matsuike, B.-J. Ryu, K. Katayama, S. Kinoi, and N. Enomoto. Effect of concentrated mass on stability of cantilevers under rocket thrust. *AIAA Journal*, 33:499–503, 1995.

Y. Sugiyama, K. Katayama, K. Kiriyama, and B.-J. Ryu. Experimental verification of dynamic stability of vertical cantilevered columns subjected to a sub-tangential force. *Journal of Sound and Vibration*, 236:193–207, 2000.

M. A. Langthjem and Y. Sugiyama. Dynamic stability of columns subjected to follower loads: A survey. *Journal of Sound and Vibration*, 238:809–851, 2000.

Index

For Product Safety Concerns and Information please contact our EU representative GPSR@taylorandfrancis.com Taylor & Francis Verlag GmbH, Kaufingerstraße 24, 80331 München, Germany